STATISTICS THROUGH APPLICATIONS

STATISTICS THROUGH APPLICATIONS

Daniel S. Yates
Statistics Consultant

Daren S. Starnes
The Webb Schools

David S. Moore
Purdue University

W. H. Freeman and Company
New York

Senior Acquisitions Editor: Patrick Farace

Development Editor: Danielle Derbenti

Directors of High School Sales and Marketing: Mike Saltzman and Cindi WeissGoldner

Publisher: Craig Bleyer

Media Editors: Brian Donnellan and Victoria Anderson

Project Editor: Jane O'Neill

Photo Editor: Vikii Wong

Cover and Text Designer: Blake Logan

Illustrations: Network Graphics

Illustration Coordinator: Bill Page

Production Coordinator: Susan Wein

Composition: TechBooks

Printing and Binding: RR Donnelley & Sons Company

Photo credits: Part opening and chapter opening pages—Corbis Images; Activity icon—Image Source.

Library of Congress Cataloging-in-Publication Data

Yates, Daniel S.
    Statistics through applications / Daniel S. Yates, Daren S. Starnes, David S. Moore.
        p. cm.
    Includes index.
    ISBN 0-7167-4772-3 (cloth)
    1. Statistics. I. Starnes, Daren S. II. Moore, David S. III. Title.

QA276.12.Y38 2004
519.5—dc22   2003069611

EAN 9 780716 747727

Printed in the United States of America

Third printing
W. H. Freeman and Company
41 Madison Avenue
New York, NY 10010
Houndmills, Basingstoke RG21 6XS, England

www.whfreeman.com

CONTENTS

DANIEL S. YATES has taught statistics in the Electronic Classroom (a distance learning facility) affiliated with Henrico County Public Schools in Richmond, Virginia. Before teaching in high school, he was on the mathematics faculty at Virginia Tech and Randolph-Macon College. Dan has a Ph.D. in Mathematics Education from Florida State University, he has served as President of the Greater Richmond Council of Teachers of Mathematics and the Virginia Council of Teachers of Mathematics, and he was the recipient of the College Board/Siemens Foundation Advanced Placement Teaching Award in 2000. Although he recently retired from classroom teaching, Dan stays in step with trends in teaching by frequently conducting College Board workshops for new and experienced AP statistics teachers and by monitoring and participating in the AP statistics electronic discussion group. Dan is lead author of the popular textbook *The Practice of Statistics.*

DAREN S. STARNES is the Mathematics Department Chair at the Webb Schools in Claremont, California. He earned his M.A. in Mathematics from the University of Michigan, received a GTE GIFT Grant to integrate AP Statistics and AP Environmental Science in 1997, and was named a Tandy Technology Scholar in 1999. Daren has led many one-day and week-long AP Statistics institutes for new and experienced AP teachers. He also collaborated in the design and implementation of a course entitled Probability, Statistics, and Finite Math for non-AP high school seniors. In 2000–2002, he served as coeditor of the Technology Tips column in the NCTM journal, *The Mathematics Teacher.* In January 2004, Daren was appointed to a three-year term on the ASA/NCTM Joint Committee on the Curriculum in Statistics and Probability. Daren is a coauthor of *The Practice of Statistics,* Second Edition.

DAVID S. MOORE is Shanti S. Gupta Distinguished Professor of Statistics at Purdue University and 1998 President of the American Statistical Association. David is an elected fellow of the American Statistical Association and of the Institute of Mathematical Statistics and an elected member of the International Statistical Institute. He has served as program director for statistics and probability at the National Science Foundation.

David has devoted his attention to the teaching of statistics. He was the content developer for the Annenberg/Corporation for Public Broadcasting college-level telecourse *Against All Odds: Inside Statistics* and for the series of video modules *Statistics: Decisions Through Data,* intended to aid the teaching of statistics in schools. He is the author of influential articles on statistics education and of several leading texts, including *Introduction to the Practice of Statistics* (cowritten with George P. McCabe), *The Basic Practice of Statistics, Statistics: Concepts and Controversies, The Practice of Statistics* (with Dan Yates and Daren Starnes), and, most recently, *The Practice of Business Statistics.*

Preface

There are books on statistical theory and books on statistical methods. This is neither. *Statistics Through Applications (STA)* is a book on statistical ideas and statistical reasoning and on their relevance in such fields as medicine, education, environmental science, business, psychology, sports, politics, and entertainment. It is designed to support a first course in high school statistics that emphasizes statistical thinking.

In its *Principles and Standards for School Mathematics,* the National Council of Teachers of Mathematics (NCTM) identifies "Data Analysis and Probability" as one of five content standards. Here is the precise statement of this NCTM standard:

Instructional programs from prekindergarten through grade 12 should enable all students to

- formulate questions that can be addressed with data and collect, organize, and display relevant data to answer them
- select and use appropriate statistical methods to analyze data
- develop and evaluate inferences and predictions that are based on data
- understand and apply basic concepts of probability[1]

STA is ideally suited for a non-AP-level introduction to statistics for high school students. It can be used effectively in either a one- or two-semester course.

The joint curriculum committee of the American Statistical Association and the Mathematical Association of America recommends that any first course in statistics "emphasize the elements of statistical thinking" and feature "more data and concepts, fewer recipes and derivations."[2] *STA* does this by taking a conceptual and verbal approach, rather than a methods-oriented focus. *STA* was adapted from David Moore's popular college text, *Statistics: Concepts and Controversies,* now in its fifth edition.

The Nature of *STA*

STA was written to be read by students. It is somewhat informal, with case studies, marginal "think pieces," and cartoons interspersed throughout. Examples and exercises were selected to pique students' interest and curiosity. Activities and applications give students an opportunity to investigate, discuss, and make use of statistical ideas and methods. Teachers should be aware, however, that the text is more serious than its relaxed style and accessible mathematical level suggest. The emphasis on ideas and reasoning asks more of the student than many recipe-filled methods texts.

Students learn to think about data by working with data. Consequently, we have included many elementary graphical and numerical techniques in *STA*. We have not, however, allowed techniques to dominate concepts. Our intention is to teach verbally rather than algebraically, to invite discussion and even argument rather than mere computation (though some computation remains essential). The coverage in *STA* is considerably broader than that of traditional statistics texts, as the table of contents reveals.

The Structure of *STA*

STA contains ten chapters organized in four parts: producing data, organizing data, chance, and inference. Each chapter consists of two or three sections, each section in turn devoted to a single coherent set of ideas. Sections are further divided into small subsections of material that can be fully addressed in a standard 45- or 50-minute high school period. Exercises appear at the end of each subsection, section, and chapter. Summaries are provided at the end of each section and chapter to help students distill what they have learned. Here are the important **features** of *STA*:

- Motivational, hands-on **activities** allow students to explore statistical concepts.
- Abundant **examples** from a variety of fields illustrate important statistical ideas.

- Varied **exercises** (with descriptive titles) demonstrate the usefulness of statistics in many different subject areas.

- **Calculator Corners** provide step-by-step instructions with screen shots for performing important statistical functions on the TI-83 (Plus/Silver Edition) graphing calculator.

- Each section opens with a brief **case study** that raises important questions to be answered in the pages ahead.

- Numerous **cartoons** enliven the pages.

- Short **marginal think pieces** offer interesting and often amusing anecdotes about statistics and its impact.

- **Applications** require students to put what they have learned to use in a real-world setting.

- **Exploring the Web** boxes highlight some useful sites connected to the material in the section.

- A short **Statistics in Summary** narrative reviews the big ideas at the end of each section. Each chapter concludes with a **Chapter Review** that lists the skills students should have acquired by that point and **Chapter Review Exercises** to check understanding.

Supplements

A full range of supplements is available to help teachers and students use *Statistics Through Applications*.

Teacher's Resource Binder (TRB), 0-7167-1262-8

This indispensable resource, compiled by Dan Yates, Daren Starnes, and Steve Kokoska of Bloomsburg University gives teachers a multitude of useful tools for planning and managing their courses, for assessing student progress, and for enhancing day-to-day instruction. The TRB offers the following:

- Two sample quizzes for every section of every chapter

- Two sample chapter tests for every chapter

- Extensive teaching suggestions, including additional examples for in-class use, for each chapter

- Additional activities and applications

- Suggestions for short-term and long-term projects, with sample grading rubrics

- Expanded technology tips for using the TI-83 (Plus/Silver Edition), as well as Java applets

- An extensive list of references to books, journal articles, video series, other sources of data in both print and electronic formats, interesting statistics-related Web sites, and more.

Instructor's CD, 0-7167-8934-5

The CD contains the following:

- Microsoft Word versions of the tests and quizzes in the Teacher's Resource Binder

- An expanded version of the Electronic Encyclopedia of Statistical Examples and Exercises (EESEE) that includes solutions to all case study questions

- All text figures in an exportable presentation format, JPEG for Windows users and PICT for Macintosh users

- PowerPoint slides that can be used directly or customized to fit your needs

- Data sets from the text in several formats, including TI-83 files

Instructor's Solutions Manual, 0-7167-1266-0

This manual was written by Duane Hinders, Foothill College, and contains detailed solutions for all problems in the text.

CD Test Item File, 0-7167-1220-2

This easy-to-use CD contains more than 750 additional questions written by David Moore and compiled by Daren Starnes for use with *STA*. Questions are grouped into Microsoft Word files by chapter, which allows for easy editing, cutting, and pasting. The *Test Bank* is also available in print format (0-7167-1248-2).

STA Web Site

The Web site at www.whfreeman. com/sta for teachers *and* students features:

- Applets that allow students to manipulate data and see the results graphically
- Data sets from the text in several formats, including TI-83 files
- Rich case studies from the Electronic Encyclopedia of Statistical Examples and Exercises (EESEE) with data and exercises for students
- Online quizzes with instant feedback to help students master material and prepare for testing

ACKNOWLEDGMENTS

To Patrick Farace, senior acquisitions editor at W. H. Freeman and Company, we owe our deepest gratitude for his willingness to take a risk on *Statistics Through Applications*, for his assistance in navigating the roadblocks that threatened to impede our progress, and for his faithful support and friendship these past few years. Our development editor, Danielle Derbenti, has (gently) cracked the whip at all the right times to ensure that we met production deadlines. Kudos, Danielle! We sincerely thank Pamela Bruton, our now long-time copy editor, for her unwillingness to allow even the slightest error to go untouched in the *STA* manuscript. Jane O'Neill has done splendid work as project editor for *STA*, especially in the final stages of writing and editing. We are grateful to Blake Logan for her design work on the book, and particularly for her willingness to entertain our "creative" suggestions along the way. To all the folks at TechBooks who played a role in producing the finished product, we offer our thanks.

We appreciate the thoughtful comments and suggestions made by our colleagues who reviewed draft chapters of the *STA* manuscript:

Rick Davis, *Norfolk Collegiate School, Virginia*

Carol K. Dormuth, *Plymouth Whitemarsh High School, Pennsylvania*

Jo Lynn Hughs, *Episcopal Collegiate School, Arkansas*

Beverly K. Kimes, *Birmingham City Schools, Alabama*

Beth Ellen Lazerick, *St. Andrews School, Florida*

Jeanne Lorenson, *James Hubert Black High School, Maryland*

Philip Mallinson, *Phillips Exeter Academy, New Hampshire*

Paul L. Myers, *Woodward Academy, Georgia*

Diann Resnick, *Bellaire High School, Texas*

Sally Ziemba, *David W. Butler High School, North Carolina*

We are particularly grateful to those pioneering teachers who agreed to serve as class testers for the preliminary version of *Statistics Through Applications* during the 2003–2004 school year:

N. Atwell, *Concord High School, North Carolina*

Ann Brock, *Lewis-Palmer High School, Colorado*

Patricia Daniel, *North Atlanta High School, Georgia*

Cathy Morgan, *McNeil High School, Texas*

Acquillahs M. Mutie, *Ganesha High School, California*

Cloe O'Grady, *Archbishop Ryan High School, Pennsylvania*

Daniel R. Shuster, *Royal High School, California*

We remain indebted to David Moore for his direct and indirect influence on our view of statistics, our teaching, and, of course, our writing. We can only hope that one day we will write as succinctly, as flawlessly, and as quickly as David. We also thank Dennis Pearl for contributing several hands-on activities and investigations using applets. To the students and teachers who have inspired us to keep learning and to share our experiences with others, we say, "This book's for you!"

                                              DSY and DSS

I firmly believe that *all* students, before they graduate from high school, should have an exposure to the basic ideas of statistics. The reason is simple: we increasingly live in a data-driven world. And even though high school graduates may not *practice* statistics very much, they need to have some baseline knowledge so that they can make sense out of data, recognize when a graph is misleading, or know if a claim of causation is valid. To me, understanding statistics is a survival skill. Toward this end, I began thinking about the kinds of knowledge and skills students should gain in an introductory statistics course. This book is my vision for such a course. I am extremely pleased that Daren Starnes consented to work with me on this project. Daren brings a flair for clear exposition that is rare in mathematics and statistics texts. His influence on every page helps make the subject accessible and enjoyable. Thanks, Daren! And thanks to my wife, Betty Jo, who typed, proofread, repaired, and compiled. Her constant support, both editorial and emotional, is greatly appreciated.

DSY

A few years ago, Dan Yates crafted the initial proposal for *Statistics Through Applications,* because he realized that there was a real need for a textbook that would support high school teachers in offering a non-AP introductory statistics course. His idea was to build on David Moore's highly engaging and well-written college text *Statistics: Concepts and Controversies (SCC)*. By adding activities, applications, and Calculator Corners, revising exercises, and reorganizing the layout of *SCC,* Dan aimed to create a book that would appeal to a high school audience. When he pitched the concept to me, I was eager to collaborate on such an exciting project. Having teamed with Dan already on the second edition of *The Practice of Statistics,* I knew that David's gift for explaining combined with Dan's feel for what works in the high school classroom would make my job as a coauthor that much simpler. I am indebted to both of them for the opportunity they have given me to grow as a writer and a statistics educator. This book is dedicated to my wife, Judy, who helps make my stories worth sharing.

DSS

Prelude

Making sense of statistics

Statistics is about data. Data are numbers, but they are not "just numbers." Data are numbers with a context. The number 10.5, for example, carries no information by itself. But if we hear that a friend's new baby weighed 10.5 pounds at birth, we congratulate her on the healthy size of the child. The context engages our background knowledge and allows us to make judgments. We know that a baby weighing 10.5 pounds is quite large, and that a human baby is unlikely to weigh 10.5 ounces or 10.5 kilograms. The context makes the number informative.

Statistics uses data to gain insight and to draw conclusions. The tools are graphs and calculations, but the tools are guided by ways of thinking that amount to educated common sense. Let's begin our study of statistics with a rapid and informal guide to coping with data and statistical studies. We will examine the examples in this prelude in more detail later.

Data beat anecdotes

> *Belief is no substitute for arithmetic.*
> HENRY SPENCER

An anecdote is a striking story that sticks in our minds exactly because it is striking. Anecdotes humanize an issue, so news reports usually start (and often stop) with anecdotes. But anecdotes are weak ground for making up your mind—they are often misleading exactly because they are striking. Always ask if a claim is backed by data, not just by an appealing personal story.

Does living near power lines cause leukemia in children? The National Cancer Institute spent 5 years and $5 million gathering data on the question. Result: No connection between leukemia and exposure to magnetic fields of the kind produced by power lines. The editorial that accompanied the study report in the *New England Journal of Medicine* thundered, "It is time to stop wasting our research resources" on the question.

Now compare the impact of a television news report of a 5-year, $5 million investigation against a televised interview with an articulate mother whose child has leukemia and who happens to live near a power line. In the public mind, the anecdote wins every time. Be skeptical. Data are more reliable than anecdotes because they systematically describe an overall picture rather than focusing on a few incidents.

Where the data come from is important

> *Figures won't lie but liars will figure.*
> CHARLES GROSVENOR

Data are numbers, and numbers always seem solid. Some are and some are not. Where the data come from is the single most important fact about any statistical study. When Ann Landers asks readers of her advice column whether they would have children again and 70% of those who reply shout "No," you should just amuse yourself with Ann's excerpts from tear-stained letters describing what beasts the writers' children are. Ann Landers is in the entertainment business. Her survey attracts people who regret having their children. Most parents don't regret having children. We know this because opinion polls have asked large numbers of parents, chosen at random to avoid attracting one opinion or another. Opinion polls have their problems, as we will see, but they beat just asking upset people to write in.

Even the most reputable publications have not been immune to bad data. The *Journal of the American Medical Association* once printed an article claiming that pumping refrigerated liquid through tubes in the stomach relieves ulcers. The patients did respond well, but only because patients often respond to any treatment given with the authority of a trusted doctor. That is, placebos (dummy treatments) work. When a skeptic finally tried a properly controlled study in which some patients got the tube and some got a placebo, the placebo actually did a bit better. "No comparison, no conclusion" is a good starting point for judging medical studies. We should be skeptical about the current boom in "natural remedies," for example. Few of these have passed a comparative trial to show that they are more than just placebos sold in bottles bearing pretty pictures of plants.

Beware the lurking variable

I have enough money to last me the rest of my life, unless I buy something.
JACKIE MASON

You read that crime is higher in counties with gambling casinos. A college teacher says that students who took a course online did better than the students in the classroom. Government reports emphasize that well-educated people earn a lot more than people with less education. Don't jump to conclusions. Ask first, "What is there that they didn't tell me that might explain this?"

Crime is higher in counties with casinos, but it is also higher in urban counties and in poor counties. What kind of counties are casinos in? Did these counties have high crime rates before the casinos arrived? The online students did better in the course, but they were older and better prepared than the in-class students. No wonder their grades were higher. Well-educated people do earn a lot. But educated people have (on the average) parents with more education and more money than the parents of poorly educated people have. They grew up in nicer places and went to better schools. These advantages help them get more education and would help them earn more even without that education.

All these studies report a connection between two variables and invite us to conclude that one of these variables influences the other. "Casinos increase crime" and "Stay in school if you want to be rich" are the messages we hear. Perhaps these messages are true. But perhaps much of the connection is explained by other variables lurking in the background, such as the nature of counties that accept casinos and the advantages that highly educated people were born with. Good statistical studies look at lots of background variables. This is tricky, but you can at least find out if it was done.

Variation is everywhere

When the facts change, I change my mind. What do you do, sir?
JOHN MAYNARD KEYNES

If a thermometer under your tongue reads higher than 98.6° F, do you have a fever? Maybe not. People vary in their "normal" temperature. Your own temperature also varies—it is higher around 6 A.M. and lower around 6 P.M. The government announces that the unemployment rate rose a tenth of a percent last month and that new home starts fell by 3%. The stock market promptly jumps (or sinks). Stocks are more variable than is sensible. The government data come from samples that give good estimates but not the exact truth. Another run of samples would give slightly different answers. And economic facts jump around anyway, because of weather, strikes, holidays, and all sorts of other reasons.

Many people join the stock market in overreacting to minor changes in data that are really nothing but background noise. Here is Arthur Nielsen, head of the country's largest market research firm, describing his experience:

> *Too many business people assign equal validity to all numbers printed on paper. They accept numbers as representing Truth and find it difficult to work with the concept of probability. They do not see a number as a kind of shorthand for a range that describes our actual knowledge of the underlying condition.*[1]

Variation is everywhere. Individuals vary; repeated measurements on the same individual vary; almost everything varies over time. Ignore the pundits who try to explain the deep reasons behind each day's stock market moves, or who condemn a team's ability and character after a game decided by a last-second shot that did or didn't go in.

Conclusions are not certain

As far as the laws of mathematics refer to reality they are not certain, and as far as they are certain they do not refer to reality.

ALBERT EINSTEIN

Because variation is everywhere, statistical conclusions are not certain. Most women who reach middle age have regular mammograms to detect breast cancer. Do mammograms really reduce the risk of dying of breast cancer? High-quality statistical studies find that mammograms reduce the risk of death in women aged 50 to 64 years by 26%. That's an average over all women in the age-group. Because variation is everywhere, the results are different for different women. Some women who have mammograms every year die of breast cancer, and some who never have mammograms live to 100 and die when they crash their motorcycles.

What the summary study actually said was "mammography reduces the risk of dying of breast cancer by 26% (95% confidence interval, 17% to 34%)."[2] That 26% is, in Arthur Nielsen's words, "shorthand for a range that describes our actual knowledge of the underlying condition." The range is 17% to 34%, and we are 95% confident that the truth lies in that range. We're pretty sure, in other words, but not certain. Once you get beyond news reports, you can look for words like "95% confident" and "statistically significant" that tell us that a study did produce findings that, while not certain, are pretty sure.

Data reflect social values

It's easy to lie with statistics. But it is easier to lie without them.

FREDERICK MOSTELLER

Good data do beat anecdotes. Data are more objective than anecdotes or loud arguments about what might happen. Statistics certainly lies on the factual, scientific, rational side of public discourse. Statistical studies deserve more weight than most other evidence about controversial issues. There is, however, no such thing as perfect objectivity. Statistics shares a social context that influences what we decide to measure and how we measure it.

Suicide rates, for example, vary greatly among nations. It appears that much of the difference in the reported rates is ascribable to social attitudes rather than to actual differences in suicides. Counts of suicides come from death certificates. The officials who complete the certificates (details vary depending on the state or nation) can choose to look more or less closely at, for example, drownings and falls that lack witnesses. Where suicide is stigmatized, deaths are more often reported as accidents. Countries that are predominantly Catholic have lower reported suicide rates than others, for example. Japanese culture has a tradition of honorable suicide as a response to shame. This tradition leads to better reporting of suicide in Japan because it reduces the stigma attached to suicide. In other nations, changes in social values may lead to higher suicide counts. It is becoming more common to view depression as a medical problem rather than a weakness of character and suicide as a tragic end to the illness rather than a moral flaw. Families and doctors then become more willing to report suicide as the cause of death.[3]

Social values influence data on matters less sensitive than suicide. The percent of people who are unemployed in the United States is measured each month by the Bureau of Labor Statistics, using a large and very professionally chosen sample of people across the country. But what does it mean to be "unemployed"? It means that you don't have a job even though you want a job and have *actively looked for work in the last two weeks.* If you went two weeks without seeking work, you are not unemployed; you are "out of the labor force." This definition of unemployment reflects the value we attach to working. A different definition might give a very different unemployment rate.

Our point is not that you should mistrust the unemployment rate. The definition of "unemployment" has been stable over time, so that we can see trends. The definition is reasonably consistent across nations, so that we can make international comparisons. The data are produced by professionals free of political interference. The unemployment rate is important and useful information. Our point is that *not everything important can be reduced to numbers* and that reducing things to numbers is done by people influenced by many pressures, conscious and unconscious.

STATISTICS and YOU

What Lies Ahead in This Book

This isn't a book about the tools of statistics. It is a book about statistical ideas and their impact on everyday life, public policy, and many different fields of study. You will learn some tools, of course. Life will be easier if you have a calculator with statistical capability available. On the other hand, you need little formal mathematics. If you can read and use simple equations, you are in good shape. Be warned, however, that you will be asked to think. Thinking exercises the mind more deeply than following mathematical recipes. *Statistics Through Applications* presents statistical ideas in four parts:

I. Producing Data describes methods for data production that can give clear answers to specific questions. Where the data come from really is important—basic concepts about how to select samples and design experiments are the most influential ideas in statistics.

II. Organizing Data concerns methods and strategies for exploring, organizing, and describing data using graphs and numerical summaries. You can learn to look at data intelligently even with quite simple tools.

III. Chance is the language we use to describe probability, variation, and risk. Because variation is everywhere, probabilistic thinking helps separate reality from background noise.

IV. Inference moves beyond the data in hand to draw conclusions about some wider universe, taking into account that variation is everywhere and that conclusions are uncertain.

Statistical ideas and tools emerged only slowly from the struggle to work with data. Two centuries ago, astronomers and surveyors faced the problem of combining many observations that, despite the greatest care, did not exactly match. Their efforts to deal with variation in their data produced some of the first statistical tools. As the social sciences emerged in the nineteenth century, old statistical ideas were transformed and new ones were invented to describe the variation in individuals and societies. The study of heredity and of variable populations in biology brought more advance. The first half of the twentieth century gave birth to statistical designs for producing data and to statistical inference based on probability. By mid-century it was clear that a new discipline had been born. As all fields of study place more emphasis on data and increasingly recognize that variability in data is unavoidable, statistics has become a central intellectual method. Every educated person should be acquainted with statistical reasoning. Reading this book will enable you to make that acquaintance.

Today, health professionals need statistics to read accounts of medical research; managers need statistics because efficient crunching of numbers will find its way to the bottom line; citizens need statistics to understand opinion polls and the consumer price index. Because data and chance are all around us, you will find statistics useful, and perhaps even profitable.

Part I

Producing Data

PEANUTS reprinted by permission of United Feature Syndicate, Inc.

What kind of music do you like? Rap? Grunge? Ska? Country? Some of your friends probably share your taste in music. But you and your friends are not typical. Your parents and grandparents likely have different tastes. They are not typical either. To get a true picture of the country as a whole (or even of high school students), you must recognize that the picture may not resemble you or what you see around you. You need *data*.

Data from retail sales show that the top-selling music types are rock (24.7% of all recorded music sold in 2002) and rap (13.8%).[1] You might like rhythm and blues (11.2%) and I might like country (10.7%), but that doesn't mean we have a clue about the tastes of the music-buying public as a whole. If we are in the music business, or even if we are interested in pop culture, we must put our own tastes aside and look at the data.

You can find data in the library or on the Internet (that's where I found the music sales data). How can we know whether data can be trusted? Good data are as much a human product as shoes and DVD players. Sloppily produced data will frustrate you as much as a sloppily made pair of shoes. You examine shoes before you buy, and you don't buy if they are not well made. Neither should you use data that are not well made. The first part of this book shows how to tell if data are well made.

1

APPLICATION The Smallpox Debate

September 11, 2001, changed our world. Since then, prevention of future terrorist attacks has become a major priority. At airports, all baggage is now X-rayed. Many passengers are searched at security checkpoints. Others are selected for "random screenings" prior to boarding. Profiling is used to identify potentially dangerous individuals. Similar precautions are taken at major sporting events, office buildings, and theme parks.

In December 2002, President George W. Bush ordered smallpox vaccinations for soldiers, health-care workers, and some emergency personnel. His action resulted from the increased likelihood of a war with Iraq and concerns about possible biological attacks. Critics argued that the vaccinations were too risky. The smallpox vaccine makes some people very ill and kills one or two of every million people who receive it.

Due to the risks involved, research into a new vaccine began immediately. The U.S. government designed a study involving 105 volunteers to test whether a vaccine called MVA would protect against smallpox. MVA has been shown to be safe enough to give to HIV/AIDS victims, whose immune systems are already weak. A U.S. and a Japanese firm teamed up to investigate another vaccine, one that had been used safely on 50,000 Japanese children in the 1970s. Another private company tested a third vaccine on animals.[2]

A survey of 1006 randomly selected U.S. adults by the Harvard School of Public Health revealed many incorrect beliefs about smallpox and vaccinations. The results of the survey were published in the *New England Journal of Medicine* on January 30, 2003. Many believed that there had been cases of smallpox in the past five years (the last known case was in 1977), and 61% said they would choose to be vaccinated if it was offered as a precaution against terrorist attack.[3] The article's authors recommended educating the public about smallpox. They also cautioned against vaccinating all doctors until more research could be completed.

1. Do some research: What new government department was formed as a result of the September 11 attacks? What government agency is now responsible for airport safety?

2. If you were given the option to receive the existing smallpox vaccine, would you accept? Justify your answer.

3. Why would a company test a new vaccine on animals rather than on people?

4. Based on the Harvard School of Public Health survey, do we know that 61% of all U.S. adults would choose to be vaccinated if they were given the option? Justify your answer.

5. Do some more research: What progress has been made since January 2003 in developing a less risky smallpox vaccine and in educating the public about the virus?

Chapter 1

How Do We Get "Good" Data?

1.1 WHERE DO DATA COME FROM?

You can't see just by watching

You can hardly go a day without meeting data and statistical studies. Have you ever wondered how the numbers were produced? Consider these examples.

* In the year 2000, 14% of drivers involved in fatal crashes were 15 to 20 years old, according to the National Highway Traffic Safety Administration.[4]

* According to researchers in the Netherlands, people who are 30 or more pounds overweight could lose up to 7 years from their life expectancy.[5]

* The Gallup Poll reports (January 10, 2003) that 46% of U.S. adults engage in "vigorous exercise" at least once a week.

* A major medical study concluded that taking aspirin regularly reduces the risk of a heart attack.

Where do these data come from? Why can we trust them? Or maybe we can't trust them. It may be, as Yogi Berra says, that "you can observe a lot by watching." But you can't see just by watching that 46% of adults exercise vigorously or that aspirin reduces heart attack risk. Good data are the fruit of intelligent human effort. Bad data result from laziness or lack of understanding, or even the desire to mislead others. "Where do the data come from?" is the first question you should ask when someone throws a number at you.

ACTIVITY 1.1 Make your own measurements

Materials: *Unlined paper (8.5 × 11 inches), calculator*

1. Prepare your "special ruler" by placing a piece of unlined paper over this page. Carefully trace the ruler below.

2. Use the ruler to measure the length of this textbook. Record your measurement to the nearest tenth of a unit (for example, 6.3 or 12.7).

NOTE: Don't tell anyone else the value you found. It may influence their measurement.

3. Record the measurements for your class.

4. Make a simple graph of these measurements as follows:

   - Draw a number line.
   - Add a scale with tick marks 0.1 unit apart. Make sure you cover the range of your data.
   - Place a dot above the appropriate value for each measurement.
   - Label your axis "textbook length."

Your finished plot (called a dotplot) should look something like this.

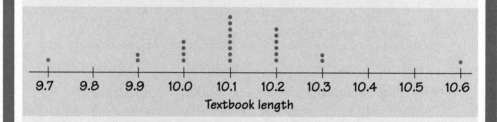

5. Now calculate the average textbook length measurement for your class. Compare it with the value provided by your teacher.

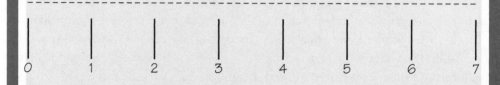

You want data on a question of interest to you. Do cell phones cause cancer? Why did pollsters incorrectly predict the 2000 presidential election? Does alternative medicine really work? Or you are writing an essay on homelessness and wonder how many people are homeless. How will you get relevant data?

It is tempting to base conclusions on your own experiences. You probably know several people who use cell phones. Do they have cancer? And polls aren't always right, are they? If one of your relatives swears that acupuncture relieves pain, you might be persuaded to try it. But our own experiences may not be typical. In fact, the incidents that stick in our memory are often unusual. We are much safer relying on data.

EXAMPLE 1.1   Counting the homeless

Homeless people are visible in any large city. They sleep where they can, trying to avoid the elements. How many homeless are there? It is difficult to tell. A few chance encounters with the homeless don't help us estimate their numbers. Before any careful data collection took place, estimates of the number of homeless in the United States ranged as high as 3 million. For the city of Chicago, estimates of 25,000 were common. These numbers, though frequently quoted by the news media, were simply guesses. The first attempts to make a careful count gave much lower estimates: about 350,000 homeless nationwide and between 2000 and 3000 in Chicago on any given night. The Chicago study was especially convincing because it involved a thorough search of several hundred city blocks chosen at random as well as visits to all shelters in the city.[6]

Talking about data: individuals and variables

Statistics is the art and science of dealing with data. Good judgment, good math, and even good taste make good statistics. A big part of good judgment is deciding what to measure in order to produce data that help answer your questions. Measurements are made on *individuals* and organized in *variables*.

Individuals and variables

Individuals are the objects described by a set of data. Individuals may be people, animals, or things.

A **variable** is any characteristic of an individual. A variable can take different values for different individuals.

For example, here is a small part of a data set from the Cyber Stat Corporation:

	A	B	C	D	E	F
1	Name	Job Type	Age	Gender	Race	Salary
2	Cedillo, Jose	Technical	27	Male	White	52,300
3	Chambers, Tonia	Management	42	Female	Black	112,800
4	Childers, Amanda	Clerical	39	Female	White	27,500
5	Chen, Huabang	Technical	51	Male	Asian	83,600
6						
7						
8						

Ready   NUM

The individuals are company employees. In addition to each employee's name, there are five variables. The first says what type of job an employee holds. The second variable gives the age, the third records gender, and the fourth reports race. The fifth variable records employee salaries.

Statistics deals with numbers, but not all variables are numerical. Of the five variables in the company's data set, only age and salary have numbers as values. To do statistics with the other variables, we use *counts or percents*. We might give the percent of employees who work in management, for example, or the percent who are Asian.

Bad judgment in choosing variables can lead to data that cost lots of time and money but that don't tell us much. What constitutes good judgment can be controversial. Here are examples of the challenges in deciding what data to collect.

EXAMPLE 1.2   Who recycles?

Who takes the trouble to recycle? Researchers spent lots of time and money weighing the stuff put out for recycling in two neighborhoods in a California city, call them Upper Crust and Lower Mid. The *individuals* here are households, because trash and recycling pickup are done for residences. The *variable* measured was the weight in pounds of the curbside recycling basket each week.

The Upper Crust households contributed more pounds per week on average than the folk in Lower Mid. Can we say that the rich are more serious about recycling? No. Someone noticed that Upper Crust recycling baskets contained lots of heavy glass wine bottles. In Lower Mid, they put out lots of light plastic soda bottles and light metal beer and soda cans. Weight tells us little about commitment to recycling.[7]

EXAMPLE 1.3 What's your race?

The U.S. census asks "What is this person's race?" for every person in every household. "Race" is a *variable,* and the Census Bureau must say exactly how to measure it. The census form does this by giving a list of races. Years of political fighting lie behind this list.

How many races should we list, and what names should we use for them? Should we have a category for people of mixed race? Asians wanted more national categories, such as Filipino and Vietnamese, for the growing Asian population. Pacific Islanders wanted to be separated from the larger Asian group. Black leaders did not want a mixed-race category, fearing that many blacks would choose it and so reduce the official count of the black population.

The 2000 census form (page 14) ended up with seven Asian groups (including "Other Asian") and four Pacific Island groups (including "Other Pacific Islander"). There is no "mixed-race" group, but you can mark more than one race. So the total of the racial group counts in 2000 is larger than the population count. Unable to decide what the proper term for blacks should be, the Census Bureau settled on "Black, African American, or Negro." What about Hispanics? That's a separate question, because Hispanics can be of any race. Again unable to choose a short name that would satisfy everyone, the Census Bureau asked if you are "Spanish/Hispanic/Latino."

The fight over "race" reminds us that data reflect society. Race is a social idea, not a biological fact. In the census, you say what race you consider yourself to be. Race is a sensitive issue in the United States, so the fight is no surprise. The Census Bureau's diplomacy seems a good compromise.

EXERCISES

1.1 Making the grade Here are a few lines from a statistics teacher's grade book:

Name	Sex	Homeroom	Gr.	Calculator no.	Test 1
Hsu, Danny	M	Blair	12	B319	81
Iris, Francine	F	Kingsley	12	B298	92
Ruiz, Ricardo	M	Alfonso	11	B304	87

(a) What individuals does this data set describe?
(b) For each individual, what variables are given? Which of these variables take numerical values?

1.2 Go Bucs! In Super Bowl XXXVII, the Tampa Bay Buccaneers defeated the Oakland Raiders. Here is a portion of Tampa Bay's team roster:

No.	Name	Pos.	Ht.	Wt.	Birth date	Exp.	College
14	Brad Johnson	QB	77	224	09/13/68	9	Florida State
99	Warren Sapp	DT	74	303	12/06/72	8	Miami
7	Martin Gramatica	K	68	170	11/27/75	4	Kansas State
9	Tom Tupa	P	76	235	02/06/66	14	Ohio

(a) What individuals does this data set describe?

(b) How many variables does the data set contain?

(c) What do you think are the units for each of the numerical variables?

1.3 Who recycles? In Example 1.2 (page 6), weight is not a good measure of the participation of households in different neighborhoods in a city recycling program. What variables would you measure in its place?

1.4 Choosing a college Popular magazines rank colleges and universities on their "academic quality." Describe five variables that you would like to see measured for each college if you were choosing where to study. Give reasons for your choices.

1.5 Activity 1.1 follow-up In Activity 1.1 (page 4),

(a) What individuals were measured?

(b) What variables did you record?

(c) Why did the measurements vary from student to student?

1.6 Chart toppers Visit the Billboard Music Web site at www.billboard.com.

(a) What are the top five songs on the Billboard Hot 100 chart?

(b) How do you think they determined these rankings?

Observational studies

Sometimes all you can do is watch. To learn how chimpanzees in the wild behave, watch. To study how a teacher and young children interact in a schoolroom, watch. It helps if the watcher knows what to look for. The chimpanzee expert may be interested in how males and females interact, in whether some chimps in the troop are dominant, in whether the chimps hunt and eat meat. In fact, chimps were thought to be vegetarians until Jane Goodall watched them carefully in Gombe National Park, Tanzania. Now it is clear that meat is a natural part of the chimpanzee diet.

At first, the observer may not know what to record. Eventually patterns seem to emerge. Then we can decide what variables we want to measure. How often do chimpanzees hunt? Alone or in groups? How large are hunting groups? Males alone, or both males and females? How much of the diet is meat? Observation that is organized and measures clearly defined variables is more convincing than just watching. Here is an example of highly organized observation.

EXAMPLE 1.4   Pain and suffering

An Associated Press news article (April 20, 2000) begins, "People hurt in traffic accidents actually recover more quickly when they cannot collect money for their pain and suffering, researchers say in a new study."[8] The Canadian province of Saskatchewan changed its insurance laws on January 1, 1995. The old system allowed lawsuits for "pain and suffering." The new no-fault system paid for medical costs and lost work but not for subjective suffering. The study looked at insurance claims filed between July 1, 1994, and December 31, 1995, on either side of the change. Sure enough, under the new system not only were fewer claims for whiplash neck injuries filed, but faster recovery with less pain was reported in the claims that were filed. This is a *comparative observational study*.

It is possible that the effect of the change in insurance system is mixed up with some other difference between the two time periods. The study gives reasonably convincing evidence that people report less pain when they can't collect money for it. But the study's design prevents us from concluding that the change in insurance law *causes* a decrease in pain. Only well-designed experiments allow us to establish causation.

Why didn't the Saskatchewan government conduct an experiment to test their hunch about "pain and suffering"? This would have required allowing some accident victims to seek money for their pain and suffering and preventing others from doing so. Ideally, each accident victim would be assigned to one of these options by a chance process, such as a coin toss. You can imagine the ethical objections that this might raise! In situations where an experiment isn't practical, a carefully designed observational study is still far better than educated guesswork.

Observational study

An **observational study** observes individuals and measures variables of interest but does not attempt to influence the responses. The purpose of an observational study is to describe some group or situation.

Sample surveys

"You don't have to eat the whole ox to know that the meat is tough." That is the idea of sampling: to gain information about the whole by examining only a part. **Sample surveys** are an important kind of observational study. They survey some group of individuals by studying only some of its members, selected not because they are of special interest but because they represent the larger group. Here is the vocabulary we use to discuss sampling.

> ## Populations and samples
>
> The **population** in a statistical study is the entire group of individuals about which we want information.
>
> A **sample** is a part of the population from which we actually collect information, which is then used to draw conclusions about the whole.

Notice that the *population* is the group we want to study. If we want information about all U.S. high school students, that is our population even if students at only one high school are available for sampling. To make sense of any sample result, you must know what population the sample represents. Did that pre-election poll, for example, ask the opinions of all adults? Citizens only? Registered voters only? Democrats only? The *sample* consists of the people we actually have information about. If the poll can't contact some of the people it selected, those people aren't in the sample.

The distinction between population and sample is basic to statistics. The following examples illustrate this distinction and also introduce some major uses of sampling. These brief descriptions also indicate the variables measured for each individual in the sample.

EXAMPLE 1.5   Public opinion polls

Polls such as those conducted by Gallup and many news organizations ask people's opinions on a variety of issues. The variables measured are responses to questions about public issues. Though most noticed at election time, these polls are conducted on a regular basis throughout the year. For a typical opinion poll:

Population: U.S. residents 18 years of age and over. Noncitizens and even illegal immigrants are included.

Sample: Between 1000 and 1500 people interviewed by telephone.

EXAMPLE 1.6   The Current Population Survey

Government economic and social data come from large sample surveys of a nation's individuals, households, or businesses. The monthly Current Population Survey (CPS) is the most important government sample survey in the United States. Many of the variables recorded by the CPS concern the employment or unemployment of everyone over 16 years old in a household. The government's monthly unemployment rate comes from the CPS. The CPS also records many other economic and social variables. For the CPS:

Population: The more than 100 million U.S. households. Notice that the individuals are households rather than people or families. A household consists of all people who share the same living quarters, regardless of how they are related to each other.

Sample: About 50,000 households interviewed each month.

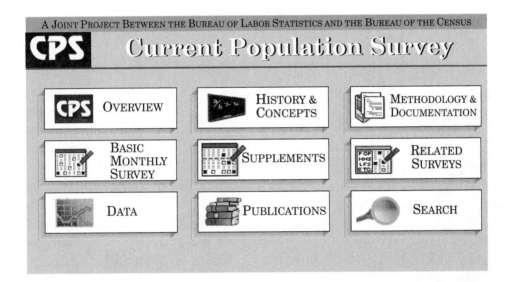

EXAMPLE 1.7   TV ratings

Market research is designed to discover what consumers want and what products they use. One example of market research is the television-rating service of Nielsen Media Research. The Nielsen ratings influence how much advertisers will pay to sponsor a program and whether or not the program stays on the air. For the Nielsen national TV ratings:

Population: The 100 million U.S. households that have a television set.

Sample: About 5000 households that agree to use a "people meter" to record the TV viewing of all people in the household.

The *variables* recorded include the number of people in the household and their age and sex, whether the TV set is in use at each time period, and, if so, what program is being watched and who is watching it.

EXAMPLE 1.8   The General Social Survey

Social science research makes heavy use of sampling. The General Social Survey (GSS), carried out every second year by the National Opinion Research Center at the University of Chicago, is the most important social science sample survey. The *variables* cover the subject's personal and family background, experiences and habits, and attitudes and opinions on subjects from abortion to war.

Population: Adults (age 18 and over) living in households in the United States. The population does not include adults in institutions such as prisons and college dormitories. It also does not include persons who cannot be interviewed in English.

Sample: About 3000 adults interviewed in person in their homes.

You just don't understand

A sample survey of journalists and scientists found quite a communications gap. Journalists think that scientists are arrogant, while scientists think that journalists are ignorant. We won't take sides, but here is one interesting result from the survey: 82% of the scientists agree that the "media do not understand statistics well enough to explain new findings" in medicine and other fields.

Most statistical studies use samples in the broad sense. We usually reserve the dignified term "sample survey" for studies that use an organized plan to choose a sample that represents some specific population. Consider the people who had traffic accidents in 1995 from Example 1.4 (page 9). They are supposed to represent all accident victims (not just those from 1995) following the change in insurance laws. Expert judgment says they are probably typical, but we can't be sure. A sample survey doesn't rely on judgment: it starts with an entire population and *chooses* a sample to represent it. Chapter 2 discusses the art and science of sample surveys.

EXERCISES

1.7 Populations and samples, I For each of the following sampling situations, identify the population and the sample as exactly as possible.

(a) A furniture maker buys hardwood in large lots. The supplier is supposed to dry the wood before shipping (wood that is not dry won't hold its size and shape). The furniture maker chooses five pieces of wood from each lot and tests their moisture content. If any piece exceeds 12% moisture content, the entire lot is sent back.

(b) An insurance company wants to monitor the quality of its procedures for handling loss claims from its auto insurance policyholders. Each month the company selects a sample from all auto insurance claims filed that month to examine them for accuracy and promptness.

1.8 Populations and samples, II For each of the following sampling situations, identify the population and the sample as exactly as possible.

(a) A business school researcher wants to know what factors affect the survival and success of small businesses. She selects a sample of 150 eating-and-drinking establishments from those listed in the Yellow Pages for a large city.

(b) Your local television station wonders if its viewers would rather watch a local college basketball team play or an NBA game scheduled at the same time. It announces that it will show the NBA game and receives 89 calls asking that it show the local game instead.

1.9 Bad apples A truckload of apples arrives at an apple juice production plant. The plant's quality control team selects three large buckets of apples from various locations within the truck. These apples are inspected carefully. Based on inspection results, the entire truckload is either accepted or rejected by the plant. Identify the population, sample, individuals, and variable(s) in this setting.

1.10 J. K. Rowling's words Different types of writing can sometimes be distinguished by the lengths of the words used. A student interested in this fact wants to study the lengths of words used by J. K. Rowling in her Harry Potter books. She opens a Harry Potter book at random and records the lengths of the first 250 words on the page.

(a) What is the population in this study? What is the sample?

(b) What variable does the student measure?

(c) Obtain a copy of one of the Harry Potter books and collect the data described above.

(d) Make a dotplot of your data. Write a few sentences describing what the graph tells you.

1.11 Nielsen ratings Find a current copy of the Nielsen television ratings. (Many local newspapers report these ratings weekly.)

(a) What individuals are being measured?

(b) What numerical variables are recorded, and in what units?

(c) How were the data produced?

Census

A sample survey looks at only a part of the population. Why not look at the entire population? A *census* tries to do this.

Census

A **census** is a sample survey that attempts to include the entire population in the sample.

The U.S. Constitution requires a census of the American population every 10 years. A census of so large a population is expensive and takes a long time. Even the federal government, which can afford a census, uses samples such as the Current Population Survey to produce timely data on employment and many other variables. If the government asked every adult in the country about his or her employment, this month's unemployment rate wouldn't be available until next year. Even the every-10-years census includes a sample survey. A census "long form" that asks many more questions than the basic census form is sent to a sample of one-sixth of all households. So time and money favor samples over a census. Samples can have other advantages as well. If you are testing fireworks or fuses, the individuals in the sample are destroyed. Moreover, a sample can produce more accurate data than a census. A careful sample of an inventory of spare parts will almost certainly give more accurate results than asking the employees to count all 500,000 parts in the warehouse. Bored people do not count accurately.

The experience of the Census Bureau reminds us that a census can only attempt to sample the entire population. The bureau estimates that the 1990 census missed 1.8% of the American population. These missing persons included an estimated 4.4% of the black population, largely in inner cities.[9] In 2000, the census overcounted the U.S. population by about 1.3 million. Millions of people who live in two places—like college students—were counted twice. About 1.8% of blacks and 0.7% of Hispanics were not counted at all. A census is not foolproof, even with the resources of the government behind it. Why take a census at all? The government needs block-by-block population figures to create election districts with equal population. The main function of the U.S. census is to provide this local information.

Experiments

Our goal in choosing a sample is a picture of the population, disturbed as little as possible by the act of gathering information. All observational studies share the principle "observe but don't disturb." When Jane Goodall first began observing chimpanzees in Tanzania, she set up a feeding station where the chimps could eat bananas. She later said that was a mistake, because it might have changed the apes' behavior.

In *experiments*, on the other hand, we want to change behavior. In doing an experiment, we don't just observe individuals or ask them questions. We actively impose some treatment in order to observe the response. Experiments can answer questions such as "Does aspirin reduce the chance of a heart attack?" and "Do a majority of high school students prefer Pepsi to Coke when they taste both without knowing which they are drinking?"

Is a census old-fashioned?

The United States has taken a census every 10 years since 1790. Technology marches on, however, and replacements for a national census look promising. Denmark has no census, and France plans to eliminate its census. Denmark has a national register of all its residents, who carry identification cards and change their register entry whenever they move. France will replace its census by a large sample survey that rotates among the nation's regions. The U.S. Census Bureau has a similar idea: the American Community Survey has already started and will eliminate the census "long form" after 2000.

"Now eat that banana. The nice statistician is watching us."

Experiments

An **experiment** deliberately imposes some treatment on individuals in order to observe their responses. The purpose of an experiment is to study whether the treatment causes a change in the response.

EXAMPLE 1.9   Curing a cold

It has been claimed that large doses of vitamin C will prevent colds. An experiment to test this claim was performed in Toronto during the winter months. About 500 volunteer subjects were assigned at random to each of two groups. Group 1 received 1 gram per day of vitamin C and 4 grams per day at the first sign of a cold. (This is a large amount of vitamin C; the recommended daily allowance of this vitamin for adults is only 60 milligrams, or 60/1000 of a gram.) Group 2 served as a control group and received a placebo pill identical in appearance to the vitamin C capsules. Both groups were regularly checked for illness during the winter.

Some of the subjects dropped out of the experiment for various reasons, but 818 completed at least two months. Groups 1 and 2 were very similar in age, occupation, smoking habits, and other extraneous variables. At the end of the winter, 26% of the subjects in Group 1 had not had a cold, compared with 18% in Group 2. Thus, vitamin C did appear to prevent colds better than the placebo, but not much better.[10]

The vitamin C example illustrates the big advantage of experiments over observational studies: *In principle, experiments can give good evidence for cause and effect.* If we design the experiment properly, we start with two very similar groups of subjects. The *individual* subjects of course differ from each other in age, occupation, smoking habits, and other respects. But the two *groups* resemble each other when we look at those variables for all subjects in each group. During the experiment, the subjects' lives differ, but there is only one systematic difference between the two groups: whether they take vitamin C or a placebo. So we should be able to say whether taking vitamin C reduces the likelihood of getting a cold.

The fact that experiments can give good evidence that a treatment causes a response is one of the big ideas of statistics. A big idea needs a big caution: statistical conclusions hold "on the average" for groups of individuals. They don't tell us much about one individual. *On the average,* the subjects taking vitamin C had fewer colds than those who were taking a placebo. That says vitamin C was somewhat effective. It doesn't say everyone who takes vitamin C will be healthy. And a big idea may also raise big questions: If we think vitamin C will prevent colds, is it ethical to offer it to some and not to others? Chapter 3 explains how to design good experiments and looks at ethical issues.

Reprinted with special permission of King Features Syndicate.

EXERCISES

1.12 Cell phones and cancer One study of cell phones and the risk of brain cancer looked at a group of 469 people who have brain cancer. The investigators matched each cancer patient with a person of the same sex, age, and race who did not have brain cancer, then asked about use of cell phones.[11] Result: "Our data suggest that use of hand-held cellular telephones is not associated with risk of brain cancer."
(a) Is this an observational study or an experiment? Why?
(b) What individuals are measured, and what variables are recorded?

1.13 Tasty muffins Before a new variety of frozen muffin is put on the market, it is subjected to extensive taste testing. People are asked to taste the new muffin and a competing brand and to say which they prefer. Is this an observational study or an experiment? Explain your answer.

1.14 The political gender gap There may be a "gender gap" in political party preference in the United States, with women more likely than men to prefer Democratic candidates. A political scientist interviews a large sample of registered voters, both men and women. She asks each voter whether they voted for the Democratic or the Republican candidate in the last congressional election. Is this study an experiment? Why or why not? What variables does the study measure?

1.15 Choose your study purpose Give an example of a question about high school students, their behavior, or their opinions that would best be answered by
(a) A sample survey
(b) An observational study that is not a sample survey
(c) An experiment

1.16 Child care In 2001, researchers announced that "children who spend most of their time in child care are three times as likely to exhibit behavioral problems in kindergarten as those who are cared for primarily by their mothers."[12]
(a) Was this an observational study or an experiment? Why?
(b) Can we conclude from this study that child care causes behavior problems? Why or why not?

1.17 Teenage drivers Go to the National Highway Traffic Safety Administration (NHTSA) Web site: www.nhtsa.dot.gov. Locate a traffic safety fact about young drivers that includes a supporting chart or graph. Print the relevant information.
(a) Summarize your safety fact with a few well-written sentences in your own words. Be sure to describe what the chart or graph tells you.
(b) How did the NHTSA obtain these data: from a census, survey, observational study, or experiment? Justify your answer.

EXPLORING THE WEB

All of the sample surveys mentioned in Examples 1.5 to 1.8 maintain Web sites:
- Gallup Poll (Example 1.5, page 10): www.gallup.com
- Current Population Survey (Example 1.6, page 10): www.bls.gov/cps/
- Nielsen Media Research (Example 1.7, page 11): www.nielsenmedia.com
- General Social Survey (Example 1.8, page 12): www.norc.org

APPLICATION 1.1   Making Sense of the Census

The U.S. Census Bureau Web site at www.census.gov contains a vast amount of data about the U.S. population. In this application, you will take a closer look at where these data come from.

1. **Population Clocks**

   * What are the current U.S. and world populations, according to the Population Clocks?

   * How is the U.S. population figure obtained?

   * When did the world population reach 6 billion?

2. **Census 2000**

   * What was the U.S. population on April 1, 2000, according to the decennial census?

   * What percent of the U.S. population was male? How many people were ages 15 to 19?

   * What percent of the U.S. population was Hispanic? How does this compare with the Hispanic population in the 1990 census?

   * How did the Census Bureau obtain its data? (Look in the Operations section of the Census 2000 page for more information.)

3. **State and County Quick Facts**

   * For your state, what was the population in 2000?

   * For your state, what was the average (mean) travel time to work in 2000? How do you think this value was obtained?

4. **The American Community Survey (ACS)**

   * Why does the Census Bureau conduct surveys, such as the ACS?

   * How many people does the ACS survey, and how are they selected?

5. **Subjects A through Z**

   * Computer use and ownership: In 2000, what percent of homes had a computer? What percent had Internet access? Where did these data come from?

   * Educational attainment: How do males and females compare in terms of bachelor's, master's, and higher degrees earned in 2000? Where do these data come from?

STATISTICS IN SUMMARY

Any statistical study records data about some **individuals** (people, animals, or things) by giving the value of one or more **variables** for each individual. Some variables, such as age and income, take numerical values. Others, such as occupation and sex, do not. Be sure the variables in a study really do tell you what you want to know.

The most important fact about any statistical study is how the data were produced. **Observational studies** try to gather information without disturbing the scene they are observing. **Sample surveys** are an important kind of observational study. A sample survey chooses a **sample** from a specific **population** and uses the sample to get information about the entire population. A **census** attempts to measure every individual in a population. **Experiments** actually do something to individuals in order to see how they respond. The goal of an experiment is usually to learn whether some treatment actually causes a certain response.

SECTION 1.1 EXERCISES

1.18 Fast food Many people eat fast food as a regular part of their diet. Is fast food unhealthy? The best answer may be, "It depends on what you eat." Here are some nutritional data about some popular fast-food burgers:

Burger	Restaurant	Calories	Fat	Cholesterol	Sodium
Quarter Pounder with cheese	McDonald's	430	30	95	1310
Classic Single with everything	Wendy's	410	19	70	890
Whopper with cheese	Burger King	706	43	113	1164
Cheeseburger	In-N-Out	480	27	60	1000

(a) What individuals and variables are recorded in this data set?
(b) In what units is each numerical variable measured?
(c) Are fast-food chicken sandwiches healthier than burgers? Collect some data for yourself. (Most fast-food restaurants will provide nutritional information on request. You can also search on the Web.)

1.19 Public housing To study the effect of living in public housing on family stability in poverty-level households, researchers obtain a list of all applicants for public housing in Chicago last year. Some applicants were accepted, while others were turned down by the housing authority. The researchers interview both groups and compare them. Is this an experiment, a sample survey, or an observational study that is not a sample survey? Explain your answer.

1.20 Baking bread A flour company wants to know what fraction of Minneapolis households bake some or all of their own bread. A sample of 500 residential addresses is taken, and interviewers are sent to those addresses. The interviewers are employed during regular working hours on weekdays and interview only during those hours.
(a) What population is the flour company interested in? What is the sample?
(b) Do you think this sample will provide accurate information to the flour company? Why or why not?

1.21 Physical fitness and leadership A study of the relationship between physical fitness and leadership uses as subjects middle-aged executives who have volunteered for an exercise program. The executives are divided into a low-fitness group and a high-fitness group on the basis of a physical examination. All subjects then take a psychological test designed to measure leadership, and the results for the two groups are compared.
(a) Is this an experiment? Explain your answer.
(b) We would prefer a sample survey to using men who volunteer for a fitness program. What population does it appear that the investigators were interested in? What variables did they measure?

1.22 Cool cars You and your friends want to find out which student at your school has the "coolest" car.
(a) What variables would you record? Which of these take numerical values?
(b) What are the individuals in your data set?
(c) Would you use a census, an observational study, a survey, or an experiment to collect your data? Explain.

1.23 Angry bees My grandmother once told me that the color red makes bees angry. Here's a method I've designed to test her claim. I'll select half of my students (by drawing names from a hat) to wear red clothes and the other half to wear white clothes. Then I'll turn a bunch of bees loose in our classroom and record how many times each student is stung.
(a) Is this an observational study or an experiment? Why?
(b) What variables are recorded?
(c) If students wearing red clothes are stung much more often than students wearing white, can we conclude that the color red *causes* bees to get angry? Why or why not?
(d) Comment on any flaws you see with my methods.

1.2 MEASURING

Got some spare time?

Do people have more or less free time now than they did a generation ago? A book titled *The Overworked Americans* says we are working more than ever. Another book, titled *Time for Life,* says that we have more free time than ever. Is somebody lying with statistics?[13]

To see if free time is increasing, we must **measure** "free time." That is, we must reduce the vague idea to a number that can go up or down. The first step is to say what we mean by free time: something like "time when you're not working, not traveling to or from work, not doing household chores, not . . ." You see that we can get different numbers by changing our list of "nots."

Once we decide what free time is, we must actually produce the numbers. We might ask a sample of people how they spent their time yesterday. Don't trust their memory? Afraid they will exaggerate how long they worked? If so, we might ask people to keep a diary of how they spend their time today. Of course, the really busy people may forget to record all their work time. Not only is it hard to say exactly what "free time" is, but it's hard to attach a number to measure whatever we say it is.

Sample surveys are wonderful. Well-designed experiments are even more wonderful. But in the end, we need to turn ideas like "free time" or "pain relief" or "income" into numbers. Don't trust the numbers until you know how this was done.

" WOW! SIX FEET LONG AND THREE FEET DEEP."

ACTIVITY 1.2A The tilted glass

Materials: Unlined paper, protractor

1. On an unlined sheet of paper, carefully trace the tilted glass.

2. The drawing represents a tilted glass of water. Draw a line to indicate the water level if the glass is as full of water as possible.

3. Use a protractor to measure the acute angle that the bottom of the cup makes with horizontal. Record your answer to the nearest degree.

4. Now measure the angle between the line you have drawn and horizontal to the nearest degree.

5. Record data from each member of your class in a table like this one.

Tilt of glass	Tilt of my line	Gender

6. Your teacher will provide you with the correct values. How do your measurements compare?

7. Make a dotplot to display the "tilt of glass" measurements for the class.

8. Write a few sentences to describe how the class's measurements compare with the correct values.

9. Calculate the average "tilt of glass" measurement for the class.

10. Now look at the "tilt of my line" measurements. Determine whether students' errors were mostly in one direction from the correct answer.

11. Compare the errors made by male and female students on the "tilt of my line" measurements. (You may want to repeat this activity with more students. Then you can see if your class's results hold true in general.)

Measurement basics

Statistics deals with numbers. Planning the production of data through a sample survey or an experiment does not by itself produce numbers. Once we have our sample respondents or our experimental subjects, we must still measure whatever characteristics interest us. First, think broadly: Are we trying to measure the right things? Are we overlooking some outcomes that are important even though they may be hard to measure?

EXAMPLE 1.10   But what about the patients?

Clinical trials tend to measure things that are easy to measure: blood pressure, tumor size, virus concentration in the blood. They often don't directly measure what matters most to patients—does the treatment really improve their lives? One study found that only 5% of trials published between 1980 and 1997 measured the effect of treatments on patients' emotional well-being or their ability to function in social settings.[14]

Once we have decided what properties we want to measure, we can think about how to do the measurements.

Measurement

We **measure** a property of a person or thing when we assign a number to represent the property.

We often use an **instrument** to make a measurement. We may have a choice of the **units** we use to record the measurements.

The result of measurement is a numerical **variable** that takes different values for people or things that differ in whatever we are measuring.

EXAMPLE 1.11   Length, college readiness, highway safety

To measure the length of my bed, I use a tape measure as the *instrument*. I can choose either inches or centimeters as the *unit of measurement*. If I choose centimeters, my *variable* is the length of the bed in centimeters.

To measure a student's readiness for college, I might ask the student to take the SAT Reasoning exam. The exam form is the *instrument*. The *variable* is the student's score in points, somewhere between 400 and 1600 if I combine the verbal and mathematics sections of the SAT.

How can I measure the safety of traveling on the highway? I might decide to count the number of people who die in motor vehicle accidents in a year. The government's Fatal Accident Reporting System collects data on all fatal traffic crashes. I can use the government count of traffic deaths as a *variable* to measure highway safety.

Here are some questions you should ask about the variables in any statistical study:

1. Exactly how is the variable defined?

2. Is the variable a valid way to describe the property it claims to measure?

3. How accurate are the measurements?

We don't often design our own measuring devices (for example, we use the results of the SAT or the Fatal Accident Reporting System), so we won't go deeply into these questions. Any consumer of numbers, however, should know a bit about them.

Know your variables

Measurement is the process of turning concepts like length or employment status into precisely defined variables. Using a tape measure to turn the idea of "length" into a number is straightforward because we know exactly what we mean by length. Measuring college readiness is controversial because it isn't clear exactly what makes a student ready for college work. Using SAT scores at least says exactly how we will get numbers. Measuring free time requires that we first say what time counts as "free." Even counting highway deaths requires us to say exactly what counts as a highway death: Pedestrians hit by cars? People in cars hit by a train at a crossing? People who die from injuries 6 months after an accident? We can simply accept the government's counts, but someone had to answer those and other questions in order to know what to count. For example, a person must die within 30 days of an accident to count as a traffic death. These details are a nuisance, but they can make a difference.

EXAMPLE 1.12   Measuring unemployment

Each month the Bureau of Labor Statistics (BLS) announces the *unemployment rate* for the previous month. People who are not available for work (retired people, for example, or students who do not want to work while in school) should not be counted as unemployed just because they don't have a job. To be unemployed, a person must first be in the labor force. That is, she must be available for work and looking for work. The unemployment rate is

$$\text{unemployment rate} = \frac{\text{number of people unemployed}}{\text{number of people in the labor force}}$$

"Unemployed? Not me, I'm out of the labor force."

To complete the exact definition of the unemployment rate, the BLS has very detailed descriptions of what it means to be "in the labor force" and what it means to be "employed." For example, if you are on strike but expect to return to the same job, you count as employed. If you are not working and did not look for work in the last two weeks, you are not in the labor force. So people who say they want to work but are too discouraged to keep looking for a job don't count as unemployed. The details matter. The official unemployment rate would be different if the government used a different definition of unemployment.

The BLS estimates the unemployment rate based on interviews with the sample in the monthly Current Population Survey. The interviewer can't simply ask "Are you in the labor force?" and "Are you employed?" Many questions are needed to classify a person as employed, unemployed, or not in the labor force. Changing the questions can change the unemployment rate. At the beginning of 1994, after several years of planning, the BLS introduced computer-assisted interviewing and improved its questions. Figure 1.1 on the facing page is a graph of the unemployment rate that appeared on the front page of the BLS monthly news release on the employment situation. There is a gap in the graph in January 1994 because of the change in the interviewing process. The unemployment rate would have been 6.3% under the old system. It was 6.7% under the new system. That's a big enough change to make politicians unhappy.

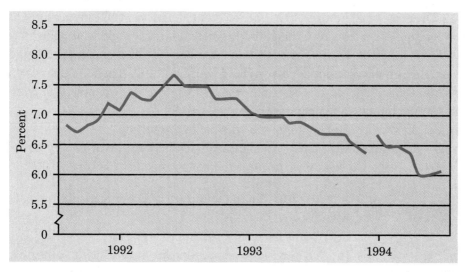

FIGURE 1.1 The unemployment rate from August 1991 to July 1994. The gap shows the effect of a change in how the government measures unemployment.

EXERCISES

1.24 Measuring pain The Department of Veterans Affairs (DVA) offers medical care to 3.4 million patients. It wants doctors and nurses to treat pain as a "fifth vital sign," to be recorded along with blood pressure, pulse, temperature, and breathing rate.
(a) For the four existing vital signs, identify the measuring instrument used, the units of measurement, and the variable.
(b) Help out the DVA: how would you measure a patient's pain?

1.25 Measuring life's quality Is life in Britain getting better or worse? The usual government data don't say. So the British government announced that it wants to add measures of such things as housing, traffic, and air pollution. "The quality of life is not simply economic," said a deputy prime minister. Help them out: How would you measure "traffic" and its impact on the quality of life? Be sure to identify the instrument and units for each variable you suggest.

1.26 Lightning strikes You can measure your distance from a lightning flash by counting the number of seconds between the flash and the thunderclap. Divide your time by 5, and you have the approximate distance in miles.
(a) Identify the instrument, units, and variable in this setting.
(b) Do you think this is a valid way to measure the distance to the lightning strike? Why or why not?
(c) How accurate do you think the measurements would be? Explain.

1.27 The heat is on According to *The Boy Scout Handbook,* you can measure the temperature of a fire using a "hand thermometer." It says, "Hold your palm where the food will be cooking. . . . Count 'one-and-one, two-and-two, . . . ,' and so on, for as many seconds as you can hold your hand still." Based on the number of seconds counted, you can determine the temperature. For example, 2 or 3 seconds would indicate a temperature of 400 to 450 °F.

(a) Identify the instrument, units, and variable in this setting.

(b) Do you think this is a valid way to measure temperature? Why or why not?

(c) How accurate do you think the measurements would be? Explain.

1.28 The BLS Go to the Bureau of Labor Statistics Web site: www.bls.gov.

(a) What is the current unemployment rate?

(b) The BLS is also responsible for measuring the consumer price index (CPI) each month. What is the current CPI? What does this number mean?

(c) Investigate at least one other variable that the BLS measures. Describe what you find in a few sentences.

Valid and invalid measurements

No one would object to using a tape measure reading in centimeters to measure the length of my bed. Many people object to using SAT scores to measure readiness for college. Let's shortcut that debate: just measure the height in inches of all applicants and accept the tallest. Bad idea, you say. Why? Because height has nothing to do with being prepared for college. In more formal language, height is not a *valid* measure of a student's academic background.

What are your units?

Not paying attention to units of measurement can get you into trouble. In 1999, the *Mars Climate Orbiter* spacecraft burned up in the Martian atmosphere. It was supposed to be 93 miles (150 kilometers) above the planet but was in fact only 35 miles (57 kilometers) up. It seems that Lockheed Martin, which built the orbiter, specified important measurements in English units (pounds, miles). The National Aeronautics and Space Administration team who operated the spacecraft thought the numbers were in metric system units (kilograms, kilometers). There went $125 million.

Valid measurement

A variable is a **valid** measure of a property if it is relevant or appropriate as a representation of that property.

It is valid to measure length with a tape measure. It isn't valid to measure a student's readiness for college by recording her height. The BLS unemployment rate is a valid measure, even though changes in the official definitions would give a

"This is our new college admissions sizer. I got the
idea from the carry-on luggage sizers at the airport."

somewhat different measure. Let's think about measures, valid and invalid, in some
other settings.

EXAMPLE 1.13   Measuring highway safety

Roads get better. Speed limits increase. Big SUVs replace cars. Enforcement campaigns
reduce drunk driving. How has highway safety changed over time in this changing environment?

We could just count deaths from motor vehicles. The Fatal Accident Reporting System
says there were 41,508 deaths in 1991 and 41,945 deaths 10 years later in 2000. But the
number of licensed drivers rose from 169 million in 1991 to 191 million in 2000. The
number of miles that people drove rose from 2172 billion to 2750 billion. If more people
drive more miles, there may be more deaths even if the roads are safer. The count of
deaths is not a valid measure of highway safety.

Rather than a *count*, we should use a *rate*. The number of deaths per mile driven takes
into account the fact that more people drive more miles than in the past. In 2000, vehicles drove 2,750,000,000,000 miles in the United States. Because this number is so large,
it is usual to measure safety by deaths per 100 million miles driven rather than deaths per
mile. For 2000, this death rate was

$$\frac{\text{motor vehicle deaths}}{\text{100s of million of miles driven}} = \frac{41,945}{27,500} = 1.53$$

The death rate fell from 1.91 deaths per 100 million miles in 1991 to 1.53 in 2000. That's a big change—there were 20% fewer deaths per mile driven in 2000 than a decade earlier. Driving has been getting safer.

Rates and counts

Often a **rate** (a fraction, proportion, or percent) at which something occurs is a more valid measure than a simple count of occurrences.

Using height to measure readiness for college, and even using counts when rates are needed, are examples of clearly invalid measures. The tougher questions concern measures that are neither clearly invalid nor obviously valid.

EXAMPLE 1.14   Achievement tests

When you take a statistics test, you hope that it will ask you about the main points of the course syllabus. If it does, the test is a valid measure of how much you know about the course material. The College Board, which administers the SAT, also offers achievement tests in a variety of disciplines (including an Advanced Placement exam in statistics). These achievement tests are not very controversial. Experts can judge validity by comparing the test questions with the syllabus of material they are supposed to cover.

EXAMPLE 1.15   IQ tests

Psychologists would like to measure aspects of the human personality that can't be observed directly, such as "intelligence" or "authoritarian personality." Does an IQ test measure intelligence? Some psychologists say "Yes" rather loudly. There is such a thing as general intelligence, they argue, and the various standard IQ tests do measure it, though not perfectly. Other experts say "No" equally loudly. There is no single intelligence, just a variety of mental abilities that no one instrument can measure.

The disagreement over the validity of IQ tests is rooted in disagreement about the nature of intelligence. If we can't agree on exactly what intelligence is, we can't agree on how to measure it.

Statistics is little help in these examples. They start with an idea like "knowledge of statistics" or "intelligence." If the idea is vague, validity becomes a matter of opinion. However, statistics can help a lot if we refine the idea of validity a bit.

EXAMPLE 1.16   The SAT again

"SAT bias will illegally cheat thousands of young women out of college admissions and scholarship aid they have earned by superior classroom performance."[15] That's what the organization FairTest said when the 1999 SAT scores were released. The gender gap was larger on the math part of the test, where women averaged 495 and men averaged 531. The federal Office of Civil Rights says that tests on which women and minorities score lower are discriminatory.

The College Board, which administers the SAT, replies that there are many reasons some groups have lower average scores than others. For example, more women than men from families with low incomes and little education sign up for the SAT. Students whose parents have low incomes and little education have, on the average, fewer advantages at home and in school than richer students. They have lower SAT scores because their backgrounds have not prepared them as well for college. The mere fact of lower scores doesn't imply that the test is not valid.

Is the SAT a valid measure of readiness for college? "Readiness for college academic work" is a vague concept that probably combines inborn intelligence (whatever we decide that is), learned knowledge, study and test-taking skills, and motivation to work at academic subjects. Opinions will always differ about whether SAT scores (or any other measure) accurately reflect this vague concept.

Instead, we ask a simpler and more easily answered question: Do SAT scores help predict students' success in college? Success in college is a clear concept, measured by whether students graduate and by their college grades. Students with high SAT scores are more likely to graduate and earn (on the average) higher grades than students with low SAT scores. We say that SAT scores have *predictive validity* as measures of readiness for college. This is the only kind of validity that data can assess directly.

Predictive validity

A measurement of a property has **predictive validity** if it can be used to predict success on tasks that are related to the property measured.

Predictive validity is the clearest and most useful form of validity from the statistical viewpoint. "Do SAT scores help predict college grades?" is a much clearer question than "Do IQ test scores measure intelligence?" However, predictive validity is not a yes-or-no idea. We must ask *how accurately* SAT scores predict college grades. Moreover, we must ask *for what groups* the SAT has predictive validity. It is possible, for example, that the SAT predicts college grades well for men but not for women. There are statistical ways to describe "how accurately." Application 1.2A uses one of these descriptions to help give you the big picture. It appears that SAT scores do have moderate predictive validity, and that they are about equally valid for different groups of students. Differences among groups in SAT scores generally reflect unequal environments that bring about unequal preparation for college.

APPLICATION 1.2A SAT Exams in College Admissions

Colleges use a variety of measures to make admissions decisions. The student's record in school is the most important, but SAT scores do matter, especially at selective colleges. The SAT has the advantage of being a national test. An A in algebra means different things in different high schools, but a math SAT score of 620 means the same thing everywhere. The SAT can't measure willingness to work hard or creativity, so it won't predict college performance exactly, but most colleges have long found it helpful.

How well do SAT scores predict first-year college grades? The table below gives some results from a sample of 48,039 students:[16]

	All students	Men only	Women only	Black students
SAT	27%	26%	31%	25%
School grades	29%	28%	28%	24%
Both together	37%	36%	38%	34%

The numbers in the table say what percent of the variation among students in college grades can be predicted by SAT scores (math and verbal combined), by high school grades, and by SAT scores and high school grades together. An entry of 0% would mean no predictive validity, and 100% would mean predictions were always exactly correct. Use the table above to answer the following questions.

1. Do SAT scores predict college grades as well as high school grades do? Explain.

2. Does combining SAT scores and high school grades help predict college grades better than using either variable alone? Justify your answer.

3. Do SAT scores and high school grades predict college grades about equally well for males, females, and black students? Explain.

APPLICATION 1.2A SAT Exams in College Admissions *(continued)*

4. How effective are SAT scores and high school grades in predicting college grades?

5. In addition to SAT scores and high school grades, what other characteristics (variables) do you think should be taken into consideration in predicting success in college?

Selective colleges are justified in paying some attention to SAT scores, but they are also justified in looking beyond SAT scores for the motivation that can bring success to students with weaker academic preparation. The SAT debate is not really about the numbers. It is about how colleges should use all the information they have in deciding whom to admit, and also about the goals colleges should have in forming their entering classes.

EXERCISES

1.29 Counting the unemployed? We could measure the extent of unemployment by a count (the number of people who are unemployed) or by giving a rate (the percent of the labor force that is unemployed). The number of people in the labor force grew from 107 million in 1980, to 126 million in 1990, and to 140 million at the beginning of 2000. Use these facts to explain why the count of unemployed people is not a valid measure of the extent of unemployment.

1.30 Measuring intelligence "Intelligence" means something like "general problem-solving ability." Explain why it is not valid to measure intelligence by a test that asks questions such as "Who wrote 'The Star-Spangled Banner'?" or "Who won the last soccer World Cup?"

1.31 School bus safety The National Highway Traffic Safety Administration says that an average of 11 children die each year in school bus accidents, and an average of 600 school-age children die each year in auto accidents during school hours. These numbers suggest that riding the bus is safer than driving to school with a parent. The counts aren't fully convincing, however. What rates would you like to know to compare the safety of buses and private autos?

1.32 Testing job applicants The law requires that tests given to job applicants must be shown to be directly job related. The Department of Labor believes that an employment test called the General Aptitude Test Battery (GATB) is valid for

What can't be measured matters

One member of the young Edmonton Oilers hockey team of 1981 finished last in almost everything one can measure: strength, speed, reflexes, eyesight. That was Wayne Gretzky, soon to be known as "the great one." He broke the National Hockey League scoring record that year, then scored yet more points in seven different seasons. Somehow the physical measurements didn't catch what made Gretzky the best hockey player ever. Not everything that matters can be measured.

a broad range of jobs. As in the case of the SATs, blacks and Hispanics get lower average scores on the GATB than do whites. Describe briefly what must be done to establish that the GATB has predictive validity as a measure of future performance on the job.

1.33 Fighting cancer Congress wants the medical establishment to show that progress is being made in fighting cancer. Some variables that might be used are the following:

(a) Total deaths from cancer. These have risen sharply over time, from 331,000 in 1970, to 505,000 in 1990, and to 553,000 in 2000.

(b) The percent of all Americans who die from cancer. The percent of deaths due to cancer rose steadily from 17.2% in 1970 to 23.5% in 1990 but then leveled off to 23.0% in 2002.

(c) The percent of cancer patients who survive for 5 years from the time the disease was discovered. These rates are rising slowly. For whites, the 5-year survival rate was 50.9% in the 1974 to 1979 period and 62.0% from 1989 to 1997.

None of these variables are fully valid as a measure of the effectiveness of cancer treatment. Explain why both (a) and (b) could increase even if treatment is getting more effective, and why (c) could increase even if treatment is getting less effective.

1.34 Online IQ tests Search for an online IQ test. When you find one that will take no more than about 15 minutes, take the test. Write a few sentences about whether you think the test provides a valid measurement of your intelligence.

Accurate and inaccurate measurements

Using a bathroom scale to measure your weight is valid. If your scale is like mine, however, the measurement may not be very accurate. Think about my bathroom scale. It measures my weight, but it may not give my true weight. My scale reads 3 pounds too high, so

$$measured\ weight = true\ weight + 3\ pounds$$

If that is the whole story, the scale will always give the same reading for the same true weight: it reads 3 pounds too high because its aim is off. But it is also erratic. Most scales vary a bit—they don't always give the same reading when you step off and step right back on. My scale is somewhat old and rusty. This morning it sticks a bit and reads one-half pound too low for that reason. So the reading is

$$measured\ weight = true\ weight + 3\ pounds - 0.5\ pound$$

When I step off and step right back on, the scale sticks in a different spot that makes it read one-quarter pound too high. The reading I get is now

$$measured\ weight = true\ weight + 3\ pounds + 0.25\ pound$$

If I have nothing better to do than keep stepping on and off the scale, I will keep getting different readings. They center on a reading 3 pounds too high, but they vary about that center.

My scale has two kinds of errors. If it didn't stick, the scale would always read 3 pounds high. That would be true every time anyone stepped on the scale. A systematic error that occurs every time we make a measurement is called *bias*. My scale also sticks—but how much this changes the reading is different every time someone steps on the scale. Sometimes stickiness pushes the scale reading up; sometimes it pulls it down. The result is that the scale weighs 3 pounds too high on the average, but its reading varies when we weigh the same thing repeatedly. We can't predict the error due to stickiness, so we call it *random error*.

Errors in measurement

We can think about errors in measurements this way:

$$measured\ value = true\ value + bias + random\ error$$

A measurement process has **bias** if it systematically overstates or understates the true value of the property it measures.

A measurement process has random error if repeated measurements on the same individual give different results. If the random error is small, we say the measurement is **reliable.**

A scale that always reads the same when it weighs the same item is perfectly reliable even if it is biased. Reliability only says that the result is repeatable. Bias and lack of reliability are different kinds of error. And don't confuse reliability with validity just because both sound like good qualities. Using a scale to measure weight is valid even if the scale is not reliable. Here's an example of a measurement that is reliable but not valid.

EXAMPLE 1.17   Do big skulls house smart brains?

In the mid-nineteenth century, it was thought that measuring the volume of a human skull would measure the intelligence of the skull's owner. It was difficult to measure a skull's volume reliably, even after it was no longer attached to its owner. Paul Broca, a professor of surgery, showed that filling a skull with small lead shot and then pouring out the shot and weighing it gave quite reliable measurements of the skull's volume. These accurate measurements do not, however, give a valid measure of intelligence. Skull volume turned out to have no relation to intelligence or achievement.

B.C.   **by johnny hart**

By permission of Johnny Hart and Creators Syndicate, Inc.

ACTIVITY 1.2B How long is a minute?

Materials: Stopwatch for each pair of students
Some people say that a minute seems like a long time. Others say, "Time flies." How well can you determine a minute?

1. You and your partner should sit as far as possible from other students in the class. Make sure you cannot see a clock from your position. Remove your watches.

2. You will take turns timing and measuring. The timer tells the measurer when to begin. When the measurer believes a minute has passed, she should quietly say, "Stop." At that point, the timer should stop the stopwatch and record the time that has passed to the nearest *tenth* of a second. Do not tell your partner how much time actually passed! Reset the stopwatch and switch roles.

3. Continue timing and measuring until each partner has measured a minute three times. Do not share your results until everyone else in the class has finished.

4. Examine your three measurements. How close did you get to a minute? Were all of your errors in the same direction? Find your average measurement.

5. Now look at your partner's data. Answer the same questions you did in Step 4.

6. Share data with your classmates. Which student was most accurate? Who was most reliable?

Extension: Analyze the entire class's measurements by making a graph. How accurate was the class as a whole? How reliable were the class's measurements?

Improving reliability, reducing bias

What time is it? Much modern technology, such as the Global Positioning System, which uses satellite signals to tell you where you are, requires very exact measurements of time. Time starts with the earth's path around the sun, which lasts one year. But the earth is much too erratic. Since 1967, time starts with the standard second, and the second is defined to be the time required for 9,192,631,770 vibrations of a cesium atom. Physical clocks are bothered by changes in temperature, humidity, and air pressure. The cesium atom doesn't care. People who need really accurate time can buy atomic clocks. The National Institute of Standards and Technology (NIST) keeps an even more accurate atomic clock and broadcasts the results (with some loss in transmission) by radio, telephone, and the Internet.

EXAMPLE 1.18 Really accurate time

NIST's atomic clock is very accurate, but not perfectly accurate. The world standard is Universal Coordinated Time, compiled by the International Bureau of Weights and Measures (BIPM) in Sèvres, France. BIPM doesn't have a better clock than NIST. It calculates the time by averaging the results of more than 200 atomic clocks around the world. NIST tells us (after the fact) by how much it misses the correct time. Here are the last 10 errors as we write, in seconds:[17]

0.000000007	0.000000000
0.000000005	−0.000000003
0.000000006	−0.000000005
0.000000000	−0.000000001
0.000000002	−0.000000001

In the long run, NIST's measurements of time are not biased. The NIST second is sometimes shorter than the BIPM second and sometimes longer, not always off in the same direction. NIST's measurements are very reliable, but the numbers above do show some variation. There is no such thing as a perfectly reliable measurement. The average (mean) of several measurements is more reliable than a single measurement. That's one reason BIPM combines the time measurements of many atomic clocks.

Scientists everywhere repeat their measurements and use the average to get more reliable results. Even students in a chemistry lab often do this. Averaging over more measurements reduces variation in the final result.

Use averages to improve reliability

No measuring process is perfectly reliable. The average of several repeated measurements of the same individual is more reliable (less variable) than a single measurement.

Unfortunately, there is no similarly straightforward way to reduce the bias of measurements. Bias depends on how good the measuring instrument is. To reduce the bias, you need a better instrument. The atomic clock at NIST (Figure 1.2) is accurate to 1 second in 6 million years and is a bit large to put beside your bed.

FIGURE 1.2 This atomic clock at the National Institute of Standards and Technology is accurate to 1 second in 6 million years. (Photo courtesy of John Wessels, NIST Time and Frequency Division.)

EXAMPLE 1.19   Measuring unemployment again

Measuring unemployment is also "measurement." The concepts of bias and reliability apply here just as they do to measuring length or time.

The Bureau of Labor Statistics checks the *reliability* of its measurements of unemployment by having supervisors reinterview about 5% of the sample. This is repeated measurement on the same individual, just as a student in a chemistry lab measures a weight several times. The BLS attacks *bias* by improving its instrument. That's what happened in 1994 when the Current Population Survey was given its biggest overhaul in more than 50 years. The old system for measuring unemployment, for example, underestimated unemployment among women because the detailed procedures had not kept up with changing patterns of women's work. The new measurement system corrected that bias—and raised the reported rate of unemployment.

EXERCISES

1.35 Measuring crime, I Crime data make headlines. We measure the amount of crime by the number of crimes committed or (better) by crime rates (crimes per 100,000 population). The FBI publishes data on crime in the United States by compiling crimes reported to police departments. The National Crime Victimization Survey publishes data based on a national sample of more than 43,000 households. The victim survey shows about two and a half times as many crimes as the FBI report. Explain why the FBI report has a large downward bias for many types of crime. (Here is a case in which bias in producing data leads to bias in measurement.)

1.36 Measuring crime, II Each year, the National Crime Victimization Survey asks a sample of more than 43,000 households whether they have been victims of crime and, if so, the details. In all, more than 80,000 people answer these questions. If other people in a household are in the room while one person is answering questions, the measurement of, for example, rape and other sexual assaults could be seriously biased. Why? Would the presence of other people lead to overreporting or underreporting of sexual assaults?

1.37 Testing job applicants A company used to give IQ tests to all job applicants. This is now illegal because IQ is not related to the performance of workers in all the company's jobs. Does the reason for the policy change involve the *reliability,* the *bias,* or the *validity* of IQ tests as a measure of future job performance? Explain your answer.

1.38 Where to live? Each year, *Money* magazine ranks the 300 largest metropolitan areas in the United States in an article on the best places to live. First place in 1997 went to Nashua, New Hampshire. Nashua was ranked 42nd in 1996 and 19th in 1995. Monmouth and Ocean Counties in New Jersey were ranked 167th in 1995, 38th in 1996, and 3rd in 1997. Are these facts evidence that *Money*'s ratings are invalid, biased, or unreliable? Explain your choice.

1.39 In the trenches Two investigators measure the depth of an oceanic trench. The first one makes 20 independent measurements and reports the average of the values obtained. The second person reports the results of a single measurement using the same process.
(a) Which investigator will have more reliability in the number reported, or will they have about the same reliability?
(b) Which investigator will have more bias in the number reported, or will they have about the same bias?

1.40 Healthy babies Apgar scores are a measurement of an infant's overall health taken a few minutes after birth. The score ranges from 0 (dead) to 10 ("perfect health") and is based on tests of the baby's heart and breathing rate, muscle tone, and other criteria. A critic gives three reasons why the Apgar score isn't a perfect measurement:
Reason I: There are many important facets of health that aren't measured by the score.
Reason II: A doctor's rating may be affected by being present at the birth; a doctor may give an unwarranted low value to a baby whose birth was difficult, for example.
Reason III: Two different doctors may give different Apgar scores, even when measuring the same baby at the same time.
(a) Which of these criticisms concerns the validity of the Apgar score? Which one concerns its reliability? Which one concerns the bias in the measurement?
(b) Suppose two doctors both judge an infant's health using the Apgar system, and the average of their two values is taken as the "official" Apgar score. Will this improve the validity of the measurement? The reliability of the measurement? Reduce the bias of the measurement?

1.41 Validity, bias, reliability Give your own example of a measurement process that is valid but has large bias. Then give your own example of a measurement process that is invalid but highly reliable.

APPLICATION 1.2B MERMAID

Mermaid is a system for monitoring pollution in coastal waters, estuaries, rivers, and lakes. It was developed by the GKSS Research Center in Geesthacht, Germany. (MERMAID stands for Marine Environmental Remote-Controlled Measuring and Integrated Detection.) Of course, the concept of "pollution" is very complex. So the MERMAID system measures dozens of different variables such as water pH, phosphate and oxygen levels, and temperature. A MERMAID system is actually a collection of modules that transmit signals to a land-based station from an ocean platform, a buoy, or even an

APPLICATION 1.2B MERMAID *(continued)*

unmanned ship. (The picture shows a MERMAID platform in the Wadden Zee, the Netherlands.)

The GKSS Research Center is very particular about the accuracy of the devices included in the MERMAID system. For example, when the phosphate-measuring mod-

Courtesy of GKSS Research Centre, Germany

ule was used many times on the same water sample, the results were so consistent that they varied from each other by only a few parts per billion. Also, the center compared their automatic measuring devices with the most modern "gold-standard" method of conducting each of the chemical analyses and found that the results were essentially identical.

1. From the above paragraph, explain in your own words how the GKSS Research Center is addressing each of the following issues in its MERMAID systems:

(a) Measurement reliability

(b) Bias

(c) Validity

2. In what units are each of the variables mentioned above measured? You may need to visit the MERMAID Web site to be sure (w3g.gkss.de/mermaid).

EXPLORING THE WEB

You can get the time directly from the atomic clock at NIST at www.time.gov. There is some error due to Internet delays, but the display even tells you roughly how accurate the time on the screen is. For a glimpse of the elaborate system that lies behind everyday measurements of length, weight, and time, visit the Web site of the International Bureau of Weights and Measures, www.bipm.fr.

STATISTICS IN SUMMARY

To **measure** something means to assign a number to some property of an individual. When we measure many individuals, we have values of a **variable** that describes them. When you work with data or read about a statistical study, ask exactly how the variables are defined and whether they leave out some things you want to know. Ask if the variables are **valid** as numerical measures of the concepts the study discusses. Validity is simple for measurements of physical properties such as length, weight, and time. When we want to measure human personality and other vague properties, **predictive validity** is the most useful way to say whether our measures are valid.

Also ask if there are **errors in measurements** that reduce the value of the data. You can think about errors in measurement like this:

$$\text{measured value} = \text{true value} + \text{bias} + \text{random error}$$

Some ways of measuring are **biased,** or systematically wrong in the same direction. To reduce bias, you must use a better **instrument** to make the measurements. Other measuring processes lack **reliability,** so that measuring the same individuals again would give quite different results due to **random error.** You can improve the reliability of a measurement by repeating it several times and using the average result.

SECTION 1.2 EXERCISES

1.42 Old trees The age of a pine tree was measured five times using a new electronic probe inserted in the tree's trunk. The measured values were 43, 40, 45, 44, and 41 years old. Later this tree was cut down and, by counting the growth rings, it was determined that the tree was really 34 years old. Does this new electronic device for measuring the age of trees have a greater problem with bias or with reliability? Explain.

1.43 Capital punishment Between 1977 and 2001, 749 convicted criminals were put to death in the United States. Here are data on the number of executions in several states during those years, as well as the 2000 population of these states:

State	Population (thousands)	Executions
Alabama	4,447	23
Arkansas	2,673	24
Florida	15,982	51
Missouri	5,595	53
Nevada	1,998	9
Texas	20,852	256
Virginia	7,079	83

Texas and Florida are among the leaders in executions. Because these are populous states, we might expect them to have many executions. Find the rate of executions for each of the states listed above, in executions per million population. Because population is given in thousands, you can find the rate per million as

$$\text{rate per million} = \frac{\text{executions}}{\text{population in thousand}} \times 1000$$

Arrange the states in order of the number of executions relative to population. Are Florida and Texas still high by this measure?

1.44 Measuring the moon The diameter of the moon is measured four times by a process that is free of bias. The individual measurements are 2157, 2166, 2162, and 2155 miles, which average out to 2160 miles. One more measurement is about to be taken using the same process. When compared with the estimate of 2160 miles, would you expect this next measurement to be more accurate, just as accurate, or less accurate as a measure of the true diameter of the moon? Explain.

1.45 Does job training work? To measure the effectiveness of government training programs, it is usual to compare workers' pay before and after training. But many workers sign up for training when their pay drops or they are laid off. So the "before" pay is unusually low and the pay gain looks large.
(a) Is this bias or random error in measuring the effect of training on pay? Why?
(b) How would you measure the success of training programs?

1.46 Measuring pain Following surgery, patients were given medication to relieve pain. They were then examined at regular intervals. The patients were asked to assess the level of pain they were experiencing at each of the exam times. A pain score was based on the patients' marks on a 10-centimeter line with a range from 0 (no discomfort) to 100 (agonizing pain).[18]
(a) Is the pain score a valid measurement of the amount of pain felt by the patients? Explain your answer.
(b) Consider the two 10-centimeter scales given below. Discuss the reliability of the data collected using each of these scales.

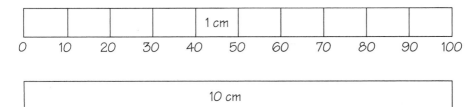

1.47 Teaching English A teacher wishes to know if a new English curriculum will increase the creativity of her students in writing poetry. At the end of a unit, a sample of each student's poetry is given to a panel of five experts who rate the creativity of the poem on a 10-point scale. For each of the statements below, tell whether it addresses the issue of measurement reliability or measurement validity.
Statement I: In reporting the results, the teacher claimed that the experts' ratings showed just how creative her students were. A critic of the new curriculum argued that the ratings were a poor way to measure creativity.
Statement II: The five experts seemed to rate each student nearly the same. For example, one student's poem got ratings of 8, 7, 8, 8, and 6 by the five experts.

1.48 Measuring good teaching You want to measure the "effectiveness" of teachers at your high school. Give an example of a clearly invalid way to measure good teaching. Then briefly describe a measurement process that you think is valid.

1.49 Activity 1.1 follow-up Refer to Activity 1.1 (page 4).
(a) How reliable are the individual measurements made by the members of the class? Explain briefly.
(b) If every measurement in the class were made by lining up one edge of the textbook with the 1 mark on the ruler rather than the 0 mark, would this affect the reliability of the measurements? Would it affect the bias of the measurements?

If the Pharaoh's architect had used the ruler from statistics class. . .

1.3 DO THE NUMBERS MAKE SENSE?

The case of the missing vans

Auto manufacturers lend their dealers money to help them keep vehicles on their lots. The loans are repaid when the vehicles are sold. A Long Island auto dealer named John McNamara borrowed over $6 billion from General Motors between 1985 and 1991. In December 1990 alone, Mr. McNamara borrowed $425 million to buy 17,000 GM vans customized by an Indiana company, allegedly for sale overseas. GM happily lent McNamara the money because he always repaid the loans.

Let's pause to consider the numbers, as GM should have done but didn't. The entire van-customizing industry produces only about 17,000 customized vans a month. So McNamara was claiming to buy an entire month's production. These large, luxurious, and gas-guzzling vehicles are designed for U.S. interstate highways. The recreational vehicle trade association says that only 1.35% were exported in 1990. It's not plausible to claim that 17,000 vans in a single month are being bought for export. McNamara's claimed purchases were large even when compared with the total production of vans. Chevrolet, for example, produced 100,067 full-sized vans in all of 1990.

Having looked at the numbers, you can guess the rest. McNamara admitted in federal court in 1992 that he was defrauding GM on a massive scale. The Indiana company was a shell set up by McNamara, its invoices were phony, and the vans didn't exist. McNamara borrowed vastly from GM, used most of each loan to pay off the previous loan (thus establishing a record as a good credit risk), and skimmed off a bit for himself. The bit he skimmed amounted to over $400 million. GM set aside $275 million to cover its losses. Two executives, who should have looked at the numbers relevant to their business, were fired.[19]

ACTIVITY 1.3 The "real facts"

Materials: Computer with Internet access
Do you drink Snapple? If so, then you have probably noticed a "Real Fact" printed on each bottle cap. Here is one we found: *Licking a stamp burns 10 calories.* How did they determine this? Does this "fact" seem reasonable? In this activity, you will learn to judge whether the numbers make sense.

1. Consider the following "Real Facts." For each one, (*a*) tell how you think the data were produced (survey, experiment, observational study) and (*b*) discuss whether the result makes sense.

ACTIVITY 1.3 The "real facts" *(continued)*

- The average American walks 18,000 steps a day.
- August has the highest percent of births.
- The average person spends about 2 years on the phone in a lifetime.
- Termites eat through wood 2 times faster when listening to rock music.
- 27% of Americans have spent at least one night in jail. (This one is actually an incorrect choice in the Real Facts game at the Snapple Web site. Do you think the correct percent is higher or lower? Why?)

2. On each Snapple bottle cap, you are encouraged to "Get all the Real Facts at snapple.com." Go to the Snapple Web site (www.snapple.com) and locate the Real Facts game. Try your hand at deciding which of two unusual statements presented is a fact.

3. Find two "Real Facts" that are different from the ones above and that involve a numerical result. Record each fact. Then tell how you think the data were produced and whether the result makes sense.

Business data, advertising claims, debate on public issues—we are assaulted daily by numbers intended to prove a point, buttress an argument, or assure us that all is well. Some, like John McNamara, feed us fake data. Others use data to argue a cause and care more for the cause than for the accuracy of the data. Others simply lack the skills needed to employ numbers carefully. We know that we should always ask:

- How were the data produced?
- What exactly was measured?

We also know quite a bit about what good answers to these questions sound like. That's wonderful, but it isn't enough. It may be that the General Motors executives taken in by John McNamara's scheme knew something about samples and reliable measurements. What they lacked was "number sense," the habit of asking if numbers make sense. To help develop number sense, we will look at how bad data, or good data used wrongly, can trick the unwary.

What didn't they tell us?

The most common way to mislead with data is to cite correct numbers that don't quite mean what they appear to say because we aren't told the full story. The numbers are not made up, so the fact that the information is a bit incomplete may be an innocent oversight. Here are some examples. You decide how innocent they are.

Reproduced by permission of Bob Schoechet.

"Sure your patients have 50% fewer cavities. That's because they have 50% fewer teeth!"

EXAMPLE 1.20   Snow! Snow! Snow!

Crested Butte attracts skiers by advertising that it has the highest average snowfall of any ski town in Colorado. That's true. But skiers want snow on the ski slopes, not in the town—and many other Colorado resorts get more snow on the slopes.[20]

EXAMPLE 1.21   Yet more snow

News reports of snowstorms say things like "A winter storm spread snow across the area, causing 28 minor traffic accidents." Eric Meyer, a reporter in Milwaukee, Wisconsin, says he often called the sheriff to gather such numbers. One day he decided to ask the sheriff how many minor accidents are typical in good weather: about 48, said the sheriff. Perhaps, says Meyer, the news should say, "Today's winter storm prevented 20 minor traffic accidents."[21]

EXAMPLE 1.22   We attract really good students

Colleges know that many prospective students look at popular guidebooks to decide where to apply for admission. The guidebooks print information supplied by the colleges themselves. Surely no college would simply lie about, say, the average SAT score of its entering students. But we do want our scores to look good. How about leaving out the scores of our international and remedial students? Northeastern University did this, making the average SAT score of its freshman class 50 points higher than if all students were included. If we admit economically disadvantaged students under a special program sponsored by the state, surely no one will complain if we leave their SAT scores out of our average? New York University did this.[22]

The point of these examples is that numbers have a context. If you don't know the context, the lonely, isolated, naked number doesn't tell you much.

Are the numbers consistent with each other?

John McNamara fooled General Motors because GM didn't compare his numbers with others. No one asked how a dealer could buy 17,000 vans in a single month for export when the entire custom van industry produces just 17,000 vans a month and only a bit over 1% are exported. Speaking of GM, here's another example in which the numbers don't line up with each other.

EXAMPLE 1.23   We won!

GM's Cadillac brand was the best-selling luxury car in the United States for 57 years in a row. In 1998, Ford's Lincoln brand seemed to be winning until the last moment. Said the *New York Times,* "After reporting almost unbelievable sales results in December, Cadillac eked out a come-from-behind victory by just 222 cars." The final count was 187,343 for Cadillac, 187,121 for Lincoln. Then GM reported that Cadillac sales dropped 38% in January. How could sales be so different in December and January? Could it be that some January sales were counted in the previous year's total? Just enough, say, to win by 222 cars? Yes, indeed. In May, GM confessed that it sold 4773 fewer Cadillacs in December than it had claimed.[23]

In the General Motors examples, we suspect something is wrong because numbers don't agree as we think they should. Here's an example where we *know* something is wrong because the numbers don't agree. This is part of an article on a cancer researcher at the Sloan-Kettering Institute who was accused of committing the ultimate scientific sin, falsifying data.

EXAMPLE 1.24   Fake data

"One thing he did manage to finish was a summary paper dealing with the Minnesota mouse experiments. . . . That paper, cleared at SKI and accepted by the *Journal of Experimental Medicine,* contains a statistical table that is erroneous in such an elementary way that a bright grammar school pupil could catch the flaw. It lists 6 sets of 20 animals

each, with the percentages of successful takes. Although any percentage of 20 has to be a multiple of 5, the percentages that Summerlin recorded were 53, 58, 63, 46, 48, and 67."[24]

Are the numbers plausible?

As the General Motors examples illustrate, you can often detect suspicious numbers simply because they don't seem plausible. Sometimes you can check an implausible number against data in reliable sources such as the annual *Statistical Abstract of the United States*. Sometimes, as the next example illustrates, you can do a calculation to show that a number can't be right.

EXAMPLE 1.25   The abundant melon field

The very respectable journal *Science*, in an article on insects that attack plants, mentioned a California field that produces 750,000 melons per acre. A reader responded, "I learned as a farm boy that an acre covers 43,560 square feet, so this remarkable field produces about 17 melons per square foot. If these are cantaloupes, with each fruit covering about 1 square foot, I guess they must grow in a stack 17 deep." Here is the calculation the reader did:

$$\text{melons per square foot} = \frac{\text{melons per acre}}{\text{square feet per acre}} = \frac{750{,}000}{43{,}560} = 17.2$$

The editor, a bit embarrassed, replied that the correct figure was about 11,000 melons per acre.[25]

EXERCISES

1.50 Advertising painkillers, I An advertisement for the pain reliever Tylenol was headlined "Why Doctors Recommend Tylenol More Than All Leading Aspirin Brands Combined." The makers of Bayer Aspirin, in a reply headlined "Makers of Tylenol, Shame on You!" accused Tylenol of misleading by giving the truth but not the whole truth. You be the detective. How is Tylenol's claim misleading even if true?

1.51 More "Real Facts" According to the Real Facts game at Snapple.com, Americans on average eat 18 acres of pizza every day.
(a) Tell how you think the data were produced.
(b) Discuss whether this result makes sense.

1.52 In the garden *Organic Gardening* magazine, describing how to improve your garden's soil, said, "Since a 6-inch layer of soil in a 100-square-foot plot weighs about 45,000 pounds, adding 230 pounds of compost will give you an instant 5% organic matter."[26]

(a) What percent of 45,000 is 230?

(b) Water weighs about 62 pounds per cubic foot. There are 50 cubic feet in a garden layer 100 square feet in area and 6 inches deep. What would 50 cubic feet of water weigh? Is it plausible that 50 cubic feet of soil weighs 45,000 pounds?

(c) It appears from (b) that the 45,000 pounds isn't right. In fact, soil weighs about 75 pounds per cubic foot. If we use the correct weight, is the "5% organic matter" conclusion roughly correct?

1.53 No eligible men? A news report quotes a sociologist as saying that for every 233 unmarried women in their 40s in the United States, there are only 100 unmarried men in their 40s. These numbers point to an unpleasant social situation for women of that age. Are the numbers plausible? (*Optional: The Statistical Abstract of the United States* has a table titled "Marital status of the population by age and sex" that gives the actual counts.)

1.54 Deer in the suburbs Westchester County is a suburban area covering 438 square miles immediately north of New York City. A garden magazine claimed that the county is home to 800,000 deer.[27] Do a calculation that shows this claim to be implausible.

Are the numbers too good to be true?

In Example 1.24, lack of consistency led to the suspicion that the data were phony. *Too much precision or regularity* can lead to the same suspicion, as when a student's lab report contains data that are exactly as the theory predicts. The teacher knows that the accuracy of the equipment and the student's laboratory technique are not good enough to give such perfect results. She suspects that the student made them up. Here is an example drawn from an article in *Science* about fraud in medical research.

EXAMPLE 1.26   More fake data

"Lasker had been asked to write a letter of support. But in reading two of Slutsky's papers side by side, he suspected that the same 'control' animals had been used in both without mention of the fact in either. Identical data points appeared in both articles, but . . . the actual number of animals cited in each case was different. This suggested at best a sloppy approach to the facts. Almost immediately after being asked about the statistical discrepancies, Slutsky resigned and left San Diego."[28]

In this case, suspicious regularity (identical data points) combined with inconsistency (different numbers of animals) led a careful reader to suspect fraud.

Is the arithmetic right?

Conclusions that are wrong or just incomprehensible are often the result of plain old-fashioned blunders. Rates and percents cause particular trouble.

EXAMPLE 1.27   Oh, those percents

Here are some examples from Australia. The *Canberra Times* reported, "Of those aged more than 60 living alone, 34% are women and only 15% are men." That's 49% of those living alone. I suppose the other 51% are neither women nor men.

Even smart people have problems with percents. A newsletter for female university teachers asked, "Does it matter that women are 550% (five and a half times) less likely than men to be appointed to a professional grade?" Now, 100% of something is all there is. If you take away 100%, there is nothing left. I have no idea what "550% less likely" might mean.[29]

It seems that few people do arithmetic once they leave school. Those who do are less likely to be taken in by meaningless numbers. A little thought and a calculator go a long way.

EXAMPLE 1.28   Summertime is burglary time

An advertisement for a home security system says, "When you go on vacation, burglars go to work. According to FBI statistics, over 26% of home burglaries take place between Memorial Day and Labor Day."[30]

This is supposed to convince us that burglars are more active in the summer vacation period. Look at your calendar. There are 14 weeks between Memorial Day and Labor Day. As a percent of the 52 weeks in the year, this is

$$\frac{14}{52} = 0.269 \quad \text{(that is, 26.9\%)}$$

So the ad claims that 26% of burglaries occur in 27% of the year. You should not be impressed.

Just a little arithmetic mistake

In 1994, an investment club of grandmotherly women wrote a best-seller, *The Beardstown Ladies' Common-Sense Investment Guide: How We Beat the Stock Market—and How You Can, Too.* On the book cover and in their many TV appearances, the downhome authors claimed a 23.4% annual return, beating the market and most professionals. Four years later, a skeptic discovered that the club treasurer had entered data incorrectly. The Beardstown ladies' true return was only 9.1%, far short of the overall stock market return of 14.9% in the same period. We all make mistakes, but most of them don't earn as much money as this one did.

EXAMPLE 1.29   The old folks are coming

A writer in *Science* claimed in 1976 that "people over 65, now numbering 10 million, will number 30 million by the year 2000, and will constitute an unprecedented 25 percent of the population."[31] Sound the alarm: the elderly were going to triple in a quarter century to become a fourth of the population.

Let's check the arithmetic. Thirty million is 25% of 120 million, because

$$\frac{30}{120} = 0.25$$

So the writer's numbers make sense only if the population in 2000 is 120 million. The U.S. population in 1975 was already 216 million. Something is wrong.

Thus alerted, we can check the *Statistical Abstract of the United States* to learn the truth. In 1975, there were 22.4 million people over age 65, not 10 million. That's more than 10% of the total population. The estimate of 30 million by the year 2000 was only about 12% of the projected population for that year. Looking back from the year 2000, we now know that people at least 65 years old are 13% of the total U.S. population. As people live longer, the numbers of the elderly are growing. But growth from 10% to 13% over 25 years is far slower than the *Science* writer claimed.

Calculating the percent increase or decrease in some quantity seems particularly prone to mistakes. The percent change in a quantity is found by

$$\text{percent change} = \frac{\text{amount of change}}{\text{starting value}} \times 100$$

EXAMPLE 1.30   Stocks go up, stocks go down

In 1999, the NASDAQ composite index of stock prices rose from 2192.69 to 4069.31. What percent increase was this?

$$\text{percent change} = \frac{\text{amount of change}}{\text{starting value}} \times 100$$

$$= \frac{4069.31 - 2192.69}{2192.69} \times 100$$

$$= \frac{1876.62}{2192.69} \times 100 = 0.856 \times 100 = 85.6\%$$

That's pretty impressive. Of course, stock prices go down as well as up. In the week ending April 14, 2000, the NASDAQ index dropped from 4446.17 to 3321.29. That's a percent decrease of

$$\frac{\text{amount of change}}{\text{starting value}} \times 100 = \frac{-1124.88}{4446.17} \times 100 = -25.3\%$$

Remember to always use the *starting* value, not the smaller value, in the denominator of your fraction.

A quantity can increase by any amount—a 100% increase just means it has doubled. But nothing can go down more than 100%—it has then lost 100% of its value, and 100% is all there is.

Is there a hidden agenda?

Lots of people feel strongly about various issues, so strongly that they would like the numbers to support their feelings. Often they can find support in numbers by choosing carefully which numbers to report or by working hard to squeeze the numbers into the shape they prefer. Here are two examples.

EXAMPLE 1.31   Heart disease in women

A highway billboard says simply, "Half of all heart disease victims are women." What might be the agenda behind this true statement? Perhaps the billboard sponsors just want to make women aware that they do face risks from heart disease. (Surveys show that many women underestimate the risk of heart disease.)

On the other hand, perhaps the sponsors want to fight what some people see as an overemphasis on male heart disease. In that case, we might want to know that although half of heart disease victims are women, they are on the average much older than male victims. Roughly 36,000 women under age 65 and 85,000 men under age 65 die from heart disease each year. The American Heart Association says, "Risk of death due to coronary heart disease in women is roughly similar to that of men 10 years younger."[32]

EXAMPLE 1.32   Income inequality

During the economic boom of the 1980s and 1990s in the United States, the gap between the highest and lowest earners widened. In 1980, the bottom fifth of households received 4.3% of all income, and the top fifth received 43.7%. By 1998, the share of the bottom fifth had fallen to 3.6% of all income, and the share of the top fifth of households had risen

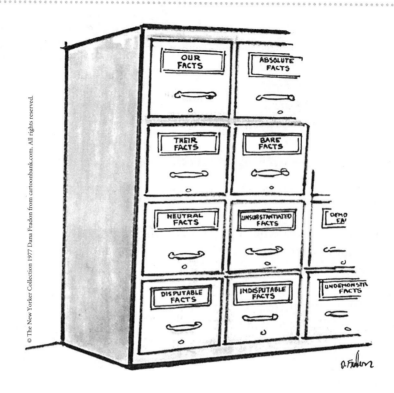

Oh, really?

In November 2002, the sensational young Atlanta Falcons quarterback, Michael Vick, was quoted, "I have two strengths: my legs, my arm, and my brains." The outspoken Mr. Vick must have attended the Yogi Berra School of Rhetoric.

to 49.2%. That is, the top fifth's share was almost 14 times the bottom fifth's share.

Can we massage the numbers to reduce the income gap? An article in *Forbes* (a magazine read mainly by rich folk) tried. First, more people live in the average rich household than in poor households, so let's change to income per person. The rich pay more taxes, so look at income after taxes. The poor receive food stamps and other assistance, so let's count that. Finally, high earners work more hours than low earners, so we should adjust for hours worked. After all this, the share of the top fifth is only 3 times that of the bottom fifth. Of course, hours worked are reduced by illness, disability, care of children and aged parents, and so on. If *Forbes*'s hidden agenda is to show that income inequality isn't important, we may not agree.

Yet other adjustments are possible. Income, in these Census Bureau figures, does not include capital gains from, for example, selling stocks that have gone up. Almost all capital gains go to the rich, so including them would widen the income gap. *Forbes* didn't make this adjustment. Making every imaginable adjustment in the meaning of "income," says the Census Bureau, gives the bottom fifth of households 4.7% of total income in 1998 and the top fifth 45.8%.[33]

EXERCISES

1.55 Drunk driving A newspaper article on drunk driving cited data on traffic deaths in Rhode Island: "Forty-two percent of all fatalities occurred on Friday, Saturday, and Sunday, apparently because of increased drinking on the weekends."[34] What percent of the week do Friday, Saturday, and Sunday make up? Are you surprised that 42% of fatalities occur on those days?

1.56 We can read, but can we count? The Census Bureau once gave a simple test of literacy in English to a random sample of 3400 people. The *New York Times* printed some of the questions under the headline "113% of Adults in U.S. Failed This Test."[35] Why is the percent in the headline clearly wrong?

1.57 Funny numbers Here's a quotation from a book review in a scientific journal:

> . . . a set of 20 studies with 57 percent reporting significant results, of which 42 percent agree on one conclusion while the remaining 15 percent favor another conclusion, often the opposite one.[36]

Do the numbers given in this quotation make sense? Can you decide how many of the 20 studies agreed on "one conclusion," how many favored another conclusion, and how many did not report significant results?

1.58 Airport delays An article in a midwestern newspaper about flight delays at major airports said:

> According to a Gannett News Service study of U.S. airlines' performance during the past five months, Chicago's O'Hare Field scheduled 114,370 flights. Nearly 10 percent, 1,136, were canceled.[37]

Check the newspaper's arithmetic. What percent of scheduled flights from O'Hare were actually canceled?

1.59 Battered women? A letter to the editor of the *New York Times* complained about a *Times* editorial that said "an American woman is beaten by her husband or boyfriend every 15 seconds." The writer of the letter claimed that "at that rate, 21 million women would be beaten by their husbands or boyfriends every year. That is simply not the case."[38] He cited the National Crime Victimization Survey, which estimated 56,000 cases of violence against women by their husbands and 198,000 by boyfriends or former boyfriends. The survey showed 2.2 million assaults against women in all, most by strangers or someone the woman knew who was not her past or present husband or boyfriend.

(a) First do the arithmetic. Every 15 seconds is 4 per minute. At that rate, how many beatings would take place in an hour? In a day? In a year? Is the letter writer's arithmetic correct?

(b) Is the letter writer correct to claim that the *Times* overstated the number of cases of domestic violence against women?

1.60 Stocks go down On August 4, 1998, the Dow Jones Industrial Average dropped 299.43 points from its opening level of 8780.94. Some newspapers headlined this as "the third-biggest point drop ever." By what percent did the Dow drop that day? Because the index had gone up so much in the previous years, this wasn't even in the top 20 of percentage drops. It's another example of rates being clearer than counts.

APPLICATION 1.3 *CHANCE News*

The CHANCE Web site at Dartmouth College contains lots of interesting stuff (at least if you are interested in statistics). In particular, the *CHANCE News* section, www.dartmouth.edu/~chance/chance_news.html, offers a monthly newsletter that keeps track of statistics in the press, including deceptive statistics.

1. Get a copy of the first five pages of *CHANCE News* 11.05 (October 11 to December 31, 2002) either from the CHANCE Web site or your teacher.

2. Look at the first two "Forsooth" items. For each one, write a sentence or two to explain why the numbers don't make sense.

3. The third "Forsooth" item shows a graph of membership in the Royal Horticultural Society. What's wrong with this picture?

4. What error was made in the quote from the *General Medical Council Newsletter*?

5. Read the article titled "Bureaucrat's Math Makes Dizzy Dozen." What mistake was made by the inspector when talking about "bad eggs"?

6. In the article "Threats and Responses: Missile Shields . . . ," how did Lt. Gen. Kadish attempt to "lie with statistics"?

EXPLORING THE WEB

The *Statistical Abstract of the United States* is an essential compilation of data. You can find it online at the Census Bureau Web site. The 1995 to 2002 editions are at www.census.gov/statab/www/. To find what you want, look in the index first.

STATISTICS IN SUMMARY

The aim of statistics is to provide insight by means of numbers. Numbers are most likely to yield their insights to those who examine them closely. Ask exactly what a number measures and decide if it is a valid measure. Look for the context of the numbers and ask if there is important **missing information.** Look for **inconsistencies,** numbers that don't agree as they should, and check for **incorrect arithmetic.** Compare numbers that are **implausible**—surprisingly large or small—with numbers you know are right. Be suspicious when numbers are **too regular or agree too well** with what their author would like to see. Look with special care if you suspect the numbers are put forward in support of some **hidden agenda.** If you form the habit of looking at numbers closely, your friends will soon think that you are brilliant. They might even be right.

SECTION 1.3 EXERCISES

1.61 Advertising painkillers, II Anacin was long advertised as containing "more of the ingredient doctors recommend most." Another over-the-counter pain reliever claimed that "doctors specify Bufferin most" over other "leading brands." Both advertising claims were literally true, but the Federal Trade Commission found them both misleading. Explain why. (*Hint:* What is the active pain reliever in both Anacin and Bufferin?)

1.62 Poverty The number of Americans living below the official poverty line increased from 27,392,580 to 32,476,000 in the 20 years between 1979 and 1998. What percent increase was this? You should not conclude from this that poverty grew more common in these years, however. Why not?   18.5 %   more people

1.63 We don't lose your baggage Continental Airlines once advertised that it had "decreased lost baggage by 100% in the past six months." Do you believe this claim?

1.64 Too good to be true? The late English psychologist Cyril Burt was known for his studies of the IQ scores of identical twins who were raised apart. The high correlation between the IQs of separated twins in Burt's studies pointed to heredity as a major factor in IQ. ("Correlation" measures how closely two variables are connected. We will meet correlation in Chapter 6.) Burt wrote several accounts of his work, adding more pairs of twins over time. Here are his reported correlations as he published them:

Publication date	Twins reared apart	Twins reared together
1955	0.771 (21 pairs)	0.944 (83 pairs)
1966	0.771 (53 pairs)	0.944 (95 pairs)

→ How could next get exact msrs to nearest thousanth

What is suspicious here?

1.65 Suicides among Vietnam veterans Did the horrors of fighting in Vietnam drive many veterans of that war to suicide? A figure of 150,000 suicides among Vietnam veterans in the 20 years following the end of the war has been widely quoted. Explain why this number is not plausible. To help you, here are some facts: about 25,000 American men commit suicide each year; about 3 million men served in Southeast Asia during the Vietnam War; there are roughly 93 million adult men in the United States.

1.66 Time off The pointy-haired boss in the *Dilbert* cartoon once noted that 40% of all sick days taken by the staff are Fridays and Mondays. Is this evidence that the staff are trying to take long weekends?   2 days = 40%

1.67 Don't dare to drive? A university sends a monthly newsletter on health to its employees. A recent issue included a column called "What Is the Chance?" that said: "Chance that you'll die in a car accident this year: 1 in 75." There are about 275 million people in the United States. About 40,000 people die each year from motor vehicle accidents. What is the chance a typical person will die in a motor vehicle accident this year?

1.68 Smoking-related deaths Below is a table from *Smoking and Health Now*, a report of the British Royal College of Physicians. It shows the number and percent of deaths among men aged 35 and over from the chief diseases related to smoking. One of the entries in the table is incorrect and an erratum slip was inserted to correct it. Which entry is wrong, and what is the correct value?

	Lung cancer	Chronic bronchitis	Coronary heart disease	All causes
Number	26,973	24,976	85,892	312,537
Percent	8.6%	8.0%	2.75%	100%

should be 27.5%

1.69 Your own example Find an example of one of the following. Explain in detail the statistical shortcomings of your example.

   Leaving out essential information

   Lack of consistency

   Implausible numbers

   Faulty arithmetic

One place to look is in the online *CHANCE News* mentioned in Application 1.3 (page 56).

CHAPTER 1 REVIEW

The first and most important question to ask about any statistical study is "Where did the data come from?" The distinction between observational and experimental data is a key part of the answer. Good statistics starts with good designs for producing data. Then, we measure the characteristics of interest to obtain numbers we can work with. We should ask, "Do the numbers make sense?" It is a valuable habit to look skeptically at numbers before accepting what they seem to say.

Here is a review list of the most important skills you should have developed from your study of this chapter.

A. DATA

1. Recognize the individuals and variables in a statistical study.

2. Identify the population and the sample in a statistical study.

3. Distinguish observational from experimental studies.

4. Identify sample surveys, censuses, and experiments.

B. MEASUREMENT

1. Explain how measuring leads to clearly defined variables in specific settings. Identify the instrument used and the units of measurement.

2. Evaluate the validity of a variable as a measure of a given characteristic, including predictive validity.

3. Explain how to reduce bias and improve reliability in measurement.

4. Distinguish issues that affect validity, bias, and reliability in a measurement process.

C. MAKING SENSE OF DATA

1. Recognize inconsistent numbers, implausible numbers, numbers so good they are suspicious, and arithmetic mistakes.

2. Calculate percent increase or decrease correctly.

CHAPTER 1 REVIEW EXERCISES

1.70 NBA All Stars For the first time in his career, Kobe Bryant of the L.A. Lakers received the most votes in 2003 NBA balloting. The table on the next page provides data about Kobe and other leading vote getters.[39]

Kobe Bryant	LAL	1,474,386	27.7	78	G
Tracey McGrady	ORL	1,316,297	30.4	80	G
Vince Carter	TOR	1,300,895	18.7	78	G-F
Yao Ming	HOU	1,286,324	12.8	89	C
Tim Duncan	SA	1,179,955	23.3	84	F-C

(a) What individuals are measured?
(b) What variables are recorded? Which are numerical? In what *units* is each numerical variable recorded?
(c) Which variable or variables do you think most influence the number of votes received by an individual player?

1.71 Keeping fit You want to measure the "physical fitness" of high school students.
(a) Give an example of a clearly invalid way to measure fitness.
(b) Then describe a measurement process that you think is valid. Define the variables you would measure, as well as the instrument(s) and units of measurement.

1.72 Trash at sea? A report on the problem of vacation cruise ships polluting the sea by dumping garbage overboard said:[40]

> On a seven-day cruise, a medium-size ship (about 1,000 passengers) might accumulate 222,000 coffee cups, 72,000 soda cans, 40,000 beer cans and bottles, and 11,000 wine bottles.

Are these numbers plausible? Do some arithmetic to back up your conclusion. Suppose, for example, that the crew is as large as the passenger list. How many cups of coffee must each person drink every day?

1.73 Multi-airline flights A passenger who takes a trip involving two or more airlines pays the first carrier, which then owes the other carrier a portion of the ticket cost. It is too expensive for the airlines to calculate exactly how much they owe each other. Instead, a sample of about 12% of tickets sold is examined and accounts are settled on that basis.
(a) What is the population of the situation? What is the sample?
(b) Will the sample results definitely, probably, probably not, or definitely not yield the correct amount of money for each airline? Explain.

1.74 An implausible number? *Newsweek* once said in a story that a woman who is not currently married at age 40 has a better chance of being killed by a terrorist than of getting married. Do you think this is plausible? What kind of data would help you check this claim?

1.75 Blood alcohol levels A researcher wants to measure the blood alcohol level of the subjects in a study of alcohol consumption and heart disease. He is not satisfied with the standard method of taking a single measurement using the "breathalyzer" machine, which measures alcohol on the breath. Three alternatives are considered:

Method 1: Check the breathalyzer machine and reset it if necessary to be sure that it gives a zero reading for the background air. Then take a single measurement using the machine.

Method 2: Instead of using the breathalyzer, take a small blood sample and measure the alcohol level in this sample.

Method 3: Use the breathalyzer 10 times and take the average of the 10 readings.

(a) Which method would improve the reliability of the standard method? Explain.

(b) Which method would have better validity than the standard method? Explain.

(c) Which method would reduce the bias associated with the standard method? Explain.

1.76 Noisy snowmobiles In spite of a study that predicted harmful environmental effects, the Bush administration lifted a ban on snowmobiles in two national parks in November 2002. The study predicted that air pollutants would be far higher with snowmobiles than with the snowcoaches that were currently being used. Noise from snowmobiles would be audible over half the time in the parks.

(a) Was this an observational study or an experiment? Why?

(b) What individuals were measured, and what variables were recorded?

(c) Why do you think the Bush administration decided to allow snowmobiles, in spite of the study results?

1.77 Measuring pulse rate All the members of a health class are asked to measure their pulse rate as they sit in the classroom. The students use a variety of methods.

Method 1: Count heartbeats for 6 seconds and multiply by 10 to get beats per minute.

Method 2: Count heartbeats for 30 seconds and multiply by 2 to get beats per minute.

(a) Which method is more reliable? Why?

(b) Is either method clearly more biased than the other? Why?

(c) One student in the class proposes a third method: starting exactly on a heartbeat, measure the time needed for 50 beats and convert this time into beats per minute. This method is more accurate than either of the two methods above. Why?

1.78 Greeks on campus The question-and-answer column of a campus newspaper was asked what percent of the campus was "Greek" (that is, members of fraternities or sororities). The answer given was that "the figures for the fall semester are approximately 13 percent for the girls and 15–18 percent for the guys, which produces a 'Greek' figure of approximately 28–31 percent of the undergraduates."[41] Discuss the campus newspaper's arithmetic.

1.79 Boating safety Data on accidents in recreational boating in the *Statistical Abstract of the United States* show that the number of deaths has dropped from 1360 in 1980 to 865 in 1990 and 819 in 1997. However, the number of injuries reported grew from 2650 in 1980 to 3822 in 1990 and 4555 in 1997.
(a) Why are there so few injuries in these government data relative to the number of deaths?
(b) Which count (deaths or injuries) is probably more accurate? Why might the injury count rise when deaths did not?
(c) By what percent did the death count decrease from 1980 to 1990? From 1990 to 1997? From 1980 to 1997?

Chapter 2

Sampling and Surveys

2.1 SAMPLES, GOOD AND BAD

The *Town Talk* takes an opinion poll

In Rapides Parish, Louisiana, only one company is allowed to provide ambulance service. The local paper, the *Town Talk,* asked readers to call in to offer their opinions on whether the company should keep its monopoly. Call-in polls are generally automated: call one telephone number to vote "Yes" and call another number to vote "No." Telephone companies often charge callers to dial these numbers.

The *Town Talk* got 3763 calls, which suggests unusual interest in ambulance service. Investigation showed that 638 calls came from the ambulance company office or from the homes of its executives. Many more no doubt came from lower-level employees. "We've got employees who are concerned about this situation, their job stability and their families and maybe called more than they should have," said a company vice-president. Other sources said employees were told to "vote early and often."[1]

We will see that a sample size of 3763 is quite respectable—if the sample is properly designed. Inviting people to call (and call, and call . . .) isn't proper sample design. We are about to learn how the pros take a sample, and why it beats the *Town Talk* method.

ACTIVITY 2.1 Random samples on the computer

In this activity, you will learn how to select "good samples" using a computer applet. You will also investigate how information obtained from a sample relates to the truth about the corresponding population.

Materials: *List of students in your class numbered 1,2,3, . . . , computer with Internet access and Java capability.*

1. Go to the textbook Web site, www.whfreeman.com/sta. Click on the applets link, then choose the *Simple Random Sample* applet. You should see a screen like Figure 2.1.

2. Enter the number 5 next to "Population = 1 to" and click the Reset button. The Population hopper now contains five balls numbered 1 to 5. Enter sample size 3 and click the Sample button. Record the numbers of the balls in the Sample bin.

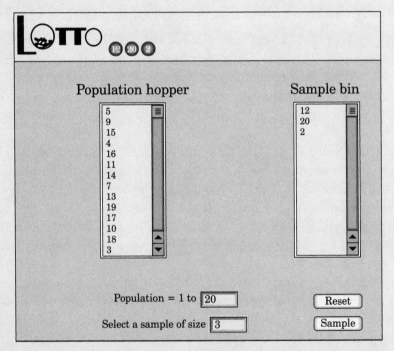

FIGURE 2.1 The *Simple Random Sample* applet at the book's Web site.

ACTIVITY 2.1 Random samples on the computer *(continued)*

3. Think: If you were to sample three balls from a population of size five, which is more likely to happen: getting the sample 1, 2, and 3 (in that order) or getting the sample 3, 1, and 5 (in that order)?

4. Think: If you were to repeatedly draw samples of size three from a population of size five, how often would the largest of the three numbers be picked first?

5. Verify: Test your answers to questions 3 and 4 by making repeated use of the applet. Record the results of each sample in a table. Did the samples behave as you thought they would?

6. Draw a sample of seven balls from a population of size twenty. After the sample is drawn, what is left in the Population hopper? Is the sample "random"? Explain your answer.

7. Every student in your class should use the applet to pick a random sample of five students from your class. Use the class list provided by your teacher. How many males and how many females were picked in your sample? Calculate the proportion of females in your sample. How does this compare to the true proportion of females in your class?

8. Pool your results with those of your classmates. Make a dotplot to display the proportion of females in each sample. How many samples contained more women than expected?

9. Find the average proportion of females found in all of the class's samples put together. How does this compare with the actual proportion of females in the class?

How to sample badly

As the *Town Talk* learned, it is easier to sample badly than to sample well. The paper relied on *voluntary response,* allowing people to call in rather than actively selecting its own sample. The result was *biased*—the sample was overweighted with people favoring the ambulance monopoly. Voluntary response samples attract people who feel strongly about the issue in question. These people, like the employees of the ambulance company, may not fairly represent the opinions of the entire population.

There are other ways to sample badly. Suppose that I sell your company several crates of oranges each week. You examine a sample of oranges from each crate to determine the quality of my oranges. It is easy to inspect a few oranges from the top of each crate, but these oranges may not be representative of the entire crate. Those on the bottom are more often damaged in shipment. If I were less than honest, I might make sure that the rotten oranges are packed on the bottom, with some good ones on top for you to inspect. If you sample from the top, your sample results are again biased—the sample oranges are systematically better than the population they are supposed to represent.

Biased sampling methods

The design of a statistical study is **biased** if it systematically favors certain outcomes.

A **voluntary response sample** chooses itself by responding to a general appeal. Write-in or call-in opinion polls are examples of voluntary response samples.

Selection of whichever individuals are easiest to reach is called **convenience sampling.**

Convenience samples and voluntary response samples are often biased.

EXAMPLE 2.1   Interviewing at the mall

Inspecting the oranges at the top of the crate is one example of convenience sampling. Mall interviews are another. Manufacturers and advertising agencies often use interviews at shopping malls to gather information about the habits of consumers and the effectiveness of ads. A sample of mall shoppers is fast and cheap. But people contacted at shopping malls are not representative of the entire U.S. population. They are richer, for example, and more likely to be teenagers or retired. Moreover, the interviewers tend to select neat, safe-looking individuals from the stream of customers. Mall samples are biased: they systematically overrepresent some parts of the population (prosperous people, teenagers, and retired people) and underrepresent others. The opinions of such a convenience sample may be very different from those of the population as a whole.

EXAMPLE 2.2   Write-in opinion polls

Ann Landers once asked the readers of her advice column, "If you had it to do over again, would you have children?" She received nearly 10,000 responses, almost 70% saying "NO!" Can it be true that 70% of parents regret having children? Not at all. This is a voluntary response sample. People who feel strongly about an issue, particularly people with strong negative feelings, are more likely to take the trouble to respond. Ann Landers's results are strongly biased—the percent of parents who would not have children again is much higher in her sample than in the population of all parents.

"Hey, Pops, what was that letter you sent off to Ann Landers yesterday?"

Write-in and call-in opinion polls are almost sure to lead to strong bias. In fact, only about 15% of the public has ever responded to a call-in poll, and these tend to be the same people who call radio talk shows. That's not a representative sample of the population as a whole.

undor covrase -> Bias

EXERCISES

2.1 Instant opinion The Harris/Excite Instant Poll can be found online at poll.excite.com. On January 25, 2000, the question was "Should female athletes be paid the same as men for the work they do?" In all, 13,147 (44%) said "Yes," another 15,182 (50%) said "No," and the remaining 1448 said "Don't know."
(a) What is the sample size for this poll?
(b) That's a much larger sample than standard sample surveys. In spite of this, we can't trust the result to give good information about any clearly defined population. Why?
(c) It is still true that more men than women use the Web. How might this fact affect the poll results?

2.2 Ann Landers takes a sample Advice columnist Ann Landers once asked her female readers whether they would be content with affectionate treatment by men, with no sex ever. Over 90,000 women wrote in, with 72% answering "Yes."

Many of the letters described unfeeling treatment by men. Explain why this sample is certainly biased. What is the likely direction of the bias? That is, is that 72% probably higher or lower than the truth about the population of all adult women?

2.3 Design your own bad sample Your school wants to gather student opinion about parking for students on campus. It isn't practical to contact all students.
(a) Give an example of a way to choose a sample of students that is poor practice because it depends on voluntary response.
(b) Give another example of a bad way to choose a sample—but this time one that doesn't use voluntary response.

2.4 Women and love In 1987, Shere Hite published a best-selling book called *Women and Love*. Hite had distributed 100,000 questionnaires through various women's groups, asking questions about love, sex, and relations between women and men. Only 4.5% of the questionnaires were returned; Hite based her book on these responses. The women who responded were fed up with men and eager to contend with them. For example, 91% of those who were divorced said that they had initiated the divorce. The anger of women toward men became the theme of the book.

Explain why Hite's sampling method is biased. Is that 91% probably higher or lower than the true percent of all divorced women who initiated the divorce themselves? Why?

2.5 SRS applet, I Explain how you would use the *Simple Random Sample* applet from Activity 2.1 (page 64) to perform each of the following tasks:
(a) Flip a coin.
(b) Shuffle a deck of 52 cards.
(c) Pick a jury of 12 people from a small town with a population of 500 adult citizens.

2.6 SRS applet, II Implement each of your ideas from the previous exercise. Can the applet be used in more than one way to carry out any of these tasks? Explain.

Simple random samples

In a voluntary response sample, people choose whether to respond. In a convenience sample, the interviewer makes the choice. In both cases, personal choice produces bias. The statistician's remedy is to allow impersonal chance to choose the sample. A sample chosen by chance allows neither favoritism by the sampler nor self-selection by respondents. Choosing a sample by chance attacks bias by giving all individuals an equal chance to be chosen. Rich and poor, young and old, black and white, all have the same chance to be in the sample.

The simplest way to use chance to select a sample is to place names in a hat (the population) and draw out a handful (the sample). This is the idea of *simple random sampling*.

Simple random sample

A **simple random sample (SRS)** of size n consists of n individuals from the population chosen in such a way that every set of n individuals has an equal chance to be the sample actually selected.

© Harley Schwadron

An SRS not only gives each individual an equal chance to be chosen (thus avoiding bias in the choice) but also gives every possible sample an equal chance to be chosen. Drawing names from a hat does this. Write 100 names on identical slips of paper and mix them in a hat. This is a population. Now draw 10 slips, one after the other. This is an SRS, because any 10 slips have the same chance as any other 10.

Drawing names from a hat makes clear what it means to give each individual and each possible set of n individuals the same chance to be chosen. That's the idea of an SRS. Of course, we would need a very big hat to get a sample of the country's 100 million households. In practice, we use computer-generated *random digits* to choose samples. If you don't use software directly, you can use a *table of random digits* to choose small samples by hand.

Random digits

A **table of random digits** is a long string of the digits 0, 1, 2, 3, 4, 5, 6, 7, 8, 9 with these two properties:

1. Each entry in the table is equally likely to be any of the 10 digits 0 through 9.

2. The entries are **independent** of each other. That is, knowledge of one part of the table gives no information about any other part.

Table A at the back of the book is a table of random digits. You can think of Table A as the result of asking an assistant (or a computer) to mix the digits 0 to 9 in a hat, draw one, then replace the digit drawn, mix again, draw a second digit, and so on. The assistant's mixing and drawing save us the work of mixing and drawing when we need to randomize. Table A begins with the digits 19223950340575628713. To make the table easier to read, the digits appear in groups of five and in numbered rows. The groups and rows have no meaning—the table is just a long list of randomly chosen digits. Here's how to use the table to choose an SRS.

EXAMPLE 2.3   Choosing an SRS: teens and the Internet

There has been much talk (and some growing concern) about teenagers' use of the Internet. In addition to countless hours spent in chat rooms and surfing the Net, teens are viewing increasing amounts of pornography. Should stricter controls be imposed on teen Internet use?

Let's select a panel of teens to help answer this question. Ideally, we would like to have teens representing each of the 50 states on the panel. Unfortunately, budget restrictions will permit us to choose only six teens for the panel. So here's what we'll do. We'll ask each state to select one teen nominee by a process of its choosing. Then, to avoid bias, we will select an SRS of size six from the 50 state nominees to form the panel.

Step 1: Label. Give each state a numerical label, using as few digits as possible. Two digits are needed to label 50 states, so we use labels

01, 02, 03, . . . , 48, 49, 50

It is also correct to use labels 00 to 49 or even another choice of 50 two-digit labels. Here is a partial list of states, with labels attached:

01 Alabama
02 Alaska
03 Arizona

.

.

.

48 Washington
49 Wisconsin
50 Wyoming

Step 2: Table. Enter Table A anywhere and read two-digit groups. Suppose we enter at line 103, which is

45467 71709 77558 00095 32863 29485 82226 90056

The first 10 two-digit groups in this line are

45 46 77 17 09 77 55 80 00 95

Each two-digit group in Table A is equally likely to be any of the 100 possible groups, 00, 01, 02, . . . , 99. So two-digit groups choose two-digit labels at random. That's just what we want.

We used only labels 01 to 50, so we ignore all other two-digit groups. The first 6 labels between 01 and 50 that we encounter in the table choose our sample. Of the first 10 labels in line 103, we ignore 6 because they are too high (over 50). The others are 45, 46, 17, and 09. The nominees from the states labeled 09, 17, 45, and 46 go into the sample. We need two more teenagers to complete the panel. The remaining 10 two-digit groups in line 103 are

32 86 32 94 85 82 22 69 00 56

The teenager from state 32 is chosen for the panel. Ignore the second 32 because that state is already in the sample. State 22's teen completes the panel.

The sample is composed of the states labeled 09, 17, 22, 32, 45, and 46. Our panel consists of the teens nominated by those states.

Are these random digits really random?

Not a chance. The random digits in Table A were produced by a computer program. Computer programs do exactly what you tell them to do. Give the program the same input and it will produce exactly the same "random" digits. Of course, clever people have devised computer programs that produce output that looks like random digits. These are called "pseudo-random numbers," and that's what Table A contains. Pseudo-random numbers work fine for statistical randomizing, but they have hidden nonrandom patterns that can mess up more refined uses.

Using the table of random digits is much quicker than drawing names from a hat. As Example 2.3 shows, choosing an SRS has two steps.

Choose an SRS in two steps

Step 1: Label. Assign a numerical label to every individual in the population. Be sure that all labels have the same number of digits.

Step 2: Table. Use random digits to select labels at random.

Golfing at random

Random drawings give everyone the same chance to be chosen, so they offer a fair way to decide who gets a scarce good—like a round of golf. Lots of golfers want to play the famous Old Course at St. Andrews, Scotland. A few can reserve in advance. Most must hope that chance favors them in the daily random drawing for tee times. At the height of the summer season, only 1 in 6 wins the right to pay over $100 for a round.

You can assign labels in any convenient manner, such as alphabetical order for names of people. As long as all labels have the same number of digits, all individuals will have the same chance to be chosen. Use the shortest possible labels: one digit for a population of up to 10 members, two digits for 11 to 100 members, three digits for 101 to 1000 members, and so on. As standard practice, we recommend that you begin with label 1 (or 01 or 001, as needed). You can read digits from Table A in any order—across a row, down a column, and so on—because the table has no order. As standard practice, we recommend reading across rows.

EXAMPLE 2.4   Choosing an SRS: Congress joins the teenagers

To add some clout to the teenagers' recommendations about Internet use in Example 2.3, we decide to add four members of the House of Representatives to the panel. Again, we decide to take an SRS.

Step 1: Label. Obtain an alphabetical listing of the 435 members of the House of Representatives. Assign numbers 001 to 435 as labels to the alphabetized list.

Step 2: Table. Enter Table A at line 134 (say) and read groups of three digits. The first 10 three-digit groups are

278	167	841	618	329	213	373	521	337	741
✓	✓	too high	too high	✓	✓				

Our sample consists of the representatives labeled 278, 167, 329, and 213. These individuals will join the teens on our panel.

Sample surveys use computer software to choose an SRS, but the software just automates the steps in Examples 2.3 and 2.4. The computer doesn't look in a table of random digits because it can generate them on the spot. You can also use a calculator to choose an SRS, as the Calculator Corner illustrates.

CALCULATOR CORNER   How to choose an SRS

The TI-83 has built-in commands for generating "random" numbers. Just how "random" are they?

1. Press MATH, then select 4:PRB and 5:randInt(. Complete the command randInt(1,5) and press ENTER. Determine by a show of hands how many students in the class got each of the numbers 1, 2, 3, 4, and 5. Do the results seem random?

```
randInt(1,5)
                   3
```

2. Press ENTER four more times. Record the sequence of five numbers you obtain. Compare your results with those of your classmates. Do the results seem random?

```
randInt(1,5)
                   3
                   4
                   5
                   1
                   3
```

3. Calculators and computers use algorithms written by humans to generate "random" numbers. The TI-83's randInt command uses a starting value, called a "seed," to determine the next "random" integer between 1 and 5. To illustrate this, everyone in your class should type 1234, press → MATH, choose 4:PRB and 1:rand, then press ENTER to generate the command 1234 → rand. Now execute the command randInt(1,5). You should all get the number 1.

```
1234→rand
                1234
randInt(1,5)
                   1
```

4. In Step 3, you saw that starting with the same seed yields the same random integer. That's hardly random! We can improve on this situation by having each student input a different seed to the random number generator. Here's one method: type the last four digits of your home phone number and store to rand, like you did in Step 3.

CALCULATOR CORNER How to choose an SRS *(continued)*

Now do `randInt(1,100)`. See if anyone else in the class got the same number.

```
0594→rand
                594
randInt(1,100)
                 87
```

5. Now you're ready to choose an SRS of students from your class. Use a numbered list, like the one described in Activity 2.1 (page 64), for reference. Execute the command `randInt(1,n)` where *n* is the number of students in your class. Note who is selected. Press ENTER four more times. Did you get five different numbers? If not, press ENTER a few more times until you have selected five people in your sample. Note who is selected.

```
randInt(1,30)
                13
                26
                22
                27
                14
```

6. An alternative is to execute the command `randInt(1,n,5)` to generate five "random" numbers between 1 and *n*. Do this now. Are there any repeated numbers among the five values? If so, press ENTER again to generate a few more. Note who is selected this time.

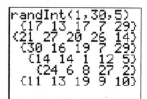

```
randInt(1,30,5)
  {17 13 17 7 29}
 {21 27 20 26 14}
  {30 16 19 7 29}
    {14 14 1 12 5}
     {24 6 8 27 2}
  {11 13 19 9 10}
```

7. Obtain eight more samples of size 5 from your class using what you learned in Steps 5 and 6. Record who is in your sample each time. How often did you choose yourself as one of the sample?

EXERCISES

2.7 Apartment living You are planning a report on apartment living in a college town. You decide to select three apartment complexes at random for in-depth interviews with residents. Use Table A, starting at line 117, to select a simple random sample of three of the following apartment complexes. Explain your method clearly enough for a classmate to obtain your results.

Ashley Oaks	Chauncey Village	Franklin Park	Richfield
Bay Pointe	Country Squire	Georgetown	Sagamore Ridge
Beau Jardin	Country View	Greenacres	Salem House
Bluffs	Country Villa	Lahr House	Village Manor
Brandon Place	Crestview	Mayfair Village	Waterford Court
Briarwood	Del-Lynn	Nobb Hill	Williamsburg
Brownstone	Fairington	Pemberly Courts	
Burberry	Fairway Knolls	Peppermill	
Cambridge	Fowler	Pheasant Run	

2.8 How do random digits behave? Which of the following statements are true of a table of random digits, and which are false? Explain your answers.
(a) There are exactly four 0s in each row of 40 digits.
(b) Each pair of digits has chance 1/100 of being 00.
(c) The digits 0000 can never appear as a group, because this pattern is not random.

2.9 An election day sample You want to choose an SRS of 25 of a city's 440 voting precincts for special voting-fraud surveillance on election day.
(a) Explain clearly how you would label the 440 precincts. How many digits make up each of your labels? What is the greatest number of precincts you could label using this number of digits?
(b) Use Table A to choose the SRS, and list the labels of the precincts you selected. Enter Table A at line 117.

2.10 Is this an SRS? A university has 1000 male and 500 female faculty members. A survey of faculty opinion selects 100 of the 1000 men at random and then separately selects 50 of the 500 women at random. The 150 faculty members chosen make up the sample.
(a) Explain why this sampling method gives each member of the faculty an equal chance to be chosen.
(b) Nonetheless, this is not an SRS. Why not?

2.11 Drug testing Use the table of random digits (Table A) to select an SRS of 3 of the following 25 volunteers for a drug test. Be sure to say where you entered the table and how you used it.

Agarwal	Fuest	Milhalko	Shen
Andrews	Fuhrmann	Moser	Smith
Baer	Garcia	Musselman	Sundheim
Berger	Healy	Pavnica	Wilson
Brockman	Hixson	Petrucelli	
Chen	Lee	Reda	
Frank	Lynch	Roberts	

Can you trust a sample?

The *Town Talk,* Ann Landers, and mall interviews produce samples. We can't trust results from these samples, because they are chosen in ways that invite bias. We have more confidence in results from an SRS, because it uses impersonal chance to avoid bias. The first question to ask of any sample is whether it was chosen at random. Opinion polls and other sample surveys carried out by people who know what they are doing use random sampling.

EXAMPLE 2.5   A Gallup Poll

A Gallup Poll on smoking began with the question "Have you, yourself, smoked any cigarettes in the past week?" The press release reported that "just 23% of Americans smoke." Can we trust this fact? Ask first how Gallup selected its sample. Later in the press release we read this: "The results are based on telephone interviews with a randomly selected national sample of 1,039 adults, 18 years and older, conducted September 23–26, 1999."[2]

This is a good start toward gaining our confidence. Gallup tells us what population it has in mind (people at least 18 years old living anywhere in the United States). We know that the sample from this population was of size 1039 and, most important, that it was chosen at random. There is more to say, and we will soon say it, but we have at least heard the comforting words "randomly selected."

What should a nation do if there are not enough volunteers for military service but only a small fraction of eligible youth are needed by the military? That question last arose during the Vietnam era. Beginning in 1970, a draft lottery was used to choose draftees by random selection. Although the draft was ended in 1976, young men must still register at age 18. The draft lottery may return if the military cannot attract enough volunteers.

Actually making a random selection in the public arena isn't always easy. In principle, it is just like choosing an SRS. In practice, random digits can't be used. Because few people understand random digits, a lottery looks fairer if a dignitary chooses capsules from a glass bowl in front of the TV cameras. This also prevents cheating: no one can check the table of random digits in advance to see how Cousin Joe will make out in the selection.

Physical mixing and drawing *look* random, but it can be hard to achieve a mixing that really *is* random. There is no better illustration of this than the first Vietnam-era draft lottery, held in 1970.

APPLICATION 2.1 The Draft Lottery

Because taking an SRS of all eligible men would be hopelessly awkward, the draft lottery selected birth dates in a random order. Men born on the date chosen first would be drafted first, then those born on the second date chosen, and so on. All men aged 19 to 25 were included in the first lottery, and there were 366 birth dates. The 366 dates were placed into identical small capsules, which were put in a bowl and publicly drawn one by one.

Here is an account of the 1970 draft lottery:

They started out with 366 cylindrical capsules, one and a half inches long and one inch in diameter. The caps at the end were round.

The men counted out 31 capsules and inserted in them slips of paper with the January dates. The January capsules were then placed in a large square wooden box and pushed to one side with a cardboard divider, leaving part of the box empty.

The 29 February capsules were then poured into the empty portion of the box, counted again, and then scraped with the divider into the January capsules. Thus, according to Captain Pascoe, the January and February capsules were thoroughly mixed.

The same process was followed with each subsequent month, counting the capsules into the empty side of the box and then pushing them with the divider into the capsules of the previous months.

"So you were born in December too, eh?"

APPLICATION 2.1 The Draft Lottery *(continued)*

Thus, the January capsules were mixed with the other capsules 11 times, the February capsules 10 times and so on with the November capsules intermingled with others only twice and the December ones only once.

The box was then shut, and Colonel Fox shook it several times. He then carried it up three flights of stairs, a process that Captain Pascoe says further mixed the capsules.

The box was carried down the three flights shortly before the drawing began. In public view, the capsules were poured from the black box into the two-foot deep bowl.[3]

1. Do you think that men are equally likely to be born on each of the 366 days? Why or why not?

2. Explain why the method does not choose an SRS of all men eligible for the draft.

3. Was this method likely to produce an SRS of all possible birth dates? Why or why not?

Did you guess what happened? Dates in January tended to be on the bottom, while birth dates in December were put in last and tended to be on top. News reporters noticed at once that men born later in the year seemed to receive lower draft numbers, and statisticians soon showed that this trend was so strong that it would occur less than once in a thousand truly random annual lotteries. An inquiry was made, which provided the account quoted.

What's done is done, and off to Vietnam went too many men born in December. But for 1971, the captain and the colonel were given other duties, and statisticians from the National Bureau of Standards were asked to design the lottery. Their design was a bit more complicated. The numbers 1 to 365 (no leap year this time) were placed in capsules in a random order determined by a table of random digits. The dates of the year were placed in another set of capsules in a random order determined again by the random digit table. Then the date capsules were put into a drum in a random order determined by a third use of the table of random digits. The number capsules went into a second drum, again in random order. The drums were rotated for an hour. The TV cameras were turned on, and the dignitary reached into the date drum: out came September 16. He reached into the number drum: out came 139. So men born on September 16 received draft number 139. Back to both drums: out came April 27 and draft number 235. And so on. It's awful, but it's random.[4] You can rejoice that in choosing samples we have Table A to do the randomization for us.

EXERCISES

2.12 A call-in opinion poll Should the United Nations continue to have its headquarters in the United States? A television program asked its viewers to call in with their opinions on that question. There were 186,000 callers, 67% of whom said "No." A nationwide random sample of 500 adults found that 72% answered "Yes" to the same question.[5] Explain to someone who knows no statistics why the opinions of only 500 randomly chosen respondents are a better guide to what all Americans think than the opinions of 186,000 callers.

2.13 Rating the police The Miami Police Department wants to know how black residents of Miami feel about police service. A sociologist prepares several questions about the police. The police department chooses an SRS of 300 mailing addresses in predominantly black neighborhoods and sends a uniformed black police officer to each address to ask the questions of an adult living there.
(a) What are the population and the sample?
(b) Why are the results likely to be biased even though the sample is an SRS?

2.14 Random selection? Choosing at random is a "fair" way to decide who gets some scarce good, in the sense that everyone has the same chance to win. Random choice isn't always a good idea—sometimes we don't want to treat everyone the same, because some people have a better claim. In each of the following situations, would you support choosing at random? Give your reasons in each case.
(a) The basketball arena has 4000 student seats, and 7000 students want tickets. Shall we choose 4000 of the 7000 at random?
(b) The list of people waiting for liver transplants is much larger than the number of available livers. Shall we let impersonal chance decide who gets a transplant?
(c) During the Vietnam War, young men were chosen for army service at random, by a "draft lottery." Is this the best way to decide who goes and who stays home?

2.15 Local draft boards Prior to 1970, young men were selected for military service by local draft boards. There was a complex system of exemptions and quotas that enabled, for example, farmers' sons and married young men with children to avoid the draft. Do you think that random selection among all men of the same age is preferable to making distinctions based on occupation and marital status? Give your reasons.

2.16 Random selection or not? Give an example of a situation where you definitely would *approve* random selection. Give an example of a situation where you would definitely *disapprove* random selection.

EXPLORING THE WEB

The Internet brings voluntary response polls to the computer nearest you. Visit www.misterpoll.com to become part of the sample in any of dozens of online polls. As the site says, "None of these polls are 'scientific,' but do represent the collective opinion of everyone who participates."

To see how software speeds up choosing an SRS, go to the Research Randomizer at www.randomizer.org. Click on "Randomizer" and fill in the boxes. You can even ask the Randomizer to arrange your sample in order.

STATISTICS IN SUMMARY

We select a **sample** in order to get information about some **population.** How can we choose a sample that represents the population fairly? **Convenience samples** and **voluntary response samples** are common but do not produce trustworthy data because these sampling methods are usually **biased.** That is, they systematically favor some parts of the population over others in choosing the sample.

The deliberate use of chance in producing data is one of the big ideas of statistics. Random samples use chance to choose a sample, thus avoiding bias due to personal choice. The basic type of random sample is the **simple random sample,** which gives all samples of the same size the same chance to be the sample we actually choose. To choose an SRS by hand, use a **table of random digits** such as Table A in the back of the book.

SECTION 2.1 EXERCISES

2.17 How much do students earn? A university's financial aid office wants to know how much it can expect students to earn from summer employment. This information will be used in setting the level of financial aid. The population contains 3478 students who have completed at least one year of study but have not yet graduated. The university will send a questionnaire to an SRS of 100 of these students, drawn from an alphabetized list.
(a) Describe how you will label the students in order to select the sample.
(b) Use Table A, beginning at line 105, to select the first 5 students in the sample.

2.18 More randomization Most sample surveys call residential telephone numbers at random. They do not, however, always ask their questions of the person who picks up the phone. Instead, they ask about the adults who live in the residence and choose one at random to be in the sample. Why is this a good idea?

2.19 A biased sample You see a female student standing in front of the cafeteria, now and then stopping other students to ask them questions. She says that she is collecting student opinions for a class assignment. Explain why this sampling method is almost certainly biased.

2.20 Sampling mushrooms A food processor has 50 large lots of canned mushrooms ready for shipment, each labeled with one of the lot numbers below.

A1109 *01*	A2056	A2219	A2381	B0001 *1*
A1123 *02*	A2083	A2336	A2382	B0012 *2*
A1186 *03*	A2084	A2337	A2383	B0046 *3*
A1197 *04*	A2100	A2338	A2384	B1195
A1198 *5*	A2108	A2339	A2385	B1196
A2016 *6*	A2113	A2340	A2390	B1197
A2017 *7*	A2119	A2351	A2396	B1198
A2020 *8*	A2124	A2352	A2410	B1199
A2029 *9*	A2125	A2367	A2411	B1200
A2032 *10*	A2130 *20*	A2372 *30*	A2500 *41*	B1201 *50*

An SRS of 5 lots must be chosen for inspection. Use Table A to do this, beginning at line 139.

2.21 Gun control Since the beginning of opinion polls over 50 years ago, two-thirds of those sampled have consistently favored stronger controls on firearms. Specific gun control proposals have often been favored by 80 to 85% of respondents. Yet little national gun control legislation has passed, and no major national restrictions on firearms exist.

Why do you think this has occurred? Does this mean that opinion polls on issues do not really offer the public a means of offsetting special-interest groups?

2.22 A lottery for AIDS drugs? Only limited supplies of some experimental drugs to treat AIDS are available because the drugs are very difficult to make. It is now usual to use a lottery to decide at random which AIDS patients can receive the drug. In 1995, for example, Hoffman–La Roche had enough doses of the drug Invirase for 2880 patients. Shortly after announcing this, the company received 10,000 calls from patients who wanted to enter the lottery.

Discuss this practice. Do you favor random selection? If not, how should recipients be chosen?

2.23 Read anything lately? The Denver Public Library wishes to estimate the percent of Denver households with an adult who has read at least one book in the last month. The homes of 400 people who have library cards are sampled, and it turns out that 90% of these households have an adult who has read a book in the past month. Is 90% likely to be a biased estimate for the true percent of Denver households with an adult who has read at least one book in the last month? Explain why or why not.

2.24 Women count! On Saturday, October 15, 1997, broadcast and print media across the country reported on the government's release of a report on the status of working women called "Working Women Count." This report gives findings from a survey of 250,000 women. According to one news report:

> More than 1600 businesses, unions, newspapers, magazines, and community service organizations helped distribute the survey to their members, subscribers, and patrons, which the White House announced with much fanfare. It sought women's opinions on job satisfaction, pay, benefits, and opportunities for advancement.[6]

A second survey, asking the same questions, was conducted at the same time. However, this survey interviewed only 1200 working women chosen at random.

Which survey is likely to give a more accurate view of "working women's opinions on job satisfaction, pay, benefits, and opportunities for advancement"? Is this improved accuracy likely to be the result of lower bias or of greater precision? Explain briefly.

2.2 WHAT DO SAMPLES TELL US?

Do you lotto?

You know that lotteries are popular. How popular? Here's what the Gallup Poll says: "They can offer massive jackpots—and a ticket just costs a dollar at your neighborhood store. For many Americans, picking up a lottery ticket has become routine, despite the massive odds against striking it rich. A new Gallup Poll Social Audit on gambling shows that 57% of Americans have bought a lottery ticket in the last 12 months, making lotteries by far the favorite choice of gamblers."

Reading further, we find that Gallup talked with 1523 randomly selected adults to reach these conclusions. We're happy Gallup chooses at random—we wouldn't get unbiased information about gambling by asking people who are standing in line to buy lotto tickets. However, the Census Bureau says that there are about 200 million adults in the United States. How can 1523 people, even a random sample of 1523 people, tell us about the habits of 200 million people?

A sample won't tell us the exact truth about the population. So Gallup gives us a margin of error. Here's Gallup's claim: "For results based on a sample of this size, one can say with 95% confidence that the error attributable to sampling and other random effects could be plus or minus 3 percentage points for adults." In the news, this gets reduced to "The margin of error for the poll was plus or minus 3 percentage points."[7] Either way, it isn't exactly obvious what the statement means. To find out, read on.

ACTIVITY 2.2 Should gambling be legal?

Figure 2.2 (next page) is a small population. Each circle represents an adult. The colored circles are people who disapprove of legal gambling, and the white circles are people who approve. You can check that 60 of the 100 circles are white, so in this population the proportion who approve of gambling is $p = 60/100 = 0.6$.

1. The circles are labeled 00, 01, . . . , 99. Use line 101 of Table A to draw an SRS of size 10. Explain your method clearly enough for a classmate to repeat the process. What is the proportion $\hat{p}$ of the people in your sample who approve of gambling?

2. Take 9 more SRSs of size 10 (10 samples in all). Be sure to use a different starting point in Table A for each sample. You now have 10 values of the sample proportion $\hat{p}$.

3. Because your samples have only 10 people, the only values $\hat{p}$ can take are 0/10, 1/10, 2/10, . . ., 9/10, and 10/10. That is, $\hat{p}$ is always 0, 0.1, 0.2, . . ., 0.9, or 1. Construct a dotplot of your 10 results.

4. Taking samples of size 10 from a population of size 100 is not a setting that occurs often in the real world, but let's look at your results anyway. How many of your 10 samples estimated the population proportion $p = 0.6$ exactly correctly? Is the true value 0.6 roughly in the center of your sample values? Explain why 0.6 would be in the center of the sample values if you took a large number of samples.

5. Calculate the average of your 10 sample proportions. How does this value compare with the proportion of adults in the population who favor legal gambling?

6. Pool results with your classmates. Make a large dotplot on the board. Describe what you see.

7. Calculate the average of the class's sample proportions. Compare with the true population proportion.

ACTIVITY 2.2 Should gambling be legal? *(continued)*

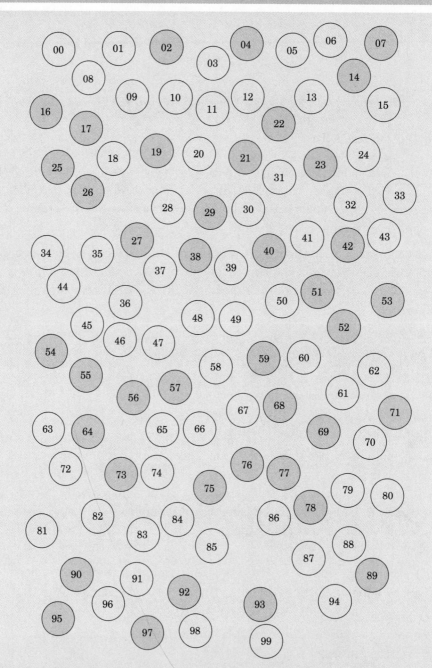

FIGURE 2.2 A population of 100 individuals for Activity 2.2. Some individuals (white circles) approve of legal gambling, and the others disapprove.

From sample to population

Gallup's finding that "57% of Americans have bought a lottery ticket in the last 12 months" makes a claim about the population of 200 million adults. But Gallup doesn't know the truth about this population. The poll contacted 1523 people and found that 57% of them said they had bought a lottery ticket in the past year. Because the sample of 1523 people was chosen at random, it's reasonable to think that they represent the entire population pretty well. So Gallup turns the *fact* that 57% of the *sample* bought lottery tickets into an *estimate* that about 57% of *all adults* bought tickets. That's a basic move in statistics: use a fact about a sample to estimate the truth about the whole population. To think about such moves, we must keep straight whether a number describes a sample or a population. Here is the vocabulary we use.

Parameters and statistics

A **parameter** is a number that describes the **population.** A parameter is a fixed number, but in practice we don't know its value.

A **statistic** is a number that describes a **sample.** The value of a statistic is known when we have taken a sample, but it can change from sample to sample. We often use a statistic to estimate an unknown parameter.

So "parameter" is to "population" as "statistic" is to "sample." Want to estimate an unknown parameter? Choose a sample from the population and use a sample statistic as your estimate. That's what Gallup did.

EXAMPLE 2.6   Do you lotto?

The proportion of all adults who bought a lottery ticket in the past 12 months is a *parameter* describing the population of 200 million adults. Call it *p*, for "proportion." Alas, we do not know the numerical value of *p*. To estimate *p*, Gallup took a sample of 1523 adults. The proportion of the sample who bought lottery tickets is a *statistic*. Call it $\hat{p}$ (read as "p-hat"). It happens that 868 of this sample of size 1523 bought tickets, so for this sample,

$$\hat{p} = \frac{868}{1523} = 0.57 \quad \text{(that is, 57\%)}$$

Because all adults had the same chance to be among the chosen 1523, it seems reasonable to use the statistic $\hat{p} = 0.57$ as an estimate of the unknown parameter *p*. It's a fact that 57% of the sample bought lottery tickets—we know because we asked them. We don't know what percent of all adults bought tickets, but we *estimate* that about 57% did.

EXERCISES

2.25 Stop smoking! A random sample of 1000 people who signed a card saying they intended to quit smoking on November 20, 1995 (the day of the "Great American Smoke-out"), were contacted in June 1996. It turned out that 210 (21%) of the sampled individuals had not smoked over the past six months. Specify the population of interest, the parameter of interest, the sample, and the sample statistic in this problem.

Each boldface number in Exercises 2.26 to 2.28 is the value of either a **parameter** *or a* **statistic.** *In each case, state which it is.*

2.26 Ketchup bottles On Tuesday, the bottles of tomato ketchup filled in a plant were supposed to contain an average of **14** ounces of ketchup. Quality control inspectors sampled 50 bottles at random from the day's production. These bottles contained an average of **13.8** ounces of ketchup.

2.27 Flight safety On a New York–to–Denver flight, **8%** of the 125 passengers were selected for random security screening prior to boarding. According to the Transportation Security Administration, **10%** of airline passengers are chosen for random screening.

2.28 Blind ducks, I A recent report in the journal *Nature* examined whether ducks keep an eye out for predators while they sleep. The researchers, from Indiana State University, put four ducks in each of four plastic boxes, which were arranged in a row. Ducks in the two end boxes slept with one eye open **31.8%** of the time, compared to only **12.4%** of the time for the ducks in the two center boxes.

2.29 Blind ducks, II Is the study described in the previous exercise an example of an observational study or a comparative experiment? Explain briefly.

Sampling variability

If Gallup took a second random sample of 1523 adults for Example 2.6, the new sample would have different people in it. It is almost certain that there would not be exactly 868 positive responses. That is, the value of the statistic $\hat{p}$ will *vary* from sample to sample. Could it happen that one random sample finds that 57% of adults recently bought a lottery ticket and a second random sample finds that only 37% had done so? Random samples eliminate *bias* from the act of choosing a sample, but they can still estimate a population proportion badly because of the *variability* that results when we choose at random. If the variation when we take repeated samples from the same population is too great, we can't trust the results of any one sample.

We are saved by the second great advantage of random samples. The first advantage is that choosing at random eliminates favoritism. That is, random sampling attacks bias. The second advantage is that if we took lots of random samples of the same size from the same population, the variation from sample to sample would follow a predictable pattern. This predictable pattern shows that results of bigger samples are less variable than the results of smaller samples.

EXAMPLE 2.7   Lots and lots of samples

Here's another big idea of statistics: to see how trustworthy one sample is likely to be, ask what would happen if we took many samples from the same population. Let's try it and see. Suppose that in fact (unknown to Gallup), exactly 60% of all adults have bought a lottery ticket in the past 12 months. That is, the truth about the population is that $p = 0.6$. What if Gallup used the sample proportion $\hat{p}$ from an SRS of size 100 to estimate the unknown value of the population proportion p?

Figure 2.3 illustrates the process of choosing many samples and finding $\hat{p}$ for each one. In the first sample, 56 of the 100 people had bought lottery tickets, so $\hat{p} = 56/100 = 0.56$. Only 46 in the next sample had bought tickets, so for that sample $\hat{p} = 0.46$. Choose 1000 samples and make a histogram of the 1000 values of $\hat{p}$. That's the graph at the right of Figure 2.3. The different values of $\hat{p}$ run along the horizontal axis. The heights of the bars show how many of our 1000 samples gave each group of values.

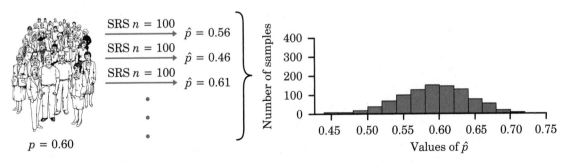

FIGURE 2.3 The results of many SRSs have a regular pattern. Here, we draw 1000 SRSs of size 100 from the same population. The population proportion is $p = 0.60$. The sample proportions vary from sample to sample, but their values center at the truth about the population.

Of course, Gallup interviewed 1523 people, not just 100. Figure 2.4 (next page) shows the results of 1000 SRSs, each of size 1523, drawn from a population in which the true proportion is $p = 0.6$. Figures 2.3 and 2.4 are drawn on the same scale. Comparing them shows what happens when we increase the size of our samples from 100 to 1523.

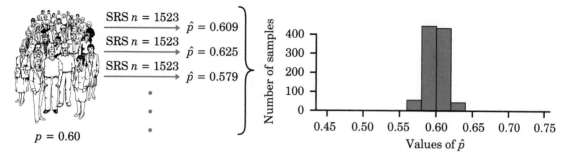

FIGURE 2.4 Draw 1000 SRSs of size 1523 from the same population as in Figure 2.3. The 1000 values of the sample proportion are much less spread out than was the case for smaller samples.

Look carefully at Figures 2.3 and 2.4. We flow from the population, to many samples from the population, to the many values of $\hat{p}$ from these many samples. Gather these values together and study the histograms that display them.

- In both cases, the values of the sample proportion $\hat{p}$ vary from sample to sample, but the values are centered at 0.6. Recall that $p = 0.6$ is the true population parameter. Some samples have a $\hat{p}$ less than 0.6 and some greater, but there is no tendency to be always low or always high. That is, $\hat{p}$ has no *bias* as an estimator of p. This is true for both large and small samples.

- The values of $\hat{p}$ from samples of size 100 are much more spread out than the values from samples of size 1523. In fact, 95% of our 1000 samples of size 1523 have a $\hat{p}$ lying between 0.576 and 0.624. That's within 0.024 on either side of the population truth 0.6. Our samples of size 100, on the other hand, spread the middle 95% of their values between 0.50 and 0.69. That goes out 0.1 from the truth, about four times as far as the larger samples. So larger random samples have less *variability* than smaller samples.

The upshot is that we can rely on a sample of size 1523 to almost always give an estimate $\hat{p}$ that is close to the truth about the population. Figure 2.4 illustrates this fact for just one value of the population proportion, but it is true for any population. Samples of size 100, on the other hand, might give an estimate of 50% or 70% when the truth is 60%.

Thinking about Figures 2.3 and 2.4 helps us restate the idea of bias when we use a statistic like $\hat{p}$ to estimate a parameter like p. It also reminds us that variability matters as much as bias.

Two types of error in estimation

Bias is consistent, repeated deviation of the sample statistic from the population parameter in the same direction when we take many samples.

Variability describes how spread out the values of the sample statistic are when we take many samples. Large variability means that the result of sampling is not repeatable.

A good sampling method has both small bias and small variability.

We can think of the true value of the population parameter as the bull's-eye on a target, and of the sample statistic as an arrow fired at the bull's-eye. Bias and variability describe what happens when an archer fires many arrows at the target. *Bias* means that the aim is off, and the arrows land consistently off the bull's-eye in the same direction. The sample values do not center about the population value. Large *variability* means that repeated shots are widely scattered on the target. Repeated samples do not give similar results but differ widely among themselves. Figure 2.5 shows this target illustration of the two types of error.

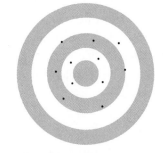

(a) Large bias, small variability   (b) Small bias, large variability

(c) Large bias, large variability   (d) Small bias, small variability

FIGURE 2.5 Bias and variability in shooting arrows at a target. Bias means the archer systematically misses in the same direction. Variability means that the arrows are scattered.

Notice that small variability (repeated shots are close together) can accompany large bias (the arrows are consistently away from the bull's-eye in one direction). And small bias (the arrows center on the bull's-eye) can accompany large variability (repeated shots are widely scattered). A good sampling scheme, like a good archer, must have both small bias and small variability. Here's how we do this.

Managing bias and variability

To reduce bias, use random sampling. When we start with a list of the entire population, simple random sampling produces unbiased estimates— the values of a statistic computed from an SRS neither consistently overestimate nor consistently underestimate the value of the population parameter.

To reduce the variability of an SRS, use a larger sample. You can make the variability as small as you want by taking a large enough sample.

In practice, Gallup takes only one sample. We don't know how close to the truth an estimate from this one sample is, because we don't know what the truth about the population is. But *large random samples almost always give an estimate that is close to the truth.* Looking at the pattern of many samples shows that we can trust the result of one sample.

EXERCISES

2.30 Sampling variability In thinking about Gallup's sample of size 1523, we asked, "Could it happen that one random sample finds that 57% of adults recently bought a lottery ticket and a second random sample finds that only 37% had done so?" Look at Figure 2.4, which shows the results of 1000 samples of this size when the population truth is 60%. Would you be surprised if a sample from this population gave 57%? Would you be surprised if a sample gave 37%?

2.31 Bias and variability Figure 2.6 on the facing page shows the behavior of a sample statistic in many samples in four situations. These graphs are like those in Figures 2.3 and 2.4. That is, the heights of the bars show how often the sample statistic took various values in many samples from the same population. The true value of the population parameter is marked by an arrow on each graph. Label each of the graphs in Figure 2.6 as showing high or low bias and as showing high or low variability.

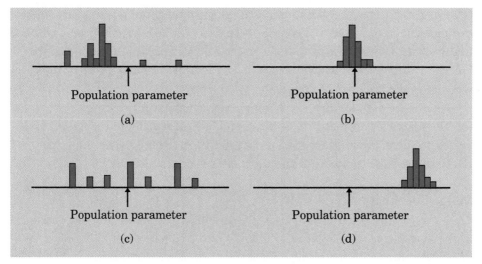

FIGURE 2.6 Take many samples from the same population and make a histogram of the values taken by a sample statistic. Here are the results for four different sampling methods, for Exercise 2.31.

2.32 Predict the election Just before a presidential election, a national opinion poll increases the size of its weekly sample from the usual 1500 people to 4000 people.
(a) Does the larger random sample reduce the bias of the poll result? Explain.
(b) Does it reduce the variability of the result? Explain.

2.33 A sampling experiment Let us illustrate sampling variability in a small sample from a small population. Ten of the 25 club members listed below are female. Their names are marked with asterisks in the list. The club chooses 5 members at random to receive free trips to the national convention.

Alonso	Darwin	Herrnstein	Myrdal	Vogt*
Binet*	Epstein	Jimenez*	Perez*	Went
Blumenbach	Ferri	Luo	Spencer*	Wilson
Chase*	Gonzales	Moll*	Thomson	Yerkes
Chen*	Gupta*	Morales*	Toulmin	Zimmer

(a) Draw 20 SRSs of size 5, using a different part of Table A each time. Record the number of females in each of your samples.
(b) Make a dotplot to display your results. What is the average number of females in your 20 samples?
(c) Do you think the club members should suspect discrimination if none of the 5 tickets go to women?

2.34 Canada's national health care The Ministry of Health in the Canadian province of Ontario wants to know whether the national health care system is achieving its goals in the province. Much information about health care comes from patient records, but that source doesn't allow us to compare people who use health services with those who don't. So the Ministry of Health conducted the Ontario Health Survey, which interviewed a random sample of 61,239 people who live in the Province of Ontario.[8]

(**a**) What is the population for this sample survey? What is the sample?

(**b**) The survey found that 76% of males and 86% of females in the sample had visited a general practitioner at least once in the past year. Do you think these estimates are close to the truth about the entire population? Why?

Margin of error and all that

The "margin of error" that sample surveys announce translates sampling variability of the kind pictured in Figures 2.3 and 2.4 into a statement of how much confidence we can have in the results of a survey. Let's start with the kind of language we hear so often in the news.

What margin of error means

"Margin of error plus or minus 3 percentage points" is shorthand for this statement:

If we took many samples using the same method we used to get this one sample, 95% of the samples would give a result within plus or minus 3 percentage points of the truth about the population.

Take this step-by-step. A sample chosen at random will usually not estimate the truth about the population exactly. We need a margin of error to tell us how close our estimate comes to the truth. But we can't be *certain* that the truth differs from the estimate by no more than the margin of error. Ninety-five percent of all samples come this close to the truth, but 5% miss by more than the margin of error. We don't know the truth about the population, so we don't know if our sample is one of the 95% that hit or one of the 5% that miss. We say we are **95% confident** that the truth lies within the margin of error.

EXAMPLE 2.8   Understanding the news

Here's what the TV news announcer says: "A new Gallup Poll finds that 57% of American adults bought a lottery ticket in the last 12 months. The margin of error for the poll was 3 percentage points." Plus or minus 3 percent starting at 57% is 54% to 60%. Most people think Gallup claims that the truth about the entire population lies in that range.

This is what Gallup actually said: "For results based on a sample of this size, one can say with 95% confidence that the error attributable to sampling and other random effects could be plus or minus 3 percentage points for adults." That is, Gallup tells us that the margin of error only works for 95% of all its samples. "95% confidence" is shorthand for that. The news report left out the "95% confidence."

Finding the margin of error exactly is a job for statisticians. You can, however, use a simple formula to get a rough idea of the size of a sample survey's margin of error.

A quick method for the margin of error

Use the sample proportion $\hat{p}$ from a simple random sample of size n to estimate an unknown population proportion p. The margin of error for 95% confidence is roughly equal to $1/\sqrt{n}$.

EXAMPLE 2.9   What is the margin of error?

The Gallup Poll in Example 2.6 interviewed 1523 people. The margin of error for 95% confidence will be about

$$\frac{1}{\sqrt{1523}} = \frac{1}{39.03} = 0.026 \quad \text{(that is, 2.6\%)}$$

Gallup actually announced a margin of error of 3%. Our result differs a bit from Gallup's for two reasons. First, polls usually round their announced margin of error to the nearest whole percent to keep their press releases simple. Second, our rough formula works for an SRS. We will see in the next section that most national samples are more complicated than an SRS in ways that tend to slightly increase the margin of error. Nonetheless, our quick method comes pretty close.

Our quick method also reveals an important fact about how margins of error behave. Because the sample size n appears in the denominator of the fraction, larger samples have smaller margins of error. We knew that. Because the formula uses the square root of the sample size, however, *to cut the margin of error in half, we must use a sample four times as large.*

" A new poll shows 67% of Americans think polls are inaccurate 52% of the time, and 84% think the margin of error of 3% is inaccurate 71% of the time…"

EXAMPLE 2.10   Estimating the margin of error

In Example 2.7 we compared the results of taking many SRSs of size $n = 100$ and many SRSs of size $n = 1523$ from the same population. We found that the spread of the middle 95% of the sample results was about four times as large for the smaller samples.

Our quick formula estimates the margin of error for SRSs of size 1523 to be about 2.6%. The margin of error for SRSs of size 100 is about

$$\frac{1}{\sqrt{100}} = \frac{1}{10} = 0.1 \quad \text{(that is, 10\%)}$$

Because 1523 is roughly 16 times 100 and the square root of 16 is 4, the margin of error is about four times larger for samples of 100 people than for samples of 1523 people.

EXERCISES

2.35 Take a bigger sample A management student is planning a project on student attitudes toward part-time work while attending college. She develops a questionnaire and plans to ask 25 randomly selected students to fill it out. Her faculty advisor approves the questionnaire but suggests that the sample size be increased to at least 100 students. Why is the larger sample helpful? Back up your statement by using the quick method to estimate the margin of error for samples of size 25 and for samples of size 100.

2.36 Gun violence The Harris Poll asked a sample of 1009 adults which causes of death they thought would become more common in the future. Topping the list was gun violence—70% of the sample thought deaths from guns would increase.
(a) How many of the 1009 people interviewed thought deaths from gun violence would increase?
(b) Harris says that the margin of error for this poll is plus or minus 3 percentage points. Explain to someone who knows no statistics what "margin of error plus or minus 3 percentage points" means.

2.37 Find the margin of error The previous exercise concerns a Harris Poll sample of 1009 people. Use the quick method to estimate the margin of error for statements about the population of all adults. Is your result close to the 3% margin of error announced by Harris?

2.38 Find the margin of error Exercise 2.34 (page 91) describes a sample survey of 61,239 adults living in Ontario. About what is the margin of error for conclusions having 95% confidence about the entire adult population of Ontario?

2.39 Smaller margin of error Exercise 2.36 describes an opinion poll that interviewed 1009 people. Suppose you want a margin of error half as large as the one you found in that exercise. How many people must you plan to interview?

Confidence statements

Here is Gallup's conclusion about buying lottery tickets in short form: "The poll found that 57% of adults bought a lottery ticket in the past 12 months. We are 95% confident that the truth about all adults is within plus or minus 3 percentage points of this sample result." Here is an even shorter form: "We are 95% confident that between 54% and 60% of all adults bought a lottery ticket in the last 12 months." These are *confidence statements*.

> ## Confidence statements
>
> A **confidence statement** has two parts: a **margin of error** and a **level of confidence.** The margin of error says how close the sample statistic lies to the population parameter. The level of confidence says what percent of all possible samples satisfy the margin of error.

A confidence statement is a fact about what happens in all possible samples, used to state how much we can trust the result of one sample. The phrase "95% confidence" means "We used a sampling method that gives a result this close to the truth 95% of the time." Here are some hints for interpreting confidence statements:

- *The conclusion of a confidence statement always applies to the population, not to the sample.* We know exactly how the 1523 people in the sample acted, because Gallup interviewed them. The confidence statement uses the sample result to say something about the population of all adults.

- *Our conclusion about the population is never completely certain.* Gallup's sample might be one of the 5% that miss by more than 3 percentage points.

- *A sample survey can choose to use a confidence level other than 95%.* We pay for higher confidence with a larger margin of error. For the same sample, a 99% confidence level requires a larger margin of error than 95% confidence. If you are content with 90% confidence, you get in return a smaller margin of error. Remember that our quick method gives the margin of error only for 95% confidence.

- *It is usual to report the margin of error for 95% confidence.* If a news report gives a margin of error but leaves out the confidence level, it's pretty safe to assume 95% confidence.

- *Want a smaller margin of error with the same confidence? Take a larger sample.* Remember that larger samples have less variability. You can get as small a margin of error as you want and still have high confidence by paying for a large enough sample.

EXAMPLE 2.11   Do you approve of gambling?

Gallup began its poll on gambling with this question: "First, generally speaking, do you approve or disapprove of legal gambling or betting?" In addition to the 1523 adults (ages 18 and older) whose lottery habits we have explored, Gallup took a random sample of 501 teenagers (ages 13 to 17). The sample results were

Adults:	959 out of 1523 approve	$\hat{p} = 959/1523 = 0.63$
Teens:	261 out of 501 approve	$\hat{p} = 261/501 = 0.52$

After reporting these and other results, Gallup says: "For results based on a sample of this size, one can say with 95 percent confidence that the error attributable to sampling and other random effects could be plus or minus 3 percentage points for adults (18+), and plus or minus 5 percentage points for teens (13–17 year olds)."

There you have it: the sample of teens is smaller, so the margin of error for conclusions about teens is wider. We are 95% confident that between 47% (that's 52% minus 5%) and 57% (that's 52% plus 5%) of all teenagers approve of legal gambling.

EXERCISES

2.40 Polling women A *New York Times* Poll on women's issues interviewed 1025 women randomly selected from the United States, excluding Alaska and Hawaii. One question was "Many women have better jobs and more opportunities than they did 20 years ago. Do you think women had to give up too much in the process, or not?" Forty-eight percent of women in the sample said "Yes."[9]

(a) The poll announced a margin of error of ±3 percentage points for 95% confidence in its conclusions. Make a 95% confidence statement about the percent of all adult women who feel that women had to give up too much.

(b) Explain to someone who knows no statistics why we can't just say that 48% of all adult women feel that women had to give up too much.

(c) Then explain clearly what "95% confidence" means.

2.41 Polling men and women The sample survey described in Exercise 2.40 interviewed 472 randomly selected men as well as 1025 women. The poll announced a margin of error of ±3 percentage points for 95% confidence in conclusions about women. The margin of error for results concerning men was ±5 percentage points. Why is this larger than the margin of error for women?

2.42 Is there a hell? A news article reports that in a recent opinion poll, 78% of a sample of 1108 adults said they believe there is a heaven. Only 60% said they believe there is a hell.

(a) Use the quick method to estimate the margin of error for a sample of this size.

(b) Make a confidence statement about the percent of all adults who believe there is a hell.

2.43 The Current Population Survey Though opinion polls usually make 95% confidence statements, some sample

The telemarketer's pause
People who do sample surveys hate telemarketing. We all get so many unwanted sales pitches by phone that many people hang up before learning that the caller is conducting a survey rather than selling vinyl siding. Here's a tip. Both sample surveys and telemarketers dial telephone numbers at random. Telemarketers automatically dial many numbers, and their sellers only come on the line after you pick up the phone. Once you know this, the telltale "telemarketer's pause" gives you a chance to hang up before the seller arrives. Sample surveys have a live interviewer on the line when you answer.

surveys use other confidence levels. The monthly unemployment rate, for example, is based on the Current Population Survey of about 50,000 households. The margin of error in the unemployment rate is announced as about two-tenths of 1 percentage point with 90% confidence. Would the margin of error for 95% confidence be smaller or larger? Why?

2.44 Find the margin of error A poll of 586 adults who had used the Internet in the past week asked whether "the Internet has made your life much better, somewhat better, somewhat worse, much worse, or has it not affected your life either way." In all, 152 of the 586 subjects said "much better."[10]
(a) What is the population for this sample survey?
(b) Use the quick method to find a margin of error. Then give a complete confidence statement for a conclusion about the population.

Sampling from large populations

Gallup's sample of 1523 adults is only 1 out of every 130,000 adults in the United States. Does it matter whether 1523 is 1 in 100 individuals in the population or 1 in 130,000?

Population size doesn't matter

The variability of a statistic from a random sample does not depend on the size of the population, as long as the population is at least 10 times larger than the sample.

Why does the size of the population have little influence on the behavior of statistics from random samples? Imagine sampling harvested corn by thrusting a scoop into a lot of corn kernels. The scoop doesn't know whether it is surrounded by a bag of corn or by an entire truckload. As long as the corn is well mixed (so that the scoop selects a random sample), the variability of the result depends only on the size of the scoop.

This is good news for national sample surveys like the Gallup Poll. A random sample of size 1000 or 1500 has small variability because the sample size is large. But remember that even a very large voluntary response sample or convenience sample is worthless because of bias. Taking a larger sample doesn't fix bias.

However, the fact that the variability of a sample statistic depends on the size of the sample and not on the size of the population is bad news for anyone planning a sample survey in a university or a small city. For example, it takes just as large an SRS to estimate the proportion of Ohio State University students who call themselves political conservatives as to estimate with the same margin of error the proportion of all adult U.S. residents who are conservatives. We can't use a smaller SRS at Ohio State just because there are 48,000 Ohio State students and 200 million adults in the United States.

EXERCISES

2.45 Strike three A survey is conducted in Chicago (population 2,900,000) using random-digit-dialing equipment that places calls at random to residential phones, both listed and unlisted. The purpose of the survey is to determine the percent of Chicago voters who support the "three strikes and you're out" provision of the recent crime bill passed by Congress. One-tenth of 1% of the adult population is interviewed and 53% of them favor the proposal. A second survey is taken in Dayton, Ohio (population 166,000), using the same techniques and asking the same question of one-tenth of 1% of the adults living in Dayton. Will the accuracy of the Chicago survey be greater than, about the same as, or less than the accuracy of the Dayton survey? Explain briefly.

2.46 Presidential polls California has about four times as many voters as Ohio, but both are very important states for the outcome of the 2008 presidential election. One national polling organization plans to take random surveys of 1000 voters in each of these two states on the day before the election. In which state is the survey likely to be more accurate in predicting the state's election results? Explain.

APPLICATION 2.2 Should Election Polls Be Banned?

The 2000 presidential election was one of the closest and most hotly contested in U.S. history. CNN ran the banner "too close to call" for many days following the election. After numerous legal challenges and meticulous recounting of ballots in some areas, George W. Bush was declared the winner. Some Americans still feel that polls had a profound effect on the final outcome.

Preelection polls tell us that Senator So-and-So is the choice of 58% of Ohio voters. The media love these polls. Statisticians don't love them, because elections often don't go as forecast even when the polls use all the right statistical methods.

Exit polls, which interview voters as they leave the voting place, don't share this problem. The people in the sample have just voted. A good exit poll, based on a national sample of election precincts, can often call a presidential election correctly long before the polls close. That fact sharpens the debate over the political effects of election forecasts.

APPLICATION 2.2   Should Election Polls Be Banned? *(continued)*

Some countries have laws restricting election forecasts. In France, no poll results can be published in the week before a presidential election. Canada forbids poll results in the 72 hours before federal elections. In all, some 30 countries restrict publication of election surveys.

The argument *for* preelection polls is simple: democracies should not forbid publication of information. Voters can decide for themselves how to use the information. After all, supporters of a candidate who is far behind know that even without polls telling them so. Restricting publication of polls just invites abuses. In France, candidates continue to take private polls (less reliable than the public polls) in the week before the election. They then leak the results to reporters in the hope of influencing press reports.[11]

1. Give at least two reasons why a preelection poll might give inaccurate results.

2. Some people have argued that preelection polls influence voter behavior. Voters may decide to stay home if the polls predict a landslide—why bother to vote if the result is a foregone conclusion? Comment on this argument.

3. Research: In the days leading up to the 2000 presidential election, numerous preelection polls were taken. Which candidate was predicted to win the election?

4. Based on exit polls taken on election day, several TV networks declared a winner. Shortly thereafter, television news anchors began to withdraw their declarations. Some even declared the other candidate the winner. Which candidate was initially declared president?

5. Give at least two reasons why the exit polls might have erred in predicting the winner of the 2000 presidential election.

6. In Florida, the polls were still open in part of the state when an initial winner was declared. How might this have affected the final result in Florida?

7. What led to the recount of votes in some Florida counties?

8. How did the final outcome—winner and percent of votes received—compare with the results of the preelection polls?

9. George W. Bush won the election, but he did not win the popular vote. Has this happened before in a U.S. presidential election? If so, when?

10. Writing: Compose a brief analysis of some aspect of the Florida vote that interests you.

EXPLORING THE WEB

You can read the Gallup Organization's own explanation of why surprisingly small samples can give trustworthy results about large populations at www.gallup.com/help/FAQs/poll1.asp.

STATISTICS IN SUMMARY

The purpose of sampling is to use a sample to gain information about a population. We often use a sample **statistic** to estimate the value of a population **parameter.** This section has one big idea: to describe how trustworthy a sample is, ask "What would happen if we took a large number of samples from the same population?" If almost all samples would give a result close to the truth, we can trust our one sample even though we can't be certain that it is close to the truth.

In planning a sample survey, first aim for small **bias** by using random sampling and avoiding bad sampling methods such as voluntary response. Next, choose a large enough random sample to reduce the **variability** of the result. Using a large random sample guarantees that almost all samples will give accurate results. To say how accurate our conclusions about the population are, make a **confidence statement.** News reports often mention only the **margin of error.** Most often this margin of error is for **95% confidence.** That is, if we chose many samples, the truth about the population would be within the margin of error 95% of the time. We can estimate the margin of error for 95% confidence based on a simple random sample of size n by the formula $1/\sqrt{n}$. As this formula suggests, only the size of the sample, not the size of the population, matters. This is true as long as the population is much larger than the sample.

SECTION 2.2 EXERCISES

Each boldface number in Exercises 2.47 and 2.48 is the value of either a **parameter** *or a* **statistic.** *In each case, state which it is.*

2.47 Lefties According to Snapple.com, **13**% of adults are left-handed. At a school administrator's conference, **16**% of those attending were left-handed.

2.48 Single-sex classes In an experiment to test the effectiveness of single-sex classrooms, girls assigned at random to a coeducational chemistry class gained an average of **12.2** points from a pretest to a posttest. Girls assigned randomly to a single-sex chemistry class taught by the same teacher gained **15.1** points.

2.49 Find the margin of error Example 2.11 (page 96) tells us that Gallup asked 501 teenagers whether they approved of legal gambling; 52% said they did. Use the quick method to estimate the margin of error for conclusions about all teenagers. How does your result compare with Gallup's margin of error quoted in Example 2.11?

2.50 Explaining confidence A student reads that we are 95% confident that the average score of young men on the quantitative part of the National Assessment of Educational Progress is 267.8 to 276.2. Asked to explain the meaning of this statement, the student says, "95% of all young men have scores between 267.8 and 276.2." Is the student right? Explain your answer.

2.51 Balancing the budget A poll of 1190 adults finds that 702 prefer balancing the budget over cutting taxes. The announced margin of error for this result is plus or minus 4 percentage points. The news report does not give the confidence level. You can be quite sure that the confidence level is 95%.
(a) What is the value of the sample proportion $\hat{p}$ who prefer balancing the budget? Explain in words what the population parameter p is in this setting.
(b) Make a confidence statement about the parameter p.

2.52 Satisfying Congress The previous exercise describes a sample survey of 1190 adults, with margin of error ±4% for 95% confidence.
(a) A member of Congress thinks that 95% confidence is not enough. He wants to be 99% confident. How would the margin of error for 99% confidence based on the same sample compare with the margin of error for 95% confidence?
(b) Another member of Congress is satisfied with 95% confidence, but she wants a smaller margin of error than ±4%. How can we get a smaller margin of error, still with 95% confidence?

2.53 Find the margin of error Should the government spend money to help low-income families who want to send their children to private or religious schools? A poll of 1006 adults found that 362 said "Yes."[12]
(a) What is the population for this sample survey?

(b) Use the quick method to find a margin of error. Make a confidence statement about the opinion of the population.

2.54 Support for Cuba A random sample of 900 people was taken in Chicago to estimate the percent of Chicago voters who favor an end to the U.S. embargo against Cuba. The results of this survey are expressed as a 95% confidence interval that turns out to be 53% ± 3.4%. A second poll, using the same methods, is taken of 900 people in Columbus, Ohio, which has about one-fourth as many voters as Chicago. Will the margin of error for the Columbus poll be larger than 3.4%, smaller than 3.4%, or about 3.4%? Explain.

2.55 The millennium begins with optimism In January of the year 2000, the Gallup Poll asked a random sample of 1633 adults, "In general, are you satisfied or dissatisfied with the way things are going in the United States at this time?" It found that 1127 said that they were satisfied.[13] Write a short report on this finding, as if you were writing for a newspaper. Be sure to include a margin of error.

2.56 Simulation Random digits can be used to simulate the results of random sampling. Suppose that you are drawing simple random samples of size 25 from a large number of college students and that 20% of the students are unemployed during the summer. To simulate this SRS, let 25 consecutive digits in Table A stand for the 25 students in your sample. The digits 0 and 1 stand for unemployed students, and other digits stand for employed students. This is an accurate imitation of the SRS because 0 and 1 make up 20% of the 10 equally likely digits.

Simulate the results of 50 samples by counting the number of 0s and 1s in the first 25 entries in each of the 50 rows of Table A. Make a histogram like that in Figure 2.3 to display the results of your 50 samples. Is the truth about the population (20% unemployed, or 5 in a sample of 25) near the center of your graph? What are the smallest and largest counts of unemployed students you obtained in your 50 samples? What percent of your samples had either 4, 5, or 6 unemployed?

2.3 SAMPLE SURVEYS IN THE REAL WORLD

The whole truth about opinion polls

An opinion poll is administered to 1000 people chosen at random, announces its results, and announces a margin of error. Should we be happy? Perhaps not. Many polls don't tell the whole truth about their samples. The Pew Research Center for the People and the Press imitated the methods of the better opinion polls and did tell the whole truth. Here it is.

Most polls are taken by telephone, dialing numbers at random to get a random sample of households. After eliminating fax and business numbers, Pew had to call

2879 residential numbers to get their sample of 1000 people. Here's the breakdown:

Never answered phone	938
Answered but refused	678
Not eligible: no person aged at least 18, or language barrier	221
Incomplete interview	42
Complete interview	1000
Total called	2879

Out of 2879 good residential phone numbers, 33% never answered. Of those who answered, 35% refused to talk. The overall rate of nonresponse (people who never answered, refused, or would not complete the interview) was 1658 out of 2879, or 58%. Pew called every number five times over a five-day period, at different times of day and on different days of the week. Many polls call only once, and it is usual to find that over half those who answer refuse to talk. In the real world, a simple random sample isn't simple and may not be random. That's our issue for this section.[14]

ACTIVITY 2.3 Design your own survey

In this activity, you will work with two or three of your classmates to design a survey. By the end of this section, you will be ready to revise your survey and to administer it at your school.

1. On your own, brainstorm three questions on topics of interest to your school community that you would like to ask.

2. Pool questions with your team members. Select three to five of the questions for your survey. You may want to choose questions that have a common theme.

3. Spend a few minutes editing your questions to make them as clear as possible. Have one member of the group write the revised list of questions neatly on a clean sheet of paper.

4. Exchange survey questions with another team, as instructed by your teacher. Discuss the content and clarity of the questions. Note any changes you would recommend on their question list.

5. Review any suggestions made regarding your own list. Once you are completely satisfied with your questions, you could go ask them to everyone in your school individually. Then you would be taking a *census*. Give a few reasons why this would be difficult.

ACTIVITY 2.3 Design your own survey *(continued)*

6. Instead of taking a census, you are going to give your survey to a smaller sample from your school community. Make a list of decisions your team needs to make before you can actually administer the survey.

How sample surveys go wrong

Random sampling eliminates bias in choosing a sample and allows control of variability. So once we see the magic words "randomly selected" and "margin of error," do we know we have trustworthy information before us? It certainly beats voluntary response, but not always by as much as we might hope. Sampling in the real world is more complex and less reliable than choosing an SRS from a list of names in a textbook exercise. Confidence statements do not reflect all of the sources of error that are present in practical sampling.

Errors in sampling

Sampling errors are errors caused by the act of taking a sample. They cause sample results to be different from the results of a census.

Random sampling error is the deviation between the sample statistic and the population parameter caused by chance in selecting a random sample. The margin of error in a confidence statement includes only random sampling error.

Nonsampling errors are errors not related to the act of selecting a sample from the population. They can be present even in a census.

Most sample surveys are afflicted by errors other than random sampling errors. These errors can introduce bias that makes a confidence statement meaningless. Good sampling technique includes the art of reducing all sources of error. Part of this art is the science of statistics, with its random samples and confidence statements. In practice, however, good statistics isn't all there is to good sampling. Let's look at sources of errors in sample surveys and at how samplers combat them.

Sampling errors

Random sampling error is one kind of sampling error. The margin of error tells us how serious random sampling error is, and we can control it by choosing the size of our random sample. Another source of sampling error is the use of *bad sampling methods,* such as voluntary response. We can avoid bad methods. Other sampling errors are not so easy to handle. Sampling begins with a list of individuals from which we will draw our sample. This list is called the **sampling frame.** Ideally, the sampling frame should list every individual in the population. Because a list of the entire population is rarely available, most samples suffer from some degree of *undercoverage.*

Undercoverage

Undercoverage occurs when some groups in the population are left out of the process of choosing the sample.

If the sampling frame leaves out certain classes of people, even random samples from that frame will be biased. Using telephone directories as the frame for a telephone survey, for example, would miss everyone with an unlisted telephone number. More than half the households in many large cities have unlisted numbers, so massive undercoverage and bias against urban areas would result. In fact, telephone surveys use random-digit-dialing equipment that dials telephone numbers in selected regions at random. In effect, the sampling frame contains all residential telephone numbers.

Opting out of modern society was tiring at times, but Ted felt that increasing the undercoverage of opinion polls made it all worthwhile.

EXAMPLE 2.12   We undercover

Most opinion polls can't afford to even attempt full coverage of the population of all adult residents of the United States. The interviews are done by telephone, thus missing the 6% of households without phones. Only households are contacted, so that students in dormitories, prison inmates, and most members of the armed forces are left out. So are the homeless and people staying in shelters. Because calls to Alaska and Hawaii are expensive, most polls restrict their samples to the continental states. Many polls interview only in English, which leaves some immigrant households out of their samples.

The kinds of undercoverage found in most sample surveys are most likely to leave out people who are young or poor or who move often. Nonetheless, random digit dialing comes close to producing a random sample of households with phones outside Alaska and Hawaii. Sampling errors in careful sample surveys are usually quite small. The real problems start when someone picks up (or doesn't pick up) the phone. Now nonsampling errors take over.

Nonsampling errors

Nonsampling errors are those that can plague even a census. They include **processing errors;** these are mistakes in mechanical tasks such as doing arithmetic or entering responses into a computer. The spread of computers has made processing errors less common than in the past.[15]

EXAMPLE 2.13   Computer-assisted interviewing

The days of the interviewer with a clipboard are past. Contemporary interviewers carry a laptop computer for face-to-face interviews or watch a computer screen as they carry out a telephone interview. Computer software manages the interview. The interviewer reads questions from the computer screen and uses the keyboard to enter the responses. The computer skips irrelevant items—once a respondent says that she has no children, further questions about children never appear. The computer can check that answers to related questions are consistent with each other. It can even present questions in random order to avoid any bias due to always asking questions in the same order.

Computer software also manages the sampling process. It keeps records of who has responded and prepares a file of data from the responses. The tedious process of transferring responses from paper to computer, once a source of processing errors, has disappeared. The computer even schedules the calls in telephone surveys, taking account of the respondent's time zone and honoring appointments made by people who were willing to respond but did not have time when first called.

Another type of nonsampling error is **response error,** which occurs when a subject gives an incorrect response. A subject may lie about her age or income or about whether she has used illegal drugs. She may remember incorrectly when asked how many packs of cigarettes she smoked last week. A subject who does not understand a question may guess at an answer rather than appear ignorant. Questions that ask subjects about their behavior during a fixed time period are notoriously prone to response errors due to faulty memory.[16] For example, the National Health Survey asks people how many times they have visited a doctor in the past year. Checking their responses against health records found that they failed to remember 60% of their visits to a doctor. A survey that asks about sensitive issues can also expect response errors, as the next example illustrates.

EXAMPLE 2.14   The effect of race

In 1989, New York City elected its first black mayor and the state of Virginia elected its first black governor. In both cases, samples of voters interviewed as they left their polling places predicted larger margins of victory than the official vote counts. The polling organizations were certain that some voters lied when interviewed because they felt uncomfortable admitting that they had voted against the black candidate.

Technology and attention to detail can minimize processing errors. Skilled interviewers greatly reduce response errors, especially in face-to-face interviews. There is no simple cure, however, for the most serious kind of nonsampling error, *nonresponse.*

"You can call, you can send email, you can stand at the door all day. The answer is still NO!"

Nonresponse

Nonresponse is the failure to obtain data from an individual selected for a sample. Most nonresponse happens because some subjects can't be contacted or because some subjects who are contacted refuse to cooperate.

Nonresponse is the most serious problem facing sample surveys. People are increasingly reluctant to answer questions, particularly over the phone. The rise of telemarketing, answering machines, and caller

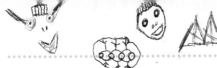

ID drives down response to telephone surveys. Gated communities and buildings guarded by doormen prevent face-to-face interviews. Nonresponse can bias sample survey results because different groups have different rates of nonresponse. Refusals are higher in large cities and among the elderly, for example. Bias due to nonresponse can easily overwhelm the random sampling error described by a survey's margin of error.

EXAMPLE 2.15 How bad is nonresponse?

The Current Population Survey has the best response rate of any poll we know: only about 6% or 7% of the households in the CPS sample don't respond. People are more likely to respond to a government survey such as the CPS, and the CPS contacts its sample in person before doing later interviews by phone.

The General Social Survey (Example 1.8, page 12) also contacts its sample in person, and it is run by a university. Despite these advantages, its recent surveys have a 24% rate of nonresponse.

What about polls done by the media and by market research and opinion-polling firms? We don't know their rates of nonresponse, because they won't say. That itself is a bad sign. The Pew study we looked at suggests how bad things are. Pew got 1221 responses (of whom 1000 were in the population they targeted) and 1658 who were never at home, refused, or would not finish the interview. That's a nonresponse rate of 1658 out of 2879, or 58%. The Pew researchers were more thorough than many polls. Insiders say that nonresponse often reaches 75% or 80% of an opinion poll's original sample.[17]

Sample surveyors know some tricks to reduce nonresponse. Carefully trained interviewers can keep people on the line if they answer at all. Calling back over longer time periods helps. So do letters sent in advance. Letters and many callbacks slow down the survey, so opinion polls that want fast answers to satisfy the media don't use them. Even the most careful surveys find that nonresponse is a problem that no amount of expertise can fully overcome. That makes this reminder even more important.

What the margin of error doesn't say

The announced margin of error for a sample survey covers only random sampling error. Undercoverage, nonresponse, and other practical difficulties can cause large bias that is not covered by the margin of error.

Careful sample surveys tell us this. Gallup, for example, says, "In addition to sampling error, question wording and practical difficulties in conducting surveys can introduce error or bias into the findings of public opinion polls." How true it is.

Does nonresponse make many sample surveys useless? Maybe not. We began this section with an account of a "standard" telephone survey done by the Pew Center. The Pew researchers also carried out a "rigorous" survey, with letters sent in advance, unlimited calls over eight weeks, letters by priority mail to people who refused, and so on. All this drove the rate of nonresponse down to 30%, compared with 58% for the standard survey. Pew then compared the answers to the same questions from the two surveys. The two samples were quite similar in age, sex, and race, though the rigorous sample was a bit more prosperous. The two samples also held similar opinions on all issues except one: race. People who at first refused to respond were less sympathetic to blacks. Overall, it appears that standard polls give reasonably accurate results. But, as in Example 2.14, race is again an exception.

EXERCISES

2.57 Not in the margin of error A recent Gallup Poll found that 68% of adult Americans favor teaching creationism along with evolution in public schools. The Gallup press release says:

> *For results based on samples of this size, one can say with 95 percent confidence that the maximum error attributable to sampling and other random effects is plus or minus 3 percentage points.*[18]

Give one example of a source of error in the poll result that is not included in this margin of error.

2.58 What kind of error? Which of the following are sources of *sampling error* and which are sources of *nonsampling error*? Explain your answers.
(a) The subject lies about past drug use.
(b) A typing error is made in recording the data.
(c) Data are gathered by asking people to mail in a coupon printed in a newspaper.

2.59 Internet users A survey of users of the Internet found that males outnumbered females by nearly 2 to 1. This was a surprise, because earlier surveys had put the ratio of men to women closer to 9 to 1. Later in the article we find this information:

> *Detailed surveys were sent to more than 13,000 organizations on the Internet; 1,468 usable responses were received. According to Mr. Quarterman, the margin of error is 2.8 percent, with a confidence level of 95 percent.*[19]

(a) What was the response rate for this survey? (The response rate is the percent of the planned sample that responded.)

(b) Use the quick method (page 93) to estimate the margin of error of this survey. Is your result close to the 2.8% claimed?

(c) Do you think that the small margin of error is a good measure of the accuracy of the survey's results? Explain your answer.

2.60 Polling students A high school chooses an SRS of 100 students from the registrar's list to interview about student life. If it selected two SRSs of 100 students at the same time, the two samples would give somewhat different results. Is this variation a source of sampling error or of nonsampling error? Will the survey's announced margin of error take this source of error into account?

2.61 Activity 2.3 follow-up Return to the survey your team constructed in Activity 2.3. Discuss how your design protects against

(a) sampling errors

(b) nonsampling errors

Wording questions

A final influence on the results of a sample survey is the exact wording of questions. It is surprisingly difficult to word questions that are completely clear. A survey that asked about "ownership of stock" found that most Texas ranchers owned stock, though probably not the kind traded on the New York Stock Exchange.

"Do I own any stock, Ma'am? Why, I've got 10,000 head out there."

EXAMPLE 2.16   A few words make a big difference

How do Americans feel about government help for the poor? Only 13% think we are spending too much on "assistance to the poor," but 44% think we are spending too much on "welfare." How do the Scots feel about the movement to become independent from England? Well, 51% would vote for "independence for Scotland," but only 34% support "an independent Scotland separate from the United Kingdom."[20]

It seems that "assistance to the poor" and "independence" are nice, hopeful words. "Welfare" and "separate" are negative words. Small changes in how a question is worded can make a big difference in the response.

The wording of questions always influences the answers. If the questions are slanted to favor one response over others, we have another source of nonsampling error. A favorite trick is to ask if the subject favors some policy as a means to a desirable end: "Do you favor banning private ownership of handguns in order to reduce the rate of violent crime?" and "Do you favor imposing the death penalty in order to reduce the rate of violent crime?" are loaded questions that draw positive responses from people who are worried about crime. Here is an example of the influence of a slanted question.

EXAMPLE 2.17   Campaign finance

The financing of political campaigns is a perennial issue. Here are two opinion poll questions on this topic:

> *Should laws be passed to eliminate all possibilities of special interests giving huge sums of money to candidates?*
>
> *Should laws be passed to prohibit interest groups from contributing to campaigns, or do groups have a right to contribute to the candidate they support?*

The first question was posed by Ross Perot, the third-party candidate in the 1992 presidential election. It drew a 99% "Yes" response in a mail-in poll. We know that voluntary response polls are worthless, so the Yankelovich survey firm asked the same question of a nationwide random sample. The result: 80% "Yes." Perot's question almost demands a "Yes" answer, so Yankelovich wrote the second question as a more neutral way of presenting the issue. Only 40% of a nationwide random sample wanted to prohibit contributions when asked this question.[21]

How to live with nonsampling errors

Nonsampling errors, especially nonresponse, are always with us. What should a careful sample survey do about this? First, **substitute other households** for the nonresponders. Because nonresponse is higher in cities, replacing nonresponders with other households in the same neighborhood may reduce bias. Once the data are in, all professional surveys use statistical methods to **weight the responses** in an attempt to correct sources of bias. If many urban households did not respond, the survey gives more weight to those that did respond. If too many women are in the sample, the survey gives more weight to the men. Here, for example, is part of a statement in the *New York Times* describing one of its sample surveys:

> *The results have been weighted to take account of household size and number of telephone lines into the residence and to adjust for variations in the sample relating to geographic region, sex, race, age and education.*[22]

The goal is to get results "as if" the sample matched the population in age, gender, place of residence, and other variables.

The practice of weighting creates job opportunities for statisticians. It also means that the results announced by a sample survey are rarely as simple as they seem to be. Gallup announces that it interviewed 1523 adults and found that 57% of them bought a lottery ticket in the last 12 months. It would seem that because 57% of 1523 is 868, Gallup found that 868 people in its sample had played the lottery. Not so. Gallup no doubt used some quite fancy statistics to weight the actual responses: 57% is Gallup's best estimate of what it would have found in the absence of nonresponse. Weighting does help correct bias. It usually also increases variability. The announced margin of error must take this into account, creating more work for statisticians.

EXERCISES

2.62 Should he go or stay? In December 1998, the House of Representatives impeached President Clinton, the first step in removing him from office. The Senate then conducted a trial and found the president not guilty. Here are two opinion poll questions asked after the House had acted:

> *What do you think President Clinton should do: fight the charges in the Senate, or resign from office?*
> *What do you think President Clinton should do: continue to serve and stand trial in the Senate, or resign from office?*

In response to the first question, 58% thought the president should resign. But only 43% of those asked the second question thought he should resign.[23] Why do you think the first wording encouraged more people to favor resignation?

2.63 Amending the Constitution You are writing an opinion poll question about a proposed amendment to the Constitution. You can ask if people are in favor of "changing the Constitution" or "adding to the Constitution" by approving the amendment. One of these choices of wording will produce a much higher percent in favor. Which one? Why?

2.64 Bad survey questions Write your own examples of bad sample survey questions.
(a) Write a biased question designed to get one answer rather than another.
(b) Write a question that is confusing, so that it is hard to answer.

2.65 Closed versus open questions Two basic types of questions are closed questions and open questions. A closed question asks the subject for one or more of a fixed set of responses. An open question allows the subject to answer in his or her own words. The interviewer writes down the responses and sorts them later. An example of an open question is

How do you feel about broccoli?

An example of a closed question is

What is your opinion about broccoli? Do you
a. Like it very much?
b. Like it somewhat?
c. Neither like nor dislike it?
d. Dislike it somewhat?
e. Dislike it very much?

What are the advantages and disadvantages of open and closed questions?

He started it!

A study of deaths in bar fights showed that in 90% of the cases, the person who died started the fight. You shouldn't believe this. If you killed someone in a fight, what would you say when the police ask you who started the fight? After all, dead men tell no tales. Now that's nonresponse.

2.66 Wording of questions A *New York Times*/CBS News Poll asked a random sample of Americans about abortion: "Do you think there should be an amendment to the Constitution prohibiting abortions, or shouldn't there be such an amendment?" The same people were later asked, "Do you believe there should be an amendment to the Constitution protecting the life of the unborn child, or shouldn't there be such an amendment?" For one of the questions, 50% were in favor and 39% were opposed. For the other, 29% were in favor and 62% were opposed. (The rest were uncertain.)[24] Which question do you think yielded each result? Explain why.

Sample design in the real world

The basic idea of sampling is straightforward: take an SRS from the population and use a statistic from your sample to estimate a parameter of the population. We now know that the sample statistic is fiddled with behind the scenes to partly correct for nonresponse. The statisticians also have their hands on our beloved SRS. In the real world, most sample surveys use more complex designs.

EXAMPLE 2.18   The Current Population Survey

The CPS population consists of all households in the United States (including Alaska and Hawaii). The sample is **chosen in stages.** The Census Bureau divides the nation into 2007 geographic areas called Primary Sampling Units (PSUs). These are generally groups of neighboring counties. At the first stage, 792 PSUs are chosen. This isn't an SRS. If all PSUs had the same chance to be chosen, the sample might miss Chicago and Los Angeles. So 432 highly populated PSUs are automatically in the sample. The other 1575 are grouped into 360 **strata** by combining PSUs that are similar in various ways. One PSU is chosen at random to represent each stratum.

Each of the 792 PSUs in the first-stage sample is divided into census blocks (smaller geographic areas). The blocks are also grouped into strata, based on such things as housing types and minority population. The households in each block are arranged in order of their location and grouped into **clusters** of about four households each. The final sample is a sample of clusters (not of individual households) from each stratum of blocks. Interviewers go to all households in the chosen clusters. The samples of clusters within each stratum of blocks are also not SRSs. To be sure that the clusters spread out geographically, the sample starts at a random cluster and then takes (say) every 10th cluster in the list.[25]

The design of the CPS illustrates several ideas that are common in real-world samples that use face-to-face interviews. Taking the sample **in several stages** with **clusters** at the final stage saves travel time for interviewers by grouping the sample households first in PSUs and then in clusters. The most important refinement mentioned in Example 2.18 is *stratified sampling.*

Stratified sample

To choose a **stratified random sample:**

    Step 1. Divide the sampling frame into distinct groups of individuals, called **strata.** Choose the strata because you have a special interest in these

> **Stratified sample** *(continued)*
>
> groups within the population or because the individuals in each stratum resemble each other.
>
> **Step 2.** Take a separate SRS in each stratum and combine these to make up the complete sample.

We must of course choose the strata using facts about the population that are known before we take the sample. You might group a university's students into undergraduate and graduate students or into those who live on campus and those who commute. Stratified samples have some advantages over an SRS. First, by taking a separate SRS in each stratum, we can set sample sizes to allow separate conclusions about each stratum. Second, a stratified sample usually has a smaller margin of error than an SRS of the same size. The reason is that the individuals in each stratum are more alike than the population as a whole, so working stratum-by-stratum eliminates some variability in the sample.

It may surprise you that stratified samples can violate one of the most appealing properties of the SRS—stratified samples need not give all individuals in the population the same chance to be chosen. Some strata may be deliberately over-represented in the sample.

EXAMPLE 2.19   Stratifying a sample of students

A large university has 30,000 students, of whom 3000 are graduate students. An SRS of 500 students gives every student the same chance to be in the sample. That chance is

$$\frac{500}{30,000} = \frac{1}{60}$$

We expect an SRS to contain only about 50 grad students—because grad students make up 10% of the population, we expect them to make up about 10% of an SRS. A sample of size 50 isn't large enough to estimate grad student opinion with reasonable accuracy. We might prefer a stratified random sample of 200 grad students and 300 undergraduates.

You know how to select such a stratified sample. Label the graduate students 0001 to 3000 and use Table A to select an SRS of 200. Then label the undergraduates 00001 to 27000 and use Table A a second time to select an SRS of 300 of them. These two SRSs together form the stratified sample.

In the stratified sample, each grad student has chance

$$\frac{200}{3000} = \frac{1}{15}$$

to be chosen. Each of the undergraduates has a smaller chance,

$$\frac{300}{27,000} = \frac{1}{90}$$

Because we have two SRSs, it is easy to estimate opinions in the two groups separately. The quick method (page 93) tells us that the margin of error for a sample proportion will be about

$$\frac{1}{\sqrt{200}} = 0.07 \qquad \text{(that is, 7%)}$$

for grad students and about

$$\frac{1}{\sqrt{300}} = 0.058 \qquad \text{(that is, 5.8%)}$$

for undergraduates.

Because the sample in Example 2.19 deliberately over-represents graduate students, the final analysis must adjust for this to get unbiased estimates of overall student opinion. Remember that our quick method only works for an SRS. In fact, a professional analysis would also take account of the fact that the population contains "only" 30,000 individuals—more job opportunities for statisticians.

EXAMPLE 2.20   The woes of telephone samples

In principle, it would seem that telephone surveys based on dialing numbers at random can really use an SRS. Telephone surveys have little need for clustering. Stratifying can still reduce variability, so telephone surveys often take samples in two stages: a stratified sample of telephone number prefixes (area code plus first three digits) followed by individual numbers (last four digits) dialed at random in each prefix.

The real problem with an SRS of telephone numbers is that too few numbers lead to households. Blame technology. Fax machines,

New York, New York

New York City, they say, is bigger, richer, faster, ruder. Maybe there's something to that. The sample survey firm Zogby International says that as a national average it takes 5 telephone calls to reach a live person. When calling to New York, it takes 12 calls. Survey firms assign their best interviewers to make calls to New York and often pay them bonuses to cope with the stress.

modems, and cell phones demand new phone numbers. Between 1988 and 1999, the number of households in the United States grew by 11%, but the number of possible residential phone numbers grew by 90%. Many new prefixes have as yet no household telephones behind them. Telephone surveys now use "list-assisted samples" that check electronic telephone directories to eliminate prefixes that have no listed numbers before random sampling begins. Fewer calls are wasted, but anyone living where all numbers are unlisted is missed. Prefixes with no listed numbers are therefore separately sampled (stratification again) to fill the gap.

It's clear from Examples 2.18, 2.19, and 2.20 that designing samples is a business for experts. Even most statisticians don't qualify. We won't worry about such details. The big idea is that good sample designs use chance to select individuals from the population. That is, all good samples are *probability samples.*

Probability sample

A **probability sample** is a sample chosen by chance. We must know what samples are possible and what chance, or probability, each possible sample has. Some probability samples, such as stratified samples, don't allow all possible samples from the population and may not give an equal chance to all the samples they do allow.

A stratified sample of 300 undergraduate students and 200 grad students, for example, only allows samples with exactly that makeup. An SRS would allow any 500 students. Both are probability samples. We need only know that estimates from any probability sample share the nice properties of estimates from an SRS. Confidence statements can be made without bias and have smaller margins of error as the size of the sample increases. Nonprobability samples such as voluntary response samples do not share these advantages and cannot give trustworthy information about a population. Now that we know that most nationwide samples are more complicated than an SRS, we will usually go back to acting as if good samples were SRSs. That keeps the big idea and hides the messy details.

EXERCISES

2.67 Genetically modified foods An article in the journal *Science* looks at differences in attitudes toward genetically modified foods in Europe and the United States. This calls for sample surveys. The European survey chose a sample of 1000 adults in each of 17 European countries. Here's part of the description: "The Eurobarometer survey is a multistage, random-probability face-to-face sample survey."[26]

(a) What does "multistage" mean?

(b) You can see that the first stage was stratified. What were the strata?

(c) What does "random-probability sample" mean?

2.68 Scholar-athletes You want to interview 10 male college-scholarship athletes at length about their attitude toward school and their future plans. Because you believe there may be large differences among athletes in different sports, you decide to interview a stratified random sample of 7 basketball players and 3 golfers. Use Table A, beginning at line 101, to select your sample from the team rosters below. Explain your method carefully enough that a classmate could obtain your results.

BASKETBALL

Abdul-Jabbar	Ewing	Robertson
Aguirre	Johnson	Robinson
Bird	Jordan	Stockton
Bryant	Malone	West
Chamberlain	McGrady	Worthy
Cousy	Miller	
Erving	O'Neal	

GOLF

Els	Mickelson	Palmer
Faldo	Nicklaus	Watson
Love	Norman	Woods

2.69 Women engineers About 20% of the engineering students at a large university are women. The school plans to poll a sample of 200 engineering students about the quality of student life.

(a) If an SRS of size 200 is selected, about how many women do you expect to find in the sample?

(b) If the poll wants to be able to report separately the opinions of male and female students, what type of sampling design would you suggest? Why?

2.70 A stratified sample A university has 2000 male and 500 female faculty members. The equal opportunity employment officer wants to poll the opinions of a random sample of faculty members. In order to give adequate attention to female faculty opinion, he decides to choose a stratified random sample of 200 males and 200 females. He has alphabetized lists of female and male faculty members.

(a) Explain how you would assign labels and use random digits to choose the desired sample. Enter Table A at line 122 and include the first 5 females and the first 5 males in your sample.

(b) What is the chance that any one of the 2000 males will be in your sample? What is the chance that any one of the 500 females will be in your sample?

(c) Each member of the sample is asked, "In your opinion, are female faculty members in general paid less than males with similar positions and qualifications?"

> 180 of the 200 females (90%) say "Yes."
> 60 of the 200 males (30%) say "Yes."

In all, 240 of the sample of 400 (60%) answered "Yes." The officer therefore reports that "Based on a sample, we can conclude that 60% of the total faculty feel that female members are underpaid relative to males." Explain why this conclusion is wrong.

(d) If we took a stratified random sample of 200 male and 50 female faculty members at this university, each member of the faculty would have the same chance of being chosen. What is that chance? Explain why this sample is *not* an SRS.

2.71 Systematic random sample The final stage in a multistage sample must choose 5 of the 500 addresses in a neighborhood. You have a list of the 500 addresses in geographical order. To choose a systematic random sample, proceed as follows:

Step 1. Choose one of the first 100 addresses on the list at random. (Label them 00, 01, . . . , 99 and use a pair of digits from Table A to make the choice.)

Step 2. The sample consists of the address from Step 1 and the addresses 100, 200, 300, and 400 positions down the list from it.

If 71 is chosen at random in Step 1, for example, the systematic random sample consists of the addresses numbered 71, 171, 271, 371, and 471.

(a) Use Table A to choose a systematic random sample of 5 from a list of 500 addresses. Enter the table at line 130.

(b) What is the chance that any specific address will be chosen? Explain your answer.

(c) Explain why this sample is *not* an SRS.

Questions to ask before you believe a poll

Opinion polls and other sample surveys can produce accurate and useful information if the pollster uses good statistical techniques and also works hard at preparing a sampling frame, wording questions, and reducing nonresponse. Many surveys, however—especially those designed to influence public opinion rather than just record it—do not produce accurate or useful information. Here are some questions to ask before you pay much attention to poll results.

- **Who carried out the survey?** Even a political party should hire a professional sample survey firm whose reputation demands that it follow good survey practices.

- **What was the population?** That is, whose opinions were being sought?

- **How was the sample selected?** Look for mention of random sampling.

- **How large was the sample?** Even better, find out both the sample size and the margin of error within which the results of 95% of all samples drawn as this one was would fall.

- **What was the response rate?** That is, what percent of the original subjects actually provided information?

- **How were the subjects contacted?** By telephone? Mail? Face-to-face interview?

- **When was the survey conducted?** Was it just after some event that might have influenced opinion?

- **What were the exact questions asked?**

Academic survey centers and government statistical offices answer these questions when they announce the results of a sample survey. National opinion polls usually don't announce their response rate (which is often low) but do give us the other information. Editors and newscasters have the bad habit of cutting out these dull facts and reporting only the sample results. Many sample surveys by interest groups and local newspapers and TV stations don't answer these questions because their polling methods are in fact unreliable. If a politician, an advertiser, or your local TV station announces the results of a poll without complete information, be skeptical.

EXERCISES

2.72 Appraising a poll The *Wall Street Journal* published an article on attitudes toward the Social Security system based on a sample survey. It found, for example, that 36% of people aged 18 to 34 years expected Social Security to pay nothing at all when they retire. News articles tend to be brief in describing sample surveys. Here is part of the *Wall Street Journal's* description of this poll:

> The Wall Street Journal/NBC News poll was based on nationwide telephone interviews of 2,012 adults, conducted Thursday through Sunday by the polling organizations of Peter Hart and Robert Teeter.

> The sample was drawn from 520 randomly selected geographic points in the continental U.S. Each region was represented in proportion to its population. Households were selected by a method that gave all telephone numbers, listed and unlisted, an equal chance of being included.[27]

Several "questions to ask" about an opinion poll are listed above. What answers does the *Wall Street Journal* give to each of these questions?

2.73 Multistage sampling The previous exercise gives part of the description of a sample survey from the *Wall Street Journal.*
(a) It appears that the sample was taken in several stages. Why can we say this?
(b) The first stage no doubt used a stratified sample, though the *Journal* does not say this. Explain why it would be bad practice to use an SRS from a large number of "geographic points" across the country rather than a stratified sample of such points.

2.74 TV ratings The method of collecting the data can influence the accuracy of sample results. The following methods have been used to collect data on television viewing in a sample household:
(a) *The diary method.* The household keeps a diary of all programs watched and who watched them for a week, then mails in the diary at the end of the week.
(b) *The roster-recall method.* An interviewer shows the subject a list of programs for the preceding week and asks which programs were watched.
(c) *The telephone-coincidental method.* The survey firm telephones the household at a specific time and asks if the television is on, which program is being watched, and who is watching it.
(d) *The automatic recorder method.* A device attached to the set records what hours the set is on and to which channel it is tuned. At the end of the week, this record is removed from the recorder.
(e) *The people meter.* Each member of the household is assigned a numbered button on a hand-held remote control. Everyone is asked to push their button whenever they start or stop watching TV. The remote control signals a device attached to the set that keeps track of what channel the set is tuned to and who is watching at all times.

Discuss the advantages and disadvantages of each of these methods, especially the possible sources of error associated with each method. The Nielsen national ratings use Method (e). Local ratings (there are more than 200 local television markets) use Method (a). Do you agree with these choices? (Do not discuss choosing the sample, just collecting the data once the sample is chosen.)

EXPLORING THE WEB

This chapter began by pointing to a study by the Pew Research Center for the People and the Press. The Center's Web site is www.people-press.org. The site contains reports on studies such as the one we have looked at and the results of the Center's own opinion polls. You might look at the Web site's "Poll Analysis" section for careful comments on what current opinion polls, both the Center's and others such as Gallup Polls, tell us about public views on important issues.

APPLICATION 2.3 Gallup Polls

Ola Babcock Miller was active in the late-nineteenth-century women's suffrage movement, was elected to three terms as Iowa's secretary of state, and in her first term founded the Iowa State Highway Patrol. Her election in 1932 surprised the political pundits of the day, as she became the first woman, and the first Democrat since the Civil War, to hold statewide political office in Iowa. However, her election did not surprise her son-in-law, George Gallup, who predicted her victory using the first scientifically sampled election poll ever. Three years later Gallup founded the American Institute of Public Opinion in Princeton, New Jersey, and began publishing the results of the Gallup Poll in his syndicated column "America Speaks."

At the Web site for the Gallup Organization at www.gallup.com you will see how the Gallup Poll provides data on the attitudes and lifestyles of people around the world. Notice that the poll is just part of a global management consulting and market research company. Many of the poll results here used to be available for public viewing but now are housed in a "subscribers only" area of the site.

Although the Gallup Poll is prominent in news stories throughout the year, it gets particular media attention every four years with its surveys prior to each presidential election. The results of the Gallup presidential election polls from 1936 to 2000 are shown in the table on the following page.

1. Did the Gallup Poll ever fail to predict the winner of a presidential election? If so, in which year(s)?

2. How far has the final Gallup Poll estimate typically deviated from the actual results of the presidential elections? Do you see any trends in the accuracy of these polls?

3. *Literary Digest* magazine made an infamous error in predicting the result of the 1936 presidential election. Do a little research. What error did they make, and how did they make it?

4. Search newspaper, magazine, or Internet sources to find the reported results of a poll. The report you find must provide answers to at least five of the eight "questions to ask before you believe a poll." Put your report on a separate sheet of paper and include the exact source (date and name of publication or Web site address). Then answer as many of the eight questions as you can.

Final Preelection Gallup Poll versus Actual Election Results (1936–2000)

Year and candidates	Democrat Gallup/ actual (%)	Republican Gallup/ actual (%)	Third party Gallup/ actual (%)	Gallup accuracy[a]
1936 **Roosevelt** (D) Landon (R)	56.0/60.8	44.0/36.5	0.0/2.7	−4.8%
1940 **Roosevelt** (D) Wilkie (R)	52.0/54.7	48.0/44.8	0.0/0.5	−2.7%
1944 **Roosevelt** (D) Wilkie (R)	51.5/53.6	48.5/46.0	0.0/0.4	−2.1%
1948 **Truman** (D) Dewey (R)	44.5/49.6	49.5/45.1	6.0/5.4	−5.1%
1952 Stevenson (D) **Eisenhower** (R)	49.0/44.4	51.0/55.1	0.0/0.5	4.6%
1956 Stevenson (D) **Eisenhower** (R)	40.5/42.1	59.5/57.6	0.0/0.3	−1.6%
1960 **Kennedy** (D) Nixon (R)	51.0/49.9	49.0/49.8	0.0/0.3	1.1%
1964 **Johnson** (D) Goldwater (R)	64.0/61.3	36.0/38.6	0.0/0.1	2.7%
1968 Humphrey (D) **Nixon** (R)	42.0/42.7	43.0/43.4	15.0/13.8	−0.7%
1972 McGovern (D) **Nixon** (R)	38.0/37.6	62.0/60.7	0.0/1.7	0.4%
1976 **Carter** (D) Ford (R)	48.0/50.1	49.0/48.0	3.0/1.8	−2.1%

1980 Carter (D) **Reagan** (R)	44.0/41.0	47.0/50.8	9.0/8.2	3.0%
1984 Mondale (D) **Reagan** (R)	41.0/40.6	59.0/58.8	0.0/0.6	0.4%
1988 Dukakis (D) **Bush** (R)	44.0/45.7	56.0/53.4	0.0/0.9	−1.7%
1992 **Clinton** (D) Bush (R)	49.0/43.0	37.0/37.5	14.0/19.5	6.0%
1996 **Clinton** (D) Dole (R)	52.0/49.2	41.0/40.9	7.0/10.0	2.8%
2000 Gore (D) **Bush** (R)	46.0/48.4	48.0/47.9	6.0/3.7	−2.4%

[a]Gallup accuracy = % Gallup predicted for Democratic party candidate − % vote received in election.

STATISTICS IN SUMMARY

Even professional sample surveys don't give exactly correct information about the population. There are many potential sources of error in sampling. The margin of error announced by a sample survey covers only **random sampling error,** the variation due to chance in choosing a random sample. Other types of error are in addition to the margin of error and can't be directly measured. **Sampling errors** come from the act of choosing a sample. Random sampling error and **undercoverage** are common types of sampling error. Undercoverage occurs when some members of the population are left out of the **sampling frame,** the list from which the sample is actually chosen.

The most serious errors in most careful surveys, however, are **nonsampling errors.** These have nothing to do with choosing a sample—they are present even in a census. The single biggest problem for sample surveys is **nonresponse:** subjects can't be contacted or refuse to answer. Mistakes in handling the data (**processing errors**) and incorrect answers by respondents (**response errors**) are other examples of nonsampling errors. Finally, the exact **wording of questions** has a big influence on the answers. People who design sample surveys use statistical techniques that help correct nonsampling errors, and they also use **probability samples** more complex than simple random samples, such as **stratified samples.**

You can assess the quality of a sample survey quite well by just looking at the basics: use of random samples, sample size and margin of error, the rate of nonresponse, and the wording of the questions.

SECTION 2.3 EXERCISES

2.75 Did you vote? When the Current Population Survey asked the adults in its sample of 50,000 households if they voted in the 1996 presidential election, 54% said they had. In fact, only 49% of the adult population voted in that election. Why do you think the CPS result missed by much more than the margin of error?

2.76 What kind of error? Each of the following is a source of error in a sample survey. Label each as *sampling error* or *nonsampling error,* and explain your answers.
(a) The telephone directory is used as a sampling frame.
(b) The subject cannot be contacted in five calls.
(c) Interviewers choose people on the street to interview.

2.77 Sampling students You want to determine whether students at your school think that teachers are sufficiently available to students outside the classroom. You have the resources to contact about 200 students.
(a) Specify the population and parameter of interest.
(b) Describe your sample design. For example, will you use a stratified sample with males and females as strata?
(c) Briefly discuss the practical difficulties that you anticipate. For example, how will you contact the students in your sample?
(d) What specific question or questions will you ask?

2.78 Telling the truth? Many subjects don't give honest answers to questions about activities that are illegal or sensitive in some other way. One study divided a large group of white adults into thirds at random. All were asked if they had ever used cocaine. The first group was interviewed by telephone: 21% said "Yes." In the group visited at home by an interviewer, 25% said "Yes." The final group was interviewed at home but answered the question on an anonymous form that they sealed in an envelope. Of this group, 28% said they had used cocaine.[28]
(a) Which result do you think is closest to the truth? Why?
(b) Give two other examples of behavior you think would be underreported in a telephone survey.

2.79 A sampling paradox? Example 2.19 compares two SRSs, of a university's undergraduate and graduate students. The sample of undergraduates contains a smaller fraction of the population, 1 out of 90, versus 1 out of 15 for graduate students. Yet sampling 1 out of 90 undergraduates gives a smaller margin of error than sampling 1 out of 15 graduate students. Explain to someone who knows no statistics why this happens.

2.80 Inspecting meat A city contains 33 supermarkets. A health inspector wants to check compliance with a new city ordinance on meat storage. Because of the time required, he can inspect only 10 markets. He decides to choose a stratified random sample and stratifies the markets by sales volume. Stratum A consists of 3 large chain stores: the inspector decides to inspect all 3. Stratum B consists of 10 smaller chain stores: 4 out of the 10 will be inspected. Stratum C consists of 20 locally owned small stores: 3 of these 20 will be inspected.

16/33

Let "Yes" mean that the store is in compliance and "No" mean that it is not. The population is as follows (unknown to the inspector, of course):

Stratum A		Stratum B		Stratum C			
Store 1	Yes	Store 1	No	Store 1	Yes	11	Yes
2	Yes	2	Yes	2	Yes	12	Yes
3	No	3	No	3	No	13	No
		4	No	4	Yes	14	Yes
		5	Yes	5	No	15	Yes
		6	No	6	No	16	No
		7	Yes	7	No	17	No
		8	No	8	Yes	18	No
		9	No	9	No	19	Yes
		10	Yes	10	No	20	Yes

(a) Use Table A to choose a stratified random sample of size 10 allotted among the strata as described above.

(b) Use your sample results to estimate the proportion of the entire population of stores in compliance with the ordinance.

(c) Use the description of the population given above to find the true proportion of stores in compliance. How accurate is the estimate from (b)?

2.81 Homosexuals in the military? Here are three opinion poll questions on the same issue, with the poll results:

Do you approve or disapprove of allowing openly homosexual men and women to serve in the armed forces of the United States? Result: 47% strongly or somewhat disapproved; 45% strongly or somewhat approved.

Do you think that gays and lesbians should be banned from the military or not?
Result: 37% said they should be banned; 57% said not.

Should President Clinton change military policy to allow gays in the military?
Result: 53% said "No"; 35% said "Yes."

Using this example, discuss the difficulty of using responses to opinion polls to understand public opinion.

2.82 Appraising a poll A *New York Times* article on attitudes toward the political parties discussed the results of a sample survey that found, for example, that 44% of adults think the Democrats "have better ideas for leading the country into the 21st century." Another 37% chose the Republicans; the others had no opinion. Here is part of the *Times*'s statement on "How the Poll Was Conducted":

> *The latest* New York Times/*CBS News poll is based on telephone interviews conducted Nov. 4 through 7 with 1,162 adults throughout the United States. . . . The sample of telephone exchanges called was randomly selected by a computer from a complete list of more than 42,000 active residential exchanges across the country.*

> *Within each exchange, random digits were added to form a complete telephone number, thus permitting access to both listed and unlisted numbers. Within each household, one adult was designated by a random procedure to be the respondent for the survey.[29]*

Page 121 lists several "questions to ask" about an opinion poll. What answers does the *Times* give to each of these questions?

CHAPTER 2 REVIEW

We have discussed sampling, the art of choosing a part of a population to represent the whole. We saw the potential for bias from voluntary response and convenience samples. This led us to the big idea of a simple random sample, which is summarized in the figure below.

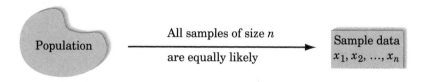

Introducing randomness into the sampling scheme means that the things we measure will vary from sample to sample. But this variation will have a predictable pattern if we repeat the procedure over and over. Increasing the size of the sample will decrease the variability of a sample statistic. But this won't help with bias.

Two types of error can plague sample surveys: sampling errors and nonsampling errors. When the sampling frame differs from the population of interest, undercoverage results. The margin of error does not account for such sampling errors. It also does not reflect nonresponse or question wording, two major causes of nonsampling error.

In many settings, an SRS would be too difficult or too costly to obtain. Many national surveys use some form of multistage sampling. If subgroups within a population differ in their opinions, a stratified random sample is appropriate.

Here are the major skills you should have developed now that you have studied this chapter.

1. Identify the population and parameter of interest.

2. Recognize bias due to voluntary response samples and other inferior sampling methods.

3. Use Table A of random digits to select a simple random sample (SRS) from a population.

4. Explain how sample surveys deal with bias and variability in their conclusions.

5. Explain in simple language what the margin of error for a sample survey result tells us and what "95% confidence" means.

6. Use the quick method to get an approximate margin of error for 95% confidence.

7. Understand the distinction between sampling errors and nonsampling errors. Recognize the presence of undercoverage and nonresponse as sources of error in a sample survey. Recognize the effect of the wording of questions on the responses.

8. Use random digits to select a stratified random sample from a population when the strata are identified.

CHAPTER 2 REVIEW EXERCISES

2.83 Baseball tickets Suppose you want to know the average amount of money spent by the fans attending opening day for the Cleveland Indians baseball season. You get permission from the team's management to conduct a survey at the stadium but they will not allow you to bother the fans in the club seating or box seat areas (the most expensive seating, where the fans have paid over $30 per ticket). Using a computer, you randomly select 500 seats from the rest of the stadium and during the game ask the fans in those seats how much they spent that day.

(a) Provide a reason why this survey might yield a biased result.

(b) Explain whether the reason you provide is a type of sampling error or a type of nonsampling error.

2.84 Choose an SRS Your class in ancient civilization is poorly taught and has decided to complain to the principal. The class decides to choose four of its members at random to carry the complaint. The class list appears below. Choose an SRS of 4 using the table of random digits, beginning at line 145. Explain your method clearly enough for a classmate to duplicate your results.

Anderson	Gutierrez	Patnaik
Aspin	Gwynn	Pirelli
Benitez	Harter	Rao
Bock	Henderson	Rider
Breiman	Hughes	Robertson
Castillo	Johnson	Rodriguez
Dixon	Kempthorn	Siegel
Edwards	Liang	Tompkins
Fernandez	Montoya	Vandegraff
Gupta	Olds	Wang

2.85 Elegant cars An advertising agency conducts a sample survey to see how adult women react to various adjectives that might be used to describe an automobile. The firm chooses 400 women from across the country. Each woman is read a list of adjectives, such as "elegant" and "prestigious." For each adjective, she is asked to indicate how desirable a car described this way seems to her. The possible responses are (1) highly desirable, (2) somewhat desirable, (3) neutral, (4) not desirable. Of the women interviewed, 76% said that a car described as "elegant" was highly desirable.

(a) What are the population and the parameter of interest in this sample survey?

(b) Is the number 76% a parameter or a statistic?

(c) Treat the sample of women as an SRS. Make a 95% confidence statement about the reaction of women to the adjective "elegant."

2.86 TV commercials The noted scientist Dr. Iconu wanted to investigate attitudes toward television advertising among American college students. He decided to use a sample of 100 students. Students in freshman psychology (PSY 001) are required to serve as subjects for experimental work. Dr. Iconu obtained a class list for PSY 001 and chose a simple random sample of 100 of the 340 students on the list. He asked each of the 100 students in the sample the following question:

Do you agree or disagree that having commercials on TV is a fair price to pay for being able to watch it?

Of the 100 students in the sample, 82 marked "Agree." Dr. Iconu announced the result of his investigation by saying "82% of American college students are in favor of TV commercials."

(a) What is the population in this example?

(b) What is the sampling frame in this example?

(c) Explain briefly why the sampling frame is or is not suitable.

(d) Discuss briefly the question Dr. Iconu asked. Is it a slanted question?

(e) Discuss briefly why Dr. Iconu's announced result is misleading.

(f) Dr. Iconu defended himself against criticism by pointing out that he had carefully selected a simple random sample from his sampling frame. Is this defense relevant? Why?

2.87 Planning a survey of students The student government plans to ask a random sample of students about their priorities for improving campus life. The college registrar provides a list of the 3500 enrolled students to serve as a sampling frame.

(a) How would you choose an SRS of 250 students?

(b) How would you choose a systematic sample of 250 students? (See Exercise 2.71 (page 120) to learn about systematic samples.)

(c) The list shows whether students live on campus (2400 students) or off campus (1100 students). How would you choose a stratified sample of 200 on-campus students and 50 off-campus students?

(d) Which of the three sampling methods would you choose? Why?

2.88 Bigger samples, please Explain in your own words the advantages of bigger random samples in a sample survey.

2.89 Exit polling Election-night television coverage includes exit polls that often allow quite precise predictions of the outcome before the polls have closed in the western part of the country. It is sometimes charged that late voters may stay home if the networks say that the national election is decided. Therefore, predictions based on samples of actual votes should not be allowed until the polls have closed everywhere in the country. Do you agree with this proposal? Explain your opinion.

2.90 Fear of crime The Gallup Poll asked a random sample of 1493 adults, "Are you afraid to go out at night within a mile of your home because of crime?" Of the sample, 672 said "Yes." Make a confidence statement about the percent of all adults who fear to go out at night because of crime. (Use the quick method to find the margin of error.)

2.91 We don't like one-way streets Highway planners decided to make a main street in West Lafayette, Indiana, a one-way street. The *Lafayette Journal and*

Courier took a one-day poll by inviting readers to call a telephone number to record their comments. The next day, the paper reported:

> Journal and Courier *readers overwhelmingly prefer two-way traffic flow in West Lafayette's Village area to one-way streets. By nearly a 7-1 margin, callers to the newspaper's Express Yourself opinion line on Wednesday complained about the one-way streets that have been in place since May. Of the 98 comments received, all but 14 said no to one-way.*

(a) What population do you think the newspaper wants information about?
(b) Is the proportion of this population who favor one-way streets almost certainly larger or smaller than the proportion 14/98 in the sample? Why?

2.92 The Harris Poll Here is the language used by the Harris Poll to explain the accuracy of its results: "In theory, with a sample of this size, one can say with 95 percent certainty that the results have a statistical precision of plus or minus 3 percentage points of what they would be if the entire adult population had been polled with complete accuracy."[30] What does Harris mean by "95 percent certainty"?

2.93 Sampling library books A group of librarians wants to estimate what fraction of books in large libraries falls in each of four height categories. This information will help them plan shelving. To obtain it, they plan to measure all of the several hundred thousand books in one library. Describe a sampling design that will save the librarians time and money.

2.94 Higher sales tax? A survey is conducted in Chicago (population 2,900,000) using random-digit-dialing equipment which places calls at random to residential phones, both listed and unlisted. The purpose of the survey is to determine the percent of adults who would favor a half-cent increase in the sales tax to help fund public transportation. Four hundred adults are interviewed and 36% of them favor the proposal. A second survey is taken in Dayton, Ohio (population 166,000), using the same techniques and asking the same question of 400 adults living in Dayton.
(a) For the Chicago survey identify the population, the sampling frame, the sample, the variable measured, the parameter of interest, and the corresponding statistic.
(b) Will the accuracy of the Chicago survey be greater than, about the same as, or less than the accuracy of the Dayton survey? Explain

Designing Experiments

3.1 Experiments, Good and Bad
3.2 Experiments in the Real World
3.3 Data Ethics

3.1 EXPERIMENTS, GOOD AND BAD

A tale of three studies

An optimistic account of learning online reports a study at Nova Southeastern University, Fort Lauderdale, Florida. The authors of the study claim that students taking courses online were "equal in learning" to students taking the same courses in class. Replacing college classes with Web sites saves colleges money, so this study suggests we should all move online.[1]

Ulcers seem to accompany modern life. "Gastric freezing" is a clever treatment for stomach ulcers. The patient swallows a deflated balloon with tubes attached; then a refrigerated solution is pumped through the balloon for an hour. The idea is that cooling the stomach will reduce its production of acid and so relieve ulcers. An experiment reported in the *Journal of the American Medical Association* claimed that gastric freezing did relieve ulcer pain.[2]

Should the government provide day care for children in low-income families? If day care helps these children stay in school and hold good jobs later in life, the government would save money by paying less welfare and collecting more taxes. The Carolina Abecedarian Project (the name suggests learning the ABCs) has followed a group of children since 1972. The results show that good day care makes a big difference in later school and work.[3]

Three big issues, three studies that claim to shed light on the issues. In fact, the first of these is an observational study that can't be trusted to give good evidence about learning in online college courses. The gastric-freezing experiment now appears to be plainly misleading. Most ulcers are caused by bacteria, and cooling the stomach is not an effective treatment. The results of the Abecedarian Project, on the other hand, are about as convincing as evidence about the long-term effects of day care can be. What makes some studies—especially some experiments— convincing? Why should we ignore others? This section shows what to look for.

ACTIVITY 3.1 Testing therapeutic touch

Materials: Blindfold and coin for each pair of students
In this Activity, you will perform an experiment to determine whether any of the students in your class has the ability to detect a "human energy field." You will work with a partner chosen by you or by your teacher.

1. With your partner, decide which of you will be the "experimenter" and which will be the "subject." You will switch roles later, so don't spend too much time deciding!

2. Blindfold the subject. Have the subject sit with both arms extended, palms up.

3. The experimenter should flip a coin. If it lands "heads," place one hand about six inches above the subject's left hand. If it lands "tails," place one hand about six inches above the subject's right hand. Ask the subject to identify the presence of the energy field by saying either "left" or "right."

4. Record the outcome of the coin toss (H or T) and whether the subject picked correctly (C) or incorrectly (I).

5. Repeat the previous two steps 19 more times, for a total of 20 trials. Organize your results in a table like this one.

Trial no.	Coin outcome (H or T)	Subject's pick (C or I)
1	T	C

6. Have the subject remove the blindfold. Share the results of the experiment.

7. Switch roles, and repeat Steps 2 through 6.

8. Pool results with your classmates. Make a dotplot that shows the percent of correct (C) identifications made by each subject.

9. Discuss what you see in the dotplot. Does any member of the class appear to be able to detect a human energy field? Why or why not?

10. Calculate the average percent of correct identifications made by all of the subjects. How does this compare with what you would expect if the subjects were just guessing?

For her fourth-grade science fair project, Emily Rosa designed an experiment to test whether an alternative healing method known as therapeutic touch actually works. Therapeutic touch requires the "healer" to detect and manipulate the patient's "human energy field." Emily doubted that this was possible. So she invited 21 people who had practiced therapeutic touch for from 1 to 27 years to participate in her experiment.

Emily's experimental design was fairly simple. She sat at one side of a table and the subject sat at the other side. A tall screen with two holes was placed in the middle of the table, so that the subject could not see Emily. The subject was asked to extend one arm, palm side up, through each of the holes. Unknown to the subject, Emily then flipped a coin to decide which of the subject's two hands she would place her hand above. She then asked the subject to identify the location of the "energy field." This process was repeated either 10 or 20 times with each of the 21 subjects.

The results of Emily Rosa's experiment were striking: the subjects correctly identified the location of Emily's hand in only 70 of 150 trials (47%). This is pretty much what we would expect if the subjects were merely guessing. With a little help from her mother (a nurse) and a statistician, Emily had an article summarizing her research published in the *Journal of the American Medical Association* on April 1, 1998.[4]

Talking about experiments

Observational studies involve passive data collection. We observe, record, or measure, but we don't interfere. Experiments involve active data production. Experimenters actively intervene by imposing some treatment in order to see what happens. All experiments and many observational studies are interested in the effect one variable has on another variable. Here is the vocabulary we use to distinguish the variable that acts from the variable that is acted upon.

The vocabulary of experiments

A **response variable** is a variable that measures an outcome or result of a study.

An **explanatory variable** is a variable that we think explains or causes changes in the response variables.

The individuals studied in an experiment are often called **subjects.**

A **treatment** is any specific experimental condition applied to the subjects. If an experiment has several explanatory variables, a treatment is a combination of specific values of these variables.

EXAMPLE 3.1   Learning on the Web and the effects of day care

College students are the subjects in the Nova Southeastern University study. The *explanatory variable* is the setting for learning (in class or online). The *response variable* is a student's score on a test at the end of the course.

The Abecedarian Project is an experiment in which the *subjects* are 111 people who in 1972 were healthy but low-income black infants in Chapel Hill, North Carolina. All the infants received nutritional supplements and help from social workers. Half, chosen at random, were also placed in an intensive preschool program. The experiment compares these two *treatments*. The *explanatory variable* is just "preschool, yes or no." There are many *response variables*, recorded over more than 20 years, including academic test scores, college attendance, and employment.

You will often see explanatory variables called *independent variables* and response variables called *dependent variables*. The idea is that the response variables depend on the explanatory variables. We won't use these older terms, partly because "independent" has other and very different meanings in statistics.

How to experiment badly

Do students who take a course via the Web learn as well as those who take the same course in a traditional classroom? The best way to find out is to assign some students to the classroom and others to the Web. That's an experiment. The Nova Southeastern University study was not an experiment, because it imposed no treat-

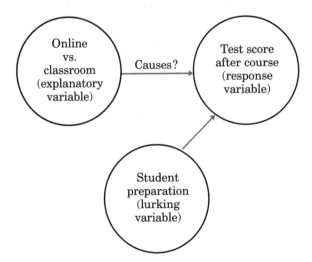

FIGURE 3.1 Confounding in the Nova Southeastern University study. The influence of course setting (the explanatory variable) can't be distinguished from the influence of student preparation (a lurking variable).

ment on the student subjects. Students chose for themselves whether to enroll in a classroom or in an online version of a course. The study simply measured their learning. The students who chose the online course were very different from the classroom students. For example, their average score on tests on the course material given before the courses started was 40.70, against only 27.64 for the classroom students. It's hard to compare in-class versus online learning when the online students have a big head start. The effect of online versus in-class instruction is hopelessly mixed up with influences lurking in the background. Figure 3.1 on the facing page shows the mixed-up influences in picture form.

Confounding

A **lurking variable** is a variable that has an important effect on the relationship among the variables in a study but is not one of the explanatory variables studied.

Two variables are **confounded** when their effects on a response variable cannot be distinguished from each other. The confounded variables may be either explanatory variables or lurking variables.

In the Nova Southeastern University study, student preparation (a lurking variable) is confounded with the explanatory variable. The study report claims that the two groups did equally well on the final test. We can't say how much of the online group's performance is due to their head start. The fact that a group that started with a big advantage did no better than the more poorly prepared classroom students is not very impressive evidence of the wonders of Web-based instruction. Here is another example, one in which a second experiment untangled the confounding.

EXAMPLE 3.2   Gastric freezing flunks

Experiments that study the effectiveness of medical treatments on actual patients are called **clinical trials.** The clinical trial that made gastric freezing a popular treatment for stomach ulcers had this "one-track" design:

Impose treatment	→	*Measure response*
Gastric freezing	→	Reduced pain?

The patients did report reduced pain, but we can't say that gastric freezing caused the reduced pain. It might be just the **placebo effect.** A **placebo** is a dummy treatment with no active ingredients. Many patients respond favorably to *any* treatment, even a placebo.

This response to a dummy treatment is the placebo effect. Perhaps the placebo effect is in our minds, based on trust in the doctor and expectations of a cure. Perhaps it is just a name for the fact that many patients improve for no visible reason. The one-track design of the experiment meant that the placebo effect was confounded with any effect gastric freezing might have.

A second clinical trial, done several years later, divided ulcer patients into two groups. One group was treated by gastric freezing as before. The other group received a placebo treatment in which the solution in the balloon was at body temperature rather than freezing. The results: 34% of the 82 patients in the treatment group improved, but so did 38% of the 78 patients in the placebo group. This and other properly designed experiments showed that gastric freezing was no better than a placebo, and doctors abandoned it.[5]

"I want to make one thing perfectly clear, Mr. Smith. The medication I prescribe will cure that run-down feeling."

Both observation and one-track experiments often yield useless data because of confounding with lurking variables. It is hard to avoid confounding when only observation is possible. Experiments offer better possibilities, as the second gastric-freezing experiment shows. This experiment included a group of subjects who received only a placebo. This allows us to see whether the treatment being tested does better than the placebo and so has more than the placebo effect going for it. Effective medical treatments pass the placebo test. Gastric freezing flunks the test.

EXERCISES

3.1 On sale A researcher studying the effect of price promotions on consumers' expectations makes up two different histories of the store price of a hypothetical brand of laundry detergent for the past year. Students in a marketing course view one or the other price history on a computer. Some students see a steady price, while others see regular promotions that temporarily cut the price. Then the students are asked what price they would expect to pay for the detergent.

(a) Is this study an experiment? Why?

(b) What are the explanatory and response variables?

3.2 Public housing A study of the effect of living in public housing on family stability and other variables in poverty-level households was carried out as follows:

A list of applicants accepted for public housing was obtained, together with a list of families who applied but were rejected by the housing authorities. A random sample was drawn from each list, and the two groups were observed for several years.
(a) Is this an experiment? Why?
(b) What are the explanatory and response variables?
(c) Does this study contain confounding that may prevent valid conclusions on the effects of living in public housing? Explain.

3.3 Nursing your baby An article in a women's magazine says that women who nurse their babies feel warmer and more receptive toward the infants than mothers who bottle-feed. The author concludes that nursing has desirable effects on the mother's attitude toward the child. But women choose whether to nurse or bottle-feed. Explain why this fact makes any conclusion about cause and effect untrustworthy. Use the language of lurking variables and confounding in your explanation, and draw a picture like Figure 3.1 to illustrate it.

3.4 Vitamin C and colds Last year only 10% of a group of adult men did not have a cold at some time during the winter. This year all the men in the group took 1 gram of vitamin C each day, and 20% had no colds. A writer claims that this shows that vitamin C helps prevent colds. Explain why this conclusion is shaky at best. Use the language of lurking variables and confounding in your explanation, and draw a picture like Figure 3.1 to illustrate it.

3.5 Activity 3.1 follow-up
(a) Identify the subjects, treatments, and response variable in the therapeutic touch experiment.
(b) After Emily's results were published, therapeutic touch practitioners argued that her results were not valid. Discuss one or two arguments they might have made.

Randomized comparative experiments

The first goal in designing an experiment is to ensure that it will show us the effect of the explanatory variables on the response variables. Confounding often prevents one-track experiments from doing this. The remedy is to compare two or more treatments. Here is an example of a new medical treatment that passes the placebo test in a direct comparison.

EXAMPLE 3.3   Sickle cell anemia

Sickle cell anemia is an inherited disorder of the red blood cells that in the United States affects mostly blacks. It can cause severe pain and many complications. The National Institutes of Health carried out a clinical trial of the drug hydroxyurea for treatment of sickle cell anemia. The subjects were 299 adult patients who had had at least three episodes of pain from sickle cell anemia in the previous year.

Simply giving hydroxyurea to all 299 subjects would confound the effect of the medication with the placebo effect and other lurking variables such as the effect of knowing that you are a subject in an experiment. Instead, half of the subjects received hydroxyurea, and the other half received a placebo that looked and tasted the same. All subjects were treated exactly the same (same schedule of medical checkups, for example) except for the content of the medicine they took. Lurking variables therefore affected both groups equally and should not cause any differences between their average responses.

The two groups of subjects must be similar in all respects before they start taking the medication. Just as in sampling, the best way to avoid bias in choosing which subjects get hydroxyurea is to allow impersonal chance to make the choice. An SRS of 152 of the subjects formed the hydroxyurea group; the remaining 147 subjects made up the placebo group. Figure 3.2 outlines the experimental design.

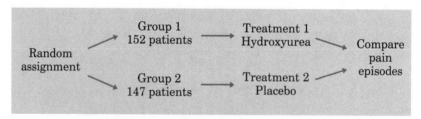

FIGURE 3.2 The design of a randomized comparative experiment to compare hydroxyurea with a placebo for treating sickle cell anemia.

The experiment was stopped ahead of schedule because the hydroxyurea group had many fewer pain episodes than the placebo group. This was compelling evidence that hydroxyurea is an effective treatment for sickle cell anemia, good news for those who suffer from this serious illness.[6]

Figure 3.2 illustrates the simplest **randomized comparative experiment,** one that compares just two treatments. The diagram outlines the essential information about the design: random assignment; one group for each treatment; the number of subjects in each group (it is generally best to keep the groups similar in size); what treatment each group gets; and the response variable we compare. You know how to carry out the random assignment of subjects to groups. Label the 299 subjects 001 to 299, then read three-digit groups from the table of random digits (Table A) until you have chosen the 152 subjects for Group 1. The remaining 147 subjects form Group 2.

The placebo group in Example 3.3 is called a **control group** because comparing the treatment and control groups allows us to control the effects of lurking variables. A control group need not receive a dummy treatment such as a placebo.

Clinical trials often compare a new treatment for a medical condition, not with a placebo, but with a treatment that is already on the market. Patients who are randomly assigned to the existing treatment form the control group. To compare more than two treatments, we can randomly assign the available experimental subjects to as many groups as there are treatments. Here is an example with three groups.

EXAMPLE 3.4 Stopping drunk drivers

Most people know that drunk driving is dangerous. Even so, some drivers are cited multiple times for driving while intoxicated. Is there anything that can be done to change their behavior? An experiment is designed to help answer this question.

The subjects are 300 people convicted of drunk driving three times in one year. The treatments, imposed after the third conviction, are a fine plus a suspended jail sentence plus one of (1) no treatment, (2) attend an alcoholism clinic, or (3) participate in Alcoholics Anonymous. The subjects assigned to Treatment 1 serve as a control group. One response variable is whether or not the subject is arrested again in the following year. Figure 3.3 outlines the design.

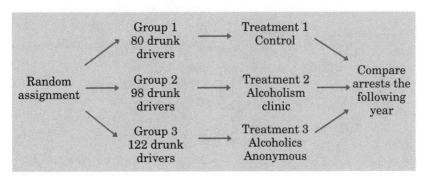

FIGURE 3.3 The design of a randomized comparative experiment to compare three programs to deter drunk drivers.

Ideally, 100 subjects would be assigned at random to each of the three treatments. This is probably not practical, since a treatment must be assigned immediately after the subject's third drunk-driving conviction. Instead, enter Table A at any row and begin reading off single digits. The digits 1, 2, and 3 will represent treatment option 1; the digits 4, 5, and 6 indicate treatment option 2; and the digits 7, 8, and 9 signal treatment option 3. Ignore the digit 0. The first subject selected for the experiment is assigned the treatment corresponding to the first nonzero digit in the row. The second subject's treatment is determined by the second nonzero digit in the row, and so on. For example, if we use row 105 in Table A, our assignment of subjects to treatments proceeds as follows:

Subject	1	2	3	4	5	6	7	8	9	10
Nonzero digit	9	5	5	9	2	9	4	7	6	9
Treatment	3	2	2	3	1	3	2	3	2	3

By the end of row 105, we have assigned 10 subjects to Treatment 1, 9 to Treatment 2, and 18 to Treatment 3. Continuing in this fashion through rows 106, 107, . . ., 113, we end up with 80 subjects receiving Treatment 1, 98 receiving Treatment 2, and 122 receiving Treatment 3.

The logic of experimental design

The randomized comparative experiment is one of the most important ideas in statistics. It is designed to allow us to draw cause-and-effect conclusions. Be sure you understand the logic:

* Randomization produces groups of subjects that should be similar in all respects before we apply the treatments.

* Comparative design ensures that influences other than the experimental treatments operate equally on all groups.

* Therefore, differences in the response variable must be due to the effects of the treatments.

We use chance to choose the groups in order to eliminate any systematic bias in assigning the subjects to groups. In the sickle cell study, for example, a doctor might subconsciously assign the most seriously ill patients to the hydroxyurea group, hoping that the untested drug will help them. That would bias the experiment against hydroxyurea. Choosing an SRS of the subjects to be Group 1 gives everyone the same chance to be in either group. We expect the two groups to be similar in all respects—age, seriousness of illness, smoker or not, and so on. Chance tends to assign equal numbers of smokers to both groups, for example, even if we don't know which subjects are smokers.

There is one important caution about randomized experiments. Like random samples, they are subject to the laws of chance. Just as an SRS of voters might by bad luck choose nearly all Republicans, a random assignment of subjects might by bad luck put nearly all the smokers in one group. We know that if we choose large

"OK, Mr. Simms. Now it's time to test your pacemaker against a control subject."

random samples, it is very likely that the sample will match the population well. In the same way, if we use many experimental subjects, it is very likely that random assignment will produce groups that match closely. More subjects means that there is less chance variation among the treatment groups and less chance variation in the outcomes of the experiment. "Use enough subjects" joins "compare two or more treatments" and "randomize" as a basic principle of statistical design of experiments.

Principles of experimental design

The basic principles of statistical design of experiments are:
1. **Control** the effects of lurking variables on the response, most simply by comparing two or more treatments.
2. **Randomize**—use impersonal chance to assign subjects to treatments.
3. **Use enough subjects** in each group to reduce chance variation in the results.

EXERCISES

3.6 Aspirin and heart attacks Can aspirin help prevent heart attacks? The Physicians' Health Study, a large medical experiment involving 22,000 male physicians, attempted to answer this question. One group of about 11,000 physicians took an aspirin every second day, while the rest took a placebo. After several years the study found that subjects in the aspirin group had significantly fewer heart attacks than subjects in the placebo group.
(a) Identify the experimental subjects, the explanatory variable and the values it can take, and the response variable.
(b) Use a diagram to outline the design of the Physicians' Health Study. (When you outline the design of an experiment, be sure to indicate the size of the treatment groups and the response variable. The diagrams in Figures 3.2 and 3.3 are models.)

3.7 Learning on the Web The discussion following Example 3.1 notes that the Nova Southeastern University study doesn't tell us much about Web versus classroom learning because the students who chose the Web version were much better prepared. Describe the design of an experiment to get better information.

3.8 Healthy turkeys Turkeys raised commercially for food are often fed the antibiotic salinomycin to prevent infections from spreading among the birds. Salinomycin can damage the birds' internal organs, especially the pancreas. A researcher believes that adding vitamin E to the diet may prevent injury. He wants to explore the effects of three levels of vitamin E added to the diet of turkeys along with the usual dose of salinomycin. There are 30 turkeys available for the study. At the end of the study, the birds will be killed and each pancreas examined under a microscope.

(a) Give a careful outline of the design of a randomized comparative experiment in this setting.

(b) Use Table A, beginning at line 125, to carry out the random assignment required by your design.

3.9 Internet phone calls It is possible to use a computer to make telephone calls over the Internet. How will the cost affect the behavior of users of this service? You will offer the service to all 200 rooms in a college dormitory. Some rooms will pay a flat rate. Others will pay higher rates at peak periods and very low rates off-peak. You are interested in the amount and time of use and in the effect on the congestion of the network.

(a) Outline the design of an experiment to study the effect of rate structure.

(b) Use Table A, starting at line 125, to assign the first 5 rooms to the flat-rate group.

3.10 Randomization at work To demonstrate how randomization reduces confounding, consider the following situation. A nutrition experimenter intends to compare the weight gain of newly weaned male rats fed Diet A with that of rats fed Diet B. To do this, she will feed each diet to 10 rats. She has available 10 rats of genetic strain 1 and 10 of strain 2. Strain 1 is more vigorous, so if the 10 rats of strain 1 were fed Diet A, the effects of strain and diet would be confounded, and the experiment would be biased in favor of Diet A.

(a) Label the rats 00, 01, . . . , 19. Use Table A to assign 10 rats to Diet A. Do this four times, using different parts of the table, and write down the four groups assigned to Diet A.

(b) Unknown to the experimenter, the rats labeled 00, 02, 04, 06, 08, 10, 12, 14, 16, and 18 are the 10 strain 1 rats. How many of these rats were in each of the four Diet A groups that you generated? What was the average number of strain 1 rats assigned to Diet A?

Statistical significance

The presence of chance variation requires us to look more closely at the logic of randomized comparative experiments. We cannot say that *any* difference in the average number of pain episodes between the hydroxyurea and control groups must be due to the effect of the drug. There will be some differences even if both treatments are the same because there will always be some differences among the individuals who are our subjects. Even though randomization eliminates systematic differences between the groups, there will still be chance differences. We should insist on a difference in the responses so large that it is unlikely to happen just because of chance variation.

Statistical significance

An observed effect so large that it would rarely occur by chance is called **statistically significant.**

[handwritten margin note: Randomization helps to reduce the effect of confounding variables.]

The difference between the average number of pain episodes for subjects in the hydroxyurea group and the average for the control group was "highly statistically significant." That means that a difference this large would almost never happen just by chance. We do indeed have strong evidence that hydroxyurea beats a placebo in helping sickle cell disease sufferers. You will often see the phrase "statistically significant" in reports of investigations in many fields of study. It tells you that the investigators found good evidence for the effect they were seeking.

Of course, the actual results of an experiment are more important than the seal of approval given by statistical significance. The treatment group in the sickle cell experiment had an average of 2.5 pain episodes per year, against 4.5 per year in the control group. That's a big enough difference to be important to people with the disease. A difference of 2.5 versus 2.8 would be much less interesting even if it were statistically significant.

How to live with observational studies

Does regular church attendance lengthen people's lives? Do doctors discriminate against women in treating heart disease? Does talking on a cell phone while driving increase the risk of having an accident? These are cause-and-effect questions, so we reach for our favorite tool, the randomized comparative experiment. Sorry. We can't randomly assign people to attend church or not, because going to religious services is an expression of beliefs or their absence. We can't use random digits to assign heart disease patients to be men or women. We are reluctant to require drivers to use cell phones in traffic, because talking while driving may be risky.

The best data we have about these and many other cause-and-effect questions come from observational studies. We know that observation is a weak second best to experiment, but good observational studies are far from worthless. What makes a good observational study?

First, good studies are **comparative** even when they are not experiments. We compare random samples of people who do and who don't attend religious services regularly. We compare how doctors treat men and women patients. We might compare drivers talking on cell phones with the same drivers when they are not on the phone. We can often combine comparison with **matching** in creating a control group.

EXAMPLE 3.5   Painkillers and pregnancy

To see the effects of taking a painkiller during pregnancy, we compare women who did so with women who did not. From a large pool of women who did not take the drug, we select individuals who match the drug group in age, education, number of children, and other lurking variables. We now have two groups that are similar in all these ways, so that these lurking variables should not affect our comparison of the groups.

Comparison does not eliminate confounding. People who attend church or synagogue or mosque take better care of themselves than nonattenders. They are less likely to smoke, more likely to exercise, and less likely to be overweight. Matching can reduce some but not all of these differences. A direct comparison of the ages at death of attenders and nonattenders would confound any effect of religion with the effects of healthy living. So a good comparative study **measures and adjusts for confounding variables.** If we measure weight, smoking, and exercise, there are statistical techniques that reduce the effects of these variables on length of life so that (we hope) only the effect of religion itself remains.

EXAMPLE 3.6   Living longer through religion

One of the better studies of the effect of regular attendance at religious services gathered data from a random sample of 3617 adults. Random sampling is a good start. The researchers then measured lots of variables, not just the explanatory variable (religious activities) and the response variable (length of life). A news article said:

Churchgoers were more likely to be nonsmokers, physically active, and at their right weight. But even after health behaviors were taken into account, those not attending religious services regularly still were about 25% more likely to have died.[7]

That "taken into account" means that the final results were adjusted for differences between the two groups. Adjustment reduced the advantage of religion but still left a large benefit.

"Statistics say that religious people live longer, so I practice a different religion every day of the week to be sure I'm covered."

EXAMPLE 3.7   Sex bias in treating heart disease?

Doctors are less likely to give aggressive treatment to women with symptoms of heart disease than to men with similar symptoms. Is this because doctors are sexist? Not necessarily. Women tend to develop heart problems much later than men, so that female heart patients are older and often have other health problems. That might explain why doctors proceed more cautiously in treating them.

This is a case for a comparative study with statistical adjustments for the effects of confounding variables. There have been several such studies, and they produce conflicting results. Some show, in the words of one doctor, "When men and women are otherwise the same and the only difference is gender, you find that treatments are very similar."[8] Other studies find that women are undertreated even after adjusting for differences between the female and male subjects.

As Example 3.7 suggests, statistical adjustment is tricky. Randomization creates groups that are similar in *all* variables known and unknown. Matching and adjustment, on the other hand, can't work with variables the study didn't think to measure. Even if you believe that the researchers thought of everything, you should be a bit skeptical about statistical adjustment. There's lots of room for cheating in deciding which variables to adjust for. And the "adjusted" conclusion is really something like this:

> *If female heart disease patients were younger and healthier than they really are, and if male patients were older and less healthy than they really are, then the two groups would get the same medical care.*

This may be the best we can get, and we should thank statistics for making such wisdom possible. But we end up longing for the clarity of a good experiment.

EXERCISES

3.11 Statistical significance, I A randomized comparative experiment examines whether a calcium supplement in the diet reduces the blood pressure of healthy men. The subjects receive either a calcium supplement or a placebo for 12 weeks. The researchers conclude that "the blood pressure of the calcium group was significantly lower than that of the placebo group." "Significant" in this conclusion means statistically significant. Explain what statistically significant means in the context of this experiment, as if you were speaking to a doctor who knows no statistics.

3.12 Statistical significance, II The financial aid office of a university asks a sample of students about their employment and earnings. The report says that "for academic year earnings, a significant difference was found between the sexes, with men earning more on the average. No significant difference was found between the earnings of black and white students." Explain the meaning of "a significant difference" and "no significant difference" in plain language.

3.13 Prayer and meditation You read in a magazine that "nonphysical treatments such as meditation and prayer have been shown to be effective in controlled scientific studies for such ailments as high blood pressure, insomnia, ulcers, and asthma." Explain in simple language what the article means by "controlled scientific studies" and why such studies might show that meditation and prayer are effective treatments for some medical problems.

3.14 TV harms children Observational studies suggest that children who watch many hours of television get lower grades in school and are more likely to commit crimes than those who watch less TV. Explain clearly why these studies do not show that watching TV causes these harmful effects. In particular, suggest some lurking variables that may be confounded with heavy TV viewing.

APPLICATION 3.1 Good for Your Heart

Heart Treatment's Value Doubted

By Michael Specter
Washington Post Staff Writer

In a study with far-reaching implications for the routine treatment of heart attack patients, researchers have found that the immediate use of special balloons to force open clogged arteries is unnecessary in the vast majority of cases if the patient is treated with a clot-dissolving drug.

Most heart specialists have assumed that the balloon treatment, an expensive and increasingly popular procedure called balloon angioplasty, should routinely follow the use of drugs, such as TPA, that dissolve the blood clots that cause most heart attacks.

But in a study of 3,262 heart attack patients that is expected to transform the standards for treatment of the nation's leading killer, researchers at medical centers across the country found the extra measures were rarely needed. Half of the randomly selected patients were treated solely with the clot-dissolving drug TPA, while the other half were given TPA followed by angioplasty.

The results, reported in today's issue of *The New England Journal of Medicine*, showed that for most people adding angioplasty was no better than relying on the less complicated and less costly drug treatment.

APPLICATION 3.1 Good for Your Heart (continued)

Only 10 percent of the group assigned solely to drug treatment died or had another heart attack in the six weeks following treatment. By comparison, 11 percent of the group assigned to receive the combination drug and angioplasty treatment suffered the same fate. The difference could have occurred by chance. But the fact that angioplasty did not prove to be the better option was a shock even to some of the nation's most renowned heart experts.

"There is no question that it bucks the trend," said Eugene Braunwald, professor of medicine at Harvard Medical School and the study chairman for the research project. "It will spare many patients unnecessary surgery and reduce the cost of medical care greatly."

1. Identify the subjects, treatments, and response variable in this study.

2. Describe the design of the experiment in as much detail as you can.

3. Explain how the principles of control, randomization, and using enough subjects were applied in this experiment.

4. What other information would you like to have to evaluate the design of the experiment?

EXPLORING THE WEB

You can find the latest medical research in the *Journal of the American Medical Association* (jama.ama-assn.org) and the *New England Journal of Medicine* (www.nejm.org). Many of the articles describe randomized comparative experiments, and even more use the language of statistical significance.

STATISTICS IN SUMMARY

Statistical studies often try to show that changing one variable (the **explanatory variable**) causes changes in another variable (the **response variable**). In an **experiment,** we actually set the explanatory variables ourselves rather than just observing them. Observational studies and one-track experiments that simply apply a single treatment often fail to produce useful data because **confounding** with **lurking variables** makes it impossible to say what the effect of the treatment was. The remedy is to use a **randomized comparative experiment.** Compare two or more treatments, use chance to decide which subjects get each treatment, and use enough subjects so that the effects of chance variations between the groups are small. Comparing two or more treatments **controls** lurking variables such as the **placebo effect** because these variables act on all the treatment groups.

Differences among the effects of the treatments so large that they would rarely happen just by chance are called **statistically significant.** Statistically significant results from randomized comparative experiments are the best available evidence that changing the explanatory variable really **causes** changes in the response. Observational studies of cause-and-effect questions are more impressive if they **compare matched groups** and measure as many lurking variables as possible to allow **statistical adjustment.** For answering questions about causation, observational studies remain a weak second best to experiments.

SECTION 3.1 EXERCISES

3.15 Does job training work? A state institutes a job-training program for manufacturing workers who lose their jobs. After five years, the state reviews how well the program works. Critics claim that because the state's unemployment rate for manufacturing workers was 6% when the program began and 10% five years later, the program is ineffective.

Explain why higher unemployment does not necessarily mean that the training program failed. In particular, identify some lurking variables whose effect on unemployment may be confounded with the effect of the training program. Draw a picture like Figure 3.1 (p. 136) to illustrate your explanation.

3.16 What a headache! Doctors identify "chronic tension-type headaches" as headaches that occur almost daily for at least six months. Can antidepressant medications or stress management training reduce the number and severity of these headaches? Are both together more effective than either alone? Investigators compared four treatments: antidepressant alone, placebo alone, antidepressant plus stress management, and placebo plus stress management.
(a) Outline the design of the experiment.
(b) The headache sufferers named below have agreed to participate in the study. Use Table A at line 130 to randomly assign the subjects to the treatments.

00 Acosta	06 Duncan	12 Han	18 Liang	24 Padilla	30 Valasco
01 Asihiro	07 Durr	13 Howard	19 Maldonado	25 Plochman	31 Vaughn
02 Bennett	08 Edwards	14 Hruska	20 Marsden	26 Rosen	32 Wei
03 Bikalis	09 Farouk	15 Imrani	21 Montoya	27 Solomon	33 Wilder
04 Chen	10 Fleming	16 James	22 O'Brien	28 Trujillo	34 Willis
05 Clemente	11 George	17 Kaplan	23 Ogle	29 Tullock	35 Zhang

3.17 Learning about markets, I An economics professor wonders if playing market games online will help students understand how markets set prices. You suggest an experiment: have some students use the online games, while others talk about markets in discussion sessions. The course has two sections, at 8:30 A.M. and 2:30 P.M. Each section has already been divided into 10 discussion groups. For practical reasons, all students in each discussion group must follow the same program. The professor says, "Let's just have the 8:30 groups do online work and the 2:30 groups do discussion." Why is this a bad idea?

3.18 Learning about markets, II
(a) Outline a better design than that of Exercise 3.17 for an experiment to compare the two methods of learning about economic markets. Treat the 20 discussion groups as individuals. What do you suggest as a response variable? (When you outline the design of an experiment, be sure to indicate the size of the treatment groups and the response variable. The diagrams in Figures 3.2 (page 140) and 3.3 (page 141) are models.)
(b) Use Table A, starting at line 116, to do the randomization your design requires.

3.19 A little alcohol is good for you The Nurses' Health Study has queried a sample of over 100,000 female registered nurses every two years since 1976. Beginning in 1980, the study asked questions about diet, including alcohol consumption. After looking at all deaths among these nurses through May 1992, the researchers concluded: "As compared with non-drinkers and heavy drinkers, light-to-moderate drinkers had a significantly lower risk of death."[9] The word "significantly" in this conclusion has a technical meaning. Explain to someone who knows no statistics what "significant" means in a statistical study.

3.20 Clumsy men? A doctor at a veterans hospital examines all of the patient records from 1985 to 1993 and finds that twice as many men as women fell out of their hospital beds during their stay. This is put forward as evidence that men are clumsier than women.
(a) Is this an experiment or an observational study? Why?
(b) Give an example of a possible confounding factor and a clear explanation of why it is a possible confounding factor.

3.21 Design an experiment A university's statistics department wants to attract more majors. It prepares two advertising brochures. Brochure A stresses the intellectual excitement of statistics. Brochure B stresses how much more money statisticians make. Which will be more attractive to first-year students? You have a questionnaire to measure interest in majoring in statistics, and you have 50 first-year students to work with. Outline the design of an experiment to decide which brochure works better.

3.2 EXPERIMENTS IN THE REAL WORLD

The case of the fickle mice

The randomized comparative experiment may be the most important idea in statistics. Experiments are the gold standard for evidence that changing one variable really causes changes in another variable. In the real world, however, experiments don't always go smoothly. Even if they do go smoothly, we can't always take a firm stand on the findings. Consider the case of the fickle mice.

That our behavior is (in part) coded into our genes is one of the big scientific issues of the day. No human experiments are allowed, so mice serve in our place. Researchers "knock out" a gene in one group of mice and compare their behavior with a control group of normal mice. All the mice have the same genetic makeup before treatment and are assigned at random to the two groups. This is an iron-clad randomized comparative experiment: if the behaviors differ, the gene that was knocked out must influence the behavior that changes. Alas, as a writer in the journal *Science* says, "No sooner has one group of researchers tied a gene to a behavior when along comes the next study, proving that the link is spurious or even that the gene in question has exactly the opposite effect."[10]

This is frustrating for the scientists, and also for news reporters who want to say "science shows" that this or that behavior is all in our genes. What goes wrong?

Some frustrated scientists tried to find out. They did the same experiments with the same genetic strain of mice in three different labs. One researcher said they "went nuts" trying to make the conditions exactly alike in Oregon, Alberta (Canada), and New York. The results were often very different. It appears that very small differences in the lab environments have big effects on the behavior of the mice. Remember this the next time you read that our genes control our behavior.

ACTIVITY 3.2A Mozart and mazes

Materials: Two mazes per student (provided by your teacher), classical music CD or cassette, contemporary music CD or cassette, CD/cassette player, stopwatch, and coin for each pair of students

In this Activity, you will take part in an experiment to determine whether listening to either classical music or contemporary music helps students complete a maze task more quickly. You will work with a partner during the data collection.

1. Your teacher will distribute two copies of Maze 1 face-down to each pair of students. With your partner, decide who will work the maze first and who will time.

ACTIVITY 3.2A Mozart and mazes *(continued)*

2. When the timer says "go," the subject should turn the paper over and connect the circles in the maze, beginning at 1 (the start) and finishing at 25 (the end). The subject should say "done" when finished. Record the time it took for the subject to complete the maze to the nearest tenth of a second. NOTE: The timer should *not* look at the maze!

3. Switch roles, and repeat Step 2.

4. Next, the members of each pair must be assigned at random to the two treatments: classical music or contemporary music. Determine this with a coin flip.

5. Your teacher will play a short selection of classical music. When the music stops, the chosen member of each pair will work Maze 2, using the procedure described in Step 2.

6. Your teacher will play a short selection of contemporary music. When the music stops, the remaining member of each pair will work Maze 2.

7. With your partner, compare your times on Maze 1 and Maze 2. Did the music appear to help? If so, did the classical music or contemporary music produce more improvement?

8. Pool results with your classmates. In one column, list the difference in times (Maze 2 – Maze 1) for each student who received the classical music treatment. In a second column, list the differences for each student who listened to contemporary music. Construct a graph to compare the results.

9. Compute the average difference for each of the two lists. Do you have reason to believe that classical music helped more than contemporary music? Explain.

In 1993, three researchers conducted an experiment to determine whether listening to Mozart's music would improve performance on spatial-reasoning tasks. Using 36 college students as subjects, Frances Rauscher and her colleagues randomly assigned 12 students to each of three treatment groups. Group 1 listened to a 10-minute selection from Mozart's Sonata for Two Pianos in D Major. Group 2 was played a relaxation tape with mixed sounds for 10 minutes. Group 3 sat in silence for 10 minutes and served as a control group. Each subject completed a pretest two days before the treatment was given and a posttest immediately following the treatment. The results of the experiment were surprising: students who listened to Mozart showed significant gains in their scores on spatial-reasoning tasks.[11]

Rauscher, Shaw, and Ky published their findings in *Nature*, and the so-called Mozart Effect was born. Parents looking for ways to improve their child's IQ scores started buying Mozart tapes. One state provided Mozart cassettes and CDs to parents of newborn babies. Another passed legislation requiring that preschools play 30 minutes of classical music a day. Meanwhile, other researchers attempted to verify the Mozart Effect in experiments of their own. A few obtained similar results. Many, however, found no evidence of a Mozart Effect. The controversy continues, and Mozart plays on.

Equal treatment for all

Probability samples are a big idea, but sampling in practice has difficulties that just using random samples doesn't solve. Randomized comparative experiments are also a big idea, but they don't solve all the difficulties of experimenting. A sampler must know exactly what information she wants and must compose questions that extract that information from her sample. An experimenter must know exactly what treatments and responses he wants information about, and he must construct the apparatus needed to apply the treatments and measure the responses. This is what psychologists or medical researchers or engineers mean when they talk about "designing an experiment." We are concerned with the *statistical* side of designing experiments, ideas that apply to experiments in psychology, medicine, engineering, and other areas as well. Even at this general level, you should understand the practical problems that can prevent an experiment from producing useful data.

The logic of a randomized comparative experiment assumes that all the subjects are treated alike except for the treatments that the experiment is designed to compare. Any other unequal treatment can cause bias. But treating subjects exactly alike is hard to do.

© Sidney Harris

EXAMPLE 3.8   Mice, rats, and rabbits

Mice, rats, and rabbits that are specially bred to be uniform in their inherited characteristics are the subjects in many experiments. As the case of the fickle mice illustrates, animals, like people, can be quite sensitive to how they are treated. Here are two amusing examples of how unequal treatment can create bias.

Does a new breakfast cereal provide good nutrition? To find out, compare the weight gains of young rats fed the new product and rats fed a standard diet. The rats are randomly assigned to diets and are housed in large racks of cages. It turns out that rats in upper cages grow a bit faster than rats in bottom cages. If the experimenters put rats fed the new product at the top and those fed the standard diet below, the experiment is biased in favor of the new product.[12] Solution: Assign the rats to cages at random.

Another study looked at the effects of human affection on the cholesterol level of rabbits. All of the rabbit subjects ate the same diet. Some (chosen at random) were regularly removed from their cages to have their furry heads scratched by friendly people. The rabbits who received affection had lower cholesterol. So affection for some but not other rabbits could bias an experiment in which the rabbits' cholesterol level is a response variable.

Double-blind experiments

Placebos work. That bare fact means that medical studies must take special care to show that a new treatment is not just a placebo. Part of equal treatment for all is to be sure that the placebo effect operates on all subjects.

EXAMPLE 3.9   The powerful placebo

Want to help balding men keep their hair? Give them a placebo—one study found that 42% of balding men maintained or increased the amount of hair on their heads when they took a placebo. Another study told 13 people who were very sensitive to poison ivy that the stuff being rubbed on one arm was poison ivy. It was a placebo, but all 13 broke out in a rash. The stuff rubbed on the other arm really was poison ivy, but the subjects were told it was harmless—and only 2 of the 13 developed a rash.[13]

When the ailment is vague and psychological, like depression, some experts think that about three-quarters of the effect of the most widely used drugs is just the placebo effect.[14] Others disagree. The strength of the placebo effect in medical treatments is hard to pin down because it depends on the exact environment, much as the behavior of the fickle mice does. How enthusiastic the doctor is seems to matter a lot. But "placebos work" is a good place to start when you think about planning medical experiments.

The strength of the placebo effect is a strong argument for randomized comparative experiments. In the baldness study, 42% of the placebo group kept or increased their hair, but 86% of the men getting a new drug to fight baldness did so. The drug beats the placebo, so it has something besides the placebo effect going for it. Of course, the placebo effect is still part of the reason this and other treatments work.

Because the placebo effect is so strong, it would be foolish to tell subjects in a medical experiment whether they are receiving a new drug or a placebo. Knowing that they are getting "just a placebo" might weaken the placebo effect and bias the experiment in favor of the other treatments. It is also foolish to tell doctors and other medical personnel what treatment each subject received. If they know that a subject is getting "just a placebo," they may expect less than if they know the subject is receiving a promising experimental drug. Doctors' expectations change how they interact with patients and even the way they diagnose a patient's condition. Whenever possible, experiments with human subjects should be *double-blind*.

Double-blind experiments

In a **double-blind experiment,** neither the subjects nor the people who work with them know which treatment each subject is receiving.

Subjects & observers are Blind

"Dr. Burns, are you sure this is what the statisticians call a double-blind experiment?"

Until the study ends and the results are in, only the study's statistician knows for sure. Reports in medical journals regularly begin with words like these, from a study of a flu vaccine given as a nose spray: "This study was a randomized, double-blind, placebo-controlled trial. Participants were enrolled from 13 sites across the continental United States between mid-September and mid-November 1997."[15] Doctors are supposed to know what this means. Now you also know.

Refusals, nonadherers, and dropouts

Sample surveys suffer from nonresponse due to failure to contact some people selected for the sample and the refusal of others to participate. Experiments with human subjects suffer from similar problems.

EXAMPLE 3.10   Minorities in clinical trials

Refusal to participate is a serious problem for medical experiments on treatments for serious diseases such as cancer. As with samples, bias can result if those who refuse are systematically different from those who cooperate.

Minorities, women, the poor, and the elderly have long been underrepresented in clinical trials. In many cases, they weren't asked. The law now requires representation of women and minorities, and data show that most clinical trials now have fair representation. But refusals remain a problem. Minorities, especially blacks, are more likely to refuse to participate. The government's Office of Minority Health says, "Though recent studies have shown that African Americans have increasingly positive attitudes toward cancer medical research, several studies corroborate that they are still cynical about clinical trials. A major impediment for lack of participation is a lack of trust in the medical establishment."[16] Some remedies for lack of trust are complete and clear information about the experiment, insurance coverage for experimental treatments, participation of black researchers, and cooperation with doctors and health organizations in black communities.

Subjects who participate but don't follow the experimental treatment, called **nonadherers,** can also cause bias. AIDS patients who participate in trials of a new drug sometimes take other treatments on their own, for example. What is more, some AIDS subjects have their medication tested privately and drop out or add other medications if they were not assigned to the new drug. This may bias the trial against the new drug.

Experiments that continue over an extended period of time also suffer **dropouts,** subjects who begin the experiment but do not complete it. If the reasons for dropping out are unrelated to the experimental treatments, no harm is done other than reducing the number of subjects. If subjects drop out because of their reaction to one of the treatments, bias can result.

EXAMPLE 3.11   Dropouts in a medical study

Orlistat is a new drug that may help reduce obesity by preventing absorption of fat from the foods we eat. As usual, the drug was compared with a placebo in a double-blind randomized trial. Here's what happened.

Start with 1187 obese subjects. First give a placebo for four weeks and drop the subjects who won't take a pill regularly. This attacks the problem of nonadherers. There are 892 subjects left. Randomly assign these subjects to Orlistat or a placebo, along with a weight-loss diet. After a year devoted to losing weight, 576 subjects are still participating. On the average, the Orlistat group lost 3.15 kilograms (about 7 pounds) more than the placebo group. Keep going for another year, now emphasizing maintaining the weight loss from the first year. At the end of the second year, 403 subjects are left. That's only 45% of the 892 who were randomized. Orlistat again beat the placebo, reducing the weight regained by an average of 2.25 kilograms (about 5 pounds).

Can we trust the results when so many subjects dropped out? The overall dropout rates were similar in the two groups: 54% of the subjects taking Orlistat and 57% of those in the placebo group dropped out. Were dropouts related to the treatments? Placebo subjects in weight-loss experiments often drop out because they aren't losing weight. This would bias the study against Orlistat because the subjects in the placebo group at the end may be those who could lose weight just by following a diet. The researchers looked carefully at the data available for subjects who dropped out. Dropouts from both groups had lost less weight than those who stayed, but careful statistical study suggested that there was little bias. Perhaps so, but the results aren't as clean as our first look at experiments promised.[17]

EXERCISES

3.22 Activity 3.2A follow-up Refer to the Mozart Effect experiment from Activity 3.2A (pages 152–153).
(a) Was the experiment double-blind? Why is this important?
(b) Did the experimental design take the placebo effect into account? Why is this important?
(c) Why was the coin flip important in this experiment?

3.23 Medical news When it was found that hydroxyurea reduced the symptoms of sickle cell anemia, the National Institutes of Health released a medical bulletin. The bulletin said, "These findings are the results of data analyzed from the

Multicenter Study of Hydroxyurea in Sickle Cell Anemia (MSH), which was a double-blind, placebo-controlled trial in which half of the patients received hydroxyurea and half received a placebo capsule." Explain to someone who knows no statistics what the terms "placebo-controlled" and "double-blind" mean here.

3.24 Emergency room care An article in a medical journal reports an experiment to see if injecting an oxygen-carrying fluid in addition to performing standard emergency room procedures would help patients in shock from loss of blood. The article describes the experiment as a "randomized, controlled, single-blinded efficacy trial conducted between February 1997 and January 1998 at 18 US trauma centers."[18] What do you think "single-blinded" means here? Why isn't a double-blind experiment possible? *The observers knowing which treatment each suspect 1st can cause bias.*

3.25 Testing a natural remedy Although the law doesn't require it, we decide to subject Dr. Moore's Old Indiana Extract to a clinical trial. We hope to show that the extract reduces pain from arthritis. Sixty patients suffering from arthritis and needing pain relief are available. We will give a pill to each patient and ask them an hour later, "About what percent of pain relief did you experience?"

(a) Why should we not simply give the extract to all 60 patients and record the responses?

(b) Outline the design of an experiment to compare the extract's effectiveness with that of aspirin and of a placebo.

(c) Should patients be told which remedy they are receiving? How might this knowledge affect their reactions?

(d) If patients are not told which treatment they are receiving, the experiment is single-blind. Should this experiment be double-blind? Explain.

The League to Mess Up Experiments meets...

"Agent B, you will scratch the heads of the lab rabbits. Agent Q, you will join a clinical trial and not take your pills. Agent K, you will sign up for an experiment, then drop out just before the end."

Can we generalize?

A well-designed experiment tells us that changes in the explanatory variable cause changes in the response variable. More exactly, it tells us that this happened for specific subjects in the specific environment of this specific experiment. No doubt we had grander things in mind. We want to proclaim that our new method of teaching math does better for high school students in general or that our new drug beats a placebo for some broad class of patients. Can we generalize our conclusions from our little group of subjects to a wider population?

The first step is to be sure that our findings are *statistically significant,* that they are too strong to often occur just by chance. That's important, but it's a technical detail that the study's statistician can reassure us about. The serious threat is that the treatments, the subjects, or the environment of our experiment may not be realistic. Let's look at some examples.

EXAMPLE 3.12   Studying frustration

A psychologist wants to study the effects of failure and frustration on the relationships among members of a work team. She forms a team of students, brings them to the psychology laboratory, and has them play a game that requires teamwork. The game is rigged so that they lose regularly. The psychologist observes the students through a one-way window and notes the changes in their behavior during an evening of game playing.

Playing a game in a laboratory for small stakes, knowing that the session will soon be over, is a long way from working for months developing a new product that never works right and is finally abandoned by your company. Does the behavior of the students in the lab tell us much about the behavior of the team whose product failed?

In Example 3.12, the subjects (students who know they are subjects in an experiment), the treatment (a rigged game), and the environment (the psychology lab) are all unrealistic if the psychologist's goal is to reach conclusions about the effects of frustration on teamwork in the workplace. Psychologists do their best to devise realistic experiments for studying human behavior, but lack of realism limits the usefulness of experiments in this area.

EXAMPLE 3.13   Center brake lights

Cars sold in the United States since 1986 have been required to have high center brake lights in addition to the usual two brake lights at the rear of the vehicle. This safety requirement was justified by randomized comparative experiments with fleets of rental and business cars. The experiments showed that the third brake light reduced rear-end collisions by as much as 50%.

After almost a decade of actual use of center brake lights, the Insurance Institute found only a 5% reduction in rear-end collisions, helpful but much less than the experiments predicted. What happened? Most cars did not have the extra brake light when the experiments were carried out, so it caught the eye of following drivers. Now that almost all cars have the third light, it no longer captures attention. The experimental conclusions did not generalize as well as safety experts hoped because the environment changed.

EXAMPLE 3.14   Are subjects treated too well?

Surely medical experiments are realistic? After all, the subjects are real patients in real hospitals really being treated for real illnesses.

Even here, there are some questions. Patients participating in medical trials get better medical care than most other patients, even if they are in the placebo group. Their doctors are specialists doing research on their specific ailment. They are watched more carefully than other patients. They are more likely to take their pills regularly because they are constantly reminded to do so. Providing "equal treatment for all" except for the experimental and control therapies translates into "provide the best possible medical care for all." The result: Ordinary patients may not do as well as the clinical trial subjects when the new therapy comes into general use. It's likely that a therapy that beats a placebo in a clinical trial will beat it in ordinary medical care, but "cure rates" or other measures of success from the trial may be optimistic.

Meta-analysis

A single study of an important issue is rarely decisive. We often find several studies in different settings, with different designs, and of different quality. Can we combine their results to get an overall conclusion? That is the idea of "meta-analysis." Of course, differences among the studies prevent us from just lumping them together. Statisticians have more sophisticated ways of combining the results. Meta-analysis has been applied to issues from the effect of secondhand smoke to whether coaching improves SAT scores.

The Carolina Abecedarian Project (Section 3.1) faces the same "too good to be realistic" question. That long and expensive experiment does show that intensive day care has substantial benefits in later life. The day care in the study was intensive indeed—lots of highly qualified staff, lots of parent participation, and detailed activities starting at a very young age, all costing about $11,000 per year for each child. It's unlikely that society will decide to offer such care to all low-income children. The unanswered question is a big one: How good must day care be to really help children succeed in life?

When experiments are not fully realistic, statistical analysis of the experimental data cannot tell us how far the results will generalize. Experimenters generalizing from students in a lab to workers in the real world must argue based on their understanding of how people function, not based just on the data. It is even harder to generalize from rats in a lab to people in the real world. This is one reason why a single experiment is rarely completely convincing, despite the compelling logic of experimental design. The true scope of a new finding must usually be explored by a number of experiments in various settings.

A convincing case that an experiment is sufficiently realistic to produce useful information is based not on statistics but on the experimenter's knowledge of the subject matter of the experiment. The attention to detail required to avoid hidden bias also rests on subject-matter knowledge. Good experiments combine statistical principles with understanding of a specific field of study.

Experimental design in the real world

The experimental designs we have met all have the same pattern: divide the subjects at random into as many groups as there are treatments, then apply each treatment to one of the groups. These are *completely randomized* designs.

Completely randomized design

In a **completely randomized** experimental design, all the experimental subjects are allocated at random among all the treatments.

What is more, our examples to this point have had only a single explanatory variable (drug versus placebo, classroom versus Web instruction). A completely randomized design can have any number of explanatory variables. Here is an example with two.

EXAMPLE 3.15   Durable fabric

A fabrics researcher is studying the durability of a fabric under repeated washings. Because the durability may depend on the water temperature and the type of cleansing agent used, the researcher decides to investigate the effect of these two explanatory variables on durability. Variable A is water temperature and has three levels: hot (145 °F), warm (100 °F), and cold (50 °F). Variable B is the cleansing agent and also has three levels: regular Tide, low-phosphate Tide, and Ivory Liquid. A treatment consists of washing a piece of fabric (a unit) 50 times in a home automatic washer with a specific combination of water temperature and cleansing agent. The response variable is strength after 50 washes, measured by a fabric-testing machine that forces a steel ball through the fabric and records the fabric's resistance to breaking. Figure 3.4 on the facing page shows the layout of the treatments.

| | Variable B Cleansing Agent | | |
	Regular Tide	Low-phosphate Tide	Ivory Liquid
Hot (145 °F)	Treatment 1	Treatment 2	Treatment 3
Variable A Temperature — Warm (100 °F)	Treatment 4	Treatment 5	Treatment 6
Cold (50 °F)	Treatment 7	Treatment 8	Treatment 9

FIGURE 3.4 The treatments in the experiment of Example 3.15. Combinations of two explanatory variables form nine treatments.

In this example there are nine possible treatments (combinations of a temperature and a cleansing agent). By using them all, the researcher obtains a wealth of information on how temperature alone, cleansing agent alone, and the two in combination affect the durability of the fabric. For example, water temperature may have no effect on the strength of the fabric when regular Tide is used, but after 50 washings in low-phosphate Tide, the fabric may be weaker when cold water is used instead of hot water. This kind of combination effect is called an interaction between cleansing agent and water temperature. Interactions can be important, as when a drug that ordinarily has no unpleasant side effects interacts with alcohol to knock out the patient who drinks a martini. Because an experiment can combine levels of several factors, interactions between the factors can be observed.

EXERCISES

3.26 Daytime running lights Canada requires that cars be equipped with "daytime running lights," headlights that automatically come on at a low level when the car is started. Some manufacturers are now equipping cars sold in the United States with running lights. Will running lights reduce accidents by making cars more visible?

(a) Briefly discuss the design of an experiment to help answer this question. In particular, what response variables will you examine?

(b) Example 3.13 discusses center brake lights. What cautions do you draw from that example that apply to an experiment on the effects of running lights?

3.27 Diet and cancer Substances that cause cancer should not appear in our food. We don't want to experiment on people to learn what substances cause cancer, so we experiment on rats instead. The rats are specially bred to have more tumors than humans do. They are fed large doses of the test chemical for most of their natural lives, about two years. Briefly discuss the questions that arise in using these experiments to decide what is safe in human diets.

3.28 Dealing with cholesterol Clinical trials have shown that reducing blood cholesterol using either drugs or diet reduces heart attacks. The first researchers followed their subjects for five to seven years. In order to see results as quickly as possible, the subjects were chosen from the group at greatest risk, middle-aged men with high cholesterol levels or existing heart disease. The experiments generally showed that reducing blood cholesterol does decrease the risk of a heart attack. Some doctors questioned whether these experimental results applied to many of their patients. Why?

3.29 Testing a natural remedy The National Institutes of Health is at last sponsoring proper clinical trials of some natural remedies. In one study at Duke University, 330 patients with mild depression are enrolled in a trial to compare Saint-John's-wort with a placebo and with Zoloft, a common prescription drug for depression. The Beck Depression Inventory is a common instrument that rates the severity of depression on a 0 to 3 scale.
(a) What would you use as the response variable to measure change in depression after treatment?
(b) Outline the design of a completely randomized clinical trial for this study.
(c) What other precautions would you take in this trial?

3.30 Baking cakes A food company is preparing to market a new cake mix. It is important that the taste of the cake not be changed by small variations in baking time or temperature. In an experiment, cakes made from the mix are baked at 300°, 320°, and 340° F, and for 1 hour and for 1 hour and 15 minutes. Ten cakes are baked at each combination of temperature and time. A panel of tasters scores each cake for texture and taste.
(a) What are the explanatory variables and the response variables for this experiment?
(b) Make a diagram like Figure 3.4 to describe the treatments. How many treatments are there? How many cakes are needed?
(c) Explain why it is a bad idea to bake all 10 cakes for one treatment at once, then bake the 10 cakes for the second treatment, and so on. Instead, the experimenters will bake the cakes in a random order determined by the randomization in your design.

Matched pairs and block designs

Completely randomized designs are the simplest statistical designs for experiments. They illustrate clearly the principles of control and randomization. However, completely randomized designs are often inferior to more elaborate statistical designs. In particular, matching the subjects in various ways can produce more precise results than simple randomization.

One common design that combines matching with randomization is the **matched pairs design.** A matched pairs design compares just two treatments. Choose pairs of subjects that are as closely matched as possible. Assign one of the treatments to each subject in a pair by tossing a coin or reading odd and even digits from Table A.

Sometimes each "pair" in a matched pairs design consists of just one subject, who gets both treatments one after the other. Each subject serves as his or her own control. The order of the treatments can influence the subject's response, so we randomize the order for each subject, again by a coin toss.

ACTIVITY 3.2B Squeeze strength

Materials: Bathroom scale, coin for each student

In this Activity, you will attempt to determine whether the right hand of right-handed people is generally stronger than the left.

1. Flip a coin to determine the order in which you will perform this experiment: "heads" means you will squeeze the scale with your left hand first; "tails" means use your right hand first.

2. Your teacher will demonstrate how to squeeze the scale. Pay close attention! It is important that all of the data are collected in the same way.

3. When it's your turn, squeeze the scale first with the hand determined in Step 1. An observer should verify the scale reading before you let go. Now squeeze the scale with your other hand, again verifying the reading. Record your left- and right-hand squeeze strengths.

4. Pool data with your classmates. Your teacher will make a large chart on the board, like this:

Subject	H or T	Left-hand reading	Right-hand reading	Difference (R – L)

Enter your data in the appropriate row.

5. Make a graph to display the difference in squeeze strength for the students in your class. Describe what you see.

6. Compute the average difference in squeeze strength. Is there good evidence that the right hand is stronger than the left? Explain.

EXAMPLE 3.16   Coke versus Pepsi

Pepsi wanted to demonstrate that Coke drinkers prefer Pepsi when they taste both colas blind. The subjects, all people who said they were Coke drinkers, tasted both colas from glasses without brand markings and said which they liked better. This is a matched pairs design in which each subject compares the two colas. Because responses may depend on which cola is tasted first, the order of tasting should be chosen at random for each subject.

When more than half the Coke drinkers chose Pepsi, Coke claimed that the experiment was biased. The Pepsi glasses were marked M and Coke glasses were marked Q. Aha, said Coke, the results could just mean that people like the letter M better than the letter Q. The matched pairs design is adequate, but a more careful experiment would avoid any distinction other than Coke versus Pepsi.[19]

Matched pairs designs use the principles of comparison of treatments and randomization. However, the randomization is not complete—we do not randomly assign all the subjects at once to the two treatments. Instead, we randomize only within each matched pair. Matching reduces the effect of variation among the subjects. Matched pairs are an example of *block designs*.

Block design

A **block** is a group of experimental subjects that are known before the experiment to be similar in some way that is expected to affect the response to the treatments. In a **block design,** the random assignment of subjects to treatments is carried out separately within each block.

A block design combines the idea of creating equivalent treatment groups by matching with the principle of forming treatment groups at random. Blocks are another form of *control*. They control the effects of some outside variables by bringing those variables into the experiment to form the blocks. Here is a typical example of a block design.

EXAMPLE 3.17   Men, women, and advertising

Women and men respond differently to advertising. An experiment to compare the effectiveness of three television commercials for the same product will want to look separately at the reactions of men and women, as well as assess the overall response to the ads.

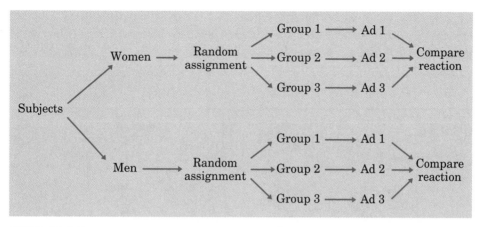

FIGURE 3.5 A block design to compare the effectiveness of three TV advertisements. Female and male subjects form two blocks.

A completely randomized design considers all subjects, both men and women, as a single pool. The randomization assigns subjects to three treatment groups without regard to their sex. This ignores the differences between men and women. A better design considers women and men separately. Randomly assign the women to three groups, one to view each commercial. Then separately assign the men at random to three groups. Figure 3.5 outlines this improved design.

A block is a group of subjects formed before an experiment starts. We reserve the word "treatment" for a condition that we impose on the subjects. We don't speak of 6 treatments in Example 3.17 even though we can compare the responses of 6 groups of subjects formed by the 2 blocks (men, women) and the 3 commercials. Block designs are similar to stratified samples. Blocks and strata both group similar individuals together. We use two different names only because the idea developed separately for sampling and experiments. The advantages of block designs are the same as the advantages of stratified samples. Blocks allow us to draw separate conclusions about each block—for example, about men and women in the advertising study in Example 3.17. Blocking also allows more precise overall conclusions, because the systematic differences between men and women can be removed when we study the overall effects of the three commercials. The idea of blocking is an important additional principle of statistical design of experiments. A wise experimenter will form blocks based on the most important unavoidable sources of variability among the experimental subjects. Randomization will then average out the effects of the remaining variation and allow an unbiased comparison of the treatments.

Like the design of samples, the design of complex experiments is a job for experts. Now that we have seen a bit of what is involved, we will usually just act as if most experiments were completely randomized.

By permission of John L. Hart FLP and Creators Syndicate, Inc.

EXERCISES

3.31 Activity 3.2B follow-up Refer to the squeeze strength experiment from Activity 3.2B (page 165).
(a) Why did we randomize the order in which each student squeezed the scale?
(b) What advantage did the matched pairs design we used have over a completely randomized design?
(c) Examine the class data from a different angle. Do you see any evidence of a "learning effect" from the first squeeze of the scale to the second? Justify your answer.

3.32 Comparing cancer treatments The progress of a type of cancer differs in women and men. A clinical experiment to compare three therapies for this cancer therefore treats sex as a blocking variable.
(a) You have 500 male and 300 female patients who are willing to serve as subjects. Use a diagram to outline a block design for this experiment. Figure 3.5 is a model.
(b) What are the advantages of a block design over a completely randomized design using these 800 subjects? What are the advantages of a block design over a completely randomized design using 800 male subjects?

3.33 In the cornfield An agronomist wants to compare the yield of 5 corn varieties. The field in which the experiment will be carried out increases in fertility from north to south. The agronomist therefore divides the field into 30 plots of equal size, arranged in 6 east-west rows of 5 plots each, and employs a block design with the rows of plots as the blocks.

(a) Draw a sketch of the field, divided into 30 plots. Label the rows Block 1 to Block 6.

(b) Do the randomization required by the block design. That is, randomly assign the 5 corn varieties A, B, C, D, and E to the 5 plots in each block. Mark on your sketch which variety is planted in each plot.

3.34 More rats In Exercise 3.10 (page 144), a nutritionist had 10 rats of each of two genetic strains. The effect of genetic strain can be controlled by treating the strains as blocks and randomly assigning 5 rats of each strain to Diet A. The remaining 5 rats of each strain receive Diet B.

(a) Describe the design of this experiment.

(b) Use Table A, beginning at line 111, to do the required randomization.

3.35 Doctors and nurses Nurse practitioners are nurses with advanced qualifications who often act much like primary-care physicians. An experiment assigned 1316 patients who had no regular source of medical care to either a doctor (510 patients) or a nurse practitioner (806 patients). All of the patients had been diagnosed with asthma, diabetes, or high blood pressure before being assigned. The response variables included measures of the patients' health and of their satisfaction with their medical care after 6 months.[20]

(a) Is the diagnosis (asthma, etc.) a treatment variable or a block? Why?

(b) Is the type of care (nurse or doctor) a treatment variable or a block? Why?

APPLICATION 3.2   Design Your Own Study

In this Application, you will design a taste-testing experiment in which subjects will be asked which of two similar items tastes better to them. However, the issue is not which item the subjects prefer, but rather *whether or not preferences are different when you know what is being tasted.* "Blinding" subjects is an important precaution in testing a new medical treatment because subjects might be influenced by the idea of being treated. We want to design a study to see if blinding is also an important issue for taste-testing experiments. For example, perhaps people will tend to state a preference for a more expensive item because they think they are supposed to like it better.

Your group will develop the protocol for a designed study of blinding in a taste-testing experiment. A protocol is a complete experimental plan. Your protocol should describe the following:

* precisely what the two items being tested are,

* how the items will be served (for example, will corn flakes be served with non-fat, low-fat, or regular milk?),

> **APPLICATION 3.2** Design Your Own Study *(continued)*
>
> - how you will decide which item to present first and which to present second,
> - how you will determine which subjects are to be blinded and which will be told what they are tasting, and
> - what other special precautions you will take to eliminate possible confounding.
>
> Incorporate these items into a written report describing your experiment.
>   Here are some examples of items you might taste-test:
>
> - a nonfat yogurt versus a regular yogurt of the same brand,
> - Kellogg's® corn flakes versus a store brand of corn flakes,
> - orange juice from a carton versus freshly squeezed orange juice, or
> - french fries cooked in a microwave versus the same brand cooked in a conventional oven.
>
> Remember the issue at hand: whether or not preferences are different when you know what is being tasted.

EXPLORING THE WEB

The Office of Minority Health newsletter *Closing the Gap* has a special issue on clinical trials. Available at www.omhrc.gov/ctg/ctg-ct.htm, the issue includes several articles on minority participation in medical experiments.
  You can find a (very) skeptical look at health claims not backed by proper clinical trials at the "Quackwatch" site, www.quackwatch.com.

STATISTICS IN SUMMARY

As with samples, experiments require a combination of good statistical design and steps to deal with practical problems. Because the **placebo effect** is strong, **clinical trials** and other experiments with human subjects should be **double-blind** whenever this is possible. The double-blind method helps achieve a basic requirement of comparative experiments: **equal treatment for all subjects** except for the actual treatments the experiment is comparing.

  Just as samples suffer from nonresponse, experiments suffer from uncooperative subjects. Some subjects refuse to participate; others drop out before the experiment is complete; others don't follow instructions, as when some subjects in a drug trial don't take their pills. The most common weakness in experiments is that we can't **generalize** the conclusions widely. Some experiments apply unrealistic treatments, some use subjects from some special group such as college students, and all take place at some specific place and time. We want to see similar experiments at other places and times confirm important findings.

Many experiments use designs that are more complex than the basic **completely randomized design** that divides all the subjects among all the treatments in one randomization. **Matched pairs designs** compare two treatments by giving one to each of a pair of similar subjects or by giving both to the same subject in random order. **Block designs** form blocks of similar subjects and assign treatments at random separately in each block. The big ideas of randomization, control, and adequate numbers of subjects remain the keys to convincing experiments.

SECTION 3.2 EXERCISES

3.36 Meditation lowers anxiety An experiment that was publicized as showing that a meditation technique lowered the anxiety level of subjects was conducted as follows: The experimenter interviewed the subjects and assessed their levels of anxiety. The subjects then learned how to meditate and did so regularly for a month. The experimenter reinterviewed them at the end of the month and assessed whether their anxiety levels had decreased or not.
(a) There was no control group in this experiment. Why is this a blunder? What extraneous variables may be confounded with the effect of meditation?
(b) The experimenter who diagnosed the effect of the treatment knew that the subjects had been meditating. Explain how this knowledge could bias the experimental conclusion.
(c) Briefly discuss a proper experimental design, with controls and blind diagnosis, to assess the effect of meditation on anxiety level.

3.37 Better air bags Car makers are interested in developing safer air bags that reduce injuries in collisions. An automobile manufacturer is trying to decide between two possible air bags for use in the company's vehicles. Six reusable crash test dummies—large male, average male, small male, large female, average female, and small female—are available for use on the company's indoor "test track." Design an experiment to help the company decide which air bag to choose.

3.38 Drug improves coordination A drug is suspected of affecting the coordination of subjects. The drug can be administered in three ways: orally, by injection under the skin, or by injection into a vein. The potency of the drug probably depends on the method of administration as well as on the dose administered. A researcher therefore wishes to study the effects of the two factors, dose at two levels and method of administration by the three methods mentioned. The response variable is the score of the subjects on a standard test of coordination. Ninety subjects are available.

(a) List the treatments that can be formed from the two factors.

(b) Describe an appropriate completely randomized design. (Just outline the design; don't do any randomization.)

(c) The researcher could study the effect of the dose in an experiment comparing two dose levels for one method of administration. He then could separately study the effect of administration by comparing the three methods for one dose level. What advantage does the two-factor experiment you designed in (a) have over these two one-factor experiments taken together?

3.39 Ugly french fries Few people want to eat discolored french fries. Potatoes are kept refrigerated before being cut for french fries to prevent spoiling and preserve flavor. But immediate processing of cold potatoes causes discoloring due to complex chemical reactions. The potatoes must therefore be brought to room temperature before processing. Design an experiment in which tasters will rate the color and flavor of french fries prepared from several groups of potatoes. The potatoes will be fresh picked or stored for a month at room temperature or stored for a month refrigerated. They will then be sliced and cooked either immediately or after an hour at room temperature.

(a) What are the factors and their levels, the treatments, and the response variables?

(b) Describe and outline the design of this experiment.

(c) It is efficient to have each taster rate fries from all treatments. How will you use randomization in presenting fries to the tasters?

3.40 Morphine and cancer Health-care providers are giving more attention to relieving the pain of cancer patients. An article in the journal *Cancer* surveyed a number of studies and concluded that controlled-release morphine tablets, which release the painkiller gradually over time, are more effective than giving standard morphine when the patient needs it.[21] The "methods" section of the article begins: "Only those published studies that were controlled (i.e., randomized, double-blind, and comparative), repeated dose studies with CR morphine tablets in cancer pain patients were considered for this review." Explain the terms in parentheses to someone who knows nothing about medical trials.

3.41 Harmful auto emissions Most motor vehicles are equipped with catalytic converters to reduce harmful emissions. The ceramic used to make the converters must be baked to a certain hardness. The manufacturer must decide which of three temperatures (500°, 750°, and 1000° F) is best. The position of the converter in the oven (front, middle, or back) also affects the hardness. So there are two experimental factors: temperature and placement.

(a) List the treatments in this experiment if all combinations of levels of the two factors are used.

(b) Design a completely randomized experiment with five units in each group.

(c) Using Table A, beginning at line 101, do the randomization for your experiment.

3.42 Bee healed! "Bee pollen is effective for combating fatigue, depression, cancer, and colon disorders." So says a Web site that offers the pollen for sale. We wonder if bee pollen really does prevent colon disorders. Here are two ways to study this question. Explain why the second design will produce more trustworthy data.

(a) Find 200 women who take bee pollen regularly. Match each with a woman of the same age, race, and occupation who does not take bee pollen. Follow both groups for five years.

(b) Find 400 women who do not have colon disorders. Assign 200 to take bee pollen capsules and the other 200 to take placebo capsules that are identical in appearance. Follow both groups for five years.

3.43 Is caffeine dependence real? Many people start their day with a jolt of caffeine from coffee or a soft drink. Most experts agree that people who take in large amounts of caffeine each day may suffer from physical withdrawal symptoms if they stop ingesting their usual amounts of caffeine. Researchers recruited 11 volunteers who were caffeine dependent and who were willing to take part in a caffeine withdrawal experiment. The experiment was conducted on two 2-day periods that occurred one week apart. During one of the 2-day periods, each subject was given a capsule containing the amount of caffeine normally ingested by that subject in one day. During the other study period, the subjects were given placebos. The order in which each subject received the two types of capsules was randomized. The subjects' diets were restricted during each of the study periods. At the end of each 2-day study period, subjects were evaluated using a tapping task in which they were instructed to press a button 200 times as fast as they could.[22]

(a) Carefully describe the design of this experiment. How was blocking incorporated, and why?

(b) Why did researchers randomize the order in which subjects received the two treatments?

(c) Could this experiment have been carried out in a double-blind manner? Explain.

3.3 DATA ETHICS

Are sham surgeries ethical?

"Randomized, double-blind, placebo-controlled trials are the gold standard for evaluating new interventions and are routinely used to assess new medical therapies."[23] So says an article in the *New England Journal of Medicine* that discusses the treatment of Parkinson's disease. The article isn't about the new treatment, which offers hope of reducing the tremors and lack of control brought on by the disease, but about the ethics of studying the treatment.

The law requires well-designed experiments to show that new drugs work and are safe. Not so with surgery—only about 7% of studies of surgery use randomized comparisons. Surgeons think their operations succeed—but innovators always think their innovations work. Even if the patients are helped, the placebo effect may deserve most of the credit. So we don't really know whether many common surgeries are worth the risk they carry. To find out, do a proper experiment. That includes a "sham surgery" to serve as a placebo. In the case of Parkinson's disease, the promising treatment involves surgery to implant new cells. The placebo subjects get the same surgery, but the cells are not implanted.

Is this ethical? One side says: Only a randomized comparative experiment can tell us if the new treatment works. If it doesn't work, thousands of patients will be spared operations that will do no good. If it does work, we can do those thousands of operations knowing that we offer more than a placebo. The other side says: Any surgery carries risks. Sham operations place patients at risk without the hope that the operation will help them. Doctors should not risk harm to some patients today just because other patients may benefit tomorrow.

Reasonable people disagree about sham surgery as part of a test of real surgery for Parkinson's disease. But there is wide agreement on some basic ethical principles for statistical studies. To think about the hard cases, we should first think about these principles.

How is heart rate affected by physical activity? The more work your body is doing, the higher your heart rate will be. An experiment was conducted by students at Ohio State University in the fall of 1993 to explore the relationship between a person's heart rate and the frequency at which that person stepped up and down on steps of various heights. Both frequency of stepping and height of the step were expected to affect heart rate.

Not surprisingly, the students found that subjects' heart rates increased with faster stepping and with higher steps. But the students also made a wise decision in designing their experiment—they used blocking to reduce unwanted variability.[24]

ACTIVITY 3.3 Stepping up your heart rate

Materials: *Stopwatch, two steps of height about 6 and 12 inches (you may be able to use some steps on your campus), metronome (if available)*

In this Activity, you will perform an experiment similar to one designed by students at Ohio State University. To make things a bit simpler, however, you will study only the effect of step height on heart rate.

1. Randomly allocate the students in your class to one of two treatment groups, with about equal numbers in each group. You can draw names from a hat or assign numbers and use the random digit table or your calculator. Students assigned to Group 1 will use the 6-inch step on the first trial and the 12-inch step on the second trial. Students assigned to Group 2 will use the 12-inch step first and then the 6-inch step.

2. When your teacher instructs you to, take your pulse for 60 seconds. Record this "resting" pulse rate.

3. Your teacher will tell you when to begin the first trial. Be sure to step up and down according to the pace set by your teacher's voice (or the metronome). You will step up and down for 3 minutes.

4. At the end of the first trial, take your pulse for 60 seconds. Record the data.

5. Your teacher will give you a few minutes to "recover." Afterward, take your pulse again prior to beginning the second trial. Record this measurement.

6. For the second trial, you will switch to the other step height for 3 minutes. When you have finished, take your pulse one more time and record the result.

7. Calculate the difference in your stepping pulse rate and your resting pulse rate for each of the two trials. Did your pulse increase more when you used the higher step?

8. Make a chart on the board that shows the differences in pulse rate for each subject, like this:

Subject   Group   6″ step diff.   12″ step diff.   Diff. (12″−6″)

9. Construct a graph to display the differences in pulse rate for the 6- and 12-inch steps (last column in the chart). Describe what you see.

10. Calculate the average difference in pulse rate for your class. Is there evidence that heart rate increases more with greater step height? Explain.

First principles

The production and use of data, like all human endeavors, raise ethical questions. We won't discuss the telemarketer who begins a telephone sales pitch with "I'm conducting a survey." Such deception is clearly unethical. It enrages legitimate survey organizations, which find the public less willing to talk with them. Neither will we discuss those few researchers who, in the pursuit of professional advancement, publish fake data. There is no ethical question here—faking data to advance your career is just wrong. It will end your career when uncovered. But just how honest must researchers be about real, unfaked data? Here is an example that suggests the answer is "More honest than they often are."

EXAMPLE 3.18   Missing details

Papers reporting scientific research are supposed to be short, with no extra baggage. Brevity can allow the researchers to avoid complete honesty about their data. Did they choose their subjects in a biased way? Did they report data on only some of their subjects? Did they try several statistical analyses and only report the ones that looked best? The statistician John Bailar screened more than 4000 medical papers in more than a decade as consultant to the *New England Journal of Medicine.* He says, "When it came to the statistical review, it was often clear that critical information was lacking, and the gaps nearly always had the practical effect of making the authors' conclusions look stronger than they should have."[25] The situation is no doubt worse in fields that screen published work less carefully.

The most complex issues of data ethics arise when we collect data from people. The ethical difficulties are more severe for experiments that impose some treatment on people than for sample surveys that simply gather information. Trials of new medical treatments, for example, can do harm as well as good to their subjects. Here are some basic standards of data ethics that must be obeyed by any study that gathers data from human subjects, whether sample survey or experiment.

Basic data ethics

The organization that carries out the study must have an **institutional review board** that reviews all planned studies in advance in order to protect the subjects from possible harm.

All individuals who are subjects in a study must give their **informed consent** before data are collected.

All individual data must be kept **confidential**. Only statistical summaries for groups of subjects may be made public.

"That's the gist of what I want to say. Now get me some statistics to base it on."

The law requires that studies funded by the federal government obey these principles. But neither the law nor the consensus of experts is completely clear about the details of their application.

Institutional review boards

The purpose of an institutional review board is not to decide whether a proposed study will produce valuable information or whether it is statistically sound. The board's purpose is, in the words of one university's board, "to protect the rights and welfare of human subjects (including patients) recruited to participate in research activities." The board reviews the plan of the study and can require changes. It reviews the consent form to be sure that subjects are informed about the nature of the study and about any potential risks. Once research begins, the board monitors its progress at least once a year.

The most pressing issue concerning institutional review boards is whether their workload has become so large that their effectiveness in protecting subjects drops. When the government temporarily stopped human subject research at Duke University Medical Center in 1999 due to inadequate protection of subjects, more than 2000 studies were going on. That's a lot of review work. There are shorter review procedures for projects that involve only minimal risks to subjects, such as most sample surveys. When a board is overloaded, there is a temptation to put more proposals in the minimal-risk category to speed the work.

Informed consent

Both words in the phrase "informed consent" are important, and both can be controversial. Subjects must be informed in advance about the nature of a study and any risk of harm it may bring. In the case of a sample survey, physical harm is not possible. The subjects should be told what kinds of questions the survey will ask and about how much of their time it will take. Experimenters must tell subjects the nature and purpose of the study and outline possible risks. Subjects must then *consent* in writing.

EXAMPLE 3.19   Who can consent?

Are there some subjects who can't give informed consent? It was once common, for example, to test new vaccines on prison inmates who gave their consent in return for good-behavior credit. Now we worry that prisoners are not really free to refuse, and the law forbids medical experiments in prisons.

Children can't give fully informed consent, so the usual procedure is to ask their parents. A study of new ways to teach reading is about to start at a local elementary school, so the study team sends consent forms home to parents. Many parents don't return the forms. Can their children take part in the study because the parents did not say "No," or should we allow only children whose parents returned the form and said "Yes"?

What about research into new medical treatments for people with mental disorders? What about studies of new ways to help emergency room patients who may be unconscious or have suffered a stroke? In most cases, there is not time even to get the consent of the family. Does the principle of informed consent bar realistic trials of new treatments for unconscious patients? These are questions without clear answers. Reasonable people differ strongly on all of them. There is nothing simple about informed consent.[26]

The difficulties of informed consent do not vanish even for capable subjects. Some researchers, especially in medical trials, regard consent as a barrier to getting patients to participate in research. They may not explain all possible risks; they may not point out that there are other therapies that might be better than those being studied; they may be too optimistic when talking with patients even when the consent form has all the right details. On the other hand, mentioning every possible risk leads to very long consent forms that really are barriers. "They are like rental car contracts," one lawyer said. Some subjects don't read forms that run five or six printed pages. Others are frightened by the large number of possible (but unlikely) disasters that might happen and so refuse to participate. Of course, unlikely disasters sometimes happen. When they do, lawsuits follow and the consent forms become even longer and more detailed.

Confidentiality

Ethical problems do not disappear once a study has been cleared by the review board, has obtained consent from its subjects, and has actually collected data about the subjects. It is important to protect the subjects' privacy by keeping all data about individuals confidential. The report of an opinion poll may say what percent of the 1500 respondents felt that legal immigration should be reduced. It may not report what you said about this or any other issue.

Confidentiality is not the same as **anonymity.** Anonymity means that subjects are anonymous—their names are not known even to the director of the study. Anonymity is rare in statistical studies. Even where anonymity is possible (mainly in surveys conducted by mail), it prevents any follow-up to improve nonresponse or inform subjects of results.

Any breach of confidentiality is a serious violation of data ethics. The best practice is to separate the identity of the subjects from the rest of the data at once. Sample surveys, for example, use the identification only to check on who did or did not respond. In an era of advanced technology, however, it is no longer enough to be sure that each individual set of data protects people's privacy.

The government, for example, maintains a vast amount of information about citizens in many separate data bases—census responses, tax returns, Social Security information, data from surveys such as the Current Population Survey, and so on. Many of these data bases can be searched by computers for statistical studies. A clever computer search of several data bases might be able, by combining information, to identify you and learn a great deal about you even if your name and other identification have been removed from the data available for search. Privacy and confidentiality of data are hot issues among statisticians in the computer age.

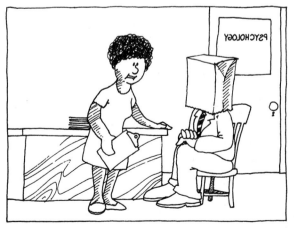

"I realize the participants in this study are to be anonymous, but you're going to have to expose your eyes."

EXAMPLE 3.20   Use of government data bases

Citizens are required to give information to the government. Think of tax returns and Social Security contributions. The government needs these data for administrative purposes—to see if you paid the right amount of tax and how large a Social Security benefit you are owed when you retire. Some people feel that individuals should be able to forbid any other use of their data, even with all identification removed. This would prevent using government records to study, say, the ages, incomes, and household sizes of Social Security recipients. But such a study could well be vital to debates on reforming Social Security.

Statisticians, honest and dishonest

Developed nations rely on government statisticians to produce honest data. We trust the monthly unemployment rate, for example, to guide both public and private decisions. Honesty can't be taken for granted everywhere, however. In 1998, the Russian government arrested the top statisticians in the State Committee for Statistics. They were accused of taking bribes to fudge data to help companies avoid taxes. "It means that we know nothing about the performance of Russian companies," said one newspaper editor.

EXERCISES

Most of the exercises in this section pose issues for discussion. There are few right or wrong answers, but there are more and less thoughtful answers.

3.44 Minimal risk? You have been invited to serve on a college's institutional review board. You must decide whether several research proposals qualify for lighter review because they involve only minimal risk to subjects. Federal regulations say that "minimal risk" means the risks are no greater than "those ordinarily encountered in daily life or during the performance of routine physical or psychological examinations or tests." That's vague. Which of these do you think qualifies as "minimal risk"?
(a) Draw a drop of blood by pricking a finger in order to measure blood sugar.
(b) Draw blood from the arm for a full set of blood tests.
(c) Insert a tube that remains in the arm, so that blood can be drawn regularly.

3.45 Is consent needed? In which of the circumstances below would you allow collecting personal information without the subjects' consent?
(a) A government agency takes a random sample of income tax returns to obtain information on the average income of people in different occupations. Only the incomes and occupations are recorded from the returns, not the names.
(b) A social psychologist attends public meetings of a religious group to study the behavior patterns of members.

(c) The social psychologist pretends to be converted to membership in a religious group and attends private meetings to study the behavior patterns of members.

3.46 Informed consent A researcher suspects that traditional religious beliefs tend to be associated with an authoritarian personality. She prepares a questionnaire that measures authoritarian tendencies and also asks many religious questions. Write a description of the purpose of this research to be read by subjects in order to obtain their informed consent. You must balance the conflicting goals of not deceiving the subjects as to what the questionnaire will tell about them and of not biasing the sample by scaring off religious people.

3.47 Anonymous or confidential? Texas A&M, like many universities, offers free screening for HIV, the virus that causes AIDS. The announcement says, "Persons who sign up for the HIV Screening will be assigned a number so that they do not have to give their name." They can learn the results of the test by telephone, still without giving their name. Does this practice offer *anonymity* or just *confidentiality*?

3.48 Not really anonymous Some common practices may appear to offer anonymity while actually delivering only confidentiality. Market researchers often use mail surveys that do not ask the respondent's identity but contain hidden codes on the questionnaire that identify the respondent. A false claim of anonymity is clearly unethical. If only confidentiality is promised, is it also unethical to say nothing about the identifying code, perhaps causing respondents to believe their replies are anonymous?

Clinical trials

Clinical trials are experiments that study the effectiveness of medical treatments on actual patients. Medical treatments can harm as well as heal, so clinical trials spotlight the ethical problems of experiments with human subjects. Here are the starting points for a discussion:

- Randomized comparative experiments are the only way to see the true effects of new treatments. Without them, risky treatments that are no better than placebos will become common.

- Clinical trials produce great benefits, but most of these benefits go to future patients. The trials also pose risks, and these risks are borne by the subjects of the trial. So we must balance future benefits against present risks.

- Both medical ethics and international human rights standards say that "the interests of the subject must always prevail over the interests of science and society."

The quoted words are from the 1964 Helsinki Declaration of the World Medical Association, the most respected international standard. The most outrageous examples of unethical experiments are those that ignore the interests of the subjects.

EXAMPLE 3.21   The Tuskegee syphilis study

In the 1930s, syphilis was common among black men in the rural South, a group that had almost no access to medical care. The Public Health Service recruited 399 poor black sharecroppers with syphilis and 201 others without the disease in order to observe how syphilis progressed when no treatment was given. Beginning in 1943, penicillin became available to treat syphilis. The study subjects were not treated. In fact, the Public Health Service prevented any treatment until word leaked out and forced an end to the study in the 1970s.

The Tuskegee study is an extreme example of investigators following their own interests and ignoring the well-being of their subjects. A 1996 review said, "It has come to symbolize racism in medicine, ethical misconduct in human research, paternalism by physicians, and government abuse of vulnerable people."[27] In 1997, President Clinton formally apologized to the surviving participants in a White House ceremony.

The Tuskegee study helps explain the lack of trust that lies behind the reluctance of many blacks to take part in clinical trials.

Because "the interests of the subject must always prevail," medical treatments can be tested in clinical trials only when there is reason to hope that they will help the patients who are subjects in the trials. Future benefits aren't enough to justify experiments with human subjects. Of course, if there is already strong evidence that a treatment works and is safe, it is unethical *not* to give it. Here are the words of Dr. Charles Hennekens of Harvard Medical School, who directed the large clinical trial that showed that aspirin reduces the risk of heart attacks:

> *There's a delicate balance between when to do or not do a randomized trial. On the one hand, there must be sufficient belief in the agent's potential to justify exposing half the subjects to it. On the other hand, there must be sufficient doubt about its efficacy to justify withholding it from the other half of subjects who might be assigned to placebos.*[28]

Why is it ethical to give a control group of patients a placebo? Well, we know that placebos often work. What is more, placebos have no harmful side effects. So in the state of balanced doubt described by Dr. Hennekens, the placebo group may be getting a better treatment than the drug group. If we *knew* which treatment was better, we would give it to everyone. When we don't know, it is ethical to try both and compare them. Here are some harder questions about placebos, with arguments on both sides.

EXAMPLE 3.22 Placebo controls?

You are testing a new drug. Is it ethical to give a placebo to a control group if an effective drug already exists?

Yes: The placebo gives a true baseline for the effectiveness of the new drug. There are three groups: new drug, best existing drug, and placebo. Every clinical trial is a bit different, and not even genuinely effective treatments work in every setting. The placebo control helps us see if the study is flawed so that even the best existing drug does not beat the placebo. Sometimes the placebo wins, so the doubt about the efficacy of the existing drug is confirmed by the use of the placebo in the trial. Placebo controls are ethical except for life-threatening conditions.

No: It isn't ethical to deliberately give patients an inferior treatment. We don't know whether the new drug is better than the existing drug, so it is ethical to give both in order to find out. If past trials showed that the existing drug is better than a placebo, it is no longer right to give patients a placebo. After all, the existing drug includes the placebo effect. A placebo group is ethical only if the existing drug is an older one that did not undergo proper clinical trials or doesn't work well or is dangerous.

EXAMPLE 3.23 Sham surgery

Placebos work. As more doctors recognize this fact, more begin to ask, "If we accept a placebo in drug trials, why don't we accept it in surgery trials?" As the Parkinson's disease case that opened this section illustrates, this is a very controversial question.

Yes: Most surgeries have not been tested in comparative experiments, and some are no doubt just placebos. Unlike placebo pills, these surgeries carry risks. Comparing real surgeries to placebo surgeries can eliminate thousands of unnecessary operations and save many lives. The placebo surgery can be made quite safe. For example, placebo patients can be given a safe drug that removes their memory of the operation rather than a more risky anesthetic. Subjects are told that they are in a placebo-controlled trial and that they must consent in order to take part. Placebo-controlled trials of surgery are ethical (except for life-threatening conditions) if the risk to the placebo group is small and there is informed consent.

No: Placebo surgery, unlike placebo drugs, always carries some risk. Remember that "the interests of the subject must always prevail." Even great future benefits can't justify risks to subjects today unless those subjects receive some benefit. We might give a patient a placebo drug as a medical therapy, because placebos work and are not risky. No doctor would do a sham surgery as ordinary therapy, because there is some risk. If we would not use it in medical practice, it isn't ethical to use it in a clinical trial.

EXERCISES

3.49 Sham surgery? Clinical trials like the Parkinson's disease study cited at the beginning of this section are becoming more common. One medical researcher says, "This is just the beginning. Tomorrow, if you have a new procedure, you will have to do a double-blind placebo trial."[29] Example 3.23 outlines the arguments for and against testing surgery just as drugs are tested. When would you allow sham surgery in a clinical trial of a new surgery?

3.50 AIDS clinical trials Now that effective treatments for AIDS are at last available, is it ethical to test treatments that may be less effective? Combinations of several powerful drugs reduce the level of the virus in the blood and at least delay illness and death from AIDS. But effectiveness depends on how damaged the patient's immune system is and what drugs he or she has previously taken. There are strong side effects, and patients must be able to take more than a dozen pills on time every day. Because AIDS is often fatal and the combination therapy works, we might argue that it isn't ethical to test any treatment for AIDS that doesn't give the full combination. But not doing further tests might prevent discovery of better treatments. This is a strong example of the conflict between doing the best we know for patients now and finding better treatments for other patients in the future. How can we ethically test new drugs for AIDS?[30]

3.51 Equal treatment Researchers on aging proposed to investigate the effect of supplemental health services on the quality of life of older people. Eligible patients on the rolls of a large medical clinic were to be randomly assigned to treatment and control groups. The treatment group would be offered hearing aids, dentures, transportation, and other services not available without charge to the control group. The review board felt that providing these services to some but not other persons in the same institution raised ethical questions. Do you agree?

3.52 Placebos At present there is no vaccine for a serious viral disease. A vaccine is developed and appears effective in animal trials. Only a comparative experiment with human subjects in which a control group receives a placebo can determine the true worth of the vaccine. Is it ethical to give some subjects the placebo, which cannot protect them against the disease?[31] *It is ethical If The subject is in no Life Threatening situation*

Behavioral and social science experiments

When we move from medicine to the behavioral and social sciences, the direct risks to experimental subjects are less acute, but so are the possible benefits to the subjects. Consider, for example, some experiments conducted by psychologists in their study of human behavior.

EXAMPLE 3.24   Group pressure

Stanley Milgram of Yale conducted a famous experiment "to see if a person will perform acts under group pressure that he would not have performed in the absence of social inducement."[32]

The subject arrives with three others who (unknown to the subject) are confederates of the experimenter. The experimenter explains that he is studying the effects of punishment on learning. The Learner (one of the confederates) is strapped into an electric chair, mentioning in passing that he has a mild heart condition. The subject and two other confederates are Teachers. They sit in front of a panel with switches with labels ranging from "Slight Shock" to "Danger: Severe Shock" and are told that they must shock the Learner whenever he fails in a memory learning task. How badly the Learner is shocked is up to the Teachers and will be the lowest level suggested by any Teacher on that trial.

All is rigged. The Learner answers incorrectly on 30 of the 40 trials. The two Teacher confederates call for a higher shock level at each failure. As the shock increases, the Learner protests, asks to be let out, shouts, complains of heart trouble, and finally screams in pain. (The "shocks" are phony and the Learner is an actor, but the subject doesn't know this.) What will the subject do? He can keep the shock at the lowest level, but the two Teacher stooges are pressuring him to torture the Learner more for each failure.

The subjects usually give in to the pressure. "While the experiment yields wide variation in performance, a substantial number of subjects submitted readily to pressure applied to them by the confederates." Milgram noted that in questioning after the experiment, the subjects often admitted that they had acted against their own principles and were upset by what they had done. "The subject was then dehoaxed carefully and had a friendly reconciliation with the victim."[33]

This experiment was acceptable in the early 1960s, when Milgram did his work. Psychologists felt it gave insight into the power of group pressure to coerce individuals into acts they would not otherwise commit. Now, however, such a study would be considered clearly unethical.

Milgram's experiment illustrates the difficulties facing those who plan and review behavioral studies.

- There is no risk of physical harm to the subjects, but they would certainly risk embarrassment and lingering emotional effects. What should we protect subjects from when physical harm is unlikely? Possible emotional harm? Undignified situations? Invasion of privacy?

- What about informed consent? Many behavioral experiments rely on hiding the true purpose of the study. Some, like Milgram's, actively deceive subjects. The subjects might change their behavior if told in advance what the investigators were looking for. Subjects are asked to consent on the basis of vague information. They receive full information only after the experiment.

"I'm doing a little study on the effects of emotional stress. Now, just take the axe from my assistant."

The "Ethical Principles" of the American Psychological Association require consent unless a study merely observes behavior in a public place. They allow deception only when it is necessary to the study, does not hide information that might influence a subject's willingness to participate, and is explained to subjects as soon as possible. Milgram's study (from the 1960s) does not meet current ethical standards because subjects did not give informed consent to take part in an experiment.

We see that the basic requirement for informed consent is understood differently in medicine and psychology. Here is an example of another setting with yet another interpretation of what is ethical. The subjects get no information and give no consent. They don't even know that an experiment may be sending them to jail for the night.

EXAMPLE 3.25 Domestic violence

How should police respond to domestic violence calls? In the past, the usual practice was to remove the offender and order him to stay out of the household overnight. Police were reluctant to make arrests because the victims rarely pressed charges. Women's groups argued that arresting offenders would help prevent future violence even if no charges were filed. Is there evidence that arrest will reduce future offenses? That's a question that experiments have tried to answer.

A typical domestic violence experiment compares two treatments: arrest the suspect and hold him overnight, or warn the suspect and release him. When police officers reach the scene of a domestic violence call, they calm the participants and investigate. Weapons or death threats require an arrest. If the facts permit an arrest but do not require it, an officer radios headquarters for instructions. The person on duty opens the next envelope in a file prepared in advance by a statistician. The envelopes contain the treatments in random order. The police either arrest the suspect or warn and release him, depending on the contents of the envelope. The researchers then watch police records and visit the victim to see if the domestic violence reoccurs.

The first such experiment appeared to show that arresting domestic violence suspects does reduce their future violent behavior. As a result of this evidence, arrest has become the common police response to domestic violence.

The domestic violence experiments shed light on an important issue of public policy. Because there is no informed consent, the ethical rules that govern clinical trials and most social science studies would forbid these experiments. They were cleared by review boards because, in the words of one domestic violence researcher, "These people became subjects by committing acts that allow the police to arrest them. You don't need consent to arrest someone."

EXERCISES

3.53 Opinion polls The presidential election campaign is in full swing, and the candidates have hired polling organizations to take regular polls to find out what the voters think about the issues. What information should the pollsters be required to give out?
(a) What does the standard of informed consent require the pollsters to tell potential respondents?
(b) The standards accepted by polling organizations also require giving respondents the name and address of the organization that carries out the poll. Why do you think this is required?
(c) The polling organization usually has a professional name such as "Samples Incorporated," so respondents don't know that the poll is being paid for by a political party or candidate. Would revealing the sponsor to respondents bias the poll? Should the sponsor always be announced whenever poll results are made public?

3.54 Students as subjects Students taking Psychology 001 are required to serve as experimental subjects. Students in Psychology 002 are not required to serve, but they are given extra credit if they do so. Students in Psychology 003 are required either to sign up as subjects or to write a term paper. Serving as an experimental subject may be educational, but current ethical standards frown on using "dependent subjects" such as prisoners or charity medical patients. Students are certainly somewhat dependent on their teachers. Do you object to any of these course policies? If so, which ones, and why?

3.55 Tempting subjects A psychologist conducts the following experiment: she measures the attitude of subjects toward cheating, then has them play a game rigged so that winning without cheating is impossible. The computer that organizes the game also records—unknown to the subjects—whether or not they cheat. Then attitude toward cheating is retested.

Subjects who cheat tend to change their attitudes to find cheating more acceptable. Those who resist the temptation to cheat tend to condemn cheating more strongly on the second test of attitude. These results confirm the psychologist's theory.

This experiment tempts subjects to cheat. The subjects are led to believe that they can cheat secretly when in fact they are observed. Is this experiment ethically objectionable? Explain your position.

3.56 Prison experiments The decision to ban medical experiments on federal prisoners followed the uncovering of experiments in the 1960s that exposed prisoners to serious harm. But experiments such as the vitamin C test of Example 1.9 (page 16) are also banned from federal prisons. Because of the difficulty of obtaining truly voluntary consent in a prison, is it necessary to ban even experiments in which all treatments appear harmless? What is your overall opinion of this ban on experimentation?

3.57 Animal welfare Many people are concerned about the ethics of experimentation with living animals. Some go as far as to regard any animal experiments as unethical, regardless of the benefits to human beings. Briefly explain your position on each of the following uses of animal subjects:

(a) Military doctors use goats that have been deliberately shot (while completely anesthetized) to study and teach the treatment of combat wounds. Assume that there is no equally effective way to prepare doctors to treat human wounds.

(b) Several states are considering legislation that would end the practice of using cats and dogs from pounds in medical research. Instead, the animals will be killed at the pounds.

(c) The cancer-causing potential of chemicals is assessed by exposing lab rats to high concentrations. The rats are bred for this specific purpose. (Would your opinion differ if dogs or monkeys were used?)

APPLICATION 3.3 Hope for Sale?

We have pointed to the ethical problems of experiments with human subjects, clinical trials in particular. *Not* doing proper experiments can also pose problems. Here is an example. Women with advanced breast cancer will eventually die. A promising but untried treatment appears. Should we wait for controlled clinical trials to show that it works, or should we make it available right now?

The promising treatment is "bone marrow transplant" (BMT for short). The idea of BMT is to harvest a patient's bone marrow cells, blast the cancer with very high doses of drugs, then return the harvested cells to keep the drugs from killing the patient. BMT has become popular. It is painful, expensive, and dangerous.

New anticancer drugs are first available through clinical trials, but there is no constraint on therapies such as BMT. When small, uncontrolled trials seemed to show success,

BMT became widely available. The economics of medicine had a lot to do with this. The early leaders in offering BMT were for-profit hospitals that advertise heavily to attract patients. Others soon jumped in. The *New York Times* reported: "Every entity offering the experimental procedure tried a different sales pitch. Some promoted the prestige of their institutions, others the convenience of their locations, others their caring attitudes and patient support." The profits for hospitals and doctors are high.

But does BMT keep patients alive longer than standard treatments? We don't know, but the answer appears to be "probably not." The patients naturally want to try anything that might keep them alive. Some doctors are willing to offer hope not backed by good evidence. Patients would not join controlled trials that might assign them to standard treatments rather than to BMT. Results from such trials were delayed for years by the difficulty in recruiting subjects. Of the first five trials reported, four found no significant difference between BMT and standard treatments. The fifth favored BMT—but the researcher soon admitted "a serious breach of scientific honesty and integrity." The *New York Times* put it more bluntly: "he falsified data."

Compassion seems to support making untested treatments available to dying patients. Reason responds that this opens the door to sellers of hope and delays development of treatments that really work. Compare children's cancer, where doctors agree not to offer experimental treatments outside controlled trials. Result: 60% of all children with cancer are in clinical trials, and progress in saving lives has been much faster for children than for adults. BMT for a rare cancer in children was tested immediately and found to be effective. In contrast, one of the pioneers in BMT for breast cancer, in the light of better evidence, now says, "We deceived ourselves and we deceived our patients."[34]

1. Describe the design of an experiment that compares BMT to an existing treatment.
2. Explain what "no significant difference between BMT and standard treatments" means.
3. Having read the background information on BMT, discuss whether BMT should be made available to advanced breast cancer patients.

EXPLORING THE WEB

For a glimpse at the work of an institutional review board, visit the site of the University of Pittsburgh's board, www.irb.pitt.edu. Look at the *Reference Manual for the Use of Human Subjects in Research* (the first item listed on this Web page) to see how elaborate the review process can be. The official ethical codes of the American Statistical Association (www.amstat.org/profession/ethicalstatistics.html) and the American Psychological Association (www.apa.org/ethics/code.html) are long documents that address many issues in addition to those discussed in this chapter.

STATISTICS IN SUMMARY

Ordinary honesty says you shouldn't make up data or claim to be taking a survey when you are really selling roofing. Data ethics begin with some principles that go beyond just being honest. Studies with human subjects must be screened in advance by an **institutional review board.** All subjects must give their **informed consent** before taking part. All information about individual subjects must be kept **confidential.** These principles are a good start, but many ethical debates remain, especially in the area of experiments with human subjects. Many of the debates concern the right balance between the welfare of the subjects and the future benefits of the experiment. Remember that randomized comparative experiments can answer questions that can't be answered without them. Also remember that "the interests of the subject must always prevail over the interests of science and society."

SECTION 3.3 EXERCISES

3.58 Who serves on the review board? Government regulations require that institutional review boards consist of at least five people, including at least one scientist, one nonscientist, and one person from outside the institution. Most boards are larger, but many contain just one outsider.
(a) Why should review boards contain people who are not scientists?
(b) Do you think that one outside member is enough? How would you choose that member? (For example, would you prefer a medical doctor? A member of the clergy? An activist for patients' rights?)

3.59 Telling the government The 2000 census long form asked 53 detailed questions, for example:

> Do you have COMPLETE plumbing facilities in this house, apartment, or mobile home; that is, 1) hot and cold piped water, 2) a flush toilet, and 3) a bathtub or shower?

The form also asked your income in dollars, broken down by source, and whether any "physical, mental, or emotional condition" causes you difficulty in "learning, remembering, or concentrating." Some members of Congress objected to these questions, even though Congress had approved them.

Give brief arguments on both sides of the debate over the long form: the government has legitimate uses for such information, but the questions seem to invade people's privacy.

3.60 Informed consent The information given to potential subjects in a clinical trial before asking them to decide whether or not to participate might include any of the following. Do you feel that all of this information is ethically required? Discuss.

(a) The basic statement that an experiment is being conducted; that is, something beyond simply treating your medical problem occurs in your therapy.

(b) A statement of any potential risks from any of the experimental treatments.

(c) An explanation that random assignment will be used to decide which treatment you get.

(d) An explanation that one "treatment" is a placebo and a statement of the probability that you will receive the placebo.

3.61 Human biological materials Long ago, doctors drew a blood specimen from you as part of treating minor anemia. Unknown to you, the sample was stored. Now researchers plan to use stored samples from you and many other people to look for genetic factors that may influence anemia. It is no longer possible to ask your consent. Modern technology can read your entire genetic make-up from the blood sample.

(a) Do you think it violates the principle of informed consent to use your blood sample if your name is on it but you were not told that it might be saved and studied later?

(b) Suppose that your identity is not attached. The blood sample is known only to come from (say) "a 20-year-old white female being treated for anemia." Is it now OK to use the sample for research?

(c) Perhaps we should use biological materials such as blood samples only from patients who have agreed to allow the material to be stored for later use in research. It isn't possible to say in advance what kind of research, so this falls short of the usual standard for informed consent. Is it nonetheless acceptable, given complete confidentiality and the fact that using the sample can't physically harm the patient?

3.62 Anonymous or confidential? One of the most important nongovernment surveys in the United States is the General Social Survey (Example 1.8, page 12). The GSS regularly monitors public opinion on a wide variety of political and social issues. Interviews are conducted in person in the subject's home. Are a subject's responses to GSS questions anonymous, confidential, or both? Explain your answer.

3.63 AIDS trials in Africa Effective drugs for treating AIDS are very expensive, so most African nations cannot afford to give them to large numbers of people. Yet AIDS is more common in parts of Africa than anywhere else. Several clinical trials are looking at ways to prevent pregnant mothers infected with HIV from passing the infection to their unborn children, a major source of HIV infections in Africa. Some people say these trials are unethical because they do not give effective AIDS drugs to their subjects, as would be required in rich nations. Others reply that the trials are looking for treatments that can work in the real world in Africa and that they promise benefits at least to the children of their subjects. What do you think?

3.64 Deceiving subjects Students sign up to be subjects in a psychology experiment. When they arrive, they are told that interviews are running late and are taken to a waiting room. The experimenters then stage a theft of a valuable object left in the waiting room. Some subjects are alone with the thief, and others are in pairs—these are the treatments being compared. Will the subject report the theft?

The students had agreed to take part in an unspecified study, and the true nature of the experiment is explained to them afterward. Do you think this study is ethically OK?

3.65 Surveys of youth A survey asked teenagers whether they had ever consumed an alcoholic beverage. Those who said "Yes" were then asked,

How old were you when you first consumed an alcoholic beverage?

Should consent of parents be required to ask minors about alcohol, drugs, and other such issues, or is consent of the minors themselves enough? Give reasons for your opinion.

CHAPTER 3 REVIEW

Sections 3.1 and 3.2 dealt with the statistical aspects of designing experiments, studies that impose some treatment in order to learn about the response. The big idea is the randomized comparative experiment. Figure 3.6 outlines the simplest design.

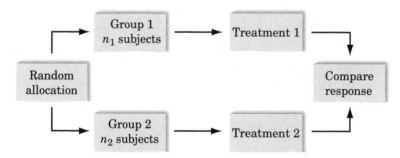

FIGURE 3.6 The idea of a randomized comparative experiment.

If the subjects in an experiment differ in some systematic way that may affect their responses to the treatments, blocking can be used to reduce unwanted variability. A matched pairs design, in which each subject serves as his or her own control, is a special form of blocking.

Well-designed experiments include random allocation of subjects to treatments, comparison, and replication. To reduce the potential for bias, many designs are double-blind—neither the subjects nor those working with them know who is receiving which treatment. Confounding can blur the relationship between the explanatory and response variables and must be actively prevented.

When we collect data about people, ethical issues can be important. Section 3.3 discussed these issues and introduced three principles that apply to any study with human subjects.

Here are the most important skills you should have developed after studying this chapter.

1. Identify the explanatory variables, treatments, response variables, and subjects in an experiment.

2. Recognize bias due to confounding of explanatory variables with lurking variables in either an observational study or an experiment.

3. Outline the design of a completely randomized experiment using a diagram like that in Figure 3.6. Such a diagram should show the sizes of the groups, the specific treatments, and the response variable.

4. Use Table A of random digits to carry out random assignment of subjects to groups in a completely randomized experiment.

5. Make use of matched pairs or other block designs when appropriate.

6. Recognize the placebo effect. Recognize when the double-blind technique should be used. Be aware of weaknesses in an experiment, especially in the ability to generalize its conclusions.

7. Explain why a randomized comparative experiment can give good evidence for cause-and-effect relationships.

8. Understand key concepts of data ethics: informed consent, confidentiality of personal data, and review of ethical considerations by an institutional review board. Discuss how these apply in specific settings.

CHAPTER 3 EXERCISES

3.66 Should I take the class? A college allows students to choose either classroom or self-paced instruction in a basic economics course. The college wants to compare the effectiveness of self-paced and regular instruction. A professor proposes giving the same final exam to all students in both versions of the course and comparing the average score of those who took the self-paced option with the average score of students in the regular sections.

(a) Explain why confounding makes the results of this study worthless.

(b) Given 30 students who are willing to use either regular or self-paced instruction, outline an experimental design to compare the two methods of instruction. Then use Table A, starting at line 108, to carry out the randomization.

3.67 Testing a new drug A drug manufacturer is studying how a new drug behaves in patients. Investigators compare 2 doses: 5 milligrams (mg) and 10 mg. The drug can be administered by injection, by a skin patch, or by intravenous drip. Concentration in the blood after 30 minutes (the response variable) may depend both on the dose and on the method of administration.
(a) Make a sketch that describes the treatments formed by combining dose and method. Then use a diagram to outline a completely randomized design for this two-factor experiment.
(b) "How many subjects?" is a tough issue. We will explain the basic idea later. What can you say now about the advantage of using larger groups of subjects?
(c) The drug may behave differently in men and women. How would you modify your experimental design from (a) to take this into account?

3.68 AIDS trials in Africa One of the most important goals of AIDS research is to find a vaccine that will protect against HIV. Because AIDS is so common in parts of Africa, that is the easiest place to test a vaccine. It is likely, however, that a vaccine would be so expensive that it could not (at least at first) be widely used in Africa. Is it ethical to test in Africa if the benefits go mainly to rich countries? The treatment group of subjects would get the vaccine, and the placebo group would later be given the vaccine if it proved effective. So the actual subjects would benefit—it is the future benefits that would go elsewhere. What do you think?

3.69 Surgery or not? To compare surgical and nonsurgical treatments of intestinal cancer, researchers examined the records of a large number of patients. Patients who received surgery survived much longer (on the average) than patients who were treated without surgery. The study concludes that surgery is more effective than nonsurgical treatment.
(a) What are the explanatory and response variables in this study?
(b) Is this study an experiment? Why or why not?
(c) Discussion with medical experts reveals that surgery is reserved for relatively healthy patients; patients who are too ill to tolerate surgery receive nonsurgical treatment. Explain why the conclusion of the study is not supported by the data.
(d) Outline the design of a randomized comparative experiment.

3.70 Reducing health-care spending Will people spend less on health care if their health insurance requires them to pay some part of the cost themselves? An experiment on this issue asked if the percent of medical costs that are paid by health insurance has an effect either on the amount of medical care that people use or on their health. The treatments were four insurance plans. Each plan paid all medical costs above a ceiling. Below the ceiling, the plans paid 100%, 75%, 50%, or 0% of costs incurred.[35]

(a) Outline the design of a randomized comparative experiment suitable for this study.

(b) Briefly describe the practical and ethical difficulties that might arise in such an experiment.

3.71 Do the twist The design of controls and instruments has a large effect on how easily people can use them. A student investigates this effect by asking right-handed students to turn a knob (with their right hands) that moves an indicator by screw action. There are two identical instruments, one with a right-hand thread (the knob turns clockwise) and the other with a left-hand thread (the knob must be turned counterclockwise). The response variable is the time required (in seconds) to move the indicator a fixed distance. Thirty right-handed students are available to serve as subjects. You have been asked to lay out the statistical design of this experiment. Describe your design, and carry out any randomization that is required.

Exercises 3.72 to 3.75 are based on an article in the *Journal of the American Medical Association* that asks if flu vaccine works.[36] The article reports a study of the effectiveness of a nasal spray vaccine called "trivalent LAIV." Here is part of the article's summary:

Design   Randomized, double-blind, placebo-controlled trial conducted from September 1997 through March 1998.

Participants   A total of 4561 healthy, working adults aged 18 to 64 years recruited through health insurance plans, at work sites, and from the general population.

Intervention   Participants were randomized 2:1 to receive intranasally administered trivalent LAIV vaccine ($n = 3041$) or placebo ($n = 1520$) in the fall of 1997.

Results   Vaccination also led to fewer days of work lost (17.9% reduction for severe febrile illnesses; 28.4% reduction for febrile upper-respiratory-tract illnesses) and fewer days with health-care-provider visits (24.8% reduction for severe febrile illnesses; 40.9% reduction for febrile upper-respiratory-tract illnesses).

3.72 Know these terms Explain in one sentence each what "randomized," "double-blind," and "placebo-controlled" mean in the description of the design of this study.

3.73 Experiment basics Identify the subjects, the explanatory variable, and several response variables in this study.

3.74 Design an experiment Use a diagram to outline the design of the experiment in this medical study.

3.75 Ethics What are the three "first principles" of data ethics? Explain briefly what the flu vaccine study must do to apply each of these principles.

Part II

Organizing Data

JUMP START reprinted by permission of United Feature Syndicate, Inc.

Words alone don't tell a story. A writer organizes words into sentences and organizes the sentences into a story line. If the words are badly organized, the story isn't clear. Data also need organizing if they are to tell a clear story. Too many words obscure a subject rather than illuminate it. Vast amounts of data are even harder to digest—we often need a brief summary to highlight essential facts. How to organize, summarize, and present data are our topics in the second part of this book.

Organizing and summarizing a large body of facts opens the door to distortions, both unintentional and deliberate. This is no less (but also no more) the case when the facts take the form of numbers rather than words. We will point out some of the traps that data presentations can set for the unwary. Those who picture statistics as primarily a piece of the liar's art concentrate on the part of statistics that deals with summarizing and presenting data. We claim that misleading summaries and selective presentations go back to that after-the-apple conversation among Adam, Eve, and God. Don't blame statistics. Do remember the saying "Figures won't lie, but liars will figure," and beware.

Chapter 4

Describing Distributions

4.1 GRAPHS, GOOD AND BAD

Getting rich

The end of the twentieth century saw a great bull market in U.S. common stocks. How great? Pictures tell the tale more clearly than words.

Look first at Figure 4.1 (next page). This shows the percent increase or decrease in stocks (measured by the Standard & Poor's 500 Index) in each year from 1971 to 1999. Until 1982, stock prices bounce up and down. Sometimes they go down a lot—stocks lost 15% of their value in 1973 and another 27% in 1974. But starting in 1982, stocks go up in 17 of the next 18 years, often by a lot.

Figure 4.2 (next page) shows how to get rich. If you had invested $1000 in stocks at the end of 1970, the graph shows how much money you had at the end of each following year. After 1974, your $1000 was down to $854, and at the end of 1981 it had grown to only $2121. That's only 7% a year. Your money would have grown faster in a bank during these years. Then the great bull market begins its work. By the end of 1999, it has turned your $1000 into $44,875.

The lesson is supposed to be that if you can wait long enough, you should put your money in stocks. Perhaps the real lesson is that you should have put your money in stocks in 1980 and taken it out of stocks in 2000—because the market took a nosedive in 2000 and stayed in the cellar for the next several years. Our statistical lesson, however, appeals to the mind rather than the pocketbook. It is the power of graphs to make clear what data say. Figure 4.1 shows how irregular the

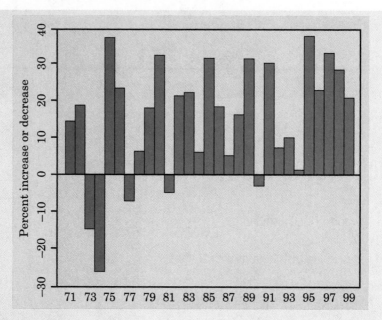

FIGURE 4.1 Percent increase or decrease in the S&P 500 Index of common stock prices, 1971 to 1999.

year-to-year behavior of stocks has been. Figure 4.2 shows the long-term growth in wealth that has rewarded patient investors. It points in particular to the effect of the 5 years of boom starting in 1995.

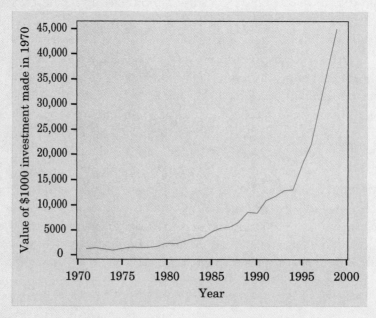

FIGURE 4.2 Value at the end of each year, 1971 to 1999, of $1000 invested in the S&P 500 Index at the end of 1970.

ACTIVITY 4.1 Looking at graphs critically

Have you ever heard that some people are "picture oriented"? Instructions for setting up a new Macintosh computer used to be provided in steps with English sentences. Dan bought an iMac in 2002 and discovered that the installation steps are given exclusively with pictures. Apple Computer, makers of the iMac, did research with a sample of its customers and found that most preferred that instructions for setting up the computer be given in a picture format.

1. Do you know of other consumer products that use pictures to show how to use the product? People who specialize in advertising and communications have known for some time that graphs can be used to convey information quickly and more efficiently than in narrative form. That's the good news. The bad news is that graphs can also be used to mislead, deceive, and obfuscate.

2. Suppose that you're looking for a summer job and you see an advertisement for a company that says that current salaries are significantly higher than they were just two years ago. The ad shows this graph:

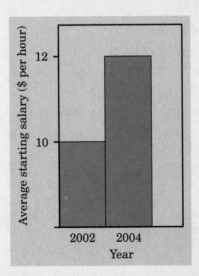

What impression do you get about the improvement in salaries at this company from this graph?

ACTIVITY 4.1 Looking at graphs critically *(continued)*

3. Here's another way to picture the same salary information:

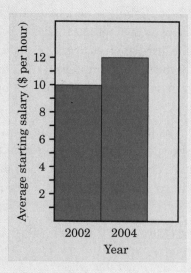

What impression do you get about the improvement in salaries from this graph?

4. Suppose that company showed the first graph and said in large type, "HUGE SALARY IMPROVEMENT." Is this a fair assessment? Is it misleading? Do you think an ad like this would be legal?

5. Here's a true story. In the latter 1990s a well-known brokerage company wanted to make the point that Americans were not saving enough money for the future and included this graph in one of its newsletters:

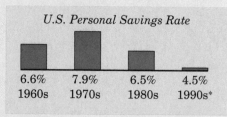

U.S. Personal Savings Rate

| 6.6% | 7.9% | 6.5% | 4.5% |
| 1960s | 1970s | 1980s | 1990s* |

*Through 1995

How much did Americans save in the 1990s, compared with the1980s?

ACTIVITY 4.1 Looking at graphs critically *(continued)*

6. Now make your own graph for "U.S. Personal Savings Rate," with the decades on the horizontal axis, like the graph at the bottom of page 202. But make your scale on the vertical axis start at 0%. Make 16 tick marks on the vertical axis with equal intervals, and label these points 0, 0.5, 1, 1.5, 2, and so forth, up to 8. Now draw the bars with the heights shown in the graph. How much taller is the 1980s bar than the 1990s bar? Was the U.S. personal savings rate in the 1980s that much more than in the 1990s?

7. Write a sentence or two to describe the impression you get about Americans' savings rate from *your* graph. Do you get the same impression from both graphs? Is your graph more "honest" than their graph? Why?

What mental picture does the word *statistics* call to mind? Very likely, images of tables crowded with numbers and graphs zigging up or zagging down. That's not a bad picture: statistics deals with data, and we use tables and graphs to present data. Random sampling, randomized comparative experiments, and valid measurement produce good data. Now we must see what the data say.

Data tables

Take a look at the *Statistical Abstract of the United States,* an annual volume packed with every variety of numerical information. Has the number of private elementary and secondary schools grown over time? What about minority enrollments in these schools? How many college degrees were given in each of the past several years, and how were these degrees divided among fields of study and by the age, race, and sex of the students? You can find all this and more in the education section of the *Statistical Abstract.* The tables *summarize* data. We don't want to see information on every college degree individually, only the counts in categories of interest to us.

EXAMPLE 4.1   What makes a clear table?

How well educated are 30-something young adults? Table 4.1 presents the data for people aged 25 to 34 years. This table illustrates some good practices for data tables. It is clearly *labeled* so that we can see the subject of the data at once. The main heading describes the general subject of the data and gives the date because these data will change over time. Labels within the table identify the variables and state the *units* in which they are measured. Notice, for example, that the counts are in thousands. The *source* of the data appears at the foot of the table. This Census Bureau publication in fact presents data from our old friend, the Current Population Survey.

TABLE 4.1 Education of People 25 to 34 Years Old, 2000

	Number of persons (thousands)	Percent
Less than high school	4,459	11.8
High school graduate	11,562	30.6
Some college	10,693	28.3
Bachelor's degree	8,577	22.7
Advanced degree	2,494	6.6
Total	37,786	100

Source: Census Bureau, *Educational Attainment in the United States: March 2000.*

Table 4.1 starts with the **counts** of young adults with each level of education. **Rates** (percents or proportions) are often clearer than counts—it is more helpful to hear that 11.8% of young adults did not finish high school than to hear that there are 4,459,000 such people. The percents also appear in Table 4.1. The two columns of the table present the *distribution* of the variable "level of education" in two alternate forms. Each column gives information about what values the variable takes and how often it takes each value.

Distribution of a variable

The **distribution** of a variable tells us what values it takes and how often it takes these values.

EXAMPLE 4.2   Roundoff errors

Did you check Table 4.1 for consistency? The total number of people should be

$$4,459 + 11,562 + 10,693 + 8,577 + 2,494 = 37,785 \text{ (thousands)}$$

The table gives the total as 37,786. What happened? Each entry is rounded to the nearest thousand. The rounded entries don't quite add to the total, which is rounded separately. Such **roundoff errors** will be with us from now on as we do more arithmetic.

Pie charts and bar graphs

The distribution in Table 4.1 is quite simple because "level of education" has only 5 possible values. To picture this distribution in a graph, we might use a **pie chart.** Figure 4.3 is a pie chart of the level of education of young adults. Pie charts show how a whole is divided into parts.

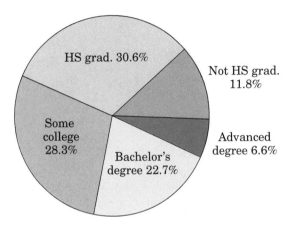

FIGURE 4.3 Pie chart of the distribution of level of education among persons aged 25 to 34 in 2000.

To make a pie chart, first draw a circle. The circle represents the whole, in this case all people aged 25 to 34 years. Wedges within the circle represent the parts, with the angle spanned by each wedge in proportion to the size of that part. For example, 22.7% of young adults have a bachelor's degree but not an advanced degree. Because there are 360 degrees in a circle, the "bachelor's degree" wedge spans an angle of

$$0.227 \times 360 = 82 \text{ degrees}$$

Pie charts force us to see that the parts do make a whole. But because angles are harder to compare than lengths, a pie chart is not the best way to compare the sizes of the various parts of the whole.

Figure 4.4 (next page) is a **bar graph** of the same data. The height of each bar shows the percent of young adults with the level of education marked at the bar's base. The bar graph makes it clear that there are more high school graduates than there are people who also have some college—the "HS grad" bar is taller. Because it is hard to see this from the wedges in the pie chart, we added labels to the wedges that include the actual percents. The bar graph is also easier to draw than the pie chart unless a computer is doing the drawing for you.

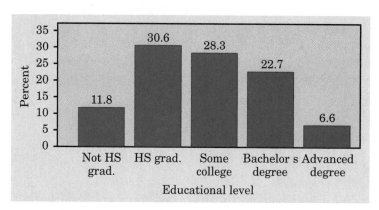

FIGURE 4.4 Bar graph of the distribution of level of education among persons aged 25 to 34 in 2000.

When we think about graphs, it is helpful to distinguish between variables whose values have a meaningful numerical scale (height in centimeters, SAT scores) and variables such as sex, occupation, or level of education that just place individuals into categories. Pie charts and bar graphs are most useful for the second kind of variable.

Categorical and quantitative variables

A **categorical variable** places an individual into one of several groups or categories.

A **quantitative variable** takes numerical values for which arithmetic operations such as adding and averaging make sense.

To display the distribution of a categorical variable, use a pie chart or a bar graph.

Although both pie charts and bar graphs can show the distribution (either counts or percents) of a categorical variable such as level of education, bar graphs have wider uses as well.

EXAMPLE 4.3   High taxes?

Figure 4.5 compares the level of taxation in eight democratic nations. The height of the bars shows the percent of each nation's gross domestic product (GDP, the total value of all goods and services produced) that is taken in taxes. Americans accustomed to complaining about high taxes may be surprised to see that the United States, at 28.5% of GDP, is near the bottom of the group.

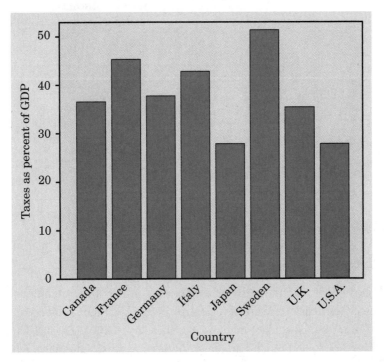

FIGURE 4.5 Total taxes as a percent of gross domestic product in eight countries in 1996.

We cannot replace Figure 4.5 by a pie chart, because it compares eight separate quantities, not the parts of some whole. A pie chart can only compare parts of a whole. Bar graphs can compare quantities that are not parts of a whole.

EXERCISES

4.1 Lottery sales States sell lots of lottery tickets. Table 4.2 (next page) shows where the money comes from. Make a bar graph that shows the distribution of lottery sales by type of game. Is it also proper to make a pie chart of these data?

4.2 Consistency? Table 4.2 shows how state lottery sales are divided among different types of games. What is the sum of the amounts spent on the five types of games? Why is this sum not exactly equal to the total given in the table?

4.3 College freshmen A survey of college freshmen in 2001 asked what field they planned to study. The results: 12.6% arts and humanities, 16.6% business, 10.1% education, 16% engineering and science, 12% professional, and 10.3% social science.[1]

TABLE 4.2 State Lottery Sales by Type of Game, 2001

	Sales (millions of dollars)
Instant games	16,420
Three-digit games	5,245
Four-digit games	2,776
Lotto	8,865
Other games	5,134
Total	38,441

Source: *Statistical Abstract of the United States*, 2002.

(a) What percent plan to study fields other than those listed?

(b) Make a graph that compares the percents of college freshmen planning to study various fields.

4.4 Girls excel Is it true that girls perform better in the study of languages and so-called soft sciences? Here are several Advanced Placement subjects and the percent of examinations taken by female candidates in 2001: English Language/Composition, 62.86%; French Language, 70.82%; Spanish Language, 64.5%; and Psychology, 65.92%.[2]

(a) Explain clearly why we cannot use a pie chart to display these data, even if we know the percent of exams taken by girls for every subject.

(b) Make a bar graph of the data. Order the bars from tallest to smallest; this will make comparisons easier.

4.5 Marital status In the *Statistical Abstract* we find these data on the marital status of adult American women as of 2000:

Marital status	Count (thousands)
Never married	22,089
Married	60,436
Widowed	11,054
Divorced	11,284

(a) How many women were not married in 2000?

(b) Make a bar graph to show the distribution of marital status.

(c) Would it also be correct to use a pie chart? If so, make a pie chart for these data.

Beware the pictogram

Bar graphs compare several quantities by comparing the heights of bars that represent the quantities. Our eyes, however, react to the *area* of the bars as well as to their height. When all bars have the same width, the area (width × height) varies in proportion to the height and our eyes receive the right impression. When you draw a bar graph, make the bars equally wide. Artistically speaking, bar graphs are a bit dull. It is tempting to replace the bars with pictures for greater eye appeal.

EXAMPLE 4.4   A misleading graph

Figure 4.6 is a pictogram. It is a bar graph in which pictures replace the bars. The graph is aimed at advertisers deciding where to spend their budgets. It shows that *Time* magazine attracts the lion's share of the advertising spending. Or does it? The numbers above the pens show that advertising spending in *Time* is 1.64 times as great as in *Newsweek*. Why does the graph suggest that *Time* is much farther ahead?

FIGURE 4.6 A pictogram. This variation of a bar graph is attractive but misleading. (Copyright © 1971 by Time, Inc. Reproduced by permission.)

To magnify a picture, the artist must increase both height and width to avoid distortion. If both the height and width of *Time*'s pen are 1.64 times larger than *Newsweek*'s, the area is 1.64 × 1.64, or 2.7 times as large. Our eyes, responding to the area of the pens, see *Time* as the big winner.

Change over time: line graphs

Many quantitative variables are measured at intervals over time. We might, for example, measure the height of a growing child or the price of a stock at the end of each month. In these examples, our main interest is change over time. To display change over time, make a *line graph*.

Line graph

A **line graph** of a variable plots each observation against the time at which it was measured. Always put time on the horizontal scale of your plot and the variable you are measuring on the vertical scale. Connect the data points by lines to display the change over time.

EXAMPLE 4.5   The price of gasoline

How has the price of gasoline at the pump changed over time? Figure 4.7 is a line graph of the average price of regular unleaded gasoline each month during the 1990s. For example, the January 1990 price was 105.1 cents per gallon. The first point in the figure is above the beginning of 1990 (January) and at 105.1 on the vertical scale.[3]

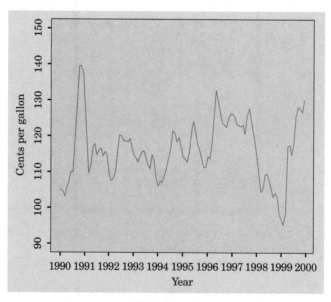

FIGURE 4.7 A line graph of the average cost of regular unleaded gasoline each month between January 1990 and December 1999.

It would be difficult to see patterns in a long table of monthly prices. Figure 4.7 makes the patterns clearer. What should we look for?

First, look for an **overall pattern.** For example, a **trend** is a long-term upward or downward movement over time. There was no consistent overall up or down trend in gasoline prices in the decade of the 1990s.

Next, look for striking **deviations** from the overall pattern. The sharp rise in prices in 1990 and the plunging prices in late 1997 and all of 1998 stand out from the overall lack of any trend. The 1990 price spike was due to the invasion of Kuwait by Iraq. The sharp fall in 1997 and 1998 reflects an Asian economic crisis (lower demand for oil products) and an oversupply of oil until the OPEC nations drove the price up in 1999 by reducing their oil production.

Change over time often has a regular pattern of **seasonal variation** that repeats each year. Gasoline prices are usually highest during the summer driving season and lowest in the winter, when there is less demand for gasoline. You can see this up-in-summer then down-in-the-fall pattern for many years in Figure 4.7.

Because seasonal variation is common, many government statistics are adjusted to remove seasonal effects. For example, the unemployment rate rises every year in January as holiday sales jobs end and outdoor work slows in the North due to winter weather. It would cause confusion (and perhaps political trouble) if the government's official unemployment rate jumped every January. The Bureau of Labor Statistics knows about how much it expects unemployment to rise in January, so it adjusts the published data for this expected change. The published unemployment rate goes up only if actual unemployment rises more than expected. We can then see the underlying changes in the employment situation without being confused by regular seasonal changes.

Seasonal variation, seasonal adjustment

A pattern that repeats itself at known regular intervals of time is called **seasonal variation.** Many series of regular measurements over time are **seasonally adjusted.** That is, the expected seasonal variation is removed before the data are published.

Watch those scales!

Because graphs speak so strongly, they can mislead the unwary. The careful reader of a line graph looks closely at the scales marked off on the axes.

EXAMPLE 4.6   Living together

The number of unmarried couples living together has increased in recent years, to the point that some people say that cohabitation is delaying or even replacing marriage. Figure 4.8 presents two line graphs of the number of unmarried-couple households in the United States. The data once again come from the Current Population Survey. The graph on the left suggests a steady but moderate increase. The right-hand graph says that cohabitation is thundering upward.

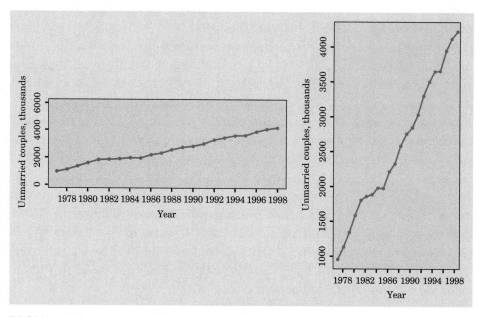

FIGURE 4.8 The effect of changing the scales in a line graph. Both graphs plot the same data, but the right-hand graph makes the increase appear much more rapid.

The secret is in the scales. You can transform the left-hand graph into the right-hand graph by stretching the vertical scale, squeezing the horizontal scale, and cutting off the vertical scale just above and below the values to be plotted. Now you know how to either exaggerate or play down a trend in a line graph.

Which graph is correct? Both are accurate graphs of the data, but both have scales chosen to create a specific effect. Because there is no one "right" scale for a line graph, correct graphs can give different impressions by their choices of scale. Watch those scales!

Making good graphs

Graphs are the most effective way to communicate using data. A good graph frequently reveals facts about the data that would be difficult or impossible to detect from a table. What is more, the immediate visual impression of a graph is much stronger than the impression made by data in numerical form. Here are some principles for making good graphs:

- Make sure **labels and legends** tell what variables are plotted, what their units are, and what is the source of the data.

- **Make the data stand out.** Be sure that the actual data, not labels, grids, or background art, catch the viewer's attention. You are drawing a graph, not a piece of creative art.

- **Pay attention to what the eye sees.** Avoid pictograms and be careful choosing scales. Avoid fancy "three-dimensional" effects that confuse the eye without adding information. Ask if a simple change in a graph would make the message clearer.

EXAMPLE 4.7   The rise in college education

Figure 4.9 shows the rise in the percent of women 25 years old and over who have at least a bachelor's degree. There are only 5 data points, so a line graph should be simple. Figure 4.9 isn't simple. The artist couldn't resist adding pictures and also cluttered the graph with grid lines. We find it harder than it should be to see the data. Grid lines on a graph serve no purpose—if your audience must know the actual numbers, supply a table along with the graph. A good graph uses no more ink than is needed to present the data clearly.

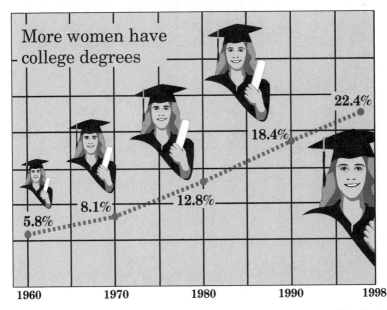

FIGURE 4.9 Chart junk: this graph is so cluttered with unnecessary ink that it is hard to see the data.

The Uietnam effect

Folklore says that during the Vietnam War many men went to college to avoid being drafted. Statistician Howard Wainer looked for traces of this "Vietnam effect" by plotting data against time. He found that scores on the Armed Forces Qualifying Test (an IQ test given to recruits) dipped sharply during the war, then rebounded. Scores on the SAT exams, taken by students applying to college, also dropped at the beginning of the war. It appears that the men who chose college over the army lowered the average test scores in both places.

EXAMPLE 4.8   High taxes, reconsidered

Figure 4.5 (page 207) is a respectable bar graph comparing taxes as a percent of GDP in eight countries. The countries are arranged alphabetically. Wouldn't it be clearer to arrange them in order of their tax burdens? Figure 4.10 does this. This simple change improves the graph by making it clearer where each country stands in the group of eight countries.

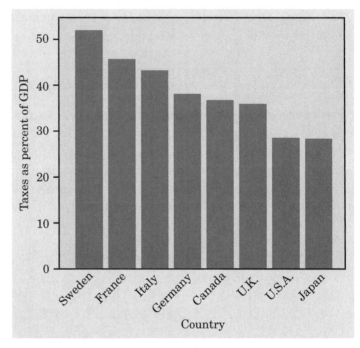

FIGURE 4.10 Taxes as a percent of gross domestic product in eight countries in 1996. Changing the order of the bars has improved the graph in Figure 4.5.

EXERCISES

4.6 We pay high interest Figure 4.11 shows a graph taken from an advertisement for an investment that promises to pay a higher interest rate than bank accounts and other competing investments. Is this graph a correct comparison of the four interest rates? Explain your answer.

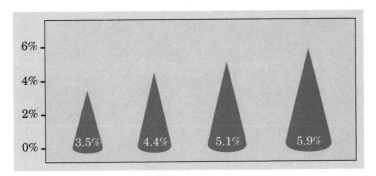

FIGURE 4.11 Comparing interest rates, for Exercise 4.6.

4.7 We sell CDs Who sold the most CDs on the Web just before listeners began to download their favorites rather than buy CDs? Figure 4.12 graphs the shares of the market leaders in 1997 and says what percent of all online sales they had.[4] Does the graph fairly represent the data? Explain your answer.

Online CD Sales Leaders, 1997

FIGURE 4.12 Market share in percent for the three leading online sellers of CDs in 1997, for Exercise 4.7.

4.8 The cost of fresh oranges Figure 4.13 (next page) is a line graph of the average cost of fresh oranges each month from January 1990 to January 2000. These data, from the Bureau of Labor Statistics' monthly survey of retail prices, are "index numbers" rather than prices in dollars and cents. That is, they give each month's price as a percent of the price in a base period (in this case, the years 1982 to 1984). So 150 means "150% of the base price."

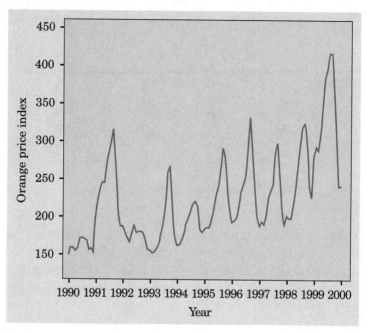

FIGURE 4.13 The price of fresh oranges, January 1990 to January 2000, for Exercise 4.8.

(a) The graph shows strong seasonal variation. How is this visible in the graph? Why would you expect the price of fresh oranges to show seasonal variation? (b) What is the overall trend in orange prices during this period, after we take account of the seasonal variation?

4.9 Electrical usage Professor Yates recorded the number of kilowatt-hours of electricity used for his house in Virginia, where there are four full seasons a year. The professor has a heat pump that doubles as an air conditioner in the summer. Here are the data for 2001 and 2002:

Month	kWh	Month	kWh	Month	kWh
1/2001	5508	9/2001	1396	5/2002	1294
2/2001	2120	10/2001	1554	6/2002	1564
3/2001	3544	11/2001	2334	7/2002	1793
4/2001	1580	12/2001	2403	8/2002	2055
5/2001	1155	1/2002	4240	9/2002	1475
6/2001	1404	2/2002	3273	10/2002	1001
7/2001	1235	3/2002	2600	11/2002	2271
8/2001	1684	4/2002	1594	12/2002	4134

(a) Make a line graph of these data. Mark intervals 1 to 12 on the horizontal axis to represent the months. In order to compare usage for the two years, make one line graph for 2001, and then overlay a second line graph for 2002.

(b) The data show seasonal variation within years. Can you suggest an explanation for the seasonal variation in the professor's electrical usage?

(c) The average temperatures for January and February are very similar, yet the electrical usage is smaller both years in February than in January. Why is that?

(d) The electricity consumed in December 2002 is much higher than the same figure for 2001. Did the professor make a mistake, or can you suggest an explanation for this discrepancy?

(e) From your plot, would you say that it is more expensive to heat the house in the winter or air-condition the house in the summer?

4.10 The Boston Marathon Women were allowed to enter the Boston Marathon in 1972. The times (in minutes, rounded to the nearest minute) for the winning woman from 1972 to 2003 appear in Table 4.3.

(a) Make a graph of the winning times.

(b) Give a brief description of the pattern of Boston Marathon winning times over these years. Have times stopped improving in recent years?

TABLE 4.3 Women's Winning Times (Minutes) in the Boston Marathon

Year	Time	Year	Time	Year	Time
1972	190	1983	143	1994	142
1973	186	1984	149	1995	145
1974	167	1985	154	1996	147
1975	162	1986	145	1997	146
1976	167	1987	146	1998	143
1977	168	1988	145	1999	143
1978	165	1989	144	2000	146
1979	155	1990	145	2001	144
1980	154	1991	144	2002	141
1981	147	1992	144	2003	145
1982	150	1993	145		

APPLICATION 4.1 College Students and Credit Cards

Nellie Mae, the federal agency that manages the college student loan program, conducted a study of credit card usage in 1998, prompted by concern over increasing credit card activity among college students. They noticed a significant increase since the early to mid-1990s in the amount of credit card balances of their student loan applicants. Indeed, the proliferation of offers on-campus, in the mail, and on the Internet, of free gifts, bonus airline miles, and low introductory rates for each new card is difficult for students to resist. The Nellie Mae study was repeated in 2000 and again in 2001. Here are some of the findings from the 2001 study, broken down by grade level.[5]

Credit card usage by grade level	Fresh	Soph	Jr	Sr/+
Percent who have credit cards	54%	92%	87%	96%
Average number of credit cards	2.5	3.67	4.5	6.13
Percent who have four or more cards	26%	44%	50%	66%
Average credit card debt	$1533	$1825	$2705	$3262
Median credit card debt	$901	$1564	$1872	$2185
Percent with balances between $3000 and $7000	8%	18%	24%	31%
Percent with balances exceeding $7000	4%	4%	7%	9%

Source: Bureau of Economic Analysis, *Survey of Current Business*, available online at www.bea.doc.gov.

1. Write a sentence that describes the usage rates and balances each year from first to final year as students progress through their four (or more) years in college.

2. What percent increase is there between the percent of freshmen who have at least one credit card and the percent of seniors who have at least one credit card?

3. Compare the amount of average credit card debt for freshmen with the average credit card debt for seniors.

4. Construct a bar graph for the percent of college students who have at least one credit card. Interpret your bar graph in a sentence.

5. Construct a line graph for the average credit card debt (*y* axis), using the scale 1, 2, 3, and 4 for the number of years in college (*x* axis). What does this graph show you?

STATISTICS IN SUMMARY

To see what data say, start with graphs. The choice of graph depends on the type of data. Do you have a **categorical variable,** such as level of education or occupation, which puts individuals into categories? Or do you have a **quantitative variable** measured in meaningful numerical units? To display the **distribution** of a categorical variable, use a **pie chart** or a **bar graph.** Pie charts always show the parts of some whole, but bar graphs can compare any set of numbers measured in the same units. To show how a quantitative variable changes over time, use a **line graph** that plots values of the variable (vertical scale) against time (horizontal scale).

Graphs can mislead the eye. Avoid **pictograms,** which replace the bars of a bar graph with pictures whose height and width both change. Look at the scales of a line graph to see if they have been stretched or squeezed to create a particular impression. Avoid clutter that makes the data hard to see.

SECTION 4.1 EXERCISES

4.11 The Border Patrol Here are the numbers of deportable aliens caught by the U.S. Border Patrol for 1971 through 1997.[6] Display these data in a graph. What are the most important facts that the data show?

Year:	1971	1973	1975	1977	1979	1981	1983
Count (1000s):	420	656	767	1042	1076	976	1251

Year:	1985	1987	1989	1991	1993	1995	1997
Count (1000s):	1349	1190	954	1198	1327	1395	1536

4.12 Line graphs Sketch line graphs of a series of observations over time having each of these characteristics. Mark your time axis in years.
(a) A strong downward trend, but no seasonal variation.
(b) Seasonal variation each year, but no clear trend.
(c) A strong downward trend with yearly seasonal variation.

4.13 Exports Figure 4.14 (next page) compares the value of exports (in billions of dollars) in 1997 from the world's leading exporters: Germany, Japan, and the United States.
(a) Explain why this is not an accurate graph.
(b) Make an accurate graph to compare the exports of these nations.

4.14 Bad habits According to the National Household Survey on Drug Abuse, 20.5% of adolescents aged 12 to 17 years used alcohol in 1997, 9.4% used marijuana, 1.0% used cocaine, and 19.9% used cigarettes. Explain why it is *not* correct to display these data in a pie chart.

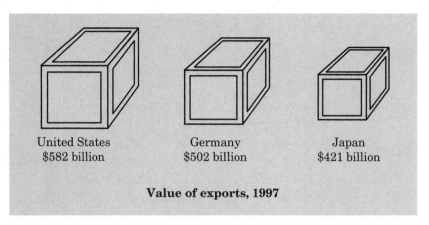

Value of exports, 1997

United States
$582 billion

Germany
$502 billion

Japan
$421 billion

FIGURE 4.14 Value of goods exported by Germany, Japan, and the United States in 1997, for Exercise 4.13. (Data from the Organization of Economic Cooperation and Development.)

4.15 Organ transplants Based on reports of procurement programs and transplant centers in the United States, in 2001, there were 2202 heart transplants, 27 heart-lung transplants, 1054 lung transplants, 5177 liver transplants, 14,152 kidney transplants, and 1464 other transplants (kidney-pancreas, pancreas, intestine, and multi-organ).[7]
(a) Find the percent of transplants for each of these categories, rounded to the nearest percent.
(b) Make a well-labeled graph of the distribution of organ transplants.

4.16 Trucks versus cars Consumers are turning to trucks, SUVs, and minivans in place of passenger cars. Here are data on sales of new cars and trucks in the United States. (The definition of "truck" includes SUVs and minivans.) Plot two line graphs on the same axes to compare the change in car and truck sales over time. Describe the trend that you see.

Year:	1981	1983	1985	1987	1989
Cars (1000s):	8536	9182	11,042	10,277	9772
Trucks (1000s):	2260	3129	4682	4912	4941

Year:	1991	1993	1995	1997	1999
Cars (1000s):	8175	8518	8636	8273	8697
Trucks (1000s):	4365	5681	6481	7226	8717

Source: Bureau of Economic Analysis, *Survey of Current Business,* available online at www.bea.doc.gov.

4.17 Seasonal variation You examine the average temperature in Chicago each month for many years. Do you expect a line graph of the data to show seasonal variation? Describe the kind of seasonal variation you expect to see.

4.2 DISPLAYING DISTRIBUTIONS WITH GRAPHS

Where does my school stand?

Tonya is a student at Grand Valley State University in her home state of Michigan. Grand Valley State charges in-state students $4108 in tuition and fees. Tonya wonders how Grand Valley State's tuition compares with the amount charged by other Michigan colleges and universities.

There are 81 colleges and universities in Michigan. Their tuition and fees for the 1999–2000 school year run from $1260 at Kalamazoo Valley Community College to $19,258 at Kalamazoo College.[8] It isn't easy to compare Grand Valley State with a list of 80 other colleges. Let's make a graph. The **histogram** in Figure 4.15 shows that many colleges charge less than $2000. The distribution extends out to the right. At the upper extreme, two colleges charge between $18,000 and $20,000. Grand Valley State is one of 5 colleges in the $4000 to $6000 group.

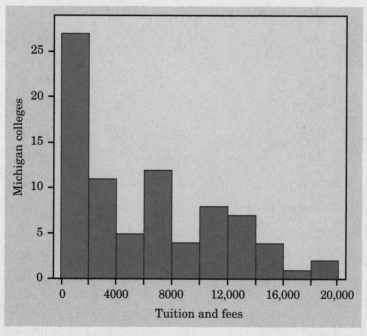

FIGURE 4.15 Histogram of the tuition and fees charged by 81 Michigan colleges and universities in 1999-2000 academic year.

Figure 4.16 is another, less familiar, graph of the same data. We have rounded the tuition charges to the nearest hundred dollars. Grand Valley State's $4108, for example, becomes just 41. Put the thousands digit or digits to the left of the vertical line, and hang the hundreds one by one on these "stems." Grand Valley State is the 1 on the 4 stem. This is a **stemplot**. The stemplot shows more detail than the histogram. Tonya can easily see, for example, that Grand Valley State's tuition and fees are 39th out of 81 colleges when we count up from the lowest charge.

```
 1 | 3555555556666667777788889999
 2 | 8
 3 | 155666899
 4 | 0125
 5 | 01
 6 | 133466
 7 | 014567
 8 | 169
 9 | 7
10 | 139
11 | 13678
12 | 39
13 | 24457
14 | 39
15 | 16
16 | 0
17 |
18 | 2
19 | 3
```

FIGURE 4.16 Stemplot of the Michigan tuition and fee data.

ACTIVITY 4.2 The nutritional value of cereals

In this activity we will make and interpret a graph called a histogram that will provide a graphical representation of the distribution of a list of numbers. We will also get some experience interpreting

ACTIVITY 4.2 The nutritional value of cereals *(continued)*

numerical descriptions of a distribution. A data file called CEREALS is stored in text format and also as an Excel file at the Web site www.whfreeman.com/sta. This file provides nutritional information for 77 brands of breakfast cereal.[9] Get a printout of this file. For each brand of cereal, we show the number of calories per serving, milligrams of sodium, grams of sugar, and grams of fat.

Your teacher will give you a copy of a histogram of the sodium content of the cereals, or you can use the *1-Variable Statistics Calculator* applet at the Web site to make your own histogram. Notice that the class intervals provided by the software are all equal in width. The vertical axis uses a frequency scale that indicates a count of the number of values falling into each class interval. Most software programs use a left endpoint convention for the histograms they produce (values at the left endpoint of the bar are included; values at the right endpoint are not).

1. Describe the shape of the histogram of the sodium content of the cereals. Use the histogram of the cereals' sodium content to answer the following questions.

2. What is the width of each bar?

3. What is the smallest amount of sodium in any of these cereals?

4. What appears to be the largest amount of sodium in any of these cereals?

5. How many cereals have no sodium at all?

6. How many brands have 200 milligrams per serving or more?

If you used statistical software to make your histogram, you could resize the display to make the histogram taller or shorter or narrower or wider by changing the starting and ending values and the "bin" width (the width of the bar). Experiment with changing the histogram characteristics. As you do this, do the features of the histogram change your impression of what the data say? For example, does changing the height or width of the histogram change your impression of the data? Explain.

7. Approximately what is the range (lowest to highest) of the sodium content of the cereals made by Ralston Purina in this data set?

Sodium tends to increase one's risk of high blood pressure. Elevated blood pressure, in turn, increases the risk of heart disease and stroke.

8. Which manufacturer sells the cereal with the highest sodium content?

9. Which manufacturers offer cereals that are essentially sodium-free? Which make high-sodium cereals? Which cereals would you recommend for people with high blood pressure?

ACTIVITY 4.2 The nutritional value of cereals *(continued)*

10. How do the different manufacturers contribute to the overall shape of the histogram of sodium content that you described in question 1? Comment about how each manufacturer contributes to the low, middle, and high areas of the sodium histogram.

Additional investigations with the other cereal data are described in Exercise 4.27 on page 235.

Histograms

Categorical variables just record group membership, such as the marital status of a woman or the race of a college student. We can use a pie chart or bar graph to display the distribution of categorical variables because they have relatively few values. What about quantitative variables such as the SAT scores of students admitted to a college or the income of families? These variables, like the sodium content of cereals in Activity 4.2, take so many values that a graph of the distribution is clearer if nearby values are grouped together. The most common graph of the distribution of a quantitative variable is a **histogram.**

EXAMPLE 4.9   How to make a histogram

Table 4.4 presents the percent of residents aged 65 years and over in each of the 50 states. To make a histogram of this distribution, proceed as follows.

Step 1: Divide the range of the data into classes of equal width. The data in Table 4.4 range from 5.7 to 17.6, so we choose as our classes

$$5.0 \leq \text{percent over } 65 < 6.0$$
$$6.0 \leq \text{percent over } 65 < 7.0$$

.

.

.

$$17.0 \leq \text{percent over } 65 < 18.0$$

Be sure to specify the classes precisely so that each individual falls into exactly one class. A state with 5.9% of its residents aged 65 or older would fall into the first class, but 6.0% falls into the second.

TABLE 4.4 Percent of Residents Aged 65 and Over in the States, 2000

State	Percent	State	Percent	State	Percent
Alabama	13.0	Louisiana	11.6	Ohio	13.3
Alaska	5.7	Maine	14.4	Oklahoma	13.2
Arizona	13.0	Maryland	11.3	Oregon	12.8
Arkansas	14.0	Massachusetts	13.5	Pennsylvania	15.6
California	10.6	Michigan	12.3	Rhode Island	14.5
Colorado	9.7	Minnesota	12.1	South Carolina	12.1
Connecticut	13.8	Mississippi	12.1	South Dakota	14.3
Delaware	13.0	Missouri	13.5	Tennessee	12.4
Florida	17.6	Montana	13.4	Texas	9.9
Georgia	9.6	Nebraska	13.6	Utah	8.5
Hawaii	13.3	Nevada	11.0	Vermont	12.7
Idaho	11.3	New Hampshire	12.0	Virginia	11.2
Illinois	12.1	New Jersey	13.2	Washington	11.2
Indiana	12.4	New Mexico	11.7	West Virginia	15.3
Iowa	14.9	New York	12.9	Wisconsin	13.1
Kansas	13.3	North Carolina	12.0	Wyoming	11.7
Kentucky	12.5	North Dakota	14.7		

Source: *Statistical Abstract of the United States*, 2000.

Step 2: Count the number of individuals in each class. Here are the counts.

Class	Count	Class	Count	Class	Count
5.0 to 5.9	1	10.0 to 10.9	1	15.0 to 15.9	2
6.0 to 6.9	0	11.0 to 11.9	8	16.0 to 16.9	0
7.0 to 7.9	0	12.0 to 12.9	13	17.0 to 17.9	1
8.0 to 8.9	1	13.0 to 13.9	14		
9.0 to 9.9	3	14.0 to 14.9	6		

Step 3: Draw the histogram. Mark on the horizontal axis the scale for the variable whose distribution you are displaying. That's "percent of residents aged 65 and over" in this example. The scale runs from 4 to 18 because that range spans the classes we chose. The vertical axis contains the scale of counts. Each bar represents a class. The base of the bar covers the class, and the bar height is the class count. *There is no horizontal space between the bars unless a class is empty,* so that its bar has height zero. Figure 4.17 (next page) is our histogram.

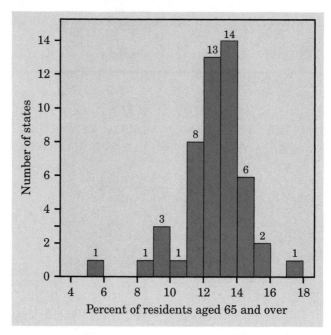

FIGURE 4.17 Histogram of the percent of residents aged 65 and older in the 50 states. Note the three outliers.

Just as with bar graphs, our eyes respond to the area of the bars in a histogram. Be sure that the classes for a histogram have equal widths. There is no one right choice of the classes. Too few classes will give a "skyscraper" histogram, with all values in a few classes with tall bars. Too many classes will produce a "pancake" graph, with most classes having one or no observations. Neither choice will give a good picture of the shape of the distribution. You must use your judgment in choosing classes to display the shape. Statistics software will choose the classes for you. The computer's choice is usually a good one, but you can change it if you want, as you saw in Activity 4.2.

EXERCISES

4.18 Poverty in the states Table 4.5 on the facing page shows the percent of people living below the poverty line in the 26 states east of the Mississippi.
(a) Make a histogram for these data. Make the classes go from 4 to 18 with 7 classes. Describe the shape of the distribution.
(b) Are there any unusual data values? Which states appear to be unusual? If there are several unusual observations, do they have any characteristic in common?

4.19 Dotplot Another way to picture a small set of data is the **dotplot.** You begin by drawing a horizontal axis and marking a scale from the smallest observation

TABLE 4.5 Percent of State Residents Living in Poverty, 1999

State	Percent	State	Percent	State	Percent
Alabama	12.5	Maryland	6.1	Pennsylvania	7.8
Connecticut	6.2	Massachusetts	6.7	Rhode Island	8.9
Delaware	6.5	Michigan	7.4	South Carolina	10.7
Florida	9.0	Mississippi	16.0	Tennessee	10.3
Georgia	9.9	New Hampshire	4.3	Vermont	6.3
Illinois	7.8	New Jersey	6.3	Virginia	7.0
Indiana	6.7	New York	11.5	West Virginia	13.9
Kentucky	12.7	North Carolina	9.0	Wisconsin	5.6
Maine	7.8	Ohio	7.8		

Source: *Statistical Abstract of the United States*, 2000.

to the largest. Of course, the intervals have to be uniform. Here is a dotplot of the percent of state residents living in poverty from the previous exercise. (The software that made this dotplot is Minitab.)

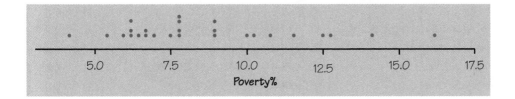

Do your answers to the previous exercise still hold when you look at this plot?

4.20 A stemplot Here is a Minitab stemplot of the poverty data from Exercise 4.18.

```
 1      4    3
 2      5    6
 9      6    1233577
(6)     7    048888
11      8    9
10      9    009
 7     10    37
 5     11    5
 4     12    57
 2     13    9
 1     14
 1     15
 1     16    0
```

(a) What is an advantage that this stemplot has over the histogram you made in Exercise 4.18?

(b) Copy the stemplot onto your paper, and then circle the two middle numbers in the stemplot. There should be as many numbers below the circled numbers as there are above them.

(c) The first column of numbers is a sort of code. Can you decipher the code? In particular, why do you think the number 6 is in parentheses?

4.21 Yankee money Table 4.6 gives the salaries of the players on the New York Yankees baseball team as of the opening day of the 2002 season. Make a histogram of these data.

TABLE 4.6 Salaries of the New York Yankees, 2002

Derek Jeter	$14,600,000	Bernie Williams	$12,357,143
Mike Mussina	11,000,000	Jason Giambi	10,428,571
Roger Clemens	10,300,000	Andy Pettitte	9,500,000
Mariano Rivera	9,450,000	Robin Ventura	8,500,000
Jorge Posada	7,000,000	Sterling Hitchcock*	4,936,719
Rondell White	4,500,000	Steve Karsay	4,000,000
Orlando Hernandez	3,200,000	Ramiro Mendoza*	2,600,000
Mike Stanton	2,500,000	David Wells	2,250,000
Gerald Williams	2,000,000	John Vander Wal	1,850,000
Shane Spencer	885,000	Ron Coomer	750,000
Enrique Wilson	720,000	Alberto Castillo	650,000
Alfonso Soriano	630,000	Ted Lilly	237,150
Randy Choate	223,350	Nick Johnson	220,650
Christian Parker*	218,600	Randy Keisler*	211,400
Jay Tessmer	~~210,000~~		

Note: Salaries are estimates from several Web sites. Most players have complex multiyear contracts.
*On the disabled list.

4.22 Setting Joe straight Joe's histogram from the poverty data in Table 4.5 is shown at the top of the next page. Write a sentence or two clearly explaining to Joe why his histogram is not legitimate.

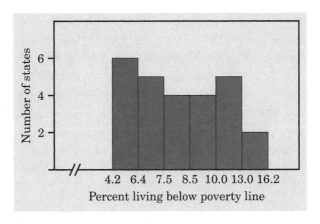

Interpreting histograms

Making a statistical graph is not an end in itself. The purpose of the graph is to help us understand the data. After you (or your computer) make a graph, always ask, "What do I see?" Here is a general strategy for looking at graphs.

> ## Pattern and deviations
>
> In any graph of data, look for an **overall pattern** and also for striking **deviations** from that pattern.

We have already applied this strategy to line graphs. Trend and seasonal variation are common overall patterns in a line graph. The upward spike in gasoline prices in Figure 4.7 (page 210) caused by the Iraqi invasion of Kuwait in 1990 is an example of a deviation from the general pattern. In the case of the histogram of Figure 4.17 (page 226), it is easiest to begin with deviations from the overall pattern of the histogram. Two states stand out. You can find them in the table once the histogram has called attention to them. Florida has 17.6% of its residents over age 65, and Alaska has only 5.7%. These states are clear *outliers.*

> ## Outliers
>
> An **outlier** in any graph of data is an individual observation that falls outside the overall pattern of the graph.

Is Utah, with 8.5% of its population over 65, an outlier? Whether an observation is an outlier is to some extent a matter of judgment. Utah is the smallest of the main body of observations, not as separated from the general pattern as Florida and Alaska are. It's not as obvious from the histogram that Utah is an outlier, but it is. Once you have spotted outliers, look for an explanation. Many outliers are due to mistakes, such as typing 4.0 as 40. Other outliers point to the special nature of some observations. Explaining outliers usually requires some background information. It is not surprising that Florida, with its many retired people, has many residents over 65 and that Alaska, the northern frontier, has few. How would you explain Uttah?

To see the *overall pattern* of a histogram, ignore any outliers. Here is a simple way to organize your thinking.

Overall pattern of a distribution

To describe the overall pattern of a distribution:
- Give the **center** and the **spread.**
- See if the distribution has a simple **shape** that you can describe in a few words.

We will learn how to describe center and spread numerically in Section 4.3. For now, we can describe the center of a distribution by its *midpoint,* the value with half the observations taking smaller values and half taking larger values. We can describe the spread of a distribution by giving the *smallest and largest values,* ignoring any outliers.

EXAMPLE 4.10   Describing distributions

Look again at the histogram in Figure 4.17 (page 226). **Shape:** The distribution has a *single peak.* It is roughly *symmetric*—that is, the pattern is similar on both sides of the peak. **Center:** The midpoint of the distribution is close to the single peak, at about 13%. **Spread:** The spread is about 9% to 16% if we ignore the three outliers.

The distribution of tuition and fees at Michigan colleges, shown in Figure 4.15 (page 221), has a quite different **shape.** There is a strong *peak* at the lowest cost class. And even though most colleges charge less than $8000, there is a long right "tail" extending up to $20,000. We call a distribution with a long tail on one side *skewed.* The **center** is roughly $4500 (half the colleges charge less than this). The **spread** is large, from less than $2000 to more than $19,000. There are no outliers—the colleges with the highest tuition just continue the long right tail that is part of the overall pattern.

When you describe a distribution, concentrate on the main features. Look for major peaks, not for minor ups and downs in the bars of the histogram like those in Figure 4.15. Look for clear outliers, not just for the smallest and largest observations. Look for rough *symmetry* or clear *skewness*.

Symmetric and skewed distributions

A distribution is **symmetric** if the right and left sides of the histogram are approximately mirror images of each other.

A distribution is **skewed to the right** if the right side of the histogram (containing the half of the observations with larger values) extends much farther out than the left side. It is **skewed to the left** if the left side of the histogram extends much farther out than the right side.

In mathematics, symmetry means that the two sides of a figure like a histogram are exact mirror images of each other. Data are almost never exactly symmetric, so we are willing to call histograms like that in Figure 4.17 roughly symmetric as an overall description. The tuition distribution in Figure 4.15, on the other hand, is clearly skewed to the right. Here are more examples.

EXAMPLE 4.11   Sampling again

The values a statistic takes in many random samples from the same population form a distribution with a regular pattern. The histogram in Figure 4.18 (next page) displays a distribution that we met in Chapter 2. Take a simple random sample of 1523 adults. Ask each whether they bought a lottery ticket in the last 12 months. The proportion who say "Yes" is the sample proportion. Do this 1000 times and collect the 1000 sample proportions from the 1000 samples. Figure 4.18 shows the distribution of 1000 sample proportions when the truth about the population is that 60% have bought lottery tickets.

This distribution is quite symmetric about a single peak in the center. The center is at 0.60, reflecting the lack of bias of the statistic. The spread from the smallest to the largest of the 1000 values is from 0.556 to 0.643.

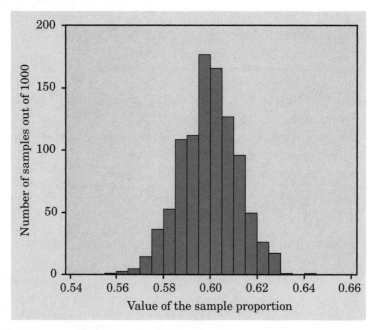

FIGURE 4.18 Histogram of the sample proportion p for 1000 simple random samples from the same population. This is a symmetric distribution.

EXAMPLE 4.12   Infant low birth weight

Figure 4.19(a) shows the distribution of percent of low birth weights for the 26 states east of the Mississippi in 2000.[10] Three states (Maine, Vermont, and New Hampshire) had the smallest percent of infants born with low birth weight, 6 to 7%. States that had 7 to 8% born with low birth weights were most numerous (11 states). We say that the distribution is *skewed right* because the right tail of the distribution extends farther out than the left tail. The center of the distribution is about 8; that is, about half of the states had percents of low birth weight that were less than 8.

Notice that the vertical scale in Figure 4.19(a) is the *number* of states in each class. We call this a **frequency histogram** because it shows counts. In Figure 4.19(b), we have calculated and displayed the *percent* of states in each low-birth-weight class. A histogram of percents rather than counts is convenient when the counts are very large or when we want to compare several distributions. There are many factors that contribute to low birth weight, such as poverty, access to health care, and education. All of these are factors for Mississippi, which had the highest percent of low birth rate, 10.1, for the eastern half of the United States.

(a)

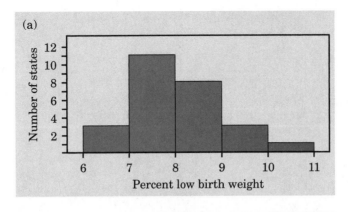

(b)

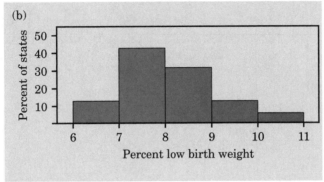

FIGURE 4.19 The distribution of infants born with low birth weight in the 26 states east of the Mississippi. (a) This distribution is skewed to the right. (b) The heights of the bars are the same, but the vertical scale has been changed to show the percent of states in each class.

EXERCISES

4.23 More about Yankee salaries Is the distribution of Yankees' salaries in Exercise 4.21 (page 228) roughly symmetrical, skewed to the left, or skewed to the right? Explain briefly. What is the spread of the distribution?

4.24 Lightning flashes Figure 4.20 (next page) comes from a study of lightning storms in Colorado.[11] It shows the distribution of the hour of the day during which the first lightning flash for that day occurred. Describe the shape, center, and spread of this distribution. Are there any outliers?

4.25 Minority students in engineering Figure 4.21 (next page) is a histogram of the number of minority students (black, Hispanic, Native American) who earned doctorate degrees in engineering from each of 115 universities in the years 1992 through 1996.[12] Briefly describe the shape, center, and spread of this distribution.

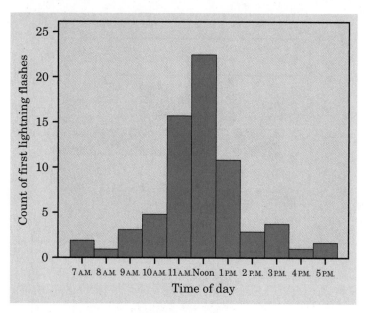

FIGURE 4.20 Histogram of time of day at which the day's first lightning flash occurred (from a study in Colorado), for Exercise 4.24.

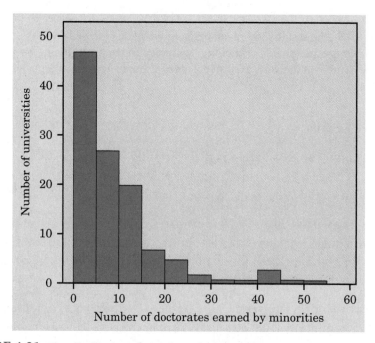

FIGURE 4.21 The distribution of number of engineering doctorates earned by minority students at 115 universities, 1992 to 1996, for Exercise 4.25.

4.26 The changing age distribution of the United States The distribution of the ages of a nation's population has a strong influence on economic and social conditions. Table 4.7 shows the age distribution of U.S. residents in 1950 and 2075, in millions of persons. The 1950 data come from that year's census. The 2075 data are projections made by the Census Bureau.

TABLE 4.7 **Age Distribution in the United States, 1950 and 2075 (in Millions of Persons)**

Age group	1950	2075
Under 10 years	29.3	34.9
10 to 19 years	21.8	35.7
20 to 29 years	24.0	36.8
30 to 39 years	22.8	38.1
40 to 49 years	19.3	37.8
50 to 59 years	15.5	37.5
60 to 69 years	11.0	34.5
70 to 79 years	5.5	27.2
80 to 89 years	1.6	18.8
90 to 99 years	0.1	7.7
100 to 109 years	—	1.7
Total	151.1	310.6

(a) Because the total population in 2075 is much larger than the 1950 population, comparing percents in each age-group is clearer than comparing counts. Make a table of the percent of the total population in each age-group for both 1950 and 2075.

(b) Make a histogram of the 1950 age distribution (in percents). Then describe the main features of the distribution. In particular, look at the percent of children relative to the rest of the population.

(c) Make a histogram of the projected age distribution for the year 2075. Use the same scales as in (b) for easy comparison. What are the most important changes in the U.S. age distribution projected for the 125-year period between 1950 and 2075?

4.27 Activity 4.2 follow-up Use the CEREALS data from Activity 4.2 and the *1-Variable Statistics Calculator* applet on the Web site to investigate the calories, sugar, or fat content of the 77 brands of cereal. Answer questions that seem interesting to you. Let the questions in Activity 4.2 be a guide.

Stemplots

Histograms are not the only graphical display of distributions. For small data sets, a *stemplot* is quicker to make and presents more detailed information.

> ## Stemplot
>
> To make a **stemplot:**
> **1.** Separate each observation into a **stem,** consisting of all but the final (rightmost) digit, and a **leaf**, the final digit. Stems may have as many digits as needed, but each leaf contains only a single digit.
> **2.** Write the stems in a vertical column with the smallest at the top, and draw a vertical line at the right of this column.
> **3.** Write each leaf in the row to the right of its stem, in increasing order out from the stem.

What the eye really sees

We make the bars in bar graphs and histograms equal in width because the eye responds to their area. That's roughly true. Careful study by statistician William Cleveland shows that our eyes "see" the size of a bar in proportion to the 0.7 power of its area. Suppose, for example, that one figure in a pictogram is both twice as high and twice as wide as another. The area of the bigger figure is 4 times that of the smaller. But we see the size of the bigger figure as 2.6 times the size of the smaller, because 2.6 is 4 to the 0.7 power.

EXAMPLE 4.13   Stemplot of the "65 and over" data

For the "65 and over" percents in Table 4.4, the whole-number part of the observation is the stem and the final digit (tenths) is the leaf. The Alabama entry, 13.0, has stem 13 and leaf 0. Stems can have as many digits as needed, but each leaf must consist of only a single digit. Figure 4.22 on the facing page shows the steps in making a stemplot for the data in Table 4.4. First, draw the stems. Then go through the table adding a leaf for each observation. Finally, arrange the leaves in increasing order out from each stem.

A stemplot looks like a histogram turned on end. The stemplot in Figure 4.22 is just like the histogram in Figure 4.17 (page 226) because the classes chosen for the histogram are the same as the stems in the stemplot. The stemplot in Figure 4.16 (page 222), on the other hand, has almost twice as many classes as the histogram of the same data in Figure 4.15 (page 221). We interpret stemplots like histograms, looking for the overall pattern and for any outliers.

You can choose the classes in a histogram. The classes (the stems) of a stemplot are given to you. You can get more

```
10 | 9
11 | 0
12 | 1344677889
13 | 0012455566789999
14 | 11222344445789
15 | 24478999
```

FIGURE 4.23 Stemplot of the percent of each state's residents who are 25 to 34 years old, for Exercise 4.28.

(a) The mountain states Montana and Wyoming have the smallest percents of young adults, perhaps because they lack job opportunities. What are the percents for these two states?

(b) Ignoring Montana and Wyoming, describe the shape, center, and spread of this distribution.

(c) Is the distribution for young adults more or less spread out than the distribution in Figure 4.22 for older adults?

4.29 More poverty Construct a stemplot of the poverty data in Table 4.5 (page 227). Is the shape of your stemplot about the same as that of the histogram you constructed in Exercise 4.18 (page 226)? Which type of display do you prefer for these data—histogram or stemplot—and why?

4.30 Babe Ruth's home runs Here are the numbers of home runs that Babe Ruth hit in his 15 years with the New York Yankees, 1920 to 1934:

   54   59   35   41   46   25   47   60   54   46   49   46   41   34   22

(a) Make a stemplot of these data. Is the distribution roughly symmetric, clearly skewed, or neither?

(b) About how many home runs did Ruth hit in a typical year?

(c) Are his famous 60 home runs in 1927 an outlier?

4.31 How many calories does a hot dog have? *Consumer Reports* magazine presented the following data on the number of calories in a hot dog for each of 17 brands of meat hot dogs:[13]

73	191	182	190	172	147	146	139	175
136	179	153	107	195	135	140	138	

Make a stemplot of the distribution of calories in meat hot dogs and briefly describe the shape of the distribution. Most brands of meat hot dogs contain a mixture of beef and pork, with up to 15% poultry allowed by government regulations. The only brand with a different makeup is *Eat Slim Veal Hot Dogs*. Which point on your stemplot do you think represents this brand?

4.32 Back-to-back stemplot The major league single-season home run record was held briefly by Mark McGwire of the St. Louis Cardinals. Here are McGwire's home run counts for 1987 to 2001, his final season:

49   32   38   39   22   42   9   9   39   52   58   70   65   32   29

A **back-to-back stemplot** helps us compare two distributions. Write the stems as usual, but with a vertical line both to their left and to their right. On the right, put leaves for Ruth (Exercise 4.30). On the left, put leaves for McGwire. Arrange the leaves on each stem in increasing order out from the stem. Now write a brief comparison of Ruth and McGwire as home run hitters. McGwire was injured in 1993 and there was a baseball strike in 1994. He had a knee injury in 2000 and was out half a season. How do these events appear in the data?

APPLICATION 4.2 Roller Coaster Safety

Roller coasters are popular with a large segment of the population, especially teens. To go fast, experience frightening drops, be turned upside down, and be subjected to several times the force of gravity are all part of the allure. Yet roller coasters can be dangerous, even deadly. Hard data on the number of accidents, injuries, and deaths can be hard to come by because there is no national requirement for reporting roller coaster injuries and deaths. And such reports are not required by many states. The data file COASTERS has information on roller coaster–connected deaths from 1980 through 2002 that have been documented by an interested layperson and posted on the Web. Because these data were not collected systematically, there may be additional deaths that were missed, and so the numbers may be higher. As is true with many sets of data that you will encounter, there are some gaps in the data. The data include the following fields:

YEAR   STATE   AGE   MALE/FEMALE   CUSTOMER/EMPLOYEE   DESCRIPTION

Print out a text version of the data to work from. Then, *based on these data*, answer the questions or carry out the instructions that follow.

APPLICATION 4.2 Roller Coaster Safety (continued)

1. Medical condition includes history of heart trouble, asthma, and brain aneurysm. What do you think "unsafe practice" refers to?

2. Why are some states listed multiple times and other states not listed at all?

3. Are boys more likely to be killed in a roller coaster accident than girls? Or are girls more likely to be killed?

4. Are children (less than 18 years old) or adults (18 and older) more likely to be killed in a roller coaster accident?

5. What percent of roller coaster fatalities are

   • employees or maintenance workers?

   • caused by operator error or equipment failure?

   • caused by unsafe practice (such as standing up or leaning out of the car)?

6. The data have been classified according to categories. Construct at least two bar graphs that plot the number of fatalities by a category of your choice. What information can you glean from your bar graphs?

7. Construct a histogram of the number of deaths per year. Describe the shape, approximate center, and spread.

8. Construct a line graph of number of deaths (y axis) by year (x axis). Is there a discernible trend in the graph? For example, perhaps the number of deaths is decreasing over time due to better safety requirements, better engineering, or better customer education. If there is a trend over time, how could this trend be explained?

9. List at least one fact that each plot tells you that the other plots don't tell you.

10. How confident are you that these data present a complete picture of roller coaster fatalities in the United States? Note, for example, that there are no entries for the period 1990 to 1993.

11. Losing a family member or friend in a roller coaster accident is devastating. Still, in a strictly statistical comparison, how would you rate the risk of a fatality in a roller coaster accident when compared with, say, the risk of dying in an automobile accident, in an airplane crash, or while working around the house?

12. Do a Web search on "roller coaster fatalities" or "roller coaster safety" to find recent articles (For example, Time, Newsweek, ABCnews.com, etc.). Things to look for: accident/injury reporting requirements (federal, state); statistics, like number of roller coaster injuries per year; trends; and proposed amusement park legislation.

EXPLORING THE WEB

Professor Webster West (University of South Carolina) has a very simple applet at www.stat.sc.edu/~west/javahtml/Histogram.html that uses a slider to let users see how the widths or the number of bins affect the shape of a histogram.

Another applet, at www.shodor.org/interactivate/activities/histogram/, does much the same thing but has colorful histograms and is more complicated. It has seven data sets to explore and also has a slider. Click-and-drag sliders are great fun to use.

A site developed by Hofstra faculty, people.hofstra.edu/faculty/Stefan_Waner/ stats/histogram.html, has an applet that lets you enter data, and then it constructs the corresponding histogram for your data.

The site www.skymark.com/resources/tools/histograms.asp has a nice summary of histograms and differently shaped distributions from a business manager's point of view.

STATISTICS IN SUMMARY

The **distribution** of a variable tells us what values the variable takes and how often it takes each value. To display the distribution of a quantitative variable, use a **histogram** or a **stemplot.** We usually favor stemplots when we have a small number of observations. For large amounts of data, histograms are usually better.

When you look at a graph, look for an **overall pattern** and for **deviations** from that pattern, such as **outliers.** We can describe the overall pattern of a histogram or stemplot by giving its **shape, center,** and **spread.** Some distributions have simple shapes such as **symmetric** or **skewed,** but others are too irregular to describe by a simple shape.

SECTION 4.2 EXERCISES

4.33 Boston Marathon

(a) Make a stemplot of the winning times by women runners in the Boston marathon (Table 4.3, page 217). Are there any unusual observations?

(b) Describe the shape of the distribution.

(c) When a stemplot is too compact, we may want to **split the stems.** This means to divide each stem in half. Drop the two high values, split the stems, and make a new stemplot. The stems are shown in the margin with several observations to help get you started. Is the shape of this new distribution similar to or different from the shape of the original stemplot?

14	
14	
15	
15	5
16	2
16	7785

4.34 Home run king Barry Bonds, the San Francisco Giants slugger, broke McGwire's home run record in 2001 with 73 home runs. Here are Bonds's home run statistics from 1986 to 2002:

16 25 24 19 33 25 34 46 37 33 42 40 37 34 49 73 46

(a) Construct a back-to-back stemplot to compare McGwire and Bonds as home run hitters. (Exercise 4.32, page 240, explains how to make a back-to-back stemplot.)
(b) Write a few sentences about this comparison.

4.35 Longevity of presidents Table 4.8 shows the ages at death of U.S. presidents.

TABLE 4.8 Age at Death of U.S. Presidents

Washington	67	Fillmore	74	T. Roosevelt	60
Adams	90	Pierce	64	Taft	72
Jefferson	83	Buchanan	77	Wilson	67
Madison	85	Lincoln	56	Harding	57
Monroe	73	A. Johnson	66	Coolidge	60
Adams	80	Grant	63	Hoover	90
Jackson	78	Hayes	70	F. Roosevelt	63
Van Buren	79	Garfield	49	Truman	88
Harrison	68	Arthur	56	Eisenhower	78
Tyler	71	Cleveland	71	Kennedy	46
Polk	53	Harrison	67	L. Johnson	64
Taylor	65	McKinley	58	Nixon	81

(a) Make a stemplot of these data. Then split the stems (see Exercise 4.33) and make another stemplot. Which stemplot do you prefer, and why? Deciding to split the stems is sometimes a judgment call.
(b) Make a histogram of these data.
(c) Which plot is better for this application, a stemplot or a histogram? Why?

4.36 Skewed left
(a) Sketch a histogram for a distribution that is skewed to the left.
(b) Suppose that you and your friends emptied your pockets of coins and recorded the year marked on each coin. The distribution of dates would be skewed to the left. Explain why.

4.37 The obesity epidemic Medical authorities describe the spread of obesity in the United States as an epidemic. Table 4.9 gives the percent of adults who are obese in each of the 45 states that participated in a study of the problem. (NA marks the states for which data are not available.) Display the distribution in a graph and briefly describe its shape, center, and spread.

TABLE 4.9 Percent of Adult Population Who Are Obese, 1998

State	Percent	State	Percent	State	Percent
Alabama	20.7	Louisiana	21.3	Ohio	19.5
Alaska	20.7	Maine	17.0	Oklahoma	18.7
Arizona	12.7	Maryland	19.8	Oregon	17.8
Arkansas	NA	Massachusetts	13.8	Pennsylvania	19.0
California	16.8	Michigan	20.7	Rhode Island	NA
Colorado	14.0	Minnesota	15.7	South Carolina	20.2
Connecticut	14.7	Mississippi	22.0	South Dakota	15.4
Delaware	16.6	Missouri	19.8	Tennessee	18.5
Florida	17.4	Montana	14.7	Texas	19.9
Georgia	18.7	Nebraska	17.5	Utah	15.3
Hawaii	15.3	Nevada	NA	Vermont	14.4
Idaho	16.0	New Hampshire	14.7	Virginia	18.2
Illinois	17.9	New Jersey	15.2	Washington	17.6
Indiana	19.5	New Mexico	14.7	West Virginia	22.9
Iowa	19.3	New York	15.9	Wisconsin	17.9
Kansas	NA	North Carolina	19.0	Wyoming	NA
Kentucky	19.9	North Dakota	18.7		

Source: Ali H. Mokdad et al., "The spread of the obesity epidemic in the United States, 1991–1998," *Journal of the American Medical Association,* 282 (1999), pp. 1519–1522.

4.38 Stock returns, I Here are the yearly returns on common stocks for the 50 years from 1950 to 1999, arranged from smallest to largest. The returns are rounded to the nearest whole percent.

−26	−15	−10	−10	−9	−8	−7	−5	−3	0
1	1	4	5	6	7	7	8	10	11
12	12	14	16	16	17	19	19	19	20
21	22	22	23	23	24	24	27	29	29
30	32	32	32	32	33	37	38	43	50

(a) Construct a histogram of these data. Make the class widths multiples of 10. Describe the shape, center, and spread of this distribution.

(b) What is the most interesting thing you discovered about returns on common stocks over this period of time?

4.39 When it rains, it pours On July 6, 1994, 21.10 inches of rain fell on Americus, Georgia. That's the most rain ever recorded in Georgia for a 24-hour period. Table 4.10 gives the maximum precipitation ever recorded in 24 hours (through 1998) at any weather station in each state. The record amount varies a great deal from state to state—hurricanes bring extreme rains on the Atlantic coast, and the mountain West is generally dry.

(a) Make a graph to display the distribution of records for the states. Mark where your state lies in this distribution.

(b) Briefly describe the distribution.

TABLE 4.10 Record 24-Hour Precipitation Amounts (Inches) by State

State	Inches	State	Inches	State	Inches
Alabama	32.52	Louisiana	22.00	Ohio	10.75
Alaska	15.20	Maine	13.32	Oklahoma	15.68
Arizona	11.40	Maryland	14.75	Oregon	11.65
Arkansas	14.06	Massachusetts	18.15	Pennsylvania	34.50
California	26.12	Michigan	9.78	Rhode Island	12.13
Colorado	11.08	Minnesota	10.84	South Carolina	17.00
Connecticut	12.77	Mississippi	15.68	South Dakota	8.00
Delaware	8.50	Missouri	18.18	Tennessee	11.00
Florida	38.70	Montana	11.50	Texas	43.00
Georgia	21.10	Nebraska	13.15	Utah	6.00
Hawaii	38.00	Nevada	7.13	Vermont	8.77
Idaho	7.17	New Hampshire	10.38	Virginia	27.00
Illinois	16.91	New Jersey	14.81	Washington	14.26
Indiana	10.50	New Mexico	11.28	West Virginia	19.00
Iowa	16.70	New York	11.17	Wisconsin	11.72
Kansas	12.59	North Carolina	22.22	Wyoming	6.06
Kentucky	10.40	North Dakota	8.10		

Source: National Oceanic and Atmospheric Administration Web site, www.ncdc.noaa.gov.

4.40 Histogram or stemplot? Explain why you might prefer a histogram to a stemplot for describing the record rainfalls for the states in Exercise 4.39.

4.3 DESCRIBING DISTRIBUTIONS WITH NUMBERS

Does education pay?

People with more education earn more on the average than people with less education. How much more? A simple way to say is to compare median incomes, the amount such that half of a group of people earn more and half earn less. Here are the median incomes of adults with four different levels of education:

High school graduate only	Some college, no degree	Bachelor's degree	Advanced degree
$16,297	$18,988	$32,581	$47,000

These numbers come from 71,512 people interviewed by the Current Population Survey in March 1999.[14] Every March, the CPS asks detailed questions about income. Boiling down 71,512 incomes into four numbers helps us get a quick picture of the relationship between education and income. The median college graduate, for example, earns about twice as much as the median high school graduate.

We know that incomes vary a lot. In fact, the highest income reported by the 31,621 people in the CPS sample who stopped with a high school diploma was $498,606. The medians compare the centers of the four income distributions. Can we also describe the spreads with just a few numbers? The largest and smallest incomes in a group of 71,512 people don't tell us much. Instead, let's give the range covered by the middle half of the incomes in each group. Here they are:

High school graduate only	Some college, no degree	Bachelor's degree	Advanced degree
$7412 to $29,000	$7803 to $33,150	$16,941 to $53,061	$28,075 to $75,162

The message of the 71,512 incomes is now clear: going to college but not getting a degree doesn't raise income much. People with bachelor's degrees, however, earn quite a bit more, and an advanced degree gives income another healthy boost. But there is a lot of variation among individuals—some very rich people never went to college. Finally, remember that this observational study says nothing about cause and effect. People who get advanced education are often smart, ambitious, and well-off to start with, so they might earn more even without their degrees.

ACTIVITY 4.3A Checking up on the candy makers

Materials: Small bags of chocolate-covered raisins or peanut M&M's (individual pieces of candy will be referred to as "items"); a weighing device such as a three-beam balance (borrowed from the science department) that weighs objects to 0.05 gram accuracy; plastic sandwich bags

RULE: The items are not to be consumed until the activity has been completed.

1. Students should be assigned to groups, with 3 to 4 students per group. Each group should have a weighing device. The same number of bags of the items will be distributed to each group.

2. Weigh the unopened bags, and record their weights on a piece of paper. You will need this information later.

3. Each group is provided enough bags of the items so that they can weigh 20 of the items. One member of the group will weigh an item and announce the weight, and another member will enter the weight into list L_1 on the calculator. Take turns with the tasks. Students not weighing or recording can monitor the weighing to ensure accuracy. It is critical that the weights be accurate, so take your time, and do a good job. Continue until the 20 items have been weighed and recorded.

4. Define Plot1 to be a histogram using the weights in list L_1. Then plot the histogram using the data in list L_1. Adjust the viewing window to leave some space at the top of the histogram.

5. Count out 30 of the items into a sandwich bag, and weigh the bag. Record these weights in a different list in the calculator. Return the 30 items to the pool, stir them up, and count out a new sample of 30 items. Weigh this new sample of 30. Continue until you have at least 5 weights for bags of 30 items. When you finish, groups should share their weights for bags of 30 so that you have at least 40 or 50 weights for bags of 30.

Keep your data in your notebook, and store your lists of weights of individual items and weights of bags of 30 in lists named CAND1 and CAND2. You will investigate these data later, after we have introduced some new concepts.

Measuring center: the median

A natural way to describe the center of a distribution is to use the "middle value" in a histogram or stemplot. That is, find the number such that half the observations are smaller and the other half are larger. This is the median of the distribution.

We will call the median M for short. Although the idea of the median as the midpoint of a distribution is simple, we need a precise rule for putting the idea into practice.

EXAMPLE 4.14   Finding the median

To find the median of the numbers

$$8 \quad 4 \quad 9 \quad 1 \quad 3$$

arrange them in increasing order as

$$1 \quad 3 \quad \mathbf{4} \quad 8 \quad 9$$

The boldface 4 is the center observation, because there are 2 observations to its left and 2 to its right. When the number of observations n is odd, there is always one observation in the center of the ordered list. This is the median, $M = 4$.

If n is even, there is no one middle observation. But there is a middle pair, and we take the median to be the mean of this middle pair, the point halfway between them. So the median of

$$8 \quad 4 \quad 1 \quad 9 \quad 1 \quad 5$$

is found by arranging these numbers in increasing order,

$$1 \quad 1 \quad \mathbf{4} \quad \mathbf{5} \quad 8 \quad 9$$

and averaging the middle pair

$$M = \frac{4 + 5}{2} = 4.5$$

Here is our rule for finding medians.

The median M

The **median** M is the midpoint of a distribution, the number such that half the observations are smaller and the other half are larger. To find the median of a distribution:

1. Arrange all observations in order of size, from smallest to largest.
2. If the number of observations n is odd, the median M is the center observation in the ordered list.
3. If the number of observations n is even, the median M is the average of the two center observations in the ordered list.

EXAMPLE 4.15   Finding another median

In Exercise 4.35, you were asked to make a stemplot of the ages at death of 36 presidents (see Table 4.8, page 243). There is an even number of data values, so to find the median age, count up from the smallest number in the stemplot or from a list ordered from 46 (John F. Kennedy) to 90 (John Adams and Herbert Hoover). Average the 18th and 19th ages. The median is

$$M = \frac{67 + 68}{2} = 67.5$$

Measuring spread with quartiles

The median income of people with a high school diploma in the CPS sample was $16,297. That's helpful but incomplete. Do most people with a high school education earn close to this amount, or are the incomes very spread out? The simplest useful description of a distribution consists of both a measure of *center* and a measure of *spread*. If we choose the median (the midpoint) to describe center, the **quartiles** are natural descriptions of spread. Again, the idea is clear: find the points one-quarter and three-quarters up the ordered list of observations. Here's an example.

EXAMPLE 4.16   Finding the quartiles

The first half of the ordered list of ages at death of U.S. presidents is

46   49   53   56   56   57   58   60 | 60   63 | 63   64   64   65   66   67   67   67

There are 18 data values *less than* the median, and the two middle numbers are boxed. The first quartile, Q_1, is

$$Q_1 = \frac{60 + 63}{2} = 61.5$$

The 18 ages *greater than* the median are

68   70   71   71   72   73   74   77 | 78   78 | 79   80   81   83   85   88   90   90

The third quartile, Q_3, is the average of the two middle numbers, which are boxed.

eye's are mest hp

$$Q_3 = \frac{78 + 78}{2} = 78$$

If there is an odd number of observations less than or greater than the median, then the middle number of that half of the data is Q_1 or Q_3. Remember that in finding the quartiles, you never include the median, whether it is a data value or not.

The quartiles get their name because, with the median, they divide the observations into quarters—one quarter lies below the first quartile, half the observations lie below the median, and three-quarters lie below the third quartile. That's the idea. Again, we need a rule to make the idea precise. The rule for calculating the quartiles uses the rule for the median.

The quartiles Q_1 and Q_3

To calculate the **quartiles**:

1. Arrange the observations in increasing order and locate the median M in the ordered list of observations.

2. The **first quartile Q_1** is the median of the observations whose position in the ordered list is to the left of the location of the overall median.

3. The **third quartile Q_3** is the median of the observations whose position in the ordered list is to the right of the location of the overall median.

EXERCISES

4.41 Median income You read that the median income of U.S. households in 1998 was $38,885. Explain in plain words what "the median income" is.

4.42 Sports statistics For each of the following, find the median and quartiles. Interpret each of these three numbers.
(a) Barry Bonds's home runs from 1986 through 2002 (the data are in Exercise 4.34, page 243)
(b) The Yankees' 2002 salaries (see Exercise 4.21, page 228)

4.43 American icon Find the median and quartiles for the number of calories in 17 brands of hot dogs (see Exercise 4.31, page 239). Interpret each of these three numbers.

4.44 Where's the median? Find the median stock return for the data given in Exercise 4.38 (page 244). Briefly explain how you determined your answer.

4.45 Median and quartiles Find the median and quartiles for the poverty data in Exercise 4.18 (pages 226–227).

The five-number summary and boxplots

The smallest and largest observations tell us little about the distribution as a whole, but they give information about the tails of the distribution that is missing if we know only the median and the quartiles. To get a quick summary of both center and spread, combine all five numbers.

> ## The five-number summary
>
> The **five-number summary** of a distribution consists of the smallest observation, the first quartile, the median, the third quartile, and the largest observation, written in order from smallest to largest. In symbols, the five-number summary is
>
> $$\text{Minimum} \quad Q_1 \quad M \quad Q_3 \quad \text{Maximum}$$

These five numbers offer a reasonably complete description of center and spread.

EXAMPLE 4.17   Poverty in the eastern states

The poverty rates for states east of the Mississippi have been divided into northern and southern states, according to the geographic divisions used by the Census Bureau.

Southern	Poverty (%)	Northern	Poverty (%)
Alabama	12.5	Connecticut	6.2
Delaware	6.5	Illinois	7.8
Florida	9.0	Indiana	6.7
Georgia	9.9	Maine	7.8
Kentucky	12.7	Massachusetts	6.7
Maryland	6.1	Michigan	7.4
Mississippi	16.0	New Hampshire	4.3
North Carolina	9.0	New Jersey	6.3
South Carolina	10.7	New York	11.5
Tennessee	10.3	Ohio	7.8
Virginia	7.0	Pennsylvania	7.8
West Virginia	13.9	Rhode Island	8.9
		Vermont	6.3
		Wisconsin	5.6

The five-number summary for the poverty rates in the southern states is

6.1   8   10.1   12.6   16.0

The five-number summary for the northern states is

4.3   6.3   7.05   7.8   11.5

The five-number summary leads to a new graph, the *boxplot*. Figure 4.24 shows boxplots for the poverty rates in southern and northern states east of the Mississippi.

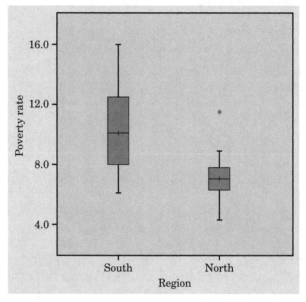

FIGURE 4.24 Boxplots comparing the poverty rates of southern and northern states east of the Mississippi in 1999.

Here are the main properties of boxplots.

Boxplot

A **boxplot** is a graph of the five-number summary.
- A central box spans the quartiles.
- A line in the box marks the median.
- Lines extend from the box out to the smallest and largest observations that are not outliers.

We will adopt the convention of identifying outliers as isolated points in the boxplot.

Side-by-side boxplots are useful for making rough comparisons among distributions. The first thing we notice in Figure 4.24 is that the poverty rate is considerably higher in the South than in the North. We also notice that there is one northern state (New York) whose poverty rate (11.5%) is an outlier among the northern states, but it wouldn't be unusual if it were in the South. We can also say that 50% of the states in the South have poverty rates higher than all of the northern states, except for the outlier, New York.

You can draw boxplots either horizontally or vertically. Be sure to include a numerical scale in the graph. When you look at a boxplot, first locate the median, which marks the center of the distribution. Then look at the spread. The quartiles show the spread of the middle half of the data, and the extremes (the smallest and largest observations) show the spread of the entire data set. We see from Figure 4.24 that the median poverty rate for the South is considerably higher than the median poverty rate for the North.

Because boxplots show less detail than histograms or stemplots, they are best used for side-by-side comparison of more than one distribution, as in Figure 4.24. For such small numbers of observations, a back-to-back stemplot is better yet (see Exercise 4.32, page 240). It would make clear, as the boxplot cannot, that Mississippi's poverty rate of 16.0 is unusual, even among the poorer southern states. Let us look at an example where boxplots are more genuinely useful.

EXAMPLE 4.18   Education and income

To see how income changes with level of education, we looked at the CPS annual sample of incomes. Figure 4.25 (next page) compares the income distributions for four levels of education, based on 71,512 observations.

This figure is a variation on the boxplot idea. The largest income among tens of thousands of people will surely be very large. The highest single income for people who stopped after high school, for example, is $498,606. Only 5% of this group, however, have incomes above $54,481. Figure 4.25 uses the 5% and 95% points in the distribution in place of the single smallest and largest incomes. So the line above the box for the "High school only" group extends only to $54,481.

Figure 4.25 gives us a clear and simple visual comparison. We see how the median and middle half move up for people with bachelor's and advanced degrees. The income of the bottom 5% stays small because there are some people in each group with no income or even negative income, perhaps due to illness or disability. The 95% point, marking off the top 5% of incomes, shoots up among people with advanced degrees, who include doctors, lawyers, and holders of MBA degrees.

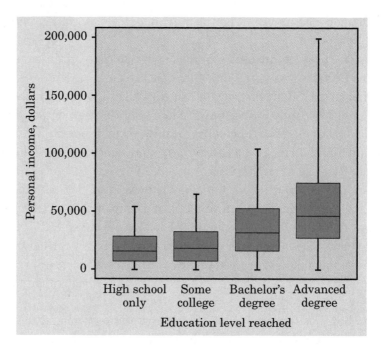

FIGURE 4.25 Boxplots comparing the distributions of income among adults with different levels of education. The ends of each plot are the 5% and 95% points in the distribution.

Figure 4.25 also illustrates how boxplots often indicate the symmetry or skewness of a distribution. In a symmetric distribution, the first and third quartiles are equally distant from the median. In most distributions that are skewed to the right, on the other hand, the third quartile will be farther above the median than the first quartile is below it. The extremes behave the same way. Even with the top 5% not present, we can see the strong right-skewness of incomes among holders of advanced degrees.

EXERCISES

4.46 Gas mileage Table 4.11 gives the highway gas mileages (in miles per gallon) for model year 2000 midsize cars.
(a) Make a stemplot of these data.
(b) Find the five-number summary of gas mileages. Which cars are in the bottom quarter of gas mileages?
(c) The stemplot shows a fact about the overall shape of the distribution that the five-number summary cannot describe. What is it?

TABLE 4.11 **Highway Gas Mileage for Model Year 2000 Midsize Cars**

Model	MPG	Model	MPG
Acura 3.5RL	24	Lexus GS300	24
Audi A6 Quattro	24	Lexus LS400	25
BMW 740I Sport M	21	Lincoln-Mercury LS	25
Buick Regal	29	Lincoln-Mercury Sable	28
Cadillac Catera	24	Mazda 626	28
Cadillac Eldorado	28	Mercedes-Benz E320	30
Chevrolet Lumina	30	Mercedes-Benz E430	24
Chrysler Cirrus	28	Mitsubishi Diamante	25
Dodge Stratus	28	Mitsubishi Galant	28
Honda Accord	29	Nissan Maxima	28
Hyundai Sonata	28	Oldsmobile Intrigue	28
Infiniti 130	28	Saab 9-3	26
Infiniti Q45	23	Saturn LS	32
Jaguar S/C	21	Toyota Camry	30
Jaguar Vanden Plas	24	Volkswagen Passat	29
Jaguar X200	26	Volvo S70	27

Source: Environmental Protection Agency, *Model Year 2000 Fuel Economy Guide*, 2000.

4.47 Home run sluggers Babe Ruth's 60 home runs in a single season in 1927 was a record that stood for 34 years as a milestone in all of sports history (Exercise 4.30, page 239). Roger Maris had these home run totals in his 10 years in the American League:

13   23   26   16   33   61   28   39   14   8

Maris broke Ruth's record with 61 home runs in 1961. Mark McGwire set a new mark at 70 home runs in 1998 (Exercise 4.32, page 240). Then in 2001, Barry Bonds hit 73 (Exercise 4.34, page 243). Construct side-by-side boxplots to compare the home run hitting performance of these exceptional players. Were any of their record-setting performances really unusual (outliers)?

4.48 Minority students in engineering Figure 4.21 (page 234) is a histogram of the number of minority students (black, Hispanic, Native American) who earned doctorate degrees in engineering from each of 115 universities in the years 1992 through 1996.

(a) What are the positions of each number in the five-number summary in a list of 115 observations arranged from smallest to largest?

(b) Even without the actual data, you can use your answer to (a) and the histogram to give the five-number summary approximately. Do this. About how many engineering doctorates must a university grant to minority students to be in the top quarter?

4.49 Stock returns, II Yearly returns on common stocks for the 50 years from 1950 to 1999 are given in Exercise 4.38 (page 244).

(a) Calculate the five-number summary.

(b) Construct a boxplot for the distribution of returns. What does the boxplot tell you about the shape of the distribution?

4.50 State SAT scores We want to compare the distributions of average SAT math and verbal scores for the states.[15] We enter these data into a computer with the names SATM for math scores and SATV for verbal scores. Here is output from the statistical software package Minitab. (Other software produces similar output. Some software uses rules for finding the quartiles that differ slightly from ours. So software may not give exactly the answer you would get by hand.)

Variable	N	Mean	Median	StDev
SATV	51	532.22	525.00	33.86
SATM	51	533.47	528.00	35.13

Variable	Minimum	Maximum	Q1	Q3
SATV	478.00	593.00	501.00	564.00
SATM	473.00	601.00	503.00	558.00

Use this output to make boxplots of SAT math and verbal scores for the states. Briefly compare the two distributions in words.

Mean and standard deviation

The five-number summary is not the most common numerical description of a distribution. That distinction belongs to the combination of the *mean* to measure center and the *standard deviation* to measure spread. The mean is familiar—it is the ordinary average of the observations. The idea of the standard deviation is to give the average distance of observations from the mean. The "average distance" in the standard deviation is found in a rather obscure way. We will give the details, but you may want to just think of the standard deviation as "average distance from the mean" and leave the details to your calculator.

Mean and standard deviation

The **mean** $\overline{x}$ (pronounced "x-bar") of a set of observations is their average. To find the mean of n observations, add the values and divide by n:

$$\overline{x} = \frac{\text{sum of the observations}}{n}$$

The **standard deviation** s measures the average distance of the observations from their mean. It is calculated by finding an average of the squared distances and then taking the square root. To find the standard deviation of n observations:

1. Find the distance of each observation from the mean and square each of these distances.
2. Average the distances by dividing their sum by $n - 1$. This average squared distance is called the **variance.**
3. The standard deviation s is the square root of this average squared distance.

"Yup, Old Bob drowned due to being ignorant of statistics. He thought it was enough to know the average depth of the river."

EXAMPLE 4.19   Finding the mean and standard deviation

A person's metabolic rate is the rate at which the body consumes energy. Metabolic rate is important in studies of weight gain, dieting, and exercise. Here are the metabolic rates of 7 men who took part in a study of dieting. (The units are calories per 24 hours. These are the same calories used to describe the energy content of foods.)

| 1792 | 1666 | 1362 | 1614 | 1460 | 1867 | 1439 |

To find the mean of these observations,

$$\bar{x} = \frac{\text{sum of observations}}{n}$$

$$= \frac{1792 + 1666 + 1362 + 1614 + 1460 + 1867 + 1439}{7}$$

$$= \frac{11,200}{7} = 1600 \text{ calories}$$

The variance and standard deviation measure spread by using the deviations of the observations from the mean. Here is a table of those deviations:

Observations x	Deviations $x - \bar{x}$	Squared deviations $(x - \bar{x})^2$
1792	$1792 - 1600 = 192$	$(192)^2 = 36,864$
1666	$1666 - 1600 = 66$	$(66)^2 = 4,356$
1362	$1362 - 1600 = -238$	$(-238)^2 = 56,644$
1614	$1614 - 1600 = 14$	$(14)^2 = 196$
1460	$1460 - 1600 = -140$	$(-140)^2 = 19,600$
1867	$1867 - 1600 = 267$	$(267)^2 = 71,289$
1439	$1439 - 1600 = -161$	$(-161)^2 = 25,921$
		sum $= 214,870$

The variance is the sum of the squared deviations divided by one less than the number of observations:

$$s^2 = \frac{214,870}{6} = 35,811.67$$

The standard deviation is the square root of the variance:

$$s = \sqrt{35,811.67} = 189.24 \text{ calories}$$

Figure 4.26 displays the data of the example as points above the number line, with their mean marked by an asterisk (*). The arrows indicate two of the deviations from the mean. These deviations show how spread out the data are about their mean. Some of the deviations will be positive and some negative because

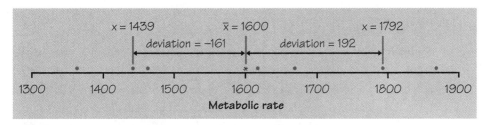

FIGURE 4.26 Metabolic rates for seven men, with their mean (*) and the deviations of two observations from the mean.

observations fall on each side of the mean. Squaring the deviations makes them all positive, so that observations far from the mean in either direction will have large positive squared deviations. The variance s^2 is the average squared deviation. The variance is large if the observations are widely spread about their mean; it is small if the observations are all close to the mean.

Because the variance involves squaring the deviations, it does not have the same unit of measurement as the original observations. If we measure energy consumed in calories, the variance is measured in calories squared. Taking the square root remedies this. The standard deviation s measures spread about the mean in the original scale.

In practice, you can key the data into your calculator and get the mean and standard deviation by asking for 1-variable statistics. Or you can enter the data into a spreadsheet or other software to find $\bar{x}$ and s. It is usual, for good but somewhat technical reasons, to average the squared distances by dividing their total by $n - 1$ rather than by n. Many calculators report two standard deviations, giving you a choice between dividing by n and dividing by $n - 1$. Be sure to choose $n - 1$.

More important than the details of the calculation are the properties that show how the standard deviation measures spread.

Properties of the standard deviation s :

- s measures spread about the mean $\bar{x}$. Use s to describe the spread of a distribution only when you use $\bar{x}$ to describe the center.
- $s = 0$ only when there is *no spread*. This happens only when all observations have the same value. So standard deviation zero means no spread at all. Otherwise, $s > 0$. As the observations become more spread out about their mean, s gets larger.

EXAMPLE 4.20   Investing 101

Enough examples about income. Here is an example about what to do with it once you've earned it. One of the first principles of investing is that taking more risk brings higher returns, at least on the average in the long run. Financial people measure risk by how unpredictable the return on an investment is. A bank account that is insured by the government and has a fixed rate of interest has no risk—its return is known exactly. Stock in a new company may soar one week and plunge the next. It has high risk because you can't predict what it will be worth when you want to sell.

Investors should think statistically. You can assess an investment by thinking about the distribution of (say) yearly returns. That means asking about both the center and the spread of the pattern of returns. Only naive investors look for a high average return without asking about risk, that is, about how spread out (or variable) the returns are. Financial experts use the mean and standard deviation to describe returns on investments. The standard deviation was long considered too complicated to mention to the public, but now you will find standard deviations appearing regularly in mutual-fund reports.

Here by way of illustration are the means and standard deviations of the yearly returns on three investments over the 50 years from 1950 to 1999:

Investment	Mean return	Standard deviation
Treasury bills	5.34%	2.96%
Treasury bonds	6.12%	10.73%
Common stocks	14.62%	16.32%

```
-2 | 6
-1 | 500
-0 | 98753
 0 | 011456778
 1 | 01224667999
 2 | 01223344799
 3 | 02222378
 4 | 3
 5 | 0
```

FIGURE 4.27 Stemplot of the yearly returns on common stocks for the 50 years 1950 to 1999. The returns are rounded to the nearest whole percent. The stems are 10s of percents and the leaves are single percents.

You can see that risk (variability) goes up as the mean return goes up, just as financial theory claims. Treasury bills and bonds are ways of loaning money to the U.S. government. Bills are paid back in one year, so their return changes from year to year depending on interest rates. Bonds are 30-year loans. They are riskier because the value of a bond you own will drop if interest rates go up. Stocks are yet riskier. They give higher returns (on the average in the long run), but at the cost of lots of sharp ups and downs on the way. As the stemplot in Figure 4.27 on the facing page shows, stocks went up by as much as 50% and down by as much as 26% in one year during the 50 years covered by our data.

CALCULATOR CORNER   Boxplots

The TI-83 can plot up to three boxplots in the same viewing window. We will compare the home run production of the current record holder, Barry Bonds, with that of Hank Aaron, whose 755 career home runs also holds the record. Bonds's numbers of home runs by season are

16	25	24	19	33
25	34	46	37	33
42	40	37	34	49
73	46			

Aaron's home run production is

13	20	24	26	27
29	30	32	34	34
38	39	40	40	44
44	44	44	45	47

- Enter Bonds's home run data into list L_1 and Aaron's into list L_2.

- Set up two statistics plots: Plot1 to show a boxplot of Bonds's data and Plot2 to show a modified boxplot of Aaron's data.

- Use the calculator's zoom feature to display the side-by-side boxplots. Press $\boxed{\text{ZOOM}}$ and select 9:ZoomStat.

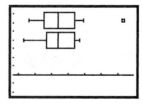

Except for the high outlier for Bonds (73 home runs in 2001), the home run production for the two players is remarkably similar.

CALCULATOR CORNER   Boxplots *(continued)*

- To see the five-number summary for each player, press the | TRACE | key. Use the left and right arrow keys to move from place to place on a boxplot. To move to a different boxplot, press the up or down arrow keys.

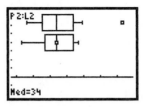

CALCULATOR CORNER   Numerical summaries

Here is how to get the calculator to produce numerical summaries for a set of data.

- With the Bonds home run data in, say, list L_1, press | STAT | | ▶ | (CALC) and select 1:1-Var Stats.
- Press | ENTER |. Now press | 2nd | | 1 | (L1). You should have 1-Var Stats L_1 showing on your screen. Then press | ENTER |.

- Notice the down arrow on the left side of the display. Press the down arrow key several times to see the rest of Bonds's statistics.

```
1-Var Stats
↑n=17
  minX=16
  Q₁=25
  Med=34
  Q₃=44
  maxX=73
```

```
1-Var Stats
 x̄=36.05882353
 Σx=613
 Σx²=24997
 Sx=13.44651715
 σx=13.04503777
↓n=17
```

EXERCISES

4.51 Barry Bonds again The number of home runs hit by Barry Bonds is shown in Exercise 4.34 (page 243). Use your calculator to find the mean and standard deviation. Then write a sentence or two explaining the meaning of each of these numbers.

4.52 Push-up king Joey is practicing push-ups for a contest. On 15 consecutive training sessions, he records the following push-ups:

48	51	46	54	51	62	55	67	55	64
62	60	71	73	74					

(a) Calculate Joey's mean number of push-ups. What is the standard deviation?
(b) Plot the data. Choose the plot type, and explain your choice.

4.53 True or false?
(a) Juan says that if the standard deviation of a list is zero, then all the numbers on the list are the same. Is Juan correct? If so, briefly explain why. If not, give a counterexample.   *The Deviations from The mean must be zero,*
(b) Letishia alleges that if the means and standard deviations of two different lists of numbers are the same, then all of the numbers in the two lists are the same. Is Letishia correct? If so, briefly explain why. If not, give a counterexample.

4.54 Imagine that! Give an example of a set of 5 data values with $\bar{x} \neq 0$ and $s = 0$.

4.55 A matchup Match the summary statistics with the histograms.
(a) mean = 6.6, median = 6.8, standard deviation = 1.3, variable = _____.
(b) mean = 6.6, median = 6.0, standard deviation = 8.65, variable = _____.
(c) mean = 6.6, median = 3.75, standard deviation = 7.4, variable = _____.

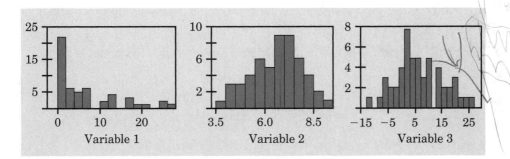

Variable 1   Variable 2   Variable 3

ACTIVITY 4.3B The *Mean and Median* applet

You can compare the behavior of the mean and median by using the *Mean and Median* applet at the *Statistics through Applications* Web site, www.whfreeman.com/sta. Click to enter data, then use the mouse to drag an outlier to the right (higher values) and watch the mean chase after it.

ACTIVITY 4.3B The *Mean and Median* applet *(continued)*

Directions: Use your mouse to click below the line and make red points (represented by black points in the figure above). This will create a dotplot (points that fall at the same place along the number line stack up to give the appearance of a histogram). Above the line, the mean is marked by a green arrow (represented by the blue arrow in the figure), and the median is marked by a red arrow (black arrow here). When the mean and median are the same, the overlapping arrows form one yellow arrow.

APPLET TIPS: To move a data point, click on it and drag it to the new position. To remove a data point, click on it and drag it to the trash can. To clear all of the points, click on the trash can.

Contemplate:

1. If you create just 1 point, where will the mean appear? How about the median?

2. If you create 2 points, where will the mean appear? How about the median?

3. If you put 2 points at the far left end of the line and 1 point at the far right (near the trash can), where will the median arrow appear? How about the mean arrow?

Verify: Use the applet to test your answers to the three questions above. Did the mean and median fall exactly where you expected in each of the three cases?

Experiment: Create a dotplot of 4 points in a symmetric pattern at the left end of the line. Notice how the mean and median overlap at the point of symmetry. Now add a 5th point just to the right of the others. Do the mean and median still overlap? Are they close? Now drag this point farther to the right and watch the behavior of the mean and the median. What happens to the mean as you move the point to the right? What happens to the median?

    Do this again, but start with 10 points in a symmetric pattern at the left end. As you move an 11th point to the right, how does the mean change? How does the median change? How do these changes compare with what happened when you started with just 5 points?

Choosing numerical descriptions

The five-number summary is easy to understand and is the best short description for most distributions. The mean and standard deviation are harder to understand but are more common. How can we decide which of these two descriptions of center and spread we should use? Let's start by comparing the mean and the median. "Midpoint" and "arithmetic average" are both reasonable ideas for describing the center of a set of data, but they are different ideas with different uses. The most important distinction is that the mean (the average) is strongly influenced by a few extreme observations and the median (the midpoint) is not.

EXAMPLE 4.21   Mean versus median

Table 4.12 gives the approximate salaries (in millions of dollars) of the 12 members of the Los Angeles Lakers basketball team for the 2002–2003 season. You can calculate that the mean is $\bar{x}$ = $5.1 million and that the median is M = $3.35 million. No wonder professional basketball players have big houses.

TABLE 4.12 Salaries of the Los Angeles Lakers, 2002–2003

Player	Salary	Player	Salary
Shaquille O'Neal	$23.6 million	Derek Fisher	$ 3.0 million
Kobe Bryant	12.4 million	Samaki Walker	1.5 million
Robert Horry	5.3 million	Kareem Rush	1.0 million
Devean George	4.6 million	Brain Shaw	1.0 million
Rick Fox	3.9 million	Mark Madsen	0.8 million
Tracy Murray	3.7 million	Jannero Pargo	0.4 million

Source: www.hoopshype.com/salaries/la_lakers.htm.

Why is the mean so much higher than the median? Figure 4.28 is a dotplot of the salaries, in millions. The distribution is skewed to the right and there are two high outliers. The very high salaries of Shaquille O'Neal and Kobe Bryant pull up the sum of the salaries and so pull up the mean. If we drop the outliers, the mean for the other 10 players is only $2.52 million. The median doesn't change nearly as much: it drops from $3.35 million to $2.25 million.

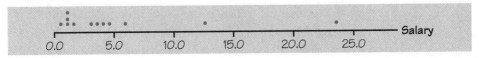

FIGURE 4.28 Dotplot of the salaries of Los Angeles Lakers players, from Table 4.12.

We can make the mean as large as we like by just increasing Shaquille O'Neal's salary. The mean will follow one outlier up and up. But to the median, Shaq's salary just counts as one observation at the upper end of the distribution. Moving it from $23.6 million to $236 million would not change the median at all.

The mean and median of a symmetric distribution are close to each other. In fact, $\bar{x}$ and M are exactly equal if the distribution is exactly symmetric. In skewed distributions, however, the mean runs away from the median toward the long tail. Many distributions of monetary values—incomes, house prices, wealth—are strongly skewed to the right. The mean may be much larger than the median. For example, we saw at the beginning of this section that the median income of people with advanced degrees in the Current Population Survey sample is $47,000. The distribution is skewed to the right, and the long right tail pulls the mean up to $65,220. Because monetary data often have a few extremely high observations, descriptions of these distributions usually employ the median.

You should think about more than symmetry versus skewness when choosing between the mean and the median. The distribution of selling prices for homes in Middletown is no doubt skewed to the right—but if the Middletown city council wants to estimate the total market value of all houses in order to set tax rates, the mean and not the median helps them out. The total market value is just the number of houses times the mean and has no connection with the median.

The standard deviation is pulled up by outliers or the long tail of a skewed distribution even more strongly than the mean. The standard deviation of the Lakers' salaries is $s = \$6.68$ million for all 12 players and only $s = \$1.79$ million when the two outliers are removed. The quartiles are much less sensitive to a few extreme observations. There is another reason to avoid the standard deviation in describing skewed distributions. Because the two sides of a strongly skewed distribution have different spreads, no single number such as s describes the spread well. The five-number summary, with its two quartiles and two extremes, does a better job. In most situations, it is wise to use $\bar{x}$ and s only for distributions that are roughly symmetric.

Choosing a summary

The mean and standard deviation are strongly affected by outliers and by the long tail of a skewed distribution. The median and quartiles are less affected.

The five-number summary is usually better than the mean and standard deviation for describing a skewed distribution or a distribution with outliers. Use $\bar{x}$ and s only for reasonably symmetric distributions that are free of outliers.

Why do we bother with the standard deviation at all? One answer appears in the next chapter: the mean and standard deviation are the natural measures of center and spread for an important kind of symmetric distribution, called normal distributions.

Do remember that a graph gives the best overall picture of a distribution. Numerical measures of center and spread report specific facts about a distribution, but they do not describe its entire shape. Numerical summaries do not disclose the presence of multiple peaks or gaps, for example. *Always start with a graph of your data.*

APPLICATION 4.3 Exploring Violent-Crime Statistics

The U.S. Department of Justice Bureau of Justice Statistics (BJS) is charged with collecting, organizing, and disseminating data on crimes and criminal activity. A visit to their Web site, www.ojp.usdoj.gov/bjs/, is a portal to a large quantity of data on this subject. One of the most surprising facts is that since 1994, the violent-crime rate has been reduced by more than half! See Figure 4.29.

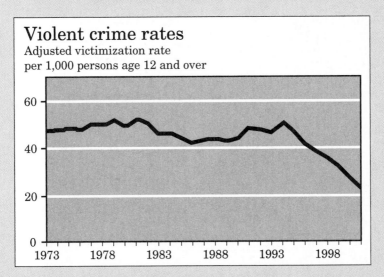

FIGURE 4.29 Violent-crime rates have declined since 1994, reaching the lowest level ever recorded in 2001.

First, we need to know what crimes are included under the heading of "violent crimes." Violent crimes include homicide, aggravated and simple assault, rape, and robbery. Next, we must ask how the data are collected. The data are from several sources. Homicide data are collected in the FBI's Uniform Crime Reports, which get their data from law enforcement agencies. Rape, robbery, and assault data are from the National

Sensitive

BS was low
rate why
mane point

APPLICATION 4.3 Exploring Violent-Crime Statistics *(continued)*

Crime Victimization Survey. This survey of households interviews about 80,000 persons aged 12 and older in 43,000 households twice each year about their victimization from crime. From our study of sample surveys, we know that anytime you conduct a survey, there is a chance that respondents will not answer some questions truthfully, especially if the questions are of a sensitive nature. In the case of rape, for example, victims may be in denial or they may be too embarrassed to answer truthfully, even in the more impersonal setting of a telephone conversation. Nevertheless, the BJS tries to take precautions to ensure data integrity.

Let's look at violent crimes state by state, and let's focus on the eastern half of the country. The number of violent crimes per 100,000 residents for the 26 states east of the Mississippi in 2001 is shown in Table 4.13.

TABLE 4.13 Number of Violent Crimes per 100,000 Residents for the 26 States East of the Mississippi in 2001

Alabama	4,546	New Hampshire	2,433
Connecticut	3,233	New Jersey	3,161
Delaware	4,478	New York	3,100
Florida	5,695	North Carolina	4,919
Georgia	4,751	Ohio	4,042
Illinois	4,286	Pennsylvania	2,995
Indiana	3,752	Rhode Island	3,476
Kentucky	2,960	South Carolina	5,221
Maine	2,620	Tennessee	4,890
Maryland	4,816	Vermont	2,987
Massachusetts	3,026	Virginia	3,028
Michigan	4,110	West Virginia	2,603
Mississippi	4,004	Wisconsin	3,209

1. Enter the data into your calculator, and determine the mean and median. Based on these two statistics, would you expect this distribution to be symmetric, skewed left, or skewed right, and why?

2. Construct a histogram for these data. Set your window dimensions as follows: X[2300,5800]$_{500}$ and Y[−2,12]$_1$. Does the shape of the distribution match your answer to (1)?

APPLICATION 4.3 Exploring Violent-Crime Statistics *(continued)*

3. Write the five-number summary for these data. Would the five-number summary or the mean and standard deviation be more appropriate to describe the distribution? Briefly explain your answer.

4. Define Plot2 to be a boxplot, and overlay the boxplot over the histogram. Does the boxplot corroborate your analysis in (a)? Briefly explain.

5. Suppose the media were pressuring you to identify the "high violent crime" states in this region. Based on your analyses of the data, how would you respond? Which states would you characterize as the "safest"?

You can learn a lot from BJS data and analyses. In 2000, 44% of murder victims were related to or acquainted with their assailants. Half of all violent crimes occurred within a mile of the victim's home. Seventy-three percent of violent crimes took place within 5 miles of the victim's home. Violence against women has decreased dramatically since 1993. In 1993, women suffered an estimated 1.1 million rapes, sexual assaults, robberies, aggravated assaults, and simple assaults at the hands of an intimate. In 2001, this number had dropped to 588,490 such crimes, a 75% decrease, if we can believe the numbers. In terms of the time of occurrence, it should come as no surprise that two-thirds of rapes/sexual assaults occurred at night (6 P.M. to 6 A.M.). In 2001, 57% percent of incidents of violent crime occurred during the day (6 A.M. to 6 P.M.). Guns, and especially handguns, were the weapon of choice in most homicides. In 2000, 52% of homicides were committed by handguns, 14% with other guns, 14% with knives, 5% with blunt objects, and 15% with other weapons. And alcohol plays a role in violent crimes. Among spouse victims, 75% of incidents were reported to have involved an offender who had been drinking.

School violence has been in the news in recent years. In 1999, about 7% of students in grades 9 to 12 reported being threatened or injured with a weapon such as a gun, knife, or club on school property in the previous 12 months. And about 7% of students carried a weapon such as a gun, knife, or club on school property in the previous 30 days. In 1999,

Poor New York?

Is New York a rich state? New York's mean income per person ranks fourth among the states, right up there with its rich neighbors Connecticut and New Jersey, which rank first and second. But while Connecticut and New Jersey rank seventh and second in median household income, New York stands 29th, well below the national average. What's going on? Just another example of mean versus median. New York has many very highly paid people, who pull up its mean income per person. But it also has a higher proportion of poor households than do New Jersey and Connecticut, and this brings the median down. New York is not a rich state—it's a state with extremes of wealth and poverty.

students aged 12 through 18 were victims of about 186,000 serious violent crimes at school and about 476,000 away from school. The good news is that between 1992 and 1999, victimization rates at school and away from school declined.

EXPLORING THE WEB

Many of our examples in this chapter have concerned income. At the Census Bureau Web site, www.census.gov, you can find the latest data on income and related issues. Look for *Money Income in the United States* (click on "Income" on the home page) and *Poverty in the United States* (click on "Poverty"). The full texts and the numerous tables in these annual publications are available online, along with many of the actual data, like the incomes and education of the 71,512 people that lie behind Figure 4.25 (page 254).

STATISTICS IN SUMMARY

To describe a set of data, always start with graphs. Then add well-chosen numbers that summarize specific aspects of the data. If we have data on a single quantitative variable, we start with a histogram or stemplot to display the distribution. Then we add numbers to describe the **center** and **spread** of the distribution.

There are two common descriptions of center and spread: **the five-number summary** and the **mean and standard deviation.** The five-number summary consists of the **median** to measure center and the two **quartiles** and the smallest and largest observations that are not outliers to describe spread. The median is the midpoint of the observations. The **mean** is the average of the observations. The **standard deviation** measures spread as a kind of average distance from the mean, so use it only with the mean.

The mean and standard deviation can be changed a lot by a few outliers. The mean and median agree for symmetric distributions, but the mean moves farther toward the long tail of a skewed distribution. In general, use the five-number summary to describe most distributions and the mean and standard deviation only for roughly symmetric distributions.

SECTION 4.3 EXERCISES

4.56 That's just wrong! Joey says that the mean and the median of a list of numbers must always appear as numbers in the list. Carla says he is wrong in both cases. Help Carla show Joey the error in his thinking by making up a very simple data set where neither the mean nor the median is in the list of data.

4.57 Reaction times Here is a sample of 100 reaction times of a subject playing a video game, in milliseconds:

10	14	11	15	7	7	20	10	14	9	8	6	12
12	10	14	11	13	9	12	13	11	12	10	8	9
14	18	12	10	10	11	7	17	12	9	9	11	7
10	14	12	12	10	9	7	11	9	18	6	12	12
10	8	14	15	12	11	9	9	11	8	11	10	13
8	11	11	13	20	6	13	13	8	9	16	15	11
10	11	20	8	17	12	19	14	17	12	18	16	15
16	10	20	11	19	20	13	11	20				

(a) Compute the mean and median of these data.

(b) Find the counts of each of the outcomes 6, 7, 8, . . ., 20. Draw a frequency histogram for these data.

(c) Explain in terms of the shape of the distribution why the mean and median fall as they do (close together or apart).

4.58 Teacher raises A school system employs teachers at salaries between $18,000 and $40,000. The teachers' union and the school board are negotiating the form of next year's increase in the salary schedule.

(a) If every teacher is given a flat $1000 raise, what will this do to the mean salary? To the median salary? To the extremes and quartiles of the salary distribution?

(b) What would a flat $1000 raise do to the standard deviation of teachers' salaries?

(c) If, instead, each teacher receives a 5% raise, the amount of the raise will vary from $900 to $2000, depending on the present salary. What will this do to the mean salary? To the median salary?

(d) A flat raise would not increase the spread of the salary distribution. What about a 5% raise? Specifically, will a 5% raise increase the distance of the quartiles from the median? Will it increase the standard deviation?

4.59 The range Another measure of the spread of a set of data is the *range*, which is the difference between the largest and smallest observations.

(a) Find the range of the video game reaction times in Exercise 4.57.

(b) The range is rarely used for any but very small samples. Can you explain why?

(c) Can you give an example in which the range is the most appropriate measure of spread?

4.60 Mean or median? You are planning a party and want to know how many cans of soda to buy. A genie offers to tell you either the mean number of cans guests will drink or the median number of cans. Which measure of center should you ask for? Why? To make your answer concrete, suppose there will be 30 guests and the genie will tell you one of $\bar{x} = 5$ cans or $M = 3$ cans. How many cans should you have on hand?

4.61 Be creative! Make up a list of numbers of which only 10% are above the average (that is, above the mean). What percent of the numbers in your list fall above the median?

4.62 Activity 4.3A follow-up, I In Activity 4.3A, "Checking Up on the Candy Makers" (page 247), you weighed individual items and bags of 30 items. You now have the tools to complete the activity.
(a) Use your TI-83 calculator to determine the mean and the standard deviation for the individual items. The standard deviation is identified as s in the 1-Var Stats screen on your calculator. Copy the mean and standard deviation from the calculator into your notebook.
(b) Define Plot2 to be a boxplot using the data from list CAND1. Compare the histogram you made in Step 4 of Activity 4.3 with the boxplot. What can you tell from the boxplot that you can't tell from the histogram? What can you tell from the histogram that you can't tell from the boxplot?
(c) Use the calculator to find the mean and standard deviation for the bags of 30 items. Compare the mean weight for the 20 individual items with the mean weight for the bags of 30 items. Is there a relationship? Would you have guessed this relationship?
(d) Compare the standard deviation for the weights of individual items with the standard deviation for the bags of 30. Are the standard deviations about the same, or is one standard deviation larger than the other? Does this make sense to you? Are you surprised? There is a relationship between these two standard deviations, but it may be hard to deduce from this investigation.

4.63 Activity 4.3A follow-up, II This exercise continues the analyses of candy from Activity 4.3A.
(a) Plot a histogram and a boxplot for the weights of the bags of 30 items (in list CAND2). Do not change the viewing window settings from before. How does this picture differ from the plots of the individual weights?
(b) Return to the weights of the individual bags. Compare the actual weights from your weighings with the advertised weight printed on the bags. What is the *proportion* of bags that have weights smaller than the listed weight? (This *proportion* is the number of bags below the stated weight divided by the total number of bags you weighed.)
(c) The FDA (Food and Drug Administration) has requirements on truth in labeling. If too many of the bags are underweight, then the company might get in trouble with the FDA or even be sued for deceptive advertising. On the other hand, if all of the bags are overweight, then executives might argue that the

company is being too generous and losing revenue. From the data you have collected and the calculations you have made, make a judgment as to whether the company is doing a good job in filling the bags that they sell. Be ready to justify your findings.

CHAPTER 4 REVIEW

Data analysis is the art of describing data using graphs and numerical summaries. The purpose of data analysis is to help us see and understand the most important features of a set of data. Sections 4.1 and 4.2 commented on basic graphs, especially pie charts, bar graphs, line graphs, histograms, and stemplots. Sections 4.2 and 4.3 showed how data analysis works by presenting statistical ideas and tools for describing the distribution of one variable. Figure 4.30 organizes the big ideas. We plot our data, and then we describe their center and spread using either the mean and standard deviation or the five-number summary.

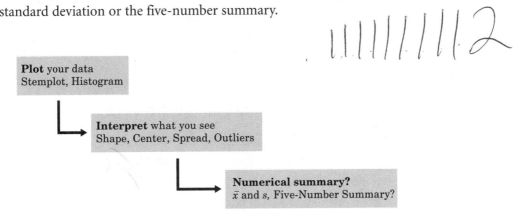

FIGURE 4.30 Plot, interpret, and summarize.

Here are the most important skills you should have acquired after reading Chapter 4.

A. DISPLAYING DISTRIBUTIONS

1. Recognize categorical and quantitative variables.

2. Recognize when a pie chart can and cannot be used.

3. Make a bar graph of the distribution of a categorical variable or, in general, to compare related quantities.

4. Interpret pie charts and bar graphs.

5. Make a line graph of a quantitative variable over time.

6. Recognize patterns such as trends and seasonal variation in line graphs.

7. Be aware of graphical abuses, especially pictograms and distorted scales in line graphs.

8. Make a histogram of the distribution of a quantitative variable.

9. Make a stemplot of the distribution of a small set of observations. Round data as needed to make an effective stemplot.

B. DESCRIBING DISTRIBUTIONS (QUANTITATIVE VARIABLE)

1. Look for the overall pattern of a histogram or stemplot and for major deviations from the pattern.

2. Assess from a histogram or stemplot whether the shape of a distribution is roughly symmetric, distinctly skewed, or neither. Assess whether the distribution has one or more major peaks.

3. Describe the overall pattern by giving numerical measures of center and spread in addition to a verbal description of shape.

4. Decide which measures of center and spread are more appropriate: the mean and standard deviation (especially for symmetric distributions) or the five-number summary (especially for skewed distributions).

5. Recognize outliers and give plausible explanations for them.

C. NUMERICAL SUMMARIES OF DISTRIBUTIONS

1. Find the median M and the quartiles Q_1 and Q_3 for a set of observations.

2. Give the five-number summary and draw a boxplot; assess center, spread, symmetry, and skewness from a boxplot.

3. Find the mean and (using a calculator) the standard deviation s for a small set of observations.

4. Understand that the median is less affected by extreme observations than the mean. Recognize that skewness in a distribution moves the mean away from the median toward the long tail.

5. Know the basic properties of the standard deviation: $s = 0$ only when all observations are identical; s increases as the spread increases; s has the same units as the original measurements; s is pulled strongly up by outliers or skewness.

CHAPTER 4 REVIEW EXERCISES

4.64 Who sells cars? Figure 4.31 on the facing page is a pie chart of the percent of passenger car sales in 1997 by various manufacturers. The artist has tried to make the graph more interesting by using the wheel of a car for the "pie." Is the graph still a correct display of the data? Explain your answer.

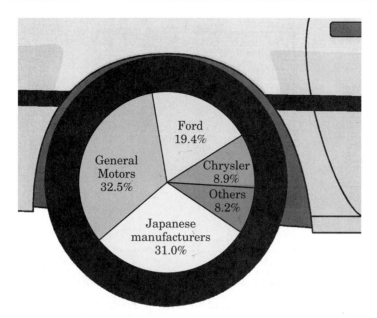

FIGURE 4.31 Passenger car sales by several manufacturers in 1997, for Exercise 4.64.

4.65 Who sells cars? Make a bar graph of the data in Exercise 4.64. What advantage does your new graph have over the pie chart in Figure 4.31?

4.66 Boys excel Is it true that boys perform better in the study of mathematics and the so-called hard sciences? Here are several Advanced Placement subjects and the percent of examinations taken by male candidates in 2001: Calculus BC, 61.45%; Computer Science AB, 89.06%; Chemistry, 55.62%; Physics—Mechanics, 73.37%; and Physics—Electricity, 77.67%.[16]
(a) Explain clearly why we cannot use a pie chart to display these data even when we know the percent of exams taken by boys for every subject.
(b) Make a bar graph of the data. To make comparisons easier, order the bars from tallest to shortest.

4.67 What about this graph? Figure 4.32 (next page) shows a graph that appeared in the *Lexington* (Kentucky) *Herald-Leader* of October 5, 1975. Discuss the correctness of this graph.

4.68 Finding the standard deviation The level of various substances in the blood influences our health. Here are measurements of the level of phosphate in the blood of a patient, in milligrams of phosphate per deciliter of blood, made on 6 consecutive visits to a clinic:

$$5.6 \qquad 5.2 \qquad 4.6 \qquad 4.9 \qquad 5.7 \qquad 6.4$$

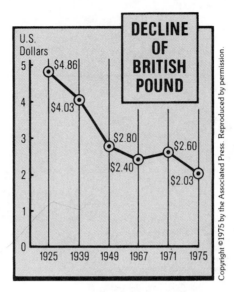

U.S. Dollars

DECLINE OF BRITISH POUND

$4.86
$4.03
$2.80
$2.40
$2.60
$2.03

1925   1939   1949   1967   1971   1975

FIGURE 4.32 A newspaper's graph of the value of the British pound, for Exercise 4.67.

(a) Make a stemplot of the data, but split the stems (see Exercise 4.33, page 242).
(b) A graph of only 6 observations gives little information, so we proceed to compute the mean and standard deviation.
(c) Find the mean from its definition. That is, find the sum of the 6 observations and divide by 6.
(d) Find the standard deviation from its definition. That is, find the distance of each observation from the mean, square the distances, then calculate the standard deviation.
(e) Now enter the data into your calculator and use the mean and standard deviation buttons to obtain $\bar{x}$ and s. Do the results agree with your hand calculations?

4.69 Roller coaster fatalities Data are provided on roller coaster accidents and fatalities in the United States in the data set COASTERS.
(a) What is the mean number of roller coaster fatalities?
(b) Plot a histogram and a boxplot in the same calculator screen. Is the mean a suitable measure of center, or should the median be used? Explain.

4.70 State SAT scores Figure 4.33 on the facing page is a histogram of the average scores on the mathematics part of the SAT exam for students in the 50 states and the District of Columbia.[17] The distinctive overall shape of this distribution implies that a single measure of center such as the mean or the median is of little value in describing the distribution. Explain why.

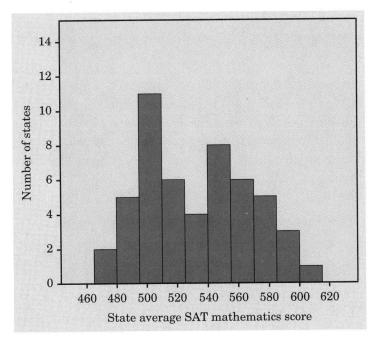

FIGURE 4.33 Histogram of the average scores on the SAT math exam for students in the 50 states and the District of Columbia in 1998, for Exercise 4.70.

4.71 Lakers' salaries Example 4.21 (page 265) gives the mean LA Lakers salary as $\bar{x} = \$5.1$ million.
(a) Why is the mean not the best measure of center for this distribution?
(b) Suppose that it's time to negotiate new salaries for the next season. Of the two groups, players and management, which would prefer to quote the mean salary, and which would prefer to quote the median? Explain.

4.72 College students and credit cards, I The Nellie Mae college student credit card studies described in Application 4.1 (page 218) looked at both the average student loan debt and the average credit card debt to determine the combined debt. Here are the numbers for 2001:

Combined Average Debt by Grade Level

	Average student loan debt	Average credit card debt	Combined debt
Freshman	$1,617	$1,825	$3,150
Sophomore	$5,596	1,825	$7,421
Junior	$10,166	$2,705	$3,262
Senior	$17,140	$3,262	$20,402

(a) Construct a bar graph of the combined debt by grade level.

(b) Write a sentence to describe the trend that you see in your bar graph.

4.73 College students and credit cards, II Student loans are designed for student borrowers and provide payment deferral during college while most students have little or no income. But college student loans and credit card balances have to be paid off eventually in order to avoid penalties for defaulting. Using the data collected in 2001, the Nellie Mae study described in Application 4.1 (page 218) made some assumptions and then, based on these assumptions, projected the following payback scenario:

Combined Average Payments after Graduation

	Student Loan	Credit card	Total
Average balance at graduation	$17,140	$3,262	$20,402
Monthly payment	$190.20	$97.86	$288.06
Total interest paid	$5,684	$3,042	$8,726

Assumptions: The student loan example assumes loans are subsidized federal Stafford Loans; borrower is making standard 10-year-term payments; the repayment interest rate remains fixed at the 2001 to 2002 rate of 5.99%; and borrower earns no repayment benefit/discounts. Total interest paid does not include any fees.

The credit card example assumes no additional purchases are made on the card; student makes minimum monthly payment of 3% of outstanding balance or $25—whichever is higher—resulting in payment duration of 11.9 years; the interest rate remains fixed at the average credit card interest rate of 18.9% as reported by American Consumer Credit Counseling, Inc. Total interest paid does not include any fees.

(a) What percent of total debt owed is due to credit card debt?

(b) Credit card payments will make up what percent of the total monthly payments students will make upon graduation?

(c) Credit card interest will make up what percent of the interest paid over time?

(d) Write a sentence that relates your answers to (a), (b), and (c).

(e) Construct a graph of your choice to visually illustrate some feature of the table above.

Chapter 5

The Normal Distributions and Government Statistics

5.1 Normal Distributions
5.2 The Consumer Price Index and Government Statistics

5.1 NORMAL DISTRIBUTIONS

Technology strikes

Bar graphs and histograms are definitely old technology. Using bars to display data goes back to William Playfair (1759–1823), an English economist who was an early pioneer of data graphics. Histograms require that we choose classes, and their appearance can change with different choices. Surely modern software offers a better way to picture distributions?

Software can replace the separate bars of a histogram with a smooth curve that represents the overall shape of a distribution. Look at Figure 5.1 (next page). The data are the numbers of minority group members who earned doctorates in engineering from 115 universities between 1992 and 1996. We met these data in Chapter 4, and the histogram in Figure 5.1 repeats Figure 4.21 (page 234). The curve is the new-technology replacement for the histogram. The software doesn't start from the histogram—it starts with the actual observations and cleverly draws a curve to describe their distribution.

In Figure 5.1, the software has caught the overall shape and shows the ripples in the long right tail more effectively than does the histogram. It struggles a bit with the peak. For example, it has extended the curve below zero in an attempt to smooth out the sharp peak. For the irregular distribution in Figure 5.1, we can't do better.

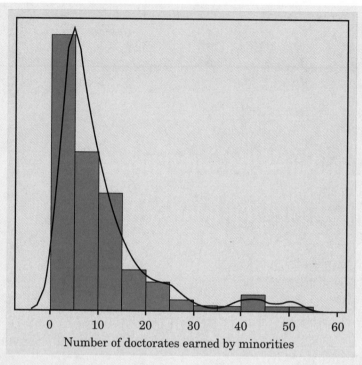

FIGURE 5.1 A histogram and a computer-drawn curve. Both picture the distribution of the number of doctorates in engineering earned by members of minority groups at 115 universities. The distribution is skewed to the right.

Figure 5.2 on the facing page shows the values of the sample proportion $\hat{p}$ for 1000 simple random samples (SRSs) of size 1523 from a population in which the population proportion is $p = 0.6$. We also met these data in Chapter 4, and the histogram here repeats Figure 4.18 (page 232). This distribution can be described by a specific kind of smooth curve called a **normal curve.** The normal curve has a distinctive, symmetric, single-peaked bell shape.

The normal curve is much easier to work with and does not require clever software. We will see that normal curves have special properties that help us use them and think about them. Only some kinds of data fit normal curves, however, so keep the technology handy for those that don't.

We now have a kit of graphical and numerical tools for describing distributions. What is more, we have a clear strategy for exploring data on a single quantitative variable:

1. Always plot your data: make a graph, usually a histogram or a stemplot.

2. Look for the overall pattern (shape, center, spread) and for striking deviations such as outliers.

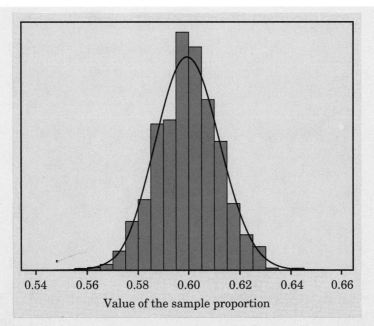

FIGURE 5.2 A normal curve used to describe the distribution of sample proportions. The histogram and the curve both picture the distribution of sample proportions in 1000 simple random samples from the same population. This distribution is quite symmetric.

3. Choose either the five-number summary or the mean and standard deviation to briefly describe center and spread in numbers.

Here is one more step to add to this strategy:

4. Sometimes the overall pattern of a large number of observations is so regular that we can describe it by a smooth curve.

ACTIVITY 5.1A Generating random numbers

In this Activity, you will use your TI-83 calculator to generate 200 random numbers between 0 and 1. Then you will divide the unit interval (0, 1) into 10 subdivisions and draw a histogram with 10 bars, where the height of each bar is the count of randomly generated numbers in that subinterval. Here's what the base of the histogram will look like:

ACTIVITY 5.1A Generating random numbers *(continued)*

1. A randomly generated number between 0 and 1 is as likely to fall into any one of these subintervals as into any other subinterval. How many of the 200 randomly generated numbers would you expect in the first (leftmost) subinterval? In each subinterval?

2. To generate 200 random numbers, enter the following WINDOW dimensions: $X[0,1]_{.1}$ and $Y[-8,30]_{5}$.

The command `rand` is found under the `MATH/PRB/1:rand` menu. Clear the home screen and enter the following command: `rand(200)→L₁`. Press ENTER . This command generates 200 random numbers between 0 and 1 and stores them in list L_1. Define STATPLOT 1 to be a histogram using the numbers in L_1, and then press the TRACE key to see the histogram. The number of observations (frequency) in the first bar, corresponding to the subinterval (0, .1), appears as $n =$ ___ in the bottom right corner of the screen.

3. Make a table like the one below and record that frequency in the table, in the box under Trial 1.

Trial number:	1	2	3	4	5	6	7	8	9	10
Frequency:										

4. Draw your histogram on a piece of grid paper. Remember to label and scale both axes.

5. How close is your result to your expected frequency from question 1? Generate a new set of 200 random numbers on your calculator by pressing 2nd [QUIT] followed by ENTER and then TRACE . Record the frequency you see this time in the box corresponding to Trial 2. Continue this sequence of calculator commands until you have filled in all 10 frequencies.

6. Looking at your frequencies, what is the greatest deviation from your expected value (question 1)?

7. Find the mean of the 10 frequencies (heights of bars).

ACTIVITY 5.1A Generating random numbers *(continued)*

8. How far is your mean from your expected height?

9. Which number would you feel more confident with as an estimate of the expected height: an individual height or the mean of 10 individual heights?

10. How could you get an estimate that you would have *lots* of confidence in?

11. Suppose that Lucky Eddie got exactly 20 for each of his 10 heights. Draw the histogram for Eddie's results.

12. Now overlay a curve that fits the histogram well. Remember that a straight line is a curve, even though it doesn't "bend."

Keep your results for this Activity handy because you will need them later in the chapter.

APPLICATION 5.1A A Student Survey

Sometimes the important features of a histogram can be summarized by a few statistics such as the mean, the median, the standard deviation, or a five-number summary. In September 1999, students in a large statistics class at a major state university responded to a survey. Students were asked their gender, birth month, the amount they spent on textbooks at the beginning of the term, the number of sodas they consumed during the previous week, and their high school grade point average (GPA). Your teacher will distribute the responses to this survey.

1. Three of the variables in the survey were birth month (1 for January, 2 for February, etc.), the amount spent on textbooks ($), and the number of sodas consumed over the previous week. Without making histograms of the data, guess which of the three rough sketches below will approximately describe each of the histograms of the three variables. Explain your guesses.

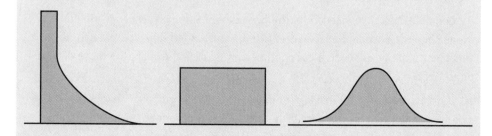

APPLICATION 5.1A   A Student Survey *(continued)*

2. After you have made your guesses at matching the histogram shape to the variable, use your calculator or software to draw the three histograms from the actual data. How do the histograms compare with what you expected?

3. Look at the three histograms for birth month, textbook costs, and sodas consumed. Guess the value of the median, the mean, and the standard deviation for each variable.

Variable	My guess at the median	My guess at the mean	My guess at the standard deviation
Month			
Textbooks			
Sodas			

4. To see how well you did at judging the descriptive statistics in question 3, use your calculator or software to compute the actual values. What are the actual values for the median, mean, and standard deviation of the textbooks and sodas variables?

Variable	Actual median	Actual mean	Actual standard deviation
Textbooks			
Sodas			

5. Compare the actual values with your guesses. Which variable's histogram was more difficult to judge? Explain.

6. Which measure of location (mean or median) would be the more appropriate in each of the following situations? Explain.

(a) You want to know how many sodas a typical student in the class drinks per week.

(b) You want to know how much soda is consumed in a week by the class as a whole.

7. Compute the values needed for the five-number summary of the distributions of the textbooks and sodas variables and use the software to make boxplots for each variable. Do the boxplots do a good job of summarizing the shapes of these distributions? Explain.

8. Boxplots are useful for making comparisons between groups in a population. Use side-by-side boxplots to compare high school GPAs for the men and women in the class. Sketch the comparative boxplots. Explain what this display shows.

Density curves

Figures 5.1 and 5.2 and Application 5.1A show curves used in place of histograms to picture the overall shape of a distribution of data. You can think of drawing a curve through the tops of the bars in a histogram and smoothing out the irregular ups and downs of the bars. There is one important distinction between histograms and these curves. Most histograms show the *counts* of observations in each class by the heights of their bars and therefore by the areas of the bars. We set up curves to show the *proportion* of observations in any region by areas under the curve. To do that, we choose the scale so that the total area under the curve is exactly 1. We then have a **density curve.**

EXAMPLE 5.1   Using a density curve

Figure 5.3 copies Figure 5.2, showing the histogram and the normal density curve that describe this data set of 1000 sample proportions. What proportion of the observations are greater than 0.61? From the actual 1000 observations, we can count that exactly 195 are greater than 0.61. So the proportion is 195/1000, or 0.195. Because 0.61 is one of the break points between the classes in the histogram, the area of the shaded bars in Figure 5.3(a) makes up 0.195 of the total area of all the bars.

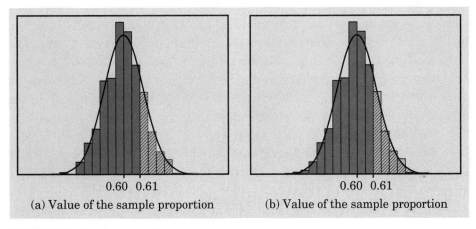

(a) Value of the sample proportion   (b) Value of the sample proportion

FIGURE 5.3 Comparing normal density curves and histograms.

Now concentrate on the density curve drawn through the histogram. The total area under this curve is 1, and the shaded area in Figure 5.3(b) represents the proportion of observations that are greater than 0.61. This area is 0.208. You can see that the density curve is a quite good approximation—0.208 is quite close to 0.195.

The area under the density curve in Example 5.1 is not exactly equal to the true proportion, because the curve is an idealized picture of the distribution. For example, the curve is exactly symmetric but the actual data are only approximately symmetric. Because density curves are smoothed-out, idealized pictures of the overall shapes of distributions, they are most useful for describing large numbers of observations.

The center and spread of a density curve

Density curves help us better understand our measures of center and spread. The median and quartiles are easy. Areas under a density curve represent proportions of the total number of observations. The median is the point with half the observations on either side. So *the median of a density curve is the equal-areas point,* the point with half the area under the curve to its left and the remaining half of the area to its right. The quartiles divide the area under the curve into quarters. One-fourth of the area under the curve is to the left of the first quartile, and three-fourths of the area is to the left of the third quartile. You can roughly locate the median and quartiles of any density curve by eye by dividing the area under the curve into four equal parts.

Because density curves are idealized patterns, a symmetric density curve is exactly symmetric. The median of a symmetric density curve is therefore at its center. Figure 5.4(a) on the facing page shows the median of a symmetric curve. We can roughly locate the equal-areas point on a skewed curve like that in Figure 5.4(b) by eye.

What about the mean? The mean of a set of observations is their arithmetic average. If we think of the observations as weights stacked on a seesaw, the mean is the point at which the seesaw would balance. This fact is also true of density curves. *The mean is the point at which the curve would balance if made of solid*

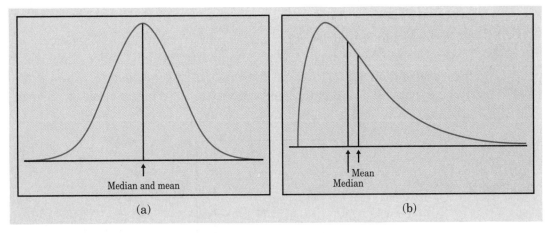

FIGURE 5.4 The median and mean for two density curves: a symmetric normal curve and a curve that is skewed to the right.

material. Figure 5.5 illustrates this fact about the mean. A symmetric curve balances at its center because the two sides are identical. *The mean and median of a symmetric density curve are equal,* as in Figure 5.4(a). We know that the mean of a skewed distribution is pulled toward the long tail. Figure 5.4(b) shows how the mean of a skewed density curve is pulled toward the long tail more than is the median.

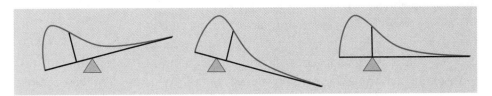

FIGURE 5.5 The mean of a density curve is the point at which it would balance.

Median and mean of a density curve

The **median** of a density curve is the equal-areas point, the point that divides the area under the curve in half.

The **mean** of a density curve is the balance point, at which the curve would balance if made of solid material.

The median and mean are the same for a symmetric density curve. They both lie at the center of the curve. The mean of a skewed curve is pulled away from the median in the direction of the long tail.

EXERCISES

5.1 Random numbers If you ask your calculator to generate random numbers between 0 and 1, as we did in Activity 5.1A, you will get observations from a **uniform distribution.** Figure 5.6 shows the density curve for a uniform distribution. This curve has height 1 over the interval from 0 to 1 and is zero outside that range.

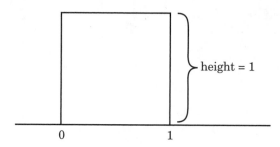

FIGURE 5.6 The density curve of a uniform distribution, for Exercise 5.1.

Use this density curve to answer these questions.
(a) Why is the height of the curve equal to 1?
(b) Recall that the mean of a density curve is the "balance point." What is the value of the mean of this distribution?
(c) What is the median? What property of this density curve tells you that the mean and median are related?
(d) What percent of the observations lie between 0 and 0.4?
(e) The birth months (1 = January, 2 = February, etc.) for a large number of people would have an approximately uniform distribution. Can you think of any other setting that would have a uniform distribution? What do all of these settings have in common (other than that they are uniformly distributed)?

5.2 More random numbers In Activity 5.1A, you obtained a histogram of the *counts* of the observations, which we arbitrarily divided into 10 equal subintervals. Generate 200 random numbers between 0 and 1 and store them in list L_1. Plot a histogram as we did in Activity 5.1A. Then find the *percent* of observations in each subinterval, and draw a new histogram of the percent of observations in each of the 10 subintervals. Copy the relative frequency distribution onto your paper and then draw a uniform density curve that best approximates your percent histogram. What is the height of your uniform density curve now?

5.3 Center and spread For each of the histograms in Figure 5.7 on the facing page, copy the distribution onto paper. Then sketch a smooth curve that describes the distribution well. Mark your best guess for the mean and median for each distribution.

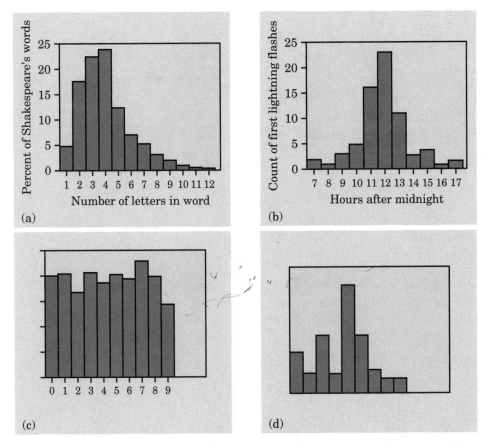

FIGURE 5.7 Four histograms for Exercise 5.3. (a) The distribution of lengths of words used in Shakespeare's plays. (b) The distribution of the time of the first lightning flash each day. (c) The distribution of random digits. (d) Unknown distribution.

5.4 Best descriptors For each of the distributions in Exercise 5.3, indicate whether it would be more appropriate to describe the distribution by the mean and standard deviation or by the median and five-number summary. Briefly explain why.

5.5 Top 98% Joey told his dad that he was in the top 98% of his statistics class. Suppose the class averages are approximately normally distributed. Sketch a normal density curve and show Joey's possible positions in the class.

Normal distributions as density curves

The density curves in Figures 5.2 (page 281) and 5.3 (page 285) belong to a particularly important family: the normal curves. Figure 5.8 (next page) presents two more normal density curves. Normal curves are symmetric, single-peaked, and bell-shaped. Their tails fall off quickly, so that we do not expect outliers. Because normal distributions are symmetric, the mean and median lie together at the peak in the center of the curve.

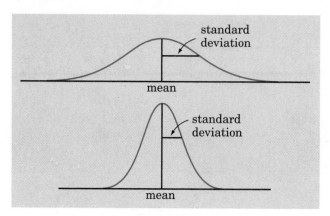

FIGURE 5.8 Two normal curves. The standard deviation fixes the spread of a normal curve.

Normal curves also have the special property that we can locate the standard deviation of the distribution by eye on the curve. This isn't true for most other density curves. Here's how to do it. Imagine that you are skiing down a mountain that has the shape of a normal curve. At first, you descend at an ever-steeper angle as you go out from the peak:

Fortunately, before you find yourself going straight down, the slope begins to grow flatter rather than steeper as you go out and down.

The points at which this change of curvature takes place are located one standard deviation on either side of the mean. The standard deviations are marked on the two curves in Figure 5.8. You can feel the change as you run a pencil along a normal curve, and so find the standard deviation.

Normal curves have the special property that giving the mean and the standard deviation completely specifies the curve. The mean fixes the center of the curve, and the standard deviation determines its shape. Changing the mean of a normal distribution does not change its shape, only its location on the axis. Changing the standard deviation does change the shape of a normal curve, as Figure 5.8 illustrates.

The distribution with the smaller standard deviation is less spread out and more sharply peaked. Here is a summary of basic facts about normal curves.

Normal density curves

The **normal curves** are symmetric, bell-shaped curves that have these properties:
- A specific normal curve is completely described by giving its mean and its standard deviation.
- The mean determines the center of the distribution. It is located at the center of symmetry of the curve.
- The standard deviation determines the shape of the curve. It is the distance from the mean to the change-of-curvature points on either side.

Why are the normal distributions important in statistics? First, normal distributions are good descriptions for some distributions of real data. Normal curves were first applied to data by the great mathematician Carl Friedrich Gauss (1777–1855), who used them to describe the small errors made by astronomers and surveyors in repeated careful measurements of the same quantity. You will sometimes see normal distributions labeled "Gaussian" in honor of Gauss. For much of the nineteenth century normal curves were called "error curves" because they were first used to describe the distribution of measurement errors. As it became clear that the distributions of some biological and psychological variables were at least roughly normal, the "error curve" terminology was dropped. The curves were first called "normal" by Francis Galton in 1889. Galton, a cousin of Charles Darwin, pioneered the statistical study of inheritance.

APPLICATION 5.1B Exploring Spread

Table 5.1 (next page) gives the ages of all U.S. presidents when they took office.

1. Enter the data into your calculator, or link them from your teacher.

2. Plot a histogram of the data. Use these window settings: X[40,70]₅ and Y[-4,15]₂. Verify that you get something like this:

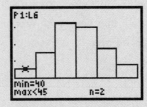

APPLICATION 5.1B Exploring Spread *(continued)*

TABLE 5.1 Ages of Presidents at Inauguration

President	Age	President	Age	President	Age
Washington	57	Lincoln	52	Hoover	54
J. Adams	61	A. Johnson	56	F. D. Roosevelt	51
Jefferson	57	Grant	46	Truman	60
Madison	57	Hayes	54	Eisenhower	61
Monroe	58	Garfield	49	Kennedy	43
J. Q. Adams	57	Arthur	51	L. B. Johnson	55
Jackson	61	Cleveland	47	Nixon	56
Van Buren	54	B. Harrison	55	Ford	61
W. H. Harrison	68	Cleveland	55	Carter	52
Tyler	51	McKinley	54	Reagan	69
Polk	49	T. Roosevelt	42	G. Bush	64
Taylor	64	Taft	51	Clinton	46
Fillmore	50	Wilson	56	G. W. Bush	54
Pierce	48	Harding	55		
Buchanan	65	Coolidge	51		

3. Describe the distribution. Include the shape, center, spread, and any deviations from the pattern. Which units of center and spread would be more appropriate: mean and standard deviation or median and five-number summary? Explain your choice.

4. There are 43 data points. What proportion (percent) of them are within one standard deviation of the mean? Here's a quick way to find out. If your data are in list L_1, then copy them into lists L_2 and L_3. Then SORT list L_2 in *ascending* order. Sort list L_3 into *descending* order. Then you can see the beginning and end of the ordered lists of data. You want to find the proportion of presidents whose ages when they took office were between (mean − 1 standard deviation) and (mean + 1 standard deviation). Verify that this interval is (48.7, 60.9). Scan lists L_2 and L_3 and count the number of data points between and including 49 and 60. You should count 28 presidents in this interval. So the proportion of ages within one standard deviation of the mean is 28/43 = 0.65. Sixty-five percent of presidents' ages are within one standard deviation of the mean.

APPLICATION 5.1B Exploring Spread *(continued)*

5. Use the method in Step 4 to find the proportion of presidents' ages within two standard deviations of the mean and the proportion of presidents' ages within three standard deviations of the mean.

6. Find another data set that is mound-shaped and symmetric, and repeat Steps 1 to 5. Do you get similar proportions with this second data set?

Commentary: Although the ages of presidents at inauguration are not a normal distribution, we can say that it is **approximately normal.** Note particularly that this data set is discrete, but the normal distribution is an idealized curve. Even though these are two quite different creatures, being able to work with a normal distribution and know some important properties of the normal distribution can be very helpful. If you inspect different data sets that appear to be approximately normal and calculate the proportion of data within one standard deviation, two standard deviations, and three standard deviations of the mean, you will get similar results, but certainly not exactly the same results for each data set. When we talk about normal distributions, however, we can be much more specific.

The 68–95–99.7 rule

There are many normal curves, each described by its mean and standard deviation. All normal curves share many properties. In particular, the standard deviation is the natural unit of measurement for normal distributions. This fact is reflected in the following rule.

The 68–95–99.7 rule

In any normal distribution, approximately
- **68%** of the observations fall within one standard deviation of the mean.
- **95%** of the observations fall within two standard deviations of the mean.
- **99.7%** of the observations fall within three standard deviations of the mean.

Figure 5.9 illustrates the 68–95–99.7 rule. By remembering these three numbers, you can think about normal distributions without constantly making detailed calculations. Remember also, though, that no set of data is exactly described by a normal curve. The 68–95–99.7 rule will be only approximately true for SAT scores or the lengths of crickets.

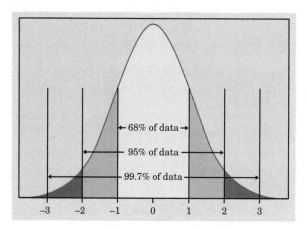

FIGURE 5.9 The 68–95–99.7 rule for normal distributions.

Standard scores

Jennie scored 600 on the verbal part of the SAT college entrance exam. How good a score is this? That depends on where a score of 600 lies in the distribution of all scores. The SAT exams are scaled so that scores should roughly follow the normal distribution with mean 500 and standard deviation 100. Jennie's 600 is one standard deviation above the mean. The 68–95–99.7 rule now tells us just where she stands (Figure 5.10 on the facing page). Half of all scores are below 500, and another 34% are between 500 and 600. So Jennie did better than 84% of the students who took the SAT. Her score report not only will say she scored 600 but will add that this is at the "84th percentile." That's statistics speak for "You did better than 84% of those who took the test."

Because the standard deviation is the natural unit of measurement for normal distributions, we restated Jennie's score of 600 as "one standard deviation above the mean." Observations expressed in standard deviations above or below the mean of a distribution are called standard scores.

Standard scores

The **standard score** for any observation is

$$\text{standard score} = \frac{\text{observation} - \text{mean}}{\text{standard deviation}}$$

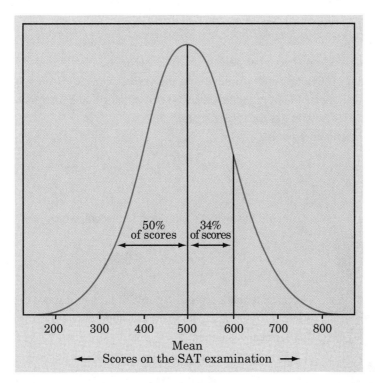

FIGURE 5.10 The 68–95–99.7 rule shows that 84% of any normal distribution lies to the left of the point one standard deviation above the mean. Here, this fact is applied to SAT scores.

We will usually let z = standard score, x = observation, μ = mean, and σ = standard deviation. Then the standard score formula looks like this:

$$z = \frac{x - \mu}{\sigma}$$

We will use the convention that x is always a nonstandardized data value and z is always a standardized value.

A standard score of 1 says that the observation in question lies one standard deviation above the mean. An observation with standard score -2 is two standard deviations below the mean. Standard scores can be used to compare values in different distributions. Of course, you should not use standard scores unless you are willing to use the standard deviation to describe the spread of the distributions. That requires that the distributions be at least roughly symmetric.

EXAMPLE 5.2   ACT versus SAT scores

Jennie scored 600 on the verbal part of the SAT. Her friend Gerald took the American College Testing (ACT) test and scored 21 on the verbal part. ACT scores are normally distributed with mean 18 and standard deviation 6. Assuming that both tests measure the same kind of ability, who has the higher score?

Jennie's standard score is

$$z = \frac{x - \mu}{\sigma} = \frac{600 - 500}{100} = \frac{100}{100} = 1.0$$

Compare this with Gerald's standard score, which is

$$z = \frac{21 - 18}{6} = \frac{3}{6} = 0.5$$

Because Jennie's score is 1 standard deviation above the mean and Gerald's is only 0.5 standard deviation above the mean, Jennie's performance is better.

The bell curve?

Does the distribution of human intelligence follow the "bell curve" of a normal distribution? Scores on IQ tests do roughly follow a normal distribution. That is because a test score is calculated from a person's answers in a way that is designed to produce a normal distribution. To conclude that intelligence follows a bell curve, we must agree that the test scores directly measure intelligence. Many psychologists don't think there is one human characteristic that we can call "intelligence" and can measure by a single test score.

EXERCISES

5.6 Uniform distribution Consider the data your calculator generated in Activity 5.1A.

(a) Should 68% of those decimal numbers fall within one standard deviation of the mean? Should that distribution of 200 numbers follow the 68–95–99.7 rule? Why or why not?

(b) Calculate the mean and standard deviation for your 200 data values. What percent of your data fall within one standard deviation of the mean? Within two standard deviations of the mean? Within three standard deviations of the mean? Do these results validate your answers in (a)?

IQ test scores Figure 5.11 on the facing page is a stemplot of the IQ test scores of 74 seventh-grade students.[1] This distribution is very close to normal with mean 111 and standard deviation 11. It includes all the seventh-graders in a rural Midwest school except for 4 low outliers who were dropped because they may have been ill or otherwise not paying attention to the test. Take the normal distribution with mean 111 and standard deviation 11 as a description of the IQ test scores of all rural

Midwest seventh-grade students. Use this distribution and the 68–95–99.7 rule to answer Exercises 5.7 to 5.9.

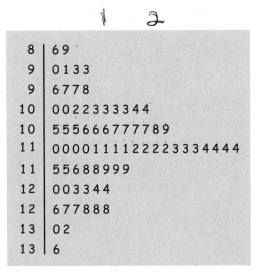

```
 8  | 6 9
 9  | 0 1 3 3
 9  | 6 7 7 8
10  | 0 0 2 2 3 3 3 3 4 4
10  | 5 5 5 6 6 6 7 7 7 8 9
11  | 0 0 0 0 1 1 1 1 2 2 2 2 3 3 3 4 4 4 4
11  | 5 5 6 8 8 9 9 9
12  | 0 0 3 3 4 4
12  | 6 7 7 8 8 8
13  | 0 2
13  | 6
```

FIGURE 5.11 Stemplot of the IQ scores of 74 seventh-grade students, for Exercises 5.7 to 5.9.

5.7 Between what values do the IQ scores of 95% of all rural Midwest seventh-graders lie?

5.8 What percent of IQ scores for rural Midwest seventh-graders are greater than 100?

5.9 What percent of all students have IQ scores 144 or higher? None of the 74 students in our sample school had scores this high. Are you surprised at this? Why?

5.10 A normal curve What are the mean and standard deviation of the normal curve in Figure 5.12 (next page)?

5.11 Horse pregnancies Bigger animals tend to carry their young longer before birth. The length of horse pregnancies from conception to birth varies according to a roughly normal distribution with mean 336 days and standard deviation 3 days. Use the 68–95–99.7 rule to answer the following questions.
(a) Almost all (99.7%) horse pregnancies fall in what range of lengths?
(b) What percent of horse pregnancies are longer than 339 days?

5.12 Eggs A truck is loaded with cartons of eggs that weigh an average of 2 pounds each with a standard deviation of 0.1 pound. A histogram of these weights looks very much like a normal distribution. What percent of the cartons weigh less than 2.1 pounds? Less than 1.8 pounds? More than 1.9 pounds?

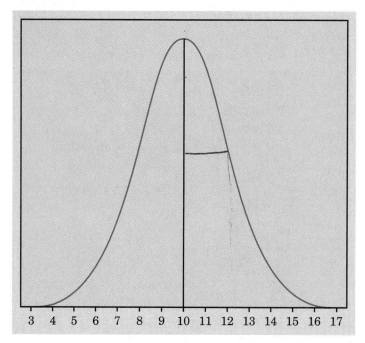

FIGURE 5.12 **What are the mean and standard deviation of this normal density curve? For Exercise 5.10.**

Percentiles of normal distributions

For normal distributions, but not for other distributions, standard scores translate directly into *percentiles*.

> ## Percentiles
>
> The **cth percentile** of a distribution is a value such that c percent of the observations lie below it and the rest lie above.

The median of any distribution is the 50th percentile, and the quartiles are the 25th and 75th percentiles. In any normal distribution, the point one standard deviation above the mean (standard score 1) is the 84th percentile.

Figure 5.10 (page 295) shows why. Every standard score for a normal distribution translates into a specific percentile, which is the same no matter what the mean and standard deviation of the original normal distribution are. Table B, at the back of this book, gives the percentiles corresponding to various standard scores. This table enables us to do calculations in greater detail than what the 68–95–99.7 rule allows.

EXAMPLE 5.3   Percentiles for college entrance exams

Jennie's score of 600 on the SAT translates into a standard score of 1.0. We saw that the 68–95–99.7 rule says that this is the 84th percentile. Table B is a bit more precise: it says that standard score 1 is the 84.13 percentile of a normal distribution. Go to Table B and verify this! Gerald's 21 on the ACT is a standard score of 0.5 (see Example 5.2 , page 296). Table B says that this is the 69.15 percentile (check this on Table B, too). Gerald did well, but not as well as Jennie. The percentile is easier to understand than either the raw score or the standard score. That's why reports of exams such as the SAT usually give both the score and the percentile.

EXAMPLE 5.4   Finding the observation that matches a percentile

How high must a student score on the SAT to fall in the top 20% of all scores? That requires a score at or above the 80th percentile. Look in the body of Table B for the percentiles closest to 80. You see that standard score 0.84 is the 79.95 percentile. We conclude that a standard score of 0.84 is approximately the 80th percentile *of any normal distribution.*

To go from the standard score back to the scale of SAT scores, substitute the known values into the standard score formula

$$z = \frac{x - \mu}{\sigma}$$

and then solve for the unknown, x:

$$\text{standard score} = 0.84 = z = \frac{x - \mu}{\sigma} = \frac{x - 500}{100}$$

Solving for x,

$$0.84\,(100) = x - 500$$
$$84 + 500 = x$$
$$x = 584$$

CALCULATOR CORNER   Normalcdf and invNorm

The command `normalcdf` can be used to find the area in the tail of a normal curve or above any interval on the horizontal axis. Here's an example.

Jennie scored 600 on the verbal part of the SAT. We want to know what her percentile is. The decimal equivalent of Jennie's percentile is the area of the shaded region shown on the next page.

CALCULATOR CORNER   Normalcdf and invNorm *(continued)*

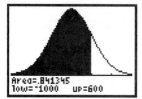

```
normalcdf(-1,1)
          .6826894809
```

- The command `normalcdf` is found by pressing [2nd] [DISTR]2:normalcdf(. Enter the following: `normalcdf(-1000,600,500,100)` and press [ENTER]. The result should be 0.8413, rounded to four decimal places.

- Example 5.4 asked how high a student needed to score on the SAT in order to fall in the top 20% of all scores. The command `invNorm(.8,500,100)` gives us the answer: 584.

```
normalcdf(-1000,
600,500,100)
          .8413447404
```

```
invNorm(.8,500,1
00)
          584.1621233
```

That agrees with what we got by using Table B. The first two numbers that you entered (−1000 and 600) specify the interval corresponding to the shaded area under the curve. The last two numbers (500 and 100) identify the particular normal curve by specifying the mean and standard deviation.

- If you have *standard* scores, you can omit the mean and standard deviation. For example, `normalcdf(-1,1)` produces 0.6827 and confirms that about 68% of the observations are within one standard deviation about the mean.

- If we leave out the mean and standard deviation, we get `invNorm(.8)=.8416`, which is the standard value of 584, but which doesn't really tell us what we want to know.

```
invNorm(.8)
          .8416212335
```

ACTIVITY 5.1B *The Normal Density Curve* applet

Directions: Go to the *Statistics Through Applications* Web site, www. whfreeman.com/sta, and select the *Normal Density Curve* applet.

Drag the green flags to have the applet compute the part of the normal curve that is in the bright yellow shaded region you create. When the 2-Tail box is checked, the applet calculates symmetric areas around the mean.

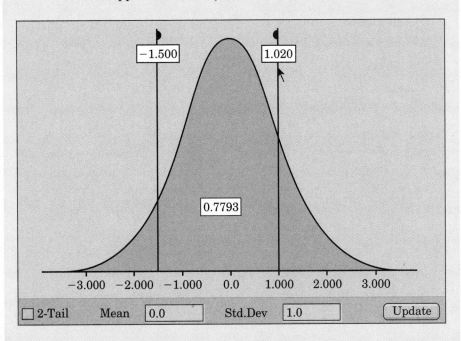

Contemplate: Answer the following without using the applet.

1. If you were to put one flag at the extreme left of the curve and the second flag exactly in the middle, what would be the proportion reported by the applet?

2. If you were to place the two flags exactly one standard deviation on either side of the mean, what would the applet say is the area between them?

3. About 99.7% of the area under the normal density curve lies within three standard deviations of the mean. Does this mean that about 99.7%/2 = 49.85% will lie within one and a half standard deviations?

Verify: Use the applet to test your answers to the three questions above. Did the shaded proportion of the curve behave as you thought it would in each case?

ACTIVITY 5.1B *The Normal Density Curve* applet *(continued)*

APPLET TIP: The direction in which the green flags point is the side of the flag poles that is shaded.

Experiment: Try comparing the values given by the applet with those given by Table B in the back of the book to see that they are the same (examine at least one positive standard score and one negative standard score).

1. Use the applet to find the quartiles of the normal distribution when the mean is 0 and the standard deviation is 1.

2. The 68–95–99.7 rule tells us that the middle 95% of values fall within about two standard deviations of the mean. Use the applet to get a more precise answer for where the middle 95% of values fall.

3. Make a small table of the areas under the normal curve within ±0.5, ±1.0, ±1.5, and ±2 of the mean. Next, change the standard deviation from 1 to 2 and note how the numbers along the horizontal axis have changed. Again make a table of the areas within ±0.5, ±1.0, ±1.5, and ±2 of the mean. Do any of the areas in this second table match any of the areas you have in your first table? Explain.

4. Solve Exercises 5.11 and 5.12 (page 297) and Examples 5.3 and 5.4 (page 299) using the applet and again using Table B. Do you find any advantage(s) to using the applet? Explain.

APPLET TIP: Remember to hit the Update button if you want to change the mean or standard deviation.

Normal curves also describe the distribution of *statistics such as sample proportions and sample means* when we take many samples from the same population. We used normal curves this way in Figures 5.2 (page 281) and 5.3 (page 285). The margins of error for the results of sample surveys are usually calculated from normal curves. However, even though many sets of data follow a normal distribution, many do not. Most income distributions, for example, are skewed to the right and so are not normal. Nonnormal data, like nonnormal people, not only are common but are sometimes more interesting than their normal counterparts.

EXERCISES

5.13 Heights of young women The distribution of the heights of young women aged 18 to 24 is approximately normal with mean 65 inches and standard deviation 2.5 inches. Sketch a picture of a normal curve and then use the 68–95–99.7 rule to show what the rule states about these women's heights.

5.14 NCAA rules for athletes The National Collegiate Athletic Association (NCAA) requires Division I athletes to score at least 820 on the combined mathematics and verbal parts of the SAT exam in order to compete in their first college year. (Higher scores are required for students with poor high school grades.) In 1999, the scores of the millions of students taking the SATs were approximately normal with mean 1017 and standard deviation 209. What percent of all students had scores less than 820?

5.15 More NCAA rules The NCAA considers a student a "partial qualifier" eligible to practice and receive an athletic scholarship, but not to compete, if the combined SAT score is at least 720. Use the information in the previous exercise to find the percent of all SAT scores that are less than 720.

5.16 800 on the SAT It is possible to score higher than 800 on either part of the SAT, but scores above 800 are reported as 800. (That is, a student can get a reported score of 800 without a perfect test.) In 1999, the scores of men on the math part of the SAT followed a normal distribution with mean 531 and standard deviation 115. What percent of scores were above 800 (and so reported as 800)?

5.17 Women's SAT scores The average performance of women on the SAT, especially the math part, is lower than that of men. The reasons for this gender gap are controversial. In 1999, women's scores on the math SAT followed a normal distribution with mean 495 and standard deviation 109. The mean for men was 531. What percent of women scored higher than the male mean?

5.18 High IQ scores Scores on the Wechsler Adult Intelligence Scale for the 20 to 34 age-group are approximately normally distributed with mean 110 and standard deviation 25. How high must a person score to be in the top 25% of all scores?

5.19 Technology to the rescue
(a) Use your TI-83 to answer the questions in Exercises 5.14 and 5.18 above.
(b) Use the *Normal Density Curve* applet to answer the questions in Exercises 5.14 and 5.18 above.

EXPLORING THE WEB

Have you ever seen a table of square roots? They were once common but have been replaced by the $\sqrt{x}$ button on a calculator. Tables of areas under a normal curve, like Table B at the back of this book, are still common but are also giving way to buttons and, better yet, to applets that let you find areas visually. You have seen our *Normal Density Curve* applet. For a variation, go to Professor Gary McClelland's site, psych.colorado.edu/~mcclella/java/normal/accurateNormal.html, and use his applet to work some representative exercises.

STATISTICS IN SUMMARY

Stemplots, histograms, and boxplots all describe the distributions of quantitative variables. **Density curves** are another kind of graph that serves the same purpose. A density curve is a curve with area exactly 1 underneath it whose shape describes the overall pattern of a distribution. An area under the curve gives the proportion of the observations that fall in an interval of values. You can roughly locate the median (equal-areas point) and the mean (balance point) by eye on a density curve.

Normal curves are a special kind of density curve that describes the overall pattern of some sets of data. Normal curves are symmetric and bell-shaped. A specific normal curve is completely described by its mean and standard deviation. You can locate the mean (center point) and the standard deviation (distance from the mean to the change-of-curvature points) on a normal curve. All normal distributions obey the **68–95–99.7 rule. Standard scores** express observations in standard deviation units about the mean, which has standard score 0. A given standard score corresponds to the same percentile in any normal distribution. Table B gives percentiles for normal distributions.

SECTION 5.1 EXERCISES

5.20 Normal curve properties, I Figure 5.13 on the facing page is a normal density curve. What are the mean and the standard deviation of this distribution?

5.21 Normal curve properties, II Explain why the point one standard deviation below the mean in a normal distribution is always the 16th percentile. Explain why the point two standard deviations above the mean is the 97.5th percentile.

The following exercises require use of your calculator or Table B of normal distribution percentiles.

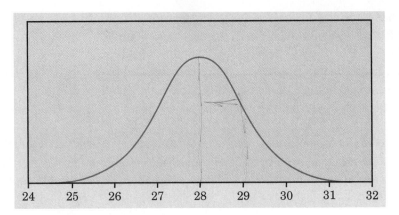

FIGURE 5.13 A normal distribution curve, for Exercise 5.20.

5.22 Am I winning? Congressman Floyd commissions a sample survey of voters to learn what percent favor him in his race for reelection. To avoid spending too much, he samples only 400 voters. Suppose that in fact only 45% of the voters support Floyd. The percent favoring Floyd in a random sample of size 400 will vary from sample to sample according to a normal distribution with mean 45% and standard deviation 2.5%. What percent of all such samples will (wrongly) show that half or more of the voters favor Floyd?

5.23 Are we getting smarter? When the Stanford-Binet "IQ test" came into use in 1932, it was adjusted so that scores for each age-group of children followed roughly the normal distribution with mean 100 and standard deviation 15. The test is readjusted from time to time to keep the mean at 100. If present-day American children took the 1932 Stanford-Binet test, their mean score would be about 120. The reasons for the increase in IQ over time are not known but probably include better childhood nutrition and more experience in taking tests.[2]
(a) IQ scores above 130 are often called "very superior." What percent of children had very superior scores in 1932?
(b) If present-day children took the 1932 test, what percent would have very superior scores? (Assume that the standard deviation 15 does not change.)

5.24 Locating the quartiles The quartiles of any distribution are the 25th and 75th percentiles. About how many standard deviations from the mean are the quartiles of any normal distribution?

5.25 Potatoes Bags of potatoes in a shipment averaged 10 pounds with a standard deviation of 0.5 pounds. A histogram of these weights followed a normal curve quite closely.

(a) What percent of the bags weighed less than 10.25 pounds?
(b) What percent weighed between 9.5 and 10.25 pounds?

5.26 Oranges Bags of oranges in a shipment averaged 8 pounds with a standard deviation of 0.5 pounds. A histogram of these weights followed a normal distribution quite closely. What percent of the bags weighed less than 8.25 pounds?

5.27 Textbook costs Students taking an introductory statistics class in fall 2000 reported spending an average of $205 on textbooks that quarter with a standard deviation of $90. A rough sketch of a curve that fitted the histogram looked like:

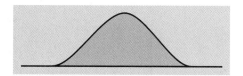

(a) Approximately what percent of the students spent between $115 and $295 on textbooks that quarter?
(b) One student spent $187 on textbooks. What was her standard score? What percent of the students spent less than she did on textbooks that quarter?
(c) Explain why the sketch of the histogram above is important to your calculations.

5.28 Ages of cars An analysis of car registrations in Texas in 2000 showed that about half of the registered cars were less than 5 years old. Another result of this analysis showed that model year 2000 cars were the most common and that there were fewer and fewer registered cars with each previous year.
(a) Describe the mean age of the registered cars in Texas.
(b) Describe the median age of the registered cars.

5.29 Sampling Suppose that the proportion of all adult Americans who are afraid to go out at night because of crime is $p = 0.4$. If we took many SRSs of size 1050, the sample proportion $\hat{p}$ would vary from sample to sample following a normal distribution with mean 0.4 and standard deviation 0.015. Use this fact and the 68–95–99.7 rule to answer the following questions.
(a) In many samples, what percent of the values of $\hat{p}$ fall above 0.4? Above 0.43?
(b) In a large number of samples, what range contains the central 95% of values of $\hat{p}$?

5.2 THE CONSUMER PRICE INDEX AND GOVERNMENT STATISTICS

The government's most important statistic

Which of the many data series produced by government statistical offices in the United States is the most important? The unemployment rate, from the monthly Current Population Survey, is certainly a candidate. Rising unemployment can influence elections and change government policy. We suggest, however, another candidate: the monthly consumer price index, or CPI.[3]

The CPI measures changes in the prices of goods and services over time, so it measures the falling buying power of the dollar over time. If the same goods and services cost more, then the dollar is worth less. A dollar in 2000 buys less than a dollar bought in 1980, so it is a different dollar even though it looks the same. In fact, the CPI tells us that a dollar at the beginning of 2000 would buy only half as much as a dollar would buy in 1980. Someone who did not earn twice as many dollars in 2000 as in 1980 had lost buying power.

Why is the CPI so important? It does at times affect elections and government policies. It also has direct links to large parts of the economy shared by no other statistic. Nobody wants to lose buying power. So groups with enough clout tie their incomes directly to the CPI. Social Security payments rise automatically as the CPI increases, and so do the pensions of people retired from the military and the federal civil service. Over 2 million union members have contracts that tie their wages to the CPI. The Bureau of Labor Statistics says that the incomes of over 80 million people are directly affected by the CPI. When the CPI rises by 1%, government spending automatically goes up by $6 billion a year. Income tax brackets go up with the CPI. You can even buy U.S. savings bonds whose value is linked to the increase in the CPI.

The CPI is also important to anyone planning for the future. Saving for education or for retirement requires that we take account of the falling buying power of the dollar. We can use the CPI to compare 1980 dollars and 2000 dollars by turning them into "real dollars" that have the same buying power over time. We'll see how in this chapter.

We all notice the high salaries paid to professional athletes. In major league baseball, for example, the mean salary rose from $143,756 in 1980 to $2,088,737 in 2002. That's a big jump. Not as big as it first appears, however. *A dollar in 2002 did not buy as much as a dollar in 1980, so 1980 salaries cannot be directly compared with 2002 salaries.* The hard fact that the dollar has steadily lost buying power over time means that we

Thought for today
"What this country really needs is a good five-cent nickel."
—Franklin P. Adams, American journalist-humorist (1881–1960)

must make an adjustment whenever we compare dollar values from different years. The adjustment is easy. What is not easy is measuring the changing buying power of the dollar. The government's CPI is the tool we need.

ACTIVITY 5.2 Exploring the consumer price index

Over the last 20 years, would you expect that the average price of fuel oil would have been increasing, decreasing, or staying about the same? What about the price of chicken? The price of orange juice? The federal Bureau of Labor Statistics (BLS) has been keeping records on a market basket of commodities such as fuel oil, chicken, and frozen orange juice concentrate. Looking at the way the price index for various commodities has changed over the recent past can be very revealing.

1. Go to the Web site www.bls.gov/data/ and scroll down to "Price and Living Conditions." Then click on the icon for "Most Requested Statistics."

2. From the list of commodities, select fuel oil, per gallon, and then click on Retrieve data. Specify the most recent 20-year period. Example: January 1983 to January 2003. Check the box for "include graphs." Click GO.

3. The graph of the monthly average price for fuel oil shows lots of variability. Is the overall trend up, down, or fairly level? What conditions might explain the peaks in the price graph? What might explain the valleys? Does this graph surprise you? Write a few sentences about the changes in the price over the last 20 years.

4. Use the BACK button to return to the list of commodities. Uncheck fuel oil, and check chicken, fresh, whole, per pound. Get a graph of the monthly prices for the last 20 years. Describe the trend.

5. Repeat Step 4, but this time select frozen orange juice concentrate. The price seemed to peak in 1990, with a high value in July. Can you think of possible explanations for this very prominent peak? (Oranges come from Florida and California, where orange trees blossom in the spring and are harvested around December.)

Index numbers

The CPI is a type of numerical description: an *index number*. We can attach an index number to any quantitative variable that we measure repeatedly over time. The idea of the index number is to give a picture of changes in a variable much like that drawn by saying, "The average cost of a day in the hospital rose 46% between 1990 and 1996." That is, an index number describes the percent change from a base period.

Index number

An **index number** measures the value of a variable relative to its value at a base period. To find the index number for any value of the variable:

$$\text{index number} = \frac{\text{value}}{\text{base value}} \times 100$$

"Now this here's a genuine 1980 dollar. They don't make 'em like that anymore."

EXAMPLE 5.5   Calculating an index number

A gallon of unleaded regular gasoline cost $1.042 in January 1990 and $1.473 in January 2003. (These are national average prices collected by the BLS.) The gasoline price index number for January 2003, with January 1990 as the base period, is

$$\text{index number} = \frac{\text{value}}{\text{base value}} \times 100$$

$$= \frac{1.473}{1.042} \times 100 = 141$$

The gasoline price index number for the base period, January 1990, is

$$\text{index number} = \frac{1.042}{1.042} \times 100 = 100$$

Knowing the base period is essential to making sense of an index number. Because the index number for the base period is always 100, it is usual to identify the base period as, for example, 1990 by writing "1990 = 100." In news reports concerning the CPI, you will notice the mysterious equation "1982–84 = 100." That's shorthand for the fact that the years 1982 to 1984 are the base period for the CPI. An index number just gives the current value as a percent of the base value. Index number 125 means 125% of the base value, or a 25% increase from the base value. Index number 80 means that the current value is 80% of the base, a 20% decrease.

Fixed market basket price indexes

It may seem that index numbers are little more than a plot to disguise simple statements in complex language. Why say, "The consumer price index (1982–84 = 100) stood at 168.8 in January 2000," instead of "Consumer prices rose 68.8% between the 1982 to 1984 average and January 2000"? In fact, the term "index number" usually means more than a measure of change relative to a base. It also tells us the kind of variable whose change we measure. That variable is a weighted average of several quantities, with fixed weights. Let's illustrate the idea by a simple price index.

EXAMPLE 5.6 The Mountain Man Price Index

Bill Smith lives in a cabin in the mountains and strives for self-sufficiency. He buys only salt, kerosene, and the services of a professional welder. Here are Bill's purchases in 1990, the base period. His cost, in the last column, is the price per unit multiplied by the number of units he purchased.

Good or service	1990 quantity	1990 price	1990 cost
Salt	100 pounds	$0.50/pound	$ 50.00
Kerosene	50 gallons	1.00/gallon	50.00
Welding	10 hours	14.00/hour	140.00
		Total cost =	$240.00

The total cost of Bill's collection of goods and services in 1990 was $240. To find the "Mountain Man Price Index" for 2000, we use 2000 prices to calculate the 2000 cost of this same collection of goods and services. Here is the calculation:

Good or service	1990 quantity	2000 price	2000 cost
Salt	100 pounds	$0.80/pound	$ 80.00
Kerosene	50 gallons	1.00/gallon	50.00
Welding	10 hours	23.00/hour	230.00
		Total cost =	$360.00

The same goods and services that cost $240 in 1990 cost $360 in 2000. So the Mountain Man Price Index (1990 = 100) for 2000 is

$$\text{index number} = \frac{360}{240} \times 100 = 150$$

The point of Example 5.6 is that we follow the cost of the same collection of goods and services over time. It may be that Bill refused to hire the welder in 2000 because his hourly rate rose sharply. No matter—the index number uses the 1990 quantities, ignoring any changes in Bill's purchases between 1990 and 2000. We call the collection of goods and services whose total cost we follow a *market basket*. The index number is then a *fixed market basket price index*.

Fixed market basket price index

A **fixed market basket price index** is an index number for the total cost of a fixed collection of goods and services.

The basic idea of a fixed market basket price index is that the weight given to each component (salt, kerosene, welding) remains fixed over time. The CPI is in essence a fixed market basket price index, with several hundred items that represent all consumer purchases. Holding the market basket fixed allows a legitimate comparison of prices because we compare the prices of exactly the same items at each time. As we will see, it also poses severe problems for the CPI.

EXERCISES

5.30 White bread The average cost of a 1-pound loaf of white bread was $0.541 in January 1983 and $1.042 in January 2003.[4] Use January 1983 as the base period.
(a) What is the white bread price index number for January 1983?
(b) Find the white bread price index number for January 2003.

5.31 Breakfast The prices of a "breakfast basket" of items shown below are given for 1983 and 2003.

Commodities	1983	2003
Eggs, large, dozen	$0.821	$1.175
Orange juice, frozen	$1.418	$1.848
Coffee, ground roast, all sizes	$2.528	$2.999

Calculate the price index number for this breakfast basket of foods in 2003.

5.32 CPI and urban areas The CPI for all items in urban areas of the United States is shown in Figure 5.14.

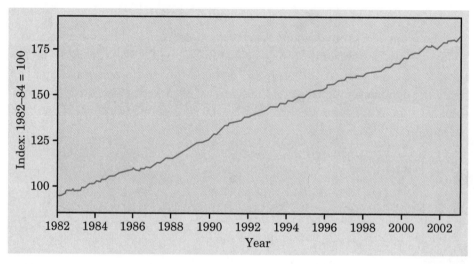

FIGURE 5.14 The consumer price index for all items in urban areas for 1982 to 2002.

(a) Describe the change in the CPI from the reference years, 1982 to 1984, until 2003.
(b) The CPI for all items in urban areas of the United States was 181.7 in January 2003. In a sentence or two, explain in simple terms what this means.
(c) Using your results from Exercise 5.30, did the price of bread rise more than, less than, or about the same as the overall CPI?

5.33 Gasoline A graph of the price of a gallon of regular unleaded gasoline for the 20-year period from 1983 to 2003 is shown in Figure 5.15 on the facing page.
(a) Describe the trend of the price of regular unleaded gasoline.
(b) There are two very noticeable peaks, the first around the end of 1990 and the second early in 2001. Can you think of any world events that might help explain these "spikes" in the cost of gasoline?

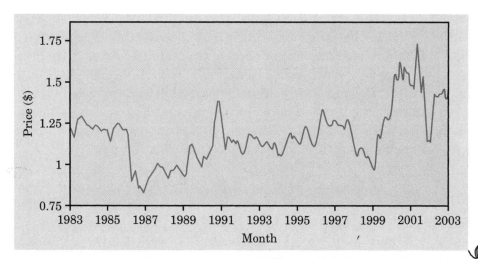

FIGURE 5.15 The consumer price index for a gallon of regular unleaded gasoline for 1983 to 2003.

(c) If the average price of a gallon of unleaded gasoline was $1.230 in January 1983 and $1.473 in January 2003, what was the price index for this commodity in 2003?

5.34 CPS In Activity 5.2, you went to the BLS Web site to explore the price indexes of several commodities. The Web site for the Current Population Survey is www.bls.gov/cps/. Go to this site and select Economic News Release. Then select Annual, and read a document on the most interesting topic you find. For example, one of the documents that has appeared in the past is entitled "Employment and Work Activity of High School Graduates." Write a short (half-page) narrative that summarizes the most important or most interesting facts you learn from the news release.

5.35 Inflation Go to the Web site www.bls.gov/home.htm and under Inflation and Consumer Spending, select the second item: Inflation Calculator. A pop-up window will appear.
(a) Enter 1.00 in the top box. $1.00 in 1980 dollars has the same buying power as what amount in 2003? Can you determine the price index for 2003 from this information? If so, what is it?
(b) Use trial and error and the inflation calculator to find out how much money you would need in 1980, to the nearest $1000, to have the buying power of $1 million in 2003.

Using the CPI

For now, think of the CPI as an index number for the cost of everything that American consumers buy. That the CPI for January 2000 was 172.2 means that we must spend $172.20 in January 2000 to buy goods and services that cost $100 in the 1982 to 1984 base period. An index number for "the cost of everything" lets us compare dollar amounts from different years by converting all the amounts into dollars of the same year. You will find tables in the *Statistical Abstract of the United States,* for example, with headings such as "Median Household Income, in Constant (2000) Dollars." That table has restated all incomes in dollars that will buy as much as the dollar would buy in 2000. Watch for the term "constant dollars" and for phrases like "real income." They mean that all dollar amounts represent the same buying power even though they may describe different years.

TABLE 5.2 **Annual Average Consumer Price Index, 1982–84 = 100**

Year	CPI	Year	CPI	Year	CPI	Year	CPI
1915	10.1	1965	31.5	1983	99.6	1993	144.5
1920	20.0	1970	38.8	1984	103.9	1994	148.2
1925	17.5	1975	53.8	1985	107.6	1995	152.4
1930	16.7	1976	56.9	1986	109.6	1996	156.9
1935	13.7	1977	60.6	1987	113.6	1997	160.5
1940	14.0	1978	65.2	1988	118.3	1998	163.0
1945	18.0	1979	72.6	1989	124.0	1999	166.6
1950	24.1	1980	82.4	1990	130.7	2000	172.2
1955	26.8	1981	90.9	1991	136.2	2001	177.1
1960	29.6	1982	96.5	1992	140.3	2002	179.9

Source: Bureau of Labor Statistics.

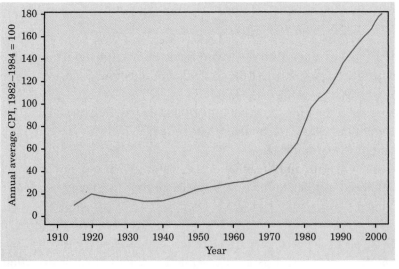

FIGURE 5.16 The consumer price index (1982–84 = 100) from 1915 to 2002.

Table 5.2 on the facing page gives annual average CPIs from 1915 to 2002. Figure 5.16 (below the table) is a line graph of the CPI values from the table. It shows that the twentieth century was a time of inflation—prices rose throughout the century and increased rapidly after 1973. Faced with this depressing fact, it would be foolish to think about dollars without adjusting for their decline in buying power. Here is the recipe for converting dollars of one year into dollars of another year.

Adjusting for changes in buying power

To convert an amount in dollars at Time A to the amount with the same buying power at Time B:

$$\text{dollars at Time B} = \text{dollars at Time A} \times \frac{\text{CPI at Time B}}{\text{CPI at Time A}}$$

Notice that the CPI for the time you are *going to* appears on the top in the ratio of CPIs in this recipe. Here are some examples.

EXAMPLE 5.7 Salaries of professional athletes

The mean salary of major league baseball players rose from $143,756 in 1980 to $1,945,712 in 2001. How big was the increase in real terms? Let's convert the 1980 average into 2001 dollars. Table 5.2 gives the annual average CPIs that we need.

$$2001 \text{ dollars} = 1980 \text{ dollars} \times \frac{2001 \text{ CPI}}{1980 \text{ CPI}}$$

$$= \$143,756 \times \frac{177.1}{82.4}$$

$$= \$308,971$$

That is, it took $308,971 in 2001 to buy what $143,756 would buy in 1980. We can now compare the 1980 mean salary of $308,971 *in 2001 dollars* with the actual 2001 mean salary, $1,945,712. Today's athletes earn much more than 1980 athletes even after adjusting for the fact that the dollar buys less now. (Of course, the mean salary is pulled up by the very high salaries paid to a few star players. The 2001 median salary was "only" $975,000.)

EXAMPLE 5.8   Rising incomes?

For a more serious example, let's leave the pampered world of professional athletes and look at the incomes of ordinary people. The median annual income of all American households was $17,710 in 1980. By 2000, the median income had risen to $42,151. Dollar income more than doubled, but we know that much of that rise is an illusion because of the dollar's declining buying power. To compare these incomes, we must express them in dollars of the same year. Let's express the 1980 median household income in 2000 dollars. The CPI in 2000 was 172.2.

$$\text{2000 dollars} = \$17,710 \times \frac{172.2}{82.4} = \$37,010$$

Real household incomes rose only from $37,010 to $42,151 in the 20 years between 1980 and 2000. That's a 14% increase.

The picture is different at the top. The 5% of households with the highest incomes earned $51,500 or more in 1980. In 2000 dollars, this is

$$\text{2000 dollars} = \$51,500 \times \frac{172.2}{82.4} = \$107,625$$

The top 5% of households earned $160,250 or more in 2000. That is, the real income of the highest earners increased by 49%.

Finally, let's look at production workers, the traditional "working men" (and women). Their average hourly earnings were $6.66 in 1980 and $13.75 in 2000. Restating the 1980 earnings in 2000 dollars,

$$\text{2000 dollars} = \$6.66 \times \frac{172.2}{82.4} = \$13.92$$

The real earnings of production workers went up only slightly (1%) between 1980 and 2000.

Example 5.8 illustrates how using the CPI to compare dollar amounts from different years brings out truths that are otherwise hidden. In this case, the truth is that the fruits of prosperity in the 1980s and 1990s went mostly to those at the top. Put another way, people with skills and education did much better than people like production workers, who generally lack special skills and college educations. Economists suggest several reasons: the "new economy," which rewards knowledge; high immigration, which leads to competition for less-skilled jobs; more competition from abroad; and so on. Exactly why the rewards of skill and education have increased so much, and what we should do about the stagnant incomes of less-educated people, are controversial questions.

Understanding the CPI

The idea of the CPI is that it is an index number for the cost of everything American consumers buy. That idea needs lots of fiddling to be practical. Much of the fiddling uses the results of large sample surveys.

Who is covered? The official name for the common version of the CPI (there are other versions, but we will ignore them) is the Consumer Price Index for All Urban Consumers. The CPI market basket represents the purchases of people living in urban areas. The official definition of "urban" is broad, so that about 80% of the U.S. population is covered. But if you live on a farm, the CPI doesn't apply to you.

How is the market basket chosen? Different households buy different things, so how can we get a single market basket? From a sample survey. The *Consumer Expenditure Survey* gathers detailed data on the spending of 29,000 households. The BLS breaks spending into categories such as "fresh fruits and vegetables," "new and used motor vehicles," and "hospital and related services." Then it chooses specific items such as "fresh oranges" to represent each group in the market basket. The items in the market basket get weights that represent their category's proportion of all spending. The weights, and even the specific market basket items, are updated regularly to keep up with changing buying habits. So the market basket isn't actually fixed.

So you think that's inflation?

Americans were unhappy when oil price increases in 1973 set off a round of inflation that saw the CPI almost double in the following decade. That's nothing. In Argentina, prices rose 127% in a single month, July 1989. The Turkish lira went from 14 to the dollar in 1970 to 579,000 to the dollar in 2000. There were 65 German marks to a dollar in January 1920 and 4,200,000,000,000 marks to a dollar in November 1923. Now *that's* inflation.

How are the prices determined? From more sample surveys. The BLS must discover the price of "fresh oranges" every month. That price differs from city to city and from store to store in the same city. The *Point of Purchase Survey* of 16,800 households keeps the BLS up-to-date on where consumers shop for each category of goods and services (supermarkets, convenience stores, discount stores, and so on). Each month, the BLS records 80,000 prices in 85 cities at a sample of stores that represents actual buying habits.

Does the CPI measure changes in the cost of living? A fixed market basket price index measures the cost of *living the same* over time, as Example 5.6 illustrated. In fact, we don't keep buying the same market basket of goods and services over time. We switch from LP records to tapes and CDs and then to DVD-audio disks. We don't buy new 1984 cars in 1995 or 2004. As prices change, we change what we buy—if beef becomes expensive, we buy less beef and more chicken or more tofu. A fixed market basket price index can't measure changes in the cost of living.

The BLS tries hard to keep its market basket up-to-date and to compensate for changes in quality. Every year, for example, the BLS must decide how much of the increase in new-car prices is paying for better quality. Only what's left counts as a genuine price increase in calculating the CPI. Between December 1967 and December 1994, actual car prices went up 313.4%, but the new-car price in the CPI went up only 172.1%. In 1995, adjustments for better quality reduced the overall rise in the prices of goods and services from 4.7% to only 2.2%. Prices of goods and services make up about 70% of the CPI. Most of the rest is the cost of shelter—renting an apartment or buying a house. House prices are another problem for the BLS. People buy houses partly to live in and partly because they think owning a house is a good investment. If we pay more for a house because we think it's a good investment, the full price should not go into the CPI.

By now it is clear that the CPI is *not* a fixed market basket price index, though that is the best way to start thinking about it. The BLS must constantly change the market basket as new products appear and our buying habits change. It must adjust the prices its sample surveys record to take account of better quality and the investment component of house prices. Yet the CPI still does not measure changes in our cost of living. It leaves out taxes, for example, which are certainly part of our cost of living.

Even if we agree that the CPI should only look at the goods and services we buy, it doesn't measure changes in our cost of living. In principle, a true "cost-of-living index" would measure the cost of the *same standard of living* over time. That's why we start with a fixed market basket price index, which also measures the cost of living the same over time but takes the simple view that "the same" means buying exactly the same things. If we are just as satisfied after switching from beef to tofu to avoid paying more for beef, our standard of living hasn't changed and a cost-of-living index should ignore the higher price of beef. If we are willing to pay more for products that keep our environment clean, we are paying for a higher standard of living, and the index should treat this just like an improvement in the quality of a new car. The BLS says that it would like the CPI to track changes in the cost of living but that a true cost-of-living index isn't possible in the real world.

EXERCISES

When you need the CPI for a year that does not appear in Table 5.2, use the table entry for the year that most closely follows the year you want.

5.36 The price of gasoline, I The year-end average price of unleaded regular gasoline has fluctuated as follows:

1985	$1.208 per gallon
1990	$1.354 per gallon
1995	$1.101 per gallon
2000	$1.489 per gallon

Give the gasoline price index numbers (1990 = 100) for 1985, 1990, 1995, and 2000.

5.37 The cost of college The part of the CPI that measures the cost of college tuition (1982–84 = 100) was 326.0 in January 2000. The overall CPI was 168.8 that month.
(a) Explain exactly what the index number 326.0 tells us about the rise in college tuition between the base period and the beginning of 2000.
(b) College tuition has risen much faster than consumer prices in general. How do you know this?

5.38 The price of gasoline, II Use your results from Exercise 5.36 to answer these questions.
(a) By how many points did the gasoline price index number change between 1985 and 2000? What percent change was this?
(b) By how many points did the gasoline price index number change between 1990 and 2000? What percent change was this?
You see that the point change and the percent change in an index number are the same if we start in the base period, but not otherwise.

5.39 Keeping up with the Joneses The Jones family had a household income of $30,000 in 1980, when the CPI (1982–84 = 100) was 82.4. The CPI at the beginning of 2000 was 168.8. How much must the Joneses earn in 2000 to have the same buying power they had in 1980?

5.40 Affording a Mercedes A Mercedes-Benz 190 cost $24,000 in 1981, when the CPI (1982–84 = 100) was 90.9. The average CPI for 2002 was 179.9. How many 2002 dollars must you earn to have the same buying power as $24,000 had in 1981?

APPLICATION 5.2 Does the CPI Overstate Inflation?

In 1995, Federal Reserve Chairman Alan Greenspan estimated that the CPI overstates inflation by somewhere between 0.5% and 1.5% per year. Mr. Greenspan was unhappy about this, because increases in the CPI automatically drive up federal spending. At the end of 1996, a group of outside experts appointed by the Senate Finance Committee

APPLICATION 5.2   Does the CPI Overstate Inflation? *(continued)*

estimated that the CPI had in the past overstated the rate of inflation by about 1.1% per year. The BLS agreed that the CPI overstates inflation but thought that the experts' guess of 1.1% per year was too high.

The reasons the CPI shows the value of a dollar falling faster than is true are partly tied to the nature of the CPI and partly due to limits on how quickly the BLS can adjust the details of the enormous machine that lies behind the CPI. Think first about the details. The prices of new products, such as digital cameras and flat-screen computer monitors, often start high and drop rapidly. The CPI market basket changes too slowly to capture the early drop in price. Discount stores with lower prices also enter the CPI sample slowly. And although the BLS tries hard to adjust for better product quality, the outside experts thought these adjustments were often too little and too late. The BLS has made many improvements in these details. The improved CPI would have grown about 0.5% per year more slowly than the actual CPI between 1978 and 1998.

The wider issue is the nature of the CPI as essentially a fixed market basket index. Such an index has an upward bias because it can't track shifts from beef to tofu and back as consumers try to get the same quality of life from whatever products are cheaper this month. At the bottom of the outside experts' criticisms of the CPI was the fact that the CPI does not track the "cost of living." Their first recommendation was, "The BLS should establish a cost of living index as its objective in measuring consumer prices." The BLS said it agreed in principle but that neither it nor anyone else knows how to do

George Chan/Tony Stone

APPLICATION 5.2 Does the CPI Overstate Inflation? *(continued)*

this in practice. It also said, "Measurement of changes in 'quality of life' may require too many subjective judgments to furnish an acceptable basis for adjusting the CPI." Nonetheless, a new kind of index that in principle comes closer to measuring changes in the cost of living appeared in 2002.

1. Digital cameras and flat-screen computer monitors are cited as examples of new products whose prices start high and then drop rapidly. Name some other examples.

2. Advocates for the elderly say that Social Security recipients would be hurt by less-rapid increases in the CPI. Why is this? Briefly explain.

Chairman Alan Greenspan's testimony on bias in the CPI before the House Committee on the Budget (March 4, 1997) began with an opening statement that can be found at the Web site www.federalreserve.gov/boarddocs/testimony/1997/19970304.htm. Read this testimony and then answer the following questions.

3. How does the Federal Reserve chairman define an unbiased cost-of-living index?

4. Does Chairman Greenspan believe that the CPI understates the cost of living or overstates it? How certain is he?

5. What does Greenspan believe to be the major reason for the bias in CPI calculation?

6. How much higher does he think the calculated inflation rate is than the actual inflation rate?

7. What two actions does Greenspan suggest to reduce bias in the CPI calculation?

8. What does the term "quality adjustment" mean, and why is this a concern?

9. What are Greenspan's suggestions for a two-track approach in dealing with bias problems in the CPI calculation?

The value of government statistics

Modern nations run on statistics. Economic data in particular guide government policy and inform the decisions of private business and individuals. Price indexes and unemployment rates, along with many other less publicized series of data, are produced by government statistical offices.

Some countries have a single statistical office, such as Statistics Canada (www.statcan.ca). Others attach smaller offices to various branches of government. The United States is an extreme case: there are 72 federal statistical offices, with relatively weak coordination among them. The Census Bureau and the Bureau of Labor Statistics are the most important, but you may at times use the products of the Bureau of Economic Analysis, the National Center for Health Statistics, the

Bureau of Justice Statistics, or others in the federal government's collection of statistical agencies.

A 1993 ranking of government statistical agencies by the heads of these agencies in several nations put Canada at the top, with the United States tied with Britain and Germany for sixth place.[5] The top spots generally went to countries with a single, independent statistical office. In 1996, Britain combined its main statistical agencies to form a new Office for National Statistics (www.ons.gov.uk). American government statistics remains fragmented.

What do citizens need from their government statistical agencies? First of all, they need data that are *accurate, timely,* and *keep up with changes in society and the economy.* Producing accurate data quickly demands considerable resources. Think of the large-scale sample surveys that produce the unemployment rate and the CPI. The major U.S. statistical offices have a good reputation for accuracy and lead the world in getting data to the public quickly. Their record for keeping up with changes is less good. The struggle to adjust the CPI for changing buying habits and changing quality is one issue. Another is the failure of U.S. economic statistics to keep up with trends such as the shift from manufacturing to services as the center of economic activity. Business organizations have expressed strong dissatisfaction with the overall state of our economic data.

Much of the difficulty stems from lack of money. In the years after 1980, reducing federal spending was a political priority. Government statistical agencies lost staff and cut programs. Lower salaries made it hard to attract the best economists and statisticians to government. The level of government spending on data also depends on our view of what data the government should produce. In particular, should the government produce data that are used mainly by private business rather than by the government's own policymakers? Perhaps such data should be either compiled by private concerns or produced only for those who are willing to pay. This is a question of political philosophy rather than statistics, but asking it helps us determine what level of government statistics taxpayers should pay for.

Freedom from political influence is as important to government statistics as accuracy and timeliness. When a statistical office is part of a government ministry, it can be influenced by the needs and desires of that ministry. The Census Bureau is in the Department of Commerce, which serves business interests. The BLS is in the Department of Labor. Thus, business and labor

"Yes sir, I know that we have to know where the economy is going. But do we have to publish the statistics so that everyone else does too?"

each have "their own" statistical office. The professionals in the statistical offices successfully resist direct political interference—a poor unemployment report is never delayed until after an election, for example. But indirect influence is clearly present. The BLS must compete with other Department of Labor activities for its budget, for example. Political interference with statistical work seems to be increasing, as when Congress refused to allow the Census Bureau to use sample surveys to correct for undercounting in the 2000 census.

The 1996 reorganization of Britain's statistical offices was prompted in part by a widespread feeling that political influence was too strong. The details of how unemployment is measured in Britain were changed many times in the 1980s, for example, and almost all the changes had the effect of reducing the reported unemployment rate—just what the government wanted to see.

Many statisticians favor a single "Statistics USA" office not attached to any other government ministry, as in Canada. Such unification might also help the money problem by eliminating duplication. It would at least allow a central decision about which programs deserve a larger share of limited resources. Unification is unlikely, but stronger coordination of the many federal statistical offices could achieve many of the same ends.

Social statistics

National economic statistics are well established with the government, the media, and the public. The government also produces many data on social issues such as education, health, housing, and crime. Social statistics are less complete than economic statistics. We have good data about how much money is spent on food but less information about how many people are poorly nourished. Social data are also less carefully produced than economic data. Economic statistics are generally based on larger samples, are compiled more often, and are published with a shorter time lag. The reason is clear: economic data are used by the government to guide economic policy month by month. Social data help us understand our society and address its problems but are not needed for short-term management.

There are other reasons the government is reluctant to produce social data. Many people don't want the government to ask about their sexual behavior or religion. Many feel that the government should avoid asking about our opinions— it's OK to ask "When did you last visit a doctor?" but not "How satisfied are you with the quality of your health care?" These hesitations reflect the American suspicion of government intrusion. Yet issues such as sexual behavior that contributes to the spread of AIDS and satisfaction with health care are important to citizens. Both facts and opinions on these issues can sway elections and influence policy. How can we get accurate information about social issues, collected consistently over time, and yet not entangle the government with sex, religion, and other touchy subjects?

The solution in the United States has been government funding of university sample surveys. After first deciding to undertake a sample survey asking people about their sexual behavior, in part to guide AIDS policy, the government backed away. Instead, it funded a much smaller survey of 3452 adults by the University of Chicago's National Opinion Research Center (NORC). NORC's General Social Survey (GSS), funded by the government's National Science Foundation, belongs with the Current Population Survey and the samples that undergird the CPI on any list of the most important sample surveys in the United States. The GSS includes both "fact" and "opinion" items. Respondents answer questions about their job security, their job satisfaction, their satisfaction with their city, their friends, and their family. They talk about race, religion, and sex. Many Americans would object if the government asked whether they had seen an X-rated movie in the past year, but they reply when the GSS asks this question.

This indirect system of government funding of a university-based sample survey fits the American feeling that the government itself should not be unduly invasive. It also insulates the survey from most political pressure. Alas, the government's budget cutting extends to the GSS, which now describes itself as an "almost annual" survey because lack of funds has prevented taking samples in some years. The GSS is, we think, a bargain.

EXPLORING THE WEB

The CPI has a home online at the BLS Web site, www.bls.gov/cpi/home.htm. Look under Frequently Asked Questions for the bureau's explanation of the CPI and its uses. The most recent CPI appears in the latest news release posted at this site.

If you like data, you can go to the BLS data page, www.bls.gov/data/, choose Most Requested Series and then Average Price Data, and see for yourself how the prices of such things as white bread and gasoline have changed over time.

U.S. government statistical agencies are fragmented, but they have cooperated in providing a single FedStats Web site that offers access to all of them. To see the variety of U.S. government statistics, go to www.fedstats.gov and click on Agencies. This is the mother lode of data about the United States.

EXERCISES

5.41 Microwaves on sale The prices of new gadgets often start high and then fall rapidly. The first home microwave oven cost $1300 in 1955. You can now buy a better microwave oven for $100. Find the latest value of the CPI (it's on the BLS Web site) and use it to restate $100 in present-day dollars in 1955 dollars. Compare this with $1300 to see how much real microwave oven prices have come down.

5.42 Saving money? One way to cut the cost of government statistics is to reduce the sizes of the samples. We might, for example, cut the CPS from 50,000 households to 20,000. Explain clearly, to someone who knows no statistics, why such cuts reduce the accuracy of the resulting data.

5.43 The General Social Survey The GSS places much emphasis on asking many of the same questions year after year. Why do you think it does this?

5.44 Measuring the effects of crime We wish to include, as part of a set of social statistics, measures of the amount of crime and of the impact of crime on people's attitudes and activities. Suggest some possible measures in each of the following categories:
(a) Statistics to be compiled from official sources such as police records.
(b) Factual information to be collected using a sample survey of citizens.
(c) Information on opinions and attitudes to be collected using a sample survey.

5.45 Statistical agencies Write a short description of the work of one of the government statistical agencies listed below. You can find information by starting at the FedStats Web site (www.fedstats.gov) and going to Agencies.
(a) Bureau of Economic Analysis (Department of Commerce).
(b) National Center for Education Statistics.
(c) National Center for Health Statistics.

STATISTICS IN SUMMARY

An **index number** describes the value of a variable relative to its value at some **base period.** A **fixed market basket price index** is an index number that describes the total cost of a collection of goods and services. Think of the government's **consumer price index** as a fixed market basket price index for the collection of all the goods and services that consumers buy. Because the CPI shows how consumer prices change over time, we can use it to change a dollar amount at one time into the amount at another time that has the same buying power. This is needed to compare dollar values from different times in **real terms.**

The details of the CPI are complex. It uses data from several large sample surveys. It is not a true fixed market basket price index because of adjustments for changing buying habits, new products, and improved quality.

Government statistical offices produce data needed for government policy and decisions by businesses and individuals. The data should be accurate, timely, and free from political interference. Citizens therefore have a stake in the competence and independence of government statistical offices.

SECTION 5.2 EXERCISES

5.46 Toxic releases The Environmental Protection Agency requires industry to report releases of any of a list of toxic chemicals. The total amounts released (in thousands of pounds) were 3,395,867 in 1988, 1,964,926 in 1995, and 1,941,870 in 1997. Give an index number for toxic chemical releases in each of these years, with 1988 as the base period. By what percent did releases increase or decrease between 1988 and 1997?

5.47 Los Angeles and New York The BLS publishes separate consumer price indexes for major metropolitan areas in addition to the national CPI. The CPI (1982–84 = 100) in January 2000 was 167.9 in Los Angeles and 179.2 in New York. (a) These numbers tell us that prices rose faster in New York than in Los Angeles between the base period and the beginning of 2000. Explain how we know this. (b) These numbers do *not* tell us that prices in January 2000 were higher in New York than in Los Angeles. Explain why.

5.48 The Food Faddist Price Index A food faddist eats only steak, rice, and ice cream. Here are his purchases in 1990:

Item	1990 quantity	1990 price
Steak	200 pounds	$5.45/pound
Rice	300 pounds	0.49/pound
Ice cream	50 gallons	5.08/gallon

After a visit from his mother, he adds oranges to his diet. Oranges cost $0.56/pound in 1990. Here are the food faddist's food purchases in 2000:

Item	2000 quantity	2000 price
Steak	175 pounds	$6.59/pound
Rice	325 pounds	0.53/pound
Ice cream	50 gallons	6.64/gallon
Oranges	100 pounds	0.61/pound

Find the fixed market basket Food Faddist Price Index (1990 = 100) for the year 2000.

5.49 Steinway A Steinway concert grand piano cost $13,500 in 1976. A similar Steinway cost $79,900 in 1999. Has the cost of the piano gone up or down in real terms? Give a calculation to justify your answer.

5.50 The price of gold Some people recommend that investors buy gold "to protect against inflation." Here are the prices of an ounce of gold at the end of the year for the years between 1983 and 1999. Make a graph that shows how the price of gold changed in real terms over this period. Would an investment in gold have protected against inflation by holding its value in real terms?

Year :	1983	1985	1987	1989	1991	1993	1995	1997	1999
Gold price :	$385	$329	$486	$403	$354	$391	$392	$368	$295

5.51 Playing major league baseball pays! Use the Inflation Calculator found at the www.bls.gov/home.htm Web site to verify that the average major league baseball player is better paid now than in 1980, even though a dollar is worth about half what it was worth then. Use the numbers from Example 5.7 (page 315).

5.52 The Guru Price Index A guru purchases only olive oil, loincloths, and copies of the *Atharva Veda,* from which he selects mantras for his disciples. Here are the quantities and prices of his purchases in 1985 and 1995:

Item	1985 quantity	1985 price	1995 quantity	1995 price
Olive oil	20 pints	$2.50/pint	18 pints	$3.80/pint
Loincloth	2	$2.75 each	3	$2.80 each
Atharva Veda	1	$10.95	1	$12.95

From these data, find the fixed market basket Guru Price Index (1985 = 100) for 1995.

5.53 The Vanguard 500 Index Fund One of the most popular mutual funds in the 1990s was the Vanguard 500 Index Fund. This fund invests in all of the Fortune 500 companies. Consequently, the fund reflects the market at large. Fund values go up when the market goes up, and the fund goes down when the market declines. Historically, the market has gained about 11% per year, on average.
(a) Would you expect this fund to have difficulties similar to those in the calculation of the CPI (see Application 5.2)? Explain briefly.
(b) Would you expect the variation in fund values for an index fund like the Vanguard 500 Index Fund to be greater than, less than, or about the same as "traditional" mutual funds, which invest in a select number of companies?

CHAPTER 5 REVIEW

Sometimes the overall pattern of a large number of observations is so regular that we can describe it by a smooth curve, called a normal distribution. Normal curves are handy models because they have properties that make them easy to use and think about. When data are approximately normally distributed, the mean and standard deviation are natural measures of center and spread. Remember, however, that the usefulness of numerical summaries and normal distributions depends on what we find when we examine graphs of our data.

In Section 5.2 we met a new kind of description, index numbers, with the consumer price index as the leading example. Section 5.2 also discussed government statistical offices, a quiet but important part of the statistical world.

Here are the most important skills you should have after studying this chapter.

1. Understand the new terminology (density curve, standard scores).

2. Interpret a density curve as a description of the distribution of a quantitative variable.

3. Recognize the shape of normal curves and estimate by eye both the mean and the standard deviation from such a curve.

4. Use the 68–95–99.7 rule and symmetry to state what percent of the observations from a normal distribution fall between two points when both points lie at the mean or one, two, or three standard deviations on either side of the mean.

5. Calculate and interpret the standard score of an observation.

6. Use Table B to find the percentile of a value from any normal distribution and the value that corresponds to a given percentile.

7. Calculate and interpret index numbers.

8. Calculate a fixed market basket price index for a small market basket.

9. Use the CPI to compare the buying power of dollar amounts from different years. Explain phrases such as "real income."

CHAPTER 5 EXERCISES

5.54 Mean and median Figure 5.17 on the facing page shows density curves of several shapes. Briefly describe the overall shape of each distribution. Two or three points are marked on each curve. The mean and the median are among these points. For each curve, which point is the median and which is the mean?

5.55 SAT scores The scale for SAT exam scores is set so that the distribution of scores is approximately normal with mean 500 and standard deviation 100. Answer these questions without using a table or a calculator.
(a) What is the median SAT score?
(b) You run a tutoring service for students who score between 400 and 600 and hope to do better. What percent of SAT scores are between 400 and 600?

5.56 Bacteria in milk A study of bacterial contamination in milk counted the number of coliform organisms (fecal bacteria) per milliliter in 100 specimens of

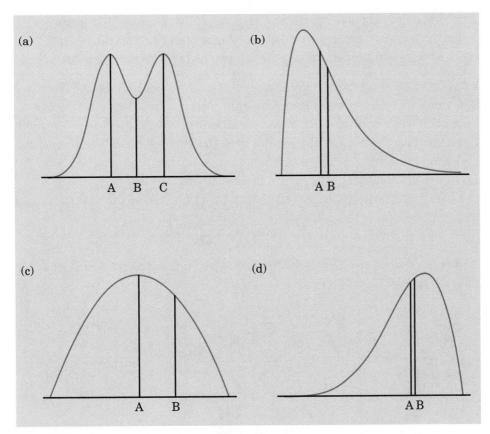

FIGURE 5.17 Four density curves of different shapes, for Exercise 5.54.

milk purchased in East Coast groceries. The U.S. Public Health Service recommends no more than 10 coliform bacteria per milliliter. Here are the data:

5	8	6	7	8	3	2	4	7	8	6	4	4	8	8	8	6	10	6	5
6	6	6	6	4	3	7	7	5	7	4	5	6	7	4	4	4	3	5	7
7	5	8	3	9	7	3	4	6	6	8	7	4	8	5	7	9	4	4	7
8	8	7	5	4	10	7	6	6	7	8	6	6	6	0	4	5	10	4	5
7	9	8	9	5	6	3	6	3	7	1	6	9	6	8	5	2	8	5	3

(a) Enter the data into your calculator and plot a histogram. Does the distribution of coliform counts appear to be approximately normal?

(b) Calculate the mean and standard deviation for the distribution.

(c) What percent of the observations fall within one, two, and three standard deviations of the mean?

5.57 Japanese IQ scores The Wechsler Intelligence Scale for Children is used (in several languages) in the United States and Europe. Scores in each case are approximately normally distributed with mean 100 and standard deviation 15. When the test was standardized in Japan, the mean was 111. To what percentile of the American-European distribution of scores does the Japanese mean correspond?

5.58 The stock market The annual rate of return on stock indexes (which combine many individual stocks) is very roughly normal. Since 1945, the Standard & Poor's 500 Index has had a mean yearly return of 12%, with a standard deviation of 16.5%. Take this normal distribution to be the distribution of yearly returns over a long period.
(a) In what range do the middle 95% of all yearly returns lie?
(b) The market is down for the year if the return on the index is less than zero. In what proportion of years is the market down?
(c) In what proportion of years does the index gain 25% or more?

5.59 Ponder this For which of the three density curves below is the median larger than the standard deviation?

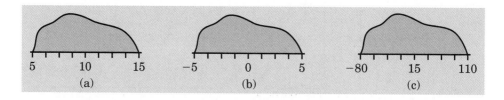

(a)   (b)   (c)

5.60 The curse of the Bambino In 1920 the Boston Red Sox sold Babe Ruth to the New York Yankees for $125,000. Between 1920 and 2000, the Yankees won 26 World Series and the Red Sox won none. How much is $125,000 in 1999 dollars?

5.61 Dream on When Julie started college in 1995, she set a goal of making $35,000 when she graduated. Julie graduated in 1999. What must Julie earn in 1999 in order to have the same buying power that $35,000 had in 1995?

5.62 Living too long? If both husband and wife are alive at age 65, in half the cases at least one will still be alive at age 93, 28 years later. This should frighten you. Myrna and Bill retired in 1970 with an income of $10,000 per year. They were quite comfortable—that was about the median family income in 1970. How much income did they need 28 years later, in 1998, to have the same buying power?

5.63 Good golfers In 1999, Tiger Woods won $6,620,970 on the Professional Golfers Association tour. The leading money winner in 1938 was Sam Snead, at $223,274. Tom Watson, the leader in 1980, won $1,041,002 that year. How do these amounts compare in real terms?

Chapter 6

Describing Relationships

6.1 Scatterplots and Correlation
6.2 Regression, Prediction, and Causation

6.1 SCATTERPLOTS AND CORRELATION

Is old faithful still faithful?

A geyser is a natural phenomenon: a type of hot spring that shoots hot water and steam into the air. A geyser erupts when some of the water in its reservoir becomes so hot that the expansion of the water into steam becomes more than the reservoir can handle. This triggers a chain reaction with the other water in the reservoir, and eventually a bit of the superhot water from the reservoir escapes to the surface and explodes into steam. This initial explosion then triggers the rest of the water in the reservoir, some or all of which is ejected through the opening in further eruptions. After the eruptions have ended, the reservoir again fills with water, and the process repeats itself.

One of the best-known geysers is Old Faithful, located in Yellowstone National Park in Wyoming. Accurate predictions of eruption times allow the National Park Service to inform visitors of the approximate time of the next eruption. Visitors can then adjust their schedules accordingly.[1] How can park geologists make predictions with any certainty? To see how, let's look at some data.

Figure 6.1 (next page) shows the distribution of intervals between eruptions of Old Faithful in July 1995. The shortest interval was 47 minutes and the longest was 113 minutes. That's quite a range! The distribution has an unusual shape: it has two clear peaks—one at about 60 minutes and the other at about 90 minutes. Let's put this in context. On many occasions, the time between eruptions of Old Faithful is about an hour. Even more often, the interval between eruptions is around 90 minutes. Can park geologists make predictions based on this double-peaked distribution?

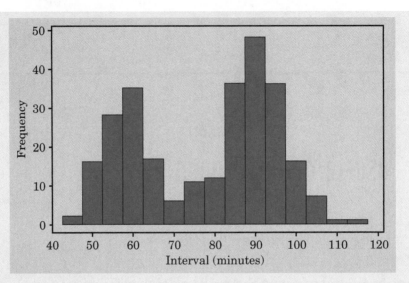

FIGURE 6.1 Histogram of the interval (in minutes) between eruptions of the Old Faithful geyser in July 1995.

If geologists predict a 60-minute gap between eruptions, but the actual interval is 90 minutes, some visitors will become impatient. If they predict a 90-minute interval, but the gap is closer to 60 minutes, some visitors may go elsewhere in the park and return after the eruption has occurred. What should the geologists do? Figure 6.2 helps

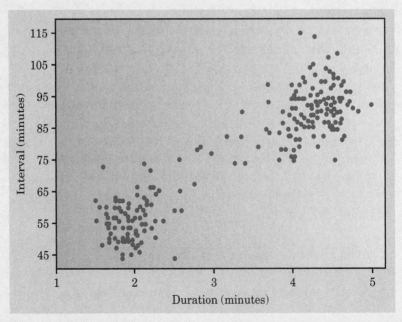

FIGURE 6.2 Scatterplot of the interval between eruptions of Old Faithful against the duration of the previous eruption.

unravel the mystery. This is a *scatterplot* that plots the interval between successive eruptions of Old Faithful on the vertical axis against the duration of the previous eruption on the horizontal axis. The plot clearly shows two groups of eruptions. In one group, the duration of the eruptions is about 2 minutes. In the other group of eruptions, most of them last at least 4 minutes. The plot as a whole shows that the interval between eruptions increases as the duration of the previous eruption increases.

Since a long eruption depletes more water from a geyser's reservoir than a short eruption, we would expect that it would take more time to refill the reservoir after a longer eruption. Consequently, we would expect a longer time interval until the next eruption. Thus, the physical properties of the geyser suggest a relationship between the duration of an eruption and the time until the next eruption. Such a relationship may be used to predict geyser eruption times.

There are general lessons to be gained from this study of Old Faithful's eruptions. To understand one variable, we must often look at how it is related to other variables. And once again, to see what data say, start by making graphs.

ACTIVITY 6.1 Tedious tasks take time

Materials: Graphing calculator, graph paper, and stopwatch for each pair of students

Many jobs require employees to perform repetitive tasks. Fast-food restaurants rely on their employees to fill customers' orders quickly and accurately. Bookkeepers and accountants enter pages of numerical data into Excel spreadsheets. Elementary school students practice their multiplication facts. Cashiers at retail stores determine the total price of goods that are purchased. Quality control inspectors examine a sample of the items produced. In each of these situations, it is important to know how long the task will take. This Activity will give you a chance to see how long a tedious task takes you.

1. Go into your Statistics/List Editor (press [STAT] then choose 1:Edit...). Clear list L_1 by highlighting L_1 with your cursor, then pressing [CLEAR][ENTER]. Clear lists L_2 through L_6 in the same way.

2. Your first task is to enter the digits 0, 1, 2, . . . , 9 in list L_1. Place your cursor just below the list name. You should see $L_1(1)=$ in the bottom left corner of the screen. Your partner will time how long it takes for you to complete the task. When your partner says "Go!" enter the digits one at a time into the list. The sequence of keystrokes will be

[0][ENTER] [1][ENTER] . . . [9][ENTER]

ACTIVITY 6.1 Tedious tasks take time *(continued)*

As soon as you have finished, say "Done!" so your partner will stop timing. If you entered all of the numbers correctly, record your time to the nearest second. If you made a mistake, repeat the task.

3. Your second task is to enter the numbers 10, 11, 12, . . . , 19 in list L_2. As in the previous task, begin when your partner says "Go!" and say "Done!" when you are finished. If you entered all the numbers correctly, record your time to the nearest second. Otherwise, repeat the task.

4. You can probably guess what the third, fourth, fifth, and sixth tasks are now. Task 3: Enter the numbers 100, 101, 102, . . . , 109 in list L_3. Task 4: Enter the numbers 1000, 1001, 1002, . . . , 1009 in list L_4. Task 5: Enter the numbers 10000, 10001, 10002, . . . , 10009 in list L_5. Task 6: Enter the numbers 100000, 100001, 100002, . . . , 100009 in list L_6. For each task, start typing when your partner says "Go!" and be sure to say "Done!" when finished. Once you enter all the numbers correctly for each task, record your time to the nearest second.

5. Switch roles with your partner and repeat Steps 2 through 4.

6. Summarize the data for your six tasks in a table like this one:

Task no.	Total no. of digits typed	Time to nearest second
1	10	
2	20	
. . .		

7. Make a *scatterplot* of your data as follows:

- On graph paper, draw x and y axes. Label the x axis "Total digits typed" and the y axis "Time." Scale the horizontal axis by adding tick marks at 0, 10, 20, . . . , 70. Scale the vertical axis in half-second increments beginning with 0.
- Plot each of your six data points on the graph.

8. Describe what the graph tells you about the relationship between the two variables.

9. Compare graphs with your partner. What similarities and differences do you see?

A medical study finds that short women are more likely to have heart attacks than women of average height, while tall women have the fewest heart attacks. An insurance group reports that heavier cars have fewer deaths per 10,000 vehicles registered than do lighter cars. These and many other statistical studies look at the relationship between two variables. To understand such a relationship, we must often examine other variables as well. To conclude that shorter women have higher risk of heart attacks, for example, the researchers had to eliminate the effect of other variables such as weight and exercise habits. Our topic in this and the following chapters is relationships between variables. One of our main themes is that the relationship between two variables can be strongly influenced by other variables that are lurking in the background. Most statistical studies examine data on more than one variable. Fortunately, statistical analysis of several-variable data builds on the tools we used to examine individual variables. The principles that guide our work also remain the same:

- First plot the data, then add numerical summaries.

- Look for overall patterns and deviations from those patterns.

- When the overall pattern is quite regular, there is sometimes a way to describe it very briefly.

Scatterplots

The most common way to display the relation between two quantitative variables is a *scatterplot*. Figure 6.2 (page 332) is a scatterplot that shows how the interval between eruptions of Old Faithful is related to the duration of the previous eruption. We think that "duration" will help explain "interval." That is, "duration" is the *explanatory variable* and "interval" is the *response variable*. We want to see how the interval changes when duration changes, so we put duration (the explanatory variable) on the horizontal axis. We can then see that as the duration of the previous eruption increases, the interval between eruptions increases. Each point on the plot represents one eruption. For one eruption that lasted 1.6 minutes, there was an interval of 74 minutes until the next eruption. Find 1.6 on the *x* (horizontal) axis to locate this eruption.

Scatterplot

A **scatterplot** shows the relationship between two quantitative variables measured on the same individuals. The values of one variable appear on the horizontal axis, and the values of the other variable appear on the vertical axis. Each individual in the data appears as the point in the plot fixed by the values of both variables for that individual.

Always plot the explanatory variable, if there is one, on the horizontal axis (the *x* axis) of a scatterplot. As a reminder, we usually call the explanatory variable *x* and the response variable *y*. If there is no explanatory/response distinction, either variable can go on the horizontal axis.

EXAMPLE 6.1   Health and wealth

Figure 6.3 is a scatterplot of data from the World Bank.[2] The individuals are all the world's nations for which data are available. The explanatory variable is a measure of how rich a country is: the gross domestic product (GDP) per person. GDP is the total value of the goods and services produced in a country, converted into dollars. The response variable is life expectancy at birth.

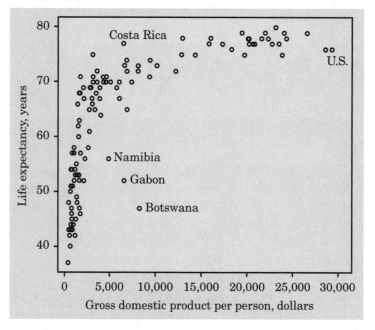

FIGURE 6.3 Scatterplot of the life expectancy of people in many nations against each nation's gross domestic product per person.

We expect people in richer countries to live longer. The overall pattern of the scatterplot does show this, but the relationship has an interesting shape. Life expectancy rises very quickly as GDP increases and then levels off. People in very rich countries such as the United States live no longer than people in poorer but not extremely poor nations. Some of these countries, such as Costa Rica, even do better than the United States.

Three African nations are outliers. Their life expectancies are similar to those of their neighbors but their GDP is higher. Gabon produces oil, and Namibia and Botswana produce diamonds. It may be that income from mineral exports goes mainly to a few people and so pulls up GDP per person without much effect on either the income or the life expectancy of ordinary citizens. That is, GDP per person is a mean, and we know that mean income can be much higher than median income. In the case of Botswana, we may also have an incorrect value for GDP. The World Bank estimates $8310 per person, and this value is used in Figure 6.3, but the CIA estimates only $3600.

EXERCISES

6.1 Explanatory or response? In each of the following situations, is it more reasonable to simply explore the relationship between the two variables or to view one of the variables as an explanatory variable and the other as a response variable? In the latter case, which is the explanatory variable and which is the response variable?
(a) The amount of time spent studying for a statistics exam and the grade on the exam.
(b) The weight in kilograms and height in centimeters of a person.
(c) Inches of rain in the growing season and the yield of corn in bushels per acre.
(d) A student's scores on the SAT math exam and the SAT verbal exam.
(e) A family's income and the years of education their eldest child completes.

6.2 Healthy breeding Often the percent of an animal species in the wild that survive to breed again is lower following a successful breeding season. This is part of nature's self-regulation, tending to keep population size stable. A study of merlins (small falcons) in northern Sweden observed the number of breeding pairs in an isolated area and the percent of males (banded for identification) who returned the next breeding season. Here are the data for nine years:[3]

Breeding pairs	Percent of males returning
28	82
29	83, 70, 61
30	69
32	58
33	43
38	50, 47

(a) Why is the response variable the *percent* of males that return rather than the *number* of males that return?
(b) Make a scatterplot. Describe the pattern. Do the data support the theory that a smaller percent of birds survive following a successful breeding season?

6.3 IQ and GPA Figure 6.4 is a scatterplot of school grade point average versus IQ score for all 78 seventh-grade students in a rural midwestern school.

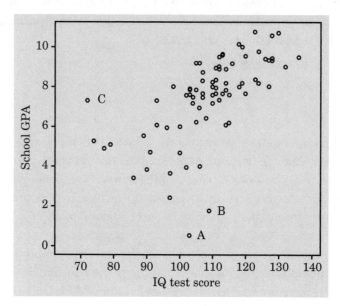

FIGURE 6.4 School grade point averages and IQ test scores for 78 seventh-grade students.

(a) Describe the overall pattern of the relationship in words. Points A, B, and C might be called outliers.
(b) About what are the IQ and GPA for Student A?
(c) For each point A, B, and C, say how it is unusual (for example, "low GPA but a moderately high IQ score").

6.4 Fast cars Interested in a sporty car? Worried that it might use too much gas? The Environmental Protection Agency lists most such vehicles in its "minicompact" or "two-seater" categories. Table 6.1 on the facing page gives city and highway gas mileages (in miles per gallon) for all model year 2001 cars in these two groups.[4]
(a) Make a scatterplot that shows the relationship between city and highway mileage for minicompact cars using city mileage as the explanatory variable. Be sure to label your axes.
(b) On the same graph, make a scatterplot that shows the relationship between city and highway mileage for two-seater cars. Use a different color or plotting symbol.
(c) Describe what you see in the scatterplot. Is the form of the relationship similar for the two types of car? What is the most important difference between the two types?

TABLE 6.1 Gas Mileages for Model Year 2001 Cars

Minicompact Cars			Two-Seater Cars		
Model	City	Highway	Model	City	Highway
Audi TT coupe	22	31	Acura NSX	17	24
BMW 325CI Convertible	19	27	Audi TT Roadster	22	30
BMW 330CI Convertible	20	28	BMW Z3 Coupe	21	28
BMW M3 Convertible	16	23	BMW Z3 Roadster	20	27
Jaguar XK8 Convertible	17	24	BMW Z8	13	21
Jaguar XKR Convertible	16	22	Chevrolet Corvette	18	26
Mercedes-Benz CLK320	20	28	Dodge Viper	11	21
Mercedes-Benz CLK430	18	24	Ferrari Maranello	8	13
Mitsubishi Eclipse	22	30	Ferrari Modena	11	16
Porsche 911 Carrera	17	25	Honda Insight	61	68
Porsche 911 Turbo	15	22	Honda S2000	20	26
			Lamborghini Diablo	10	13
			Mazda Miata	22	28
			Mercedes-Benz SL500	16	23
			Mercedes-Benz SL600	13	19
			Mercedes-Benz SLK320	21	27
			Plymouth Prowler	17	23
			Porsche Boxster	19	27
			Toyota MR2	25	30

Interpreting scatterplots

To interpret a scatterplot, apply the usual strategies of data analysis.

Examining a scatterplot

In any graph of data, look for the **overall pattern** and for striking **deviations** from that pattern.

You can describe the overall pattern of a scatterplot by the **form, direction, and strength** of the relationship.

An important kind of deviation is an **outlier,** an individual value that falls outside the overall pattern of the relationship.

Both Figures 6.2 (page 332) and 6.3 (page 336) have a clear *direction:* the time between eruptions goes up as the length of the previous eruption increases, and life expectancy generally goes up as GDP increases. We say that Figures 6.2 and 6.3 show a *positive association* between the variables.

Positive association, negative association

Two variables are **positively associated** when above-average values of one tend to accompany above-average values of the other and below-average values also tend to occur together. The scatterplot slopes upward as we move from left to right.

Two variables are **negatively associated** when above-average values of one tend to accompany below-average values of the other, and vice versa. The scatterplot slopes downward from left to right.

Each of our scatterplots has a distinctive *form.* Figure 6.2 shows two *clusters* of eruptions, and Figure 6.3 shows a *curved relationship.* The *strength* of a relationship in a scatterplot is determined by how closely the points follow a clear form. The relationships in Figures 6.2 and 6.3 are not strong. Eruptions of similar lengths show quite a bit of scatter in times until the next eruption, and nations with similar GDPs can have quite different life expectancies. Here is an example of two variables with a strong negative relationship and a simple form.

EXAMPLE 6.2   How much natural gas does a household use?

Joan is concerned about the amount of energy she uses to heat her home in the Midwest. She keeps a record of the natural gas she consumes each month over one year's heating season. Because the months are not equally long, she divides each month's consumption by the number of days in the month to get the average number of cubic feet of gas used per day. Demand for heating is strongly influenced by the outside temperature. From local weather records, Joan obtains the average temperature for each month, in degrees Fahrenheit. Here are Joan's data:

Month:	Oct.	Nov.	Dec.	Jan.	Feb.	Mar.	Apr.	May
Temperature, x:	49.4	38.2	27.2	28.6	29.5	46.4	49.7	57.1
Gas consumed, y:	520	610	870	850	880	490	450	250

A scatterplot of these data appears in Figure 6.5. Temperature is the explanatory variable *x* (plotted on the horizontal axis) because outside temperature helps explain gas consumption. The scatterplot shows *a strong negative straight-line association*. The straight-line form is important because it is common and simple. The association is strong because the points lie close to a line. It is negative because as temperature increases, gas consumption goes down because less gas is used for heating.

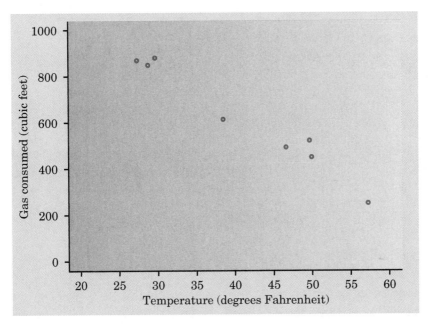

FIGURE 6.5 Home consumption of natural gas versus outdoor temperature.

CALCULATOR CORNER   How to make a scatterplot

In this Calculator Corner, you will make a scatterplot of the natural gas consumption data from Example 6.2.

• Go into your Statistics/List Editor (press STAT then choose 1:Edit...).

Clear lists L_1 and L_2 as you did in Activity 6.1.

• Enter the temperature values in list L_1 and the corresponding gas consumed values in list L_2.

CALCULATOR CORNER   How to make a scatterplot *(continued)*

- Go to the statistics plots menu (press 2nd Y=) and select 1:Plot1. Adjust your settings as shown.

- Make sure that Plot2 and Plot3 are Off, and that all functions are deactivated in your Y = menu. Then use ZoomStat (press ZOOM and choose 9:ZoomStat) to produce the graph.

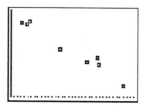

- One disadvantage of using ZoomStat is that the calculator chooses the viewing window for you. Press WINDOW . Notice the unusual choices for Xmin, Xmax, Ymin, and Ymax.

```
WINDOW
 Xmin=24.21
 Xmax=60.09
 Xscl=1
 Ymin=142.9
 Ymax=987.1
 Yscl=1
 Xres=1
```

- You can adjust the window settings to values that seem more reasonable, which can be especially useful if you want to transfer a calculator graph to paper. Change your settings to the ones shown.

```
WINDOW
 Xmin=20
 Xmax=60
 Xscl=10
 Ymin=100
 Ymax=1000
 Yscl=100
 Xres=1
```

- Press GRAPH . Now the tick marks around the edges of the window give you a better idea of each point's location.

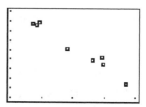

CALCULATOR CORNER   How to make a scatterplot *(continued)*

- Press ⌷TRACE⌷ and use your arrow keys to move from one point to another in the scatterplot. Notice that the *x* and *y* coordinates of each point you trace are given at the bottom of the screen.

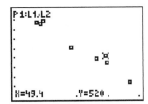

- Save the data for later use. First, go to the home screen (press ⌷2nd⌷ ⌷MODE⌷ to get there). Then press ⌷2nd⌷⌷1⌷⌷STO →⌷⌷2nd⌷ ⌷ALPHA⌷, type the letters TEMP (the letters

are in green above the keys), and press ⌷ENTER⌷. Compare with the screen shot shown.

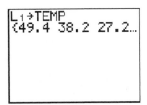

- Follow the same procedure to store the values in list L₂ to the variable GAS.

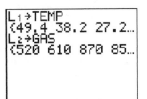

EXERCISES

6.5 Rich states, poor states One measure of a state's prosperity is the median income of its households. Another measure is the mean personal income per person in the state. Figure 6.6 (next page) is a scatterplot of these two variables, both measured in thousands of dollars. Because both variables have the same units, the plot is square with the same scales on both axes.

(a) Explain why you expect a positive association between these variables. Also, explain why you expect household income to be generally higher than income per person.

(b) Nonetheless, the mean income per person in a state can be higher than the median household income. In fact, the District of Columbia has median income $33,433 per household and mean income $38,429 per person. Explain why this can happen.

(c) Describe the overall pattern of the plot, ignoring the outliers.

(d) We have labeled some interesting states on the graph. New York has low median household income relative to its mean individual income. In Utah, median household income is high relative to mean individual income. What facts about a state could make median household income unusually high or low relative to mean individual income?

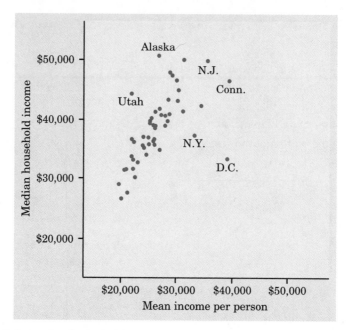

FIGURE 6.6 Median household income and mean income per person in the states, for Exercise 6.5.

6.6 Crawl before you walk At what age do babies learn to crawl? Does it take longer to learn in the winter, when babies are often bundled in clothes that restrict their movement? Perhaps there might even be an association between babies' crawling age and the average temperature during the month they first try to crawl (around six months after birth). Data were collected from parents who brought their babies into the University of Denver Infant Study Center to participate in one of a number of experiments. Parents reported the birth month and the age at which their child was first able to creep or crawl a distance of four feet within one minute. Information was obtained on 414 infants, 208 boys and 206 girls. Crawling age is given in weeks, and average temperature (in °F) refers to the month that is six months after the birth month.[5]

Birth month	Average crawling age	Average temperature
January	29.84	66
February	30.52	73
March	29.70	72
April	31.84	63
May	28.58	52
June	31.44	39
July	33.64	33
August	32.82	30

Birth month	Average crawling age	Average temperature
September	33.83	33
October	33.35	37
November	33.38	48
December	32.32	57

(a) Make a scatterplot that displays the relationship between average crawling age and average temperature. Be sure to think about which is the explanatory variable.

(b) Describe the direction, form, and strength of the relationship. What do you conclude about when babies learn to crawl?

6.7 Interested in stocks? When interest rates are high, investors may shun stocks because they can get high returns with less risk. Figure 6.7 plots the annual return on U.S. common stocks for the years 1950 to 2000 against the returns on Treasury bills for the same years.[6] (The interest paid by Treasury bills is a measure of how high interest rates were that year.)

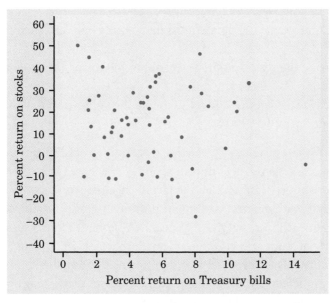

FIGURE 6.7 Percent returns on Treasury bills and common stocks for the years 1950 to 2000, for Exercise 6.7.

(a) Approximately what were the highest and lowest annual percent returns on stocks during this period? What were the highest and lowest returns for Treasury bills?

(b) Describe the pattern you see. Are high interest rates bad for stocks? Is the relationship between interest rates and stock returns strong or weak?

6.8 Teen drivers Thirty students in a high school statistics class were asked to report the age (in years) and odometer reading of their primary vehicle. Here are their data:[7]

Age	Mileage	Age	Mileage	Age	Mileage
3	40,300	6	85,000	8	110,000
2	11,912	4	20,000	4	30,323
3	30,000	3	30,000	7	100,000
4	40,000	2	17,000	10	98,000
8	98,000	3	25,000	2	12,000
3	48,000	7	150,000	5	53,000
8	120,000	2	10,000	7	40,000
11	185,000	10	110,000	4	76,000
4	40,000	5	103,000	2	3,000
1	1,050	5	66,610	7	75,000

(a) Enter the data into your Statistics/List Editor, using list L_1 for age and list L_2 for mileage.
(b) Make a scatterplot of the data. Adjust the viewing window, then transfer your graph to paper.
(c) Describe what the scatterplot tells you about the form, direction, and strength of the relationship. Are there any outliers?
(d) Save these lists for later use. Refer to the method shown in the Calculator Corner on page 341.

6.9 NAEP for fourth-graders The National Assessment of Educational Progress (NAEP) assesses what students know in several subject areas based on large representative samples. Table 6.2 on the facing page reports some findings of the NAEP year 2000 Mathematics Assessment for fourth-graders in the 40 states that participated. For each state we give the mean NAEP math score (out of 500) and also the percent of students who were at least "proficient" in the sense of being able to use math skills to solve real-world problems.[8] Nationally, about 25% of students are "proficient" by NAEP standards. We expect that average performance and percent of proficient performers will be strongly related.
(a) Make a scatterplot, using mean NAEP scores as the explanatory variable. Notice that there are several pairs of states with identical values. Use a different symbol for points that represent two states.
(b) Describe the form, direction, and strength of the relationship.
(c) Circle your home state's point in the scatterplot. Although there are no clear outliers, there are some points that you may consider interesting, perhaps because they are on the edge of the pattern. Choose one such point: Which state is this, and in what way is it interesting?

TABLE 6.2 State Performance on the NAEP Mathematics Assessment of Fourth-Graders

State	Mean NAEP score	Percent proficient	Percent poverty	State	Mean NAEP score	Percent proficient	Percent poverty
Alabama	218	14	21.8	Missouri	229	24	14.4
Arizona	219	17	23.6	Montana	230	25	21.2
Arkansas	217	14	13.1	Nebraska	226	24	14.8
California	214	15	22.3	Nevada	220	16	12.8
Connecticut	234	32	13.4	New Mexico	214	12	23.5
Georgia	220	18	24.7	New York	227	22	28.9
Hawaii	216	14	14.5	North Carolina	232	28	21.3
Idaho	227	21	17.4	North Dakota	231	25	17.2
Illinois	225	22	12.2	Ohio	231	26	16.0
Indiana	234	31	12.6	Oklahoma	225	17	19.9
Iowa	233	28	14.2	Oregon	227	24	19.4
Kansas	232	30	13.3	Rhode Island	225	23	20.5
Kentucky	221	17	16.7	South Carolina	220	18	17.6
Louisiana	218	14	29.8	Tennessee	220	13	14.5
Maine	231	24	12.0	Texas	233	27	20.1
Maryland	222	22	8.1	Utah	227	24	11.8
Massachusetts	235	33	15.0	Vermont	232	30	12.2
Michigan	231	29	14.8	Virginia	230	25	7.9
Minnesota	235	34	12.6	West Virginia	225	18	25.7
Mississippi	211	9	19.3	Wyoming	229	25	13.0

Correlation

A scatterplot displays the direction, form, and strength of the relationship between two variables. Straight-line relations are particularly important because a straight line is a simple pattern that is quite common. A straight-line relation is strong if the points lie close to a straight line and weak if they are widely scattered about a line. Our eyes are not good judges of how strong a relationship is. The two scatterplots in Figure 6.8 (next page) depict the same data, but the right-hand plot is drawn smaller in a large field. The right plot seems to show a stronger straight-line relationship. Our eyes can be fooled by changing the plotting scales or the amount of blank space around the cloud of points in a scatterplot. We need to follow our strategy for data analysis by using a numerical measure to supplement the graph. *Correlation* is the measure we use.

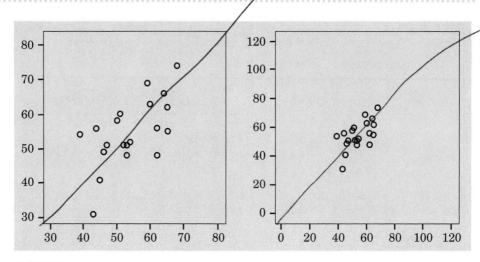

FIGURE 6.8 Two scatterplots of the same data. The right-hand plot suggests a stronger relationship between the variables because of the surrounding space.

Correlation

The **correlation** describes the direction and strength of a straight-line relationship between two quantitative variables. Correlation is usually written as *r*.

Calculating a correlation takes a bit of work. You can usually think of *r* as the result of pushing a calculator button or giving a command in software and concentrate on understanding its properties and use. Knowing how we obtain *r* from data does help us understand how correlation works, however, so here we go.

EXAMPLE 6.3   Classifying fossils and calculating correlation

Archaeopteryx is an extinct beast having feathers like a bird but teeth and a long bony tail like a reptile. Only six fossil specimens are known. Because these fossils differ greatly in size, some scientists think they are different species rather than individuals from the same species. We will examine data on the lengths in centimeters (cm) of the femur (a leg bone) and the humerus (a bone in the upper arm) for the five fossils that preserve both bones. Here are the data:[9]

Femur:	38	56	59	64	74
Humerus:	41	63	70	72	84

Because there is no explanatory/response distinction, we can put either measurement on the *x* axis of a scatterplot. We chose to use *x* for femur length and *y* for humerus length. The plot appears in Figure 6.9.

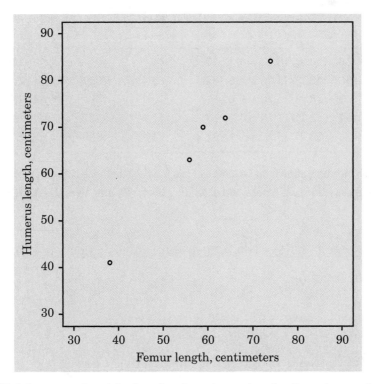

FIGURE 6.9 Scatterplot of the lengths of two bones in 5 fossil specimens of the extinct beast *Archaeopteryx*.

The plot shows a strong, positive, straight-line association. As the length of one bone increases, so does the length of the other bone. These data suggest that all five fossils belong to the same species and differ in size because some are younger than others. We expect that a different species would have a different relationship between the lengths of the two bones, so that it would appear as an outlier.

To calculate the correlation, we follow these steps.

Step 1. Find the mean and standard deviation for both *x* and *y*. For the fossil data, a calculator tells us that

Femur: $\bar{x} = 58.2$ cm $s_x = 13.20$ cm

Humerus: $\bar{y} = 66.0$ cm $s_y = 15.89$ cm

We use s_x and s_y to remind ourselves that there are two standard deviations, one for the values of x and the other for the values of y.

Step 2. Using the means and standard deviations from Step 1, find the standard scores for each x-value and for each y-value:

Value of x	Standard score $(x - \bar{x})/s_x$	Value of y	Standard score $(y - \bar{y})/s_y$
38	$(38 - 58.2)/13.20 = -1.530$	41	$(41 - 66.0)/15.89 = -1.573$
56	$(56 - 58.2)/13.20 = -0.167$	63	$(63 - 66.0)/15.89 = -0.189$
59	$(59 - 58.2)/13.20 = 0.061$	70	$(70 - 66.0)/15.89 = 0.252$
64	$(64 - 58.2)/13.20 = 0.439$	72	$(72 - 66.0)/15.89 = 0.378$
74	$(74 - 58.2)/13.20 = 1.197$	84	$(84 - 66.0)/15.89 = 1.133$

Step 3. The correlation is the average of the products of these standard scores. As with the standard deviation, we "average" by dividing by $n - 1$, one fewer than the number of individuals:

$$r = \frac{1}{4}[(-1.530)(-1.573) + (-0.167)(-0.189) + (0.061)(0.252) + (0.439)(0.378)$$

$$+ (1.197)(1.133)]$$

$$= \frac{1}{4}(2.4067 + 0.0316 + 0.0154 + 0.1659 + 1.3562)$$

$$= \frac{3.9758}{4} = 0.994$$

The algebraic shorthand for the set of calculations in Example 6.3 is

$$r = \frac{1}{n - 1} \sum \left(\frac{x - \bar{x}}{s_x} \right) \left(\frac{y - \bar{y}}{s_y} \right)$$

The symbol $\sum$ means "add them all up."

EXERCISES

6.10 How tall is your date? A student wonders if tall women tend to date taller men than do short women. She measures herself, her dormitory roommate, and the women in the adjoining rooms; then she measures the next man each woman dates. Here are the data (heights in inches):

Women (x):	66	64	66	65	70	65
Men (y):	72	68	70	68	71	65

(a) Make a scatterplot of these data. Based on the scatterplot, do you expect the correlation to be positive or negative? Near ± 1 or not?

(b) Find the correlation r between the heights of the men and women. Follow the method of Example 6.3.

(c) How would r change if all the men were 6 inches shorter than the heights given in the table? Does the correlation tell us whether women tend to date men taller than themselves?

(d) If heights were measured in centimeters rather than inches, how would the correlation change? (There are 2.54 centimeters in an inch.)

(e) If every woman dated a man exactly 3 inches taller than herself, what would be the correlation between male and female heights?

6.11 Explaining correlation You have data on the current grade point average (GPA) of students in a basic statistics course and their scores on the first examination in the course. Say as specifically as you can what the correlation r between GPA and exam score measures.

6.12 What correlation doesn't measure
(a) Make a scatterplot of the following data:

x	-5	-3	0	3	5
y	0	4	5	4	0

(b) Show that the correlation is zero. Use the method of Example 6.3.

(c) The scatterplot shows a strong association between x and y. Explain how it can happen that $r = 0$ in this case.

Understanding correlation

More important than calculating r (a task for a machine) is understanding how correlation measures association. Here are the facts:

After you plot your data, think!

Abraham Wald (1902–1950), like many statisticians, worked on war problems during World War II. Wald invented some statistical methods that were military secrets until the war ended. Here is one of his simpler ideas. Asked where extra armor should be added to airplanes, Wald studied the location of enemy bullet holes in planes returning from combat. He plotted the locations on an outline of the plane. As data accumulated, most of the outline filled up. Put the armor in the few spots with no bullet holes, said Wald. That's where bullets hit the planes that didn't make it back.

- **Positive *r* indicates positive association between the variables, and negative *r* indicates negative association.** The scatterplot in Figure 6.9 (page 349) shows a strong positive association between femur length and humerus length. Both bones are longer than average in three fossils, so their standard scores are positive for both *x* and *y*. The bones are shorter than average in the other two fossils, so both standard scores are negative. The products are all positive, giving a positive *r*.

- **The correlation *r* always falls between −1 and 1.** Values of *r* near 0 indicate a very weak straight-line relationship. The strength of the relationship increases as *r* moves away from 0 toward either −1 or 1. Values of *r* close to −1 or 1 indicate that the points lie close to a straight line. The extreme values *r* = −1 and *r* = 1 occur only when the points in a scatterplot lie exactly along a straight line.

The result *r* = 0.994 in Example 6.3 reflects the strong, positive, straight-line pattern in Figure 6.9.

"He says we've ruined his positive correlation between height and weight."

The scatterplots in Figure 6.10 on the facing page illustrate how *r* measures both the direction and strength of a straight-line relationship. Study them carefully. Note that the sign of *r* matches the direction of the slope in each plot, and that *r* approaches −1 or 1 as the pattern of the plot comes closer to a straight line.

- Because *r* uses the standard scores for the observations, **the correlation does not change when we change the units of measurement** of *x, y,* or both. Measuring length in inches rather than centimeters in Example 6.3 would not change the correlation *r* = 0.994.

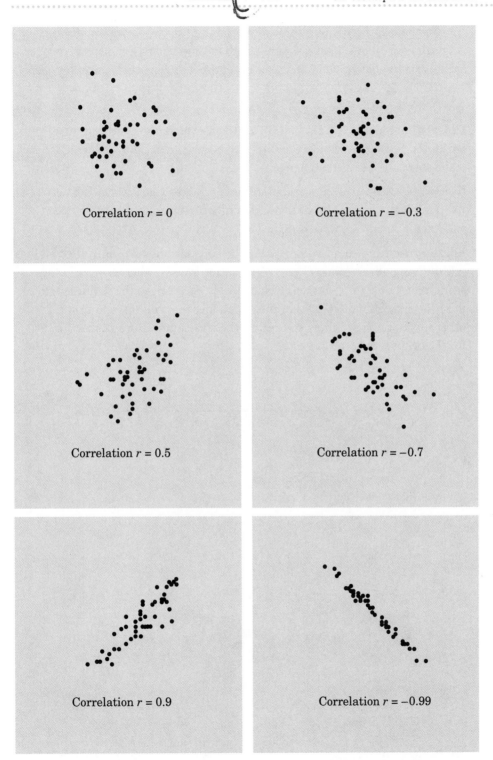

FIGURE 6.10 How correlation measures the strength of a straight-line relationship. Patterns closer to a straight line have correlations closer to 1 or −1.

Our descriptive measures for one variable all share the same units as the original observations. If we measure length in centimeters, the median, quartiles, mean, and standard deviation are all in centimeters. The correlation between two variables, however, has no unit of measurement; it is just a number between −1 and 1.

- **Correlation ignores the distinction between explanatory and response variables.** If we reverse our choice of which variable to call *x* and which to call *y*, the correlation does not change.

- **Correlation measures the strength of only straight-line association between two variables.** Correlation does not describe curved relationships between variables, no matter how strong they are.

- Like the mean and standard deviation, **the correlation is strongly affected by a few outlying observations.** Use *r* with caution when outliers appear in the scatterplot. Look, for example, at Figure 6.11. We changed the femur length of the first fossil from 38 to 60 centimeters. Rather than falling in line with the other fossils, the first is now an outlier. The correlation drops from *r* = 0.994 for the original data to *r* = 0.640.

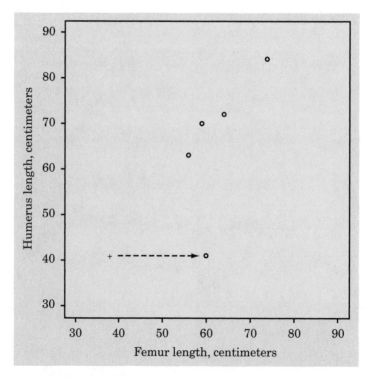

FIGURE 6.11 Moving one point reduces the correlation from *r* = 0.994 to *r* = 0.640.

There are many kinds of relationships between variables and many ways to measure them. Although correlation is very common, remember its limitations. Correlation makes sense only for quantitative variables—we can speak of the relationship between the sex of voters and the political party they prefer, but not of the correlation between these variables. Even for quantitative variables such as the length of bones, correlation measures only straight-line association.

Remember also that correlation is not a complete description of two-variable data, even when there is a straight-line relationship between the variables. You should give the means and standard deviations of both x and y along with the correlation. Because the formula for correlation uses the means and standard deviations, these measures are the proper choice to accompany a correlation.

EXERCISES

6.13 What number can I be?
(a) What are all the values that a correlation r can possibly take?
(b) What are all the values that a standard deviation s can possibly take?

6.14 Stretching a scatterplot Changing the units of measurement can greatly alter the appearance of a scatterplot. Return to the fossil data from Example 6.3:

Femur:	38	56	59	64	74
Humerus:	41	63	70	72	84

These measurements are in centimeters. Suppose a deranged scientist measured the femur in meters and the humerus in millimeters. The data would then be

Femur:	0.38	0.56	0.59	0.64	0.74
Humerus:	410	630	700	720	840

(a) Draw an x axis extending from 0 to 75 and a y axis extending from 0 to 850. Plot the original data on these axes. Then plot the new data on the same axes in a different color. The two plots look very different.
(b) Nonetheless, the correlation is exactly the same for the two sets of measurements. Why do you know that this is true without doing any calculations?

6.15 Rank the correlations Consider each of the following relationships: the heights of fathers and the heights of their adult sons, the heights of husbands and the heights of their wives, and the heights of women at age 4 and their heights at age 18. Rank the correlations between these pairs of variables from highest to lowest. Explain your reasoning.

6.16 Matching correlations Figure 6.12 displays five scatterplots. Match each to the *r* below that best describes it. (Some *r*'s will be left over.)

$r = -0.9$		$r = 0.3$
$r = -0.7$	$r = 0$	$r = 0.7$
$r = -0.3$		$r = 0.9$

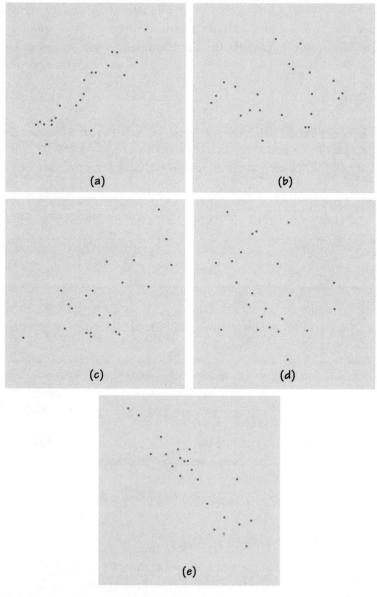

FIGURE 6.12 Match the correlations.

APPLICATION 6.1 Cars and Correlation

The scatterplots below show some data from Environmental Protection Agency (EPA) fuel economy tests on 140 different model year 1999 cars that run on unleaded gas. Variables that were measured for each car include make of car, horsepower, weight (in pounds), displacement (in cubic inches), mpg:highway (highway driving miles per gallon), mpg:city (city driving miles per gallon), and carbon dioxide emitted.

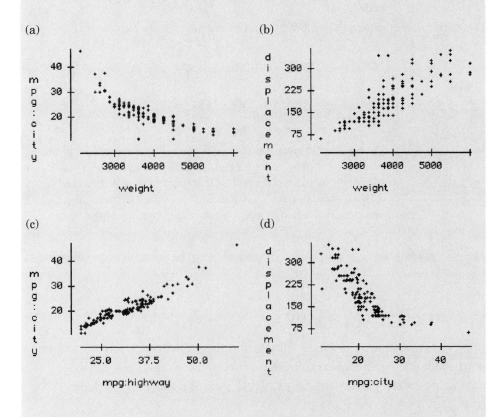

(a)

(b)

(c)

(d)

1. Describe the direction, form, and strength of each relationship in your own words.

2. Estimate the value of the correlation between the variables in each plot.

3. For each of the relationships, tell whether there are clear explanatory and response variables, or whether either variable could be viewed as the explanatory variable.

4. How would the correlation between the variables in plot (b) change if the weights were measured in kilograms instead of pounds? Explain.

APPLICATION 6.1 Cars and Correlation *(continued)*

5. How would the correlation in plot (c) change if the *x* and *y* variables were switched? Explain.

6. Predict how the following pairs of variables would be related. Include direction, form, and strength in your predictions.

- Horsepower and mpg:city
- Make of car and weight
- Weight and amount of carbon dioxide emitted

EXPLORING THE WEB

The best way to grasp how the correlation reflects the pattern of the points on a scatterplot is to use an applet that allows you to practice estimating correlation. Go to the Champaign-Urbana Web University (CUWU) statistics page www.stat.uiuc.edu/~stat100/cuwu/ and click on the Correlations icon. With enough practice you might even become one of the top scorers.

STATISTICS IN SUMMARY

Most statistical studies examine relationships between two or more variables. A **scatterplot** is a graph of the relationship between two quantitative variables. If you have an explanatory and a response variable, put the explanatory variable on the *x* (horizontal) axis of the scatterplot.

When you examine a scatterplot, look for the **direction, form, and strength** of the relationship and also for possible **outliers.** If there is a clear direction, is it positive (the scatterplot slopes upward from left to right) or negative (the plot slopes downward)? Is the form straight or curved? Are there clusters of observations? Is the relationship strong (a tight pattern in the plot) or weak (the points scatter widely)?

The **correlation *r*** measures the direction and strength of a straight-line relationship between two quantitative variables. Correlation is a number between −1 and 1. The sign of *r* shows whether the association is positive or negative. The value of *r* gets closer to −1 or 1 as the points cluster more tightly about a straight line. The extreme values −1 and 1 occur only when the scatterplot shows a perfectly straight line.

SECTION 6.1 EXERCISES

6.17 Guess the correlation For each of the following pairs of variables, would you expect a substantial negative correlation, a substantial positive correlation, or a small correlation?
(a) The age of secondhand cars and their prices.
(b) The weight of new cars and their gas mileages in miles per gallon.
(c) The heights and the weights of adult men.
(d) The heights and the IQ scores of adult men.
(e) The heights of daughters and the heights of their mothers.

6.18 Heavy backpacks Ninth-grade students at the Webb Schools go on a 30-mile backpacking trip each fall. Before leaving, students and their backpacks are weighed. The table below shows the body weights and backpack weights of the eight members of Mr. Starnes's group. (Names have been changed to conceal students' identities.)

Name	Weight (lb.)	Backpack weight (lb.)
Amble	120	26
Plodder	187	30
Slog	109	26
Cliff	103	24
Saunter	131	29
Thumper	165	35
Belay	158	31
Rocky	116	28

(a) Is there a relationship between a student's weight and the weight of his backpack? Make a scatterplot to help answer this question. Describe what you see.
(b) Calculate the correlation between body weight and backpack weight using the method of Example 6.3 (page 348).
(c) Suppose that Plodder had decided not to go on the trip. Make a prediction about the correlation between body and backpack weight for the remaining seven students. Then calculate the actual correlation. How much effect did Plodder's point have? Justify your answer.

6.19 Pollution from vehicle exhaust Auto manufacturers are required to test their vehicles for the amount of each of several pollutants in the exhaust. The amount of a pollutant varies even among identical vehicles, so that several vehicles must be tested. Figure 6.13 is a scatterplot of the amounts of two pollutants, carbon monoxide and nitrogen oxides, for 46 vehicles of the same model. Both variables are measured in grams of the pollutant per mile driven.[10]

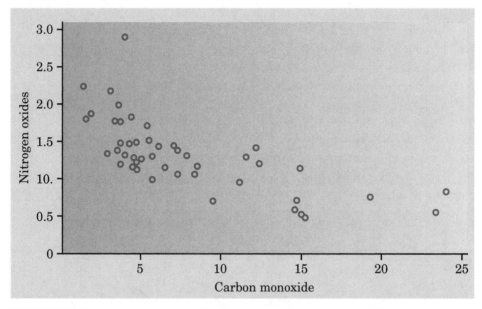

FIGURE 6.13 Amounts of two pollutants in the exhaust of 46 vehicles of the same model.

(a) Describe the nature of the relationship. Is the association positive or negative? Is the relation close to a straight line or clearly curved? Are there any outliers?
(b) A car magazine says, "When a car's engine is properly tuned, it emits few pollutants. If the engine is out of tune, it emits more of all pollutants. You can find out how badly a car is polluting the air by measuring any one pollutant. If that value is acceptable, the other emissions will also be OK." Do the data in Figure 6.13 support this claim?

6.20 IQ and siblings The correlation between a child's score on the vocabulary portion of the Wechsler Intelligence Scale for Children (a standard IQ test) and the number of siblings a child has is $r = -0.319$.[11]
(a) Explain in words what this r says.
(b) Can you suggest an explanation for this relationship?

6.21 Women can run Table 6.3 on the facing page shows the progress of world record times (in seconds) for the 10,000-meter run.[12] Concentrate first on the women's world records.

TABLE 6.3 World Record Times for the 10,000-Meter Run

Men		Men		Women	
Record year	Time (seconds)	Record year	Time (seconds)	Record year	Time (seconds)
1912	1880.8	1962	1698.2	1967	2286.4
1921	1840.2	1963	1695.6	1970	2130.5
1924	1835.4	1965	1659.3	1975	2100.4
1924	1823.2	1972	1658.4	1975	2041.4
1924	1806.2	1973	1650.8	1977	1995.1
1937	1805.6	1977	1650.5	1979	1972.5
1938	1802.0	1978	1642.4	1981	1950.8
1939	1792.6	1984	1633.8	1981	1937.2
1944	1775.4	1989	1628.2	1982	1895.3
1949	1768.2	1993	1627.9	1983	1895.0
1949	1767.2	1993	1618.4	1983	1887.6
1949	1761.2	1994	1612.2	1984	1873.8
1950	1742.6	1995	1603.5	1985	1859.4
1953	1741.6	1996	1598.1	1986	1813.7
1954	1734.2	1997	1591.3	1993	1771.8
1956	1722.8	1997	1587.8		
1956	1710.4	1997	1582.7		
1960	1698.8				

(a) Make a scatterplot with year as the explanatory variable. Describe the pattern of improvement over time that your plot displays.

(b) The correlation between these two variables is -0.97. Explain what this number tells you.

(c) Add the men's world record data to your scatterplot using a different plotting symbol. Then compare the progress of men and women.

(d) Women began running this long distance later than men, so we might expect their improvement to be more rapid. Moreover, it is often said that men have little advantage over women in distance running as opposed to sprints, where muscular strength plays a greater role. Do the data appear to support these claims?

6.22 Correlation "facts." Which of these statements are true and which are false?

(a) A correlation of 0.8 means that 80% of the points in the scatterplot lie on a line.

(b) If the correlation between two lists of numbers is zero, then there can be no relationship between them.

(c) For all books in the Library of Congress, the correlation between the thickness of the books (in inches) and their number of pages would be positive.

(d) For all of the cars registered in the state of Ohio, the correlation between their fuel efficiency (in miles per gallon) and their weight (in pounds) would be positive.

(e) If the correlation between height in inches and weight in pounds for a group of people is 0.7, then the correlation between their heights in centimeters and their weights in kilograms will still be 0.7.

6.23 Foot problems Metatarsus adductus (call it MA) is a turning in of the front part of the foot that is common in adolescents and usually corrects itself. Hallux abducto valgus (call it HAV) is a deformation of the big toe that is not common in youth and often requires surgery. Perhaps the severity of MA can help predict the severity of HAV. Table 6.4 gives data on 38 consecutive patients who came to a medical center for HAV surgery.[13] Using X-rays, doctors measured the angle of deformity for both MA and HAV. They speculated that there is a positive association: more serious MA is associated with more serious HAV.

(a) Use your calculator to help you make a scatterplot of the data. Which is the explanatory variable?

TABLE 6.4 The Severity of Metatarsus Adductus and Hallux Abducto Valgus

HAV angle	MA angle	HAV angle	MA angle	HAV angle	MA angle
28	18	21	15	16	10
32	16	17	16	30	12
25	22	16	10	30	10
34	17	21	7	20	10
38	33	23	11	50	12
26	10	14	15	25	25
25	18	32	12	26	30
18	13	25	16	28	22
30	19	21	16	31	24
26	10	22	18	38	20
28	17	20	10	32	37
13	14	18	15	21	23
20	20	26	16		

(b) Describe the form, direction, and strength of the relationship between MA angle and HAV angle. Are there any clear outliers in your graph?

(c) The correlation between HAV angle and MA angle is 0.30. What does this tell you?

(d) Do you think the data confirm the doctors' speculation?

6.2 REGRESSION, PREDICTION, AND CAUSATION

Will stocks go up or down?

Predicting the future course of the stock market could make you rich. No wonder lots of people and lots of computers pore over market data looking for patterns.

Some popular patterns are a bit bizarre. The "Super Bowl Indicator" says that the football Super Bowl, played in January, predicts how stocks will behave each year. The current National Football League (NFL) was formed by merging the original NFL with the American Football League (AFL). The indicator claims that stocks go up in years when a team from the old NFL wins and down when an AFL team wins. The indicator was right in 28 of 33 years between the first Super Bowl in 1967 and 1999. Sounds impressive. But stocks went down only 6 times in those 33 years, so just predicting "up" every year would be right 27 times. Original NFL teams have won most Super Bowls for good reasons: there are 17 of them against only 11 old AFL teams, and the NFL was the established and stronger league. So "NFL wins" has been pretty much the same as "up every year." The great 1990s boom in stocks roared on unhindered by AFL wins in 1998 and 1999.

The Super Bowl predictor is simple-minded. There are statistical methods to predict one variable from others that go well beyond just counting ups and downs. *Regression*, which we will meet in this chapter, is the starting point for these methods. They go on quite a ways. One Web site sells stock price predictions based on "artificial neural networks, genetic algorithms, nearest-neighbor modeling, and other pattern classification techniques, combined with chaotic-, fractal-, and wavelet-based price-time series transforms."[14] Sounds impressive, but does it work? Here's a hint: if these methods really predicted stock prices, their owners wouldn't be selling them to us—they would be quietly using them to get very rich.

If a pattern were found, says the accepted financial wisdom, so many people would try to take advantage of it that the pattern would quickly disappear. If, for example, stocks regularly went up between St. Patrick's Day (March 17) and April Fool's Day (April 1), people would buy stocks around March 17 and sell them around April 1. Result: Buying would drive stock prices up at the beginning of the period and selling would push them down at the end, destroying the pattern. This logic is more convincing than claims to have discovered a way to beat the market. Prediction, as we will often remark in this chapter, is a subtle business.

ACTIVITY 6.2 Vitruvius and the ideal man

Materials: Ruler or measuring tape for each small group of 2 to 3 students, graph paper

According to the ancient architect Vitruvius, the measurements of various parts of the human body have been set by nature to follow certain ratios. Leonardo da Vinci paid tribute to these claims in his painting entitled *Vitruvian Man.* Here is a brief excerpt from the text accompanying da Vinci's painting: "The length of a man's outspread arms is equal to his height. The greatest width of the shoulders contains in itself the fourth part of the man. From the elbow to the tip of the hand will be the fifth part of the man." In this Activity, you will make some measurements to determine whether da Vinci's statements seem to accurately reflect reality.

1. Measure the distance from the elbow to the tip of the hand (end of the middle finger) and the height for each group member to the nearest inch. Record your data.

2. Your teacher will prepare a chart for recording your data. When you have finished measuring, enter your group members' values in the chart.

3. On graph paper, make a scatterplot of the class's measurements using the distance from the elbow to the tip of the hand as the explanatory variable. Be sure to label and scale your axes.

4. Describe the direction, form, and strength of the relationship between these variables. Estimate the correlation from your scatterplot.

5. Enter the data into lists L_1 and L_2 on your calculator and make a scatterplot. How does the calculator graph compare with the one you made on paper? Adjust the viewing window to make the graphs appear more similar.

6. Calculate $\bar{x}$, s_x, $\bar{y}$, s_y, and r using your calculator.

7. According to da Vinci, the distance from the elbow to the tip of the hand should be one-fifth of the height. That is, the height should be five times the distance measured on the arm. To see how close your class's data come to this prediction, define $Y_1 = 5x$ and graph this line on top of your scatterplot. Describe what you see.

8. Based on your scatterplot, make a prediction for the height of a student whose distance measured on the arm is 10 inches. How does your prediction compare with those of your classmates?

9. Save your data: $L_1 \rightarrow$ `ARM` and $L_2 \rightarrow$ `HEIT`. You will use them later in the chapter.

Regression lines

If a scatterplot shows a straight-line relationship between two quantitative variables, we would like to summarize this overall pattern by drawing a line on the graph. A *regression line* summarizes the relationship between two variables, but only in a specific setting: when one of the variables helps explain or predict the other. That is, regression describes a relationship between an explanatory variable and a response variable.

> ## Regression line
>
> A **regression line** is a straight line that describes how a response variable y changes as an explanatory variable x changes. We often use a regression line to predict the value of y for a given value of x.

EXAMPLE 6.4   Fossil bones

We saw that the lengths of two bones in fossils of the extinct beast *Archaeopteryx* closely follow a straight-line pattern. Figure 6.14 plots the lengths for the 5 available fossils. The line on the plot gives a quick summary of the overall pattern.

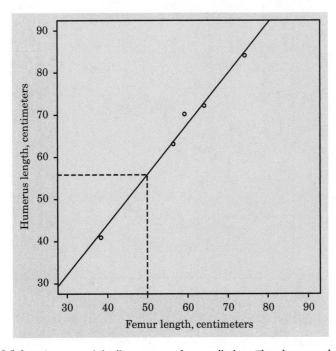

FIGURE 6.14 Using a straight-line pattern for prediction. The data are the lengths of two bones in 5 fossils of the extinct beast *Archaeopteryx*.

Another *Archaeopteryx* fossil is incomplete. Its femur is 50 centimeters long, but the humerus is missing. Can we predict how long the humerus is? The straight-line pattern connecting humerus length to femur length is so strong that we feel quite safe in using femur length to predict humerus length. Figure 6.14 shows how: starting at the femur length (50 cm) on the *x* axis, go up to the line, then over to the humerus length axis. We predict a humerus length of about 56 cm. This is the length the humerus would have if this fossil's point lay exactly on the line. All the other points are close to the line, so we think the missing point would also be close to the line. That is, we think this prediction will be quite accurate.

EXAMPLE 6.5   Presidential elections

Republican Ronald Reagan was elected president twice, in 1980 and in 1984. Figure 6.15 plots the percent in each state who voted for Reagan's Democratic opponents: Jimmy Carter in 1980 and Walter Mondale in 1984. The plot shows a positive straight-line relationship. We expect this because some states tend to vote Democratic and others tend to vote Republican. There is one outlier: Georgia, President Carter's home state, voted 56% for the Democrat Carter in 1980 but only 40% Democratic in 1984.

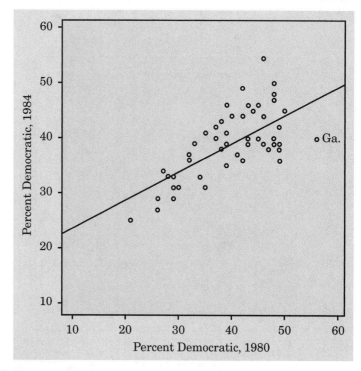

FIGURE 6.15 A weaker straight-line pattern. The data are the percent in each state who voted Democratic in the two Reagan presidential elections.

usually important for understanding the data. The slope is the rate of change, the amount of change in the predicted *y* when *x* increases by 1.

The *intercept* of the least-squares line is $a = -3.66$. This is the value of the predicted *y* when $x = 0$. Although we need the intercept to draw the line, it is statistically meaningful only when *x* can actually take values close to zero. Here, femur length 0 is impossible, so the intercept has no statistical meaning.

To use the equation for *prediction*, substitute the value of *x* and calculate *y*. The predicted humerus length for a fossil with a femur 50 cm long is

$$\text{humerus length} = -3.66 + (1.197)(50)$$
$$= 56.2 \text{ cm}$$

To *draw the line* on the scatterplot, predict *y* for two different values of *x*. This gives two points. Plot them and draw the line through them.

EXERCISES

6.24 Pizza party Figure 6.17 displays data on the number of slices of pizza consumed by pledges at a fraternity party (the explanatory variable *x*) and the number of laps around the block the pledges could run immediately afterward

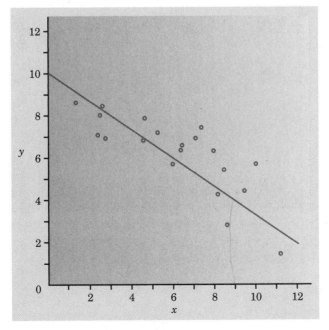

FIGURE 6.17 A fraternity pizza regression for Exercise 6.24.

(the response variable y). The line on the scatterplot is the least-squares regression line computed from these points for predicting y from x.

(a) At the next party, a pledge eats 6 slices of pizza before running. How many laps do you predict he will complete? *about 6 slices*

(b) Another pledge eats 9 slices of pizza. Predict how many laps he will complete. *4. slices*

(c) A third pledge shows off by eating 16 slices of pizza. Can you predict his performance from the scatterplot? Why or why not? *No not online but can predict*

From Rex Boggs in Australia comes an unusual data set. Before showering in the morning, he weighed the bar of soap in his shower stall. The weight goes down as the soap is used. The data appear in Table 6.5 (weights in grams). Notice that Mr. Boggs forgot to weigh the soap on some days. Exercises 6.25 to 6.27 are based on the soap data set.

Regression toward the mean

To "regress" means to go backward. Why are statistical methods for predicting a response from an explanatory variable called "regression"? Sir Francis Galton (1822–1911), who was the first to apply regression to biological and psychological data, looked at examples such as the heights of children versus the heights of their parents. He found that the taller-than-average parents tended to have children who were also taller than average but not as tall as their parents. Galton called this fact "regression toward the mean," and the name came to be applied to the statistical method.

TABLE 6.5 Weight (Grams) of a Bar of Soap Used to Shower

Day	Weight	Day	Weight	Day	Weight
1	124	8	84	16	27
2	121	9	78	18	16
5	103	10	71	19	12
6	96	12	58	20	8
7	90	13	50	21	6

Source: Rex Boggs.

6.25 Scatterplot Plot the weight of the bar of soap against day. Is the overall pattern roughly straight line? Based on your scatterplot, is the correlation between day and weight close to 1, positive but not close to 1, close to 0, negative but not close to −1, or close to −1? Explain your answer.

6.26 Regression The equation for the least-squares regression line for the data in Table 6.5 is

$$\text{weight} = 133.2 - 6.31 \times \text{day}$$

(a) Explain carefully what the slope $b = -6.31$ tells us about how fast the soap lost weight.

(b) Interpret the y intercept, 133.2.

(c) Mr. Boggs did not measure the weight of the soap on Day 4. Use the regression equation to predict that weight.

(d) Draw the regression line on your scatterplot from the previous exercise.

6.27 Prediction? Use the regression equation in the previous exercise to predict the weight of the soap after 30 days. Why is it clear that your answer makes no sense? What's wrong with using the regression line to predict weight after 30 days?

6.28 Calculating the least-squares line Like to know the details when you study something? Here is the formula for the least-squares regression line for predicting y from x. Start with the means and the standard deviations s_x and s_y of the two variables and the correlation r between them. The least-squares line has equation $y = a + bx$ with

$$b = r\frac{s_y}{s_x} \quad \text{and} \quad a = \bar{y} - b\bar{x}$$

Example 6.3 (page 348) gives the means, standard deviations, and correlation for the fossil bone length data. Use these values in the formulas just given to verify the equation of the least-squares line given in Example 6.6 (page 368).

$$\text{humerus length} = -3.66 + (1.197 \times \text{femur length})$$

Understanding prediction

Computers make prediction easy and automatic, even from very large sets of data. Anything that can be done automatically is often done thoughtlessly. Regression software will happily fit a straight line to a curved relationship, for example. Also, the computer cannot decide which is the explanatory variable and which is the response variable. This is important, because the same data give two different lines depending on which is the explanatory variable.

In practice, we often use several explanatory variables to predict a response. As part of its admissions process, a college might use SAT math and verbal scores and high school grades in English, math, and science (5 explanatory variables) to predict first-year college grades. Although the details are messy, all statistical methods of predicting a response share some basic properties of least-squares regression lines.

- **Prediction is based on fitting some "model" to a set of data.** In Figures 6.14 and 6.15, our model is a straight line that we draw through the points in a scatterplot. Other prediction methods use more elaborate models.

- **Prediction works best when the model fits the data closely.** Compare again Figure 6.14, where the data closely follow a line, with Figure 6.15, where there is more scatter about the line. Prediction is more trustworthy in Figure 6.14. It is not so easy to see patterns when there are many variables, but if the data do not have strong patterns, prediction may be very inaccurate.

"How did I get into this business? Well, I couldn't understand regression and correlation in college, so I settled for this instead."

- **Prediction outside the range of the available data is risky.** Suppose that you have data on a child's growth between 3 and 8 years of age. You find a strong straight-line relationship between age x and height y. If you fit a regression line to these data and use it to predict height at age 25 years, you will predict that the child will be 8 feet tall. Growth slows down and stops at maturity, so extending the straight line to adult ages is foolish. No one would make this mistake in predicting height. But almost all economic predictions try to tell us what will happen next quarter or next year. No wonder economic predictions are often wrong.

EXAMPLE 6.7   Predicting the national surplus

The Congressional Budget Office is required to submit annual reports that predict the federal budget and its deficit or surplus for the next 5 years. These forecasts depend on future economic trends (unknown) and on what Congress will decide about taxes and spending (also unknown). Even the prediction of the state of the budget if current policies are not changed has been wildly inaccurate. The forecast made in 1996 for 1999, for example, missed by more than $300 billion. The 1997 forecast for the very next year was $102 billion off. As Senator Everett Dirksen once said, "A billion here and a billion there and pretty soon you are talking real money." By 1999, the Budget Office was predicting a surplus (ignoring Social Security) of $996 billion over the next 10 years. Politicians debated what to do with the money, but no one else believed the prediction.

Correlation and regression

Correlation measures the direction and strength of a straight-line relationship. Regression draws a line to describe the relationship. Correlation and regression are closely connected, even though regression requires choosing an explanatory variable and correlation does not.

Both correlation and regression are strongly affected by outliers. Be wary if your scatterplot shows strong outliers. Figure 6.18 plots the record-high yearly precipitation in each state against that state's record-high 24-hour precipitation. Hawaii is a high outlier, with a yearly record of 704.83 inches of rain recorded at Kukui in

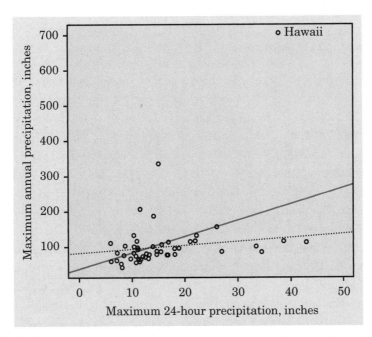

FIGURE 6.18 Least-squares regression lines are strongly influenced by outliers. The solid line is based on all 50 data points. The dotted line leaves out Hawaii.

1982. The correlation for all 50 states in Figure 6.18 is 0.408. If we leave out Hawaii, the correlation drops to $r = 0.195$. The solid line in the figure is the least-squares line for predicting the annual record from the 24-hour record. If we leave out Hawaii, the least-squares line drops down to the dotted line. This line is nearly flat—there is little relation between yearly and 24-hour record precipitation once we decide to ignore Hawaii.

The usefulness of the regression line for prediction depends on the strength of the association. That is, the usefulness of a regression line depends on the correlation between the variables. It turns out that the square of the correlation is the right measure to use.

r^2 in regression

The **square of the correlation, r^2,** is the fraction of the variation in the values of y that is explained by the least-squares regression of y on x.

The idea is that when there is a straight-line relationship, some of the variation in y is accounted for by the fact that as x changes it pulls y along with it.

EXAMPLE 6.8   Using r^2

Look again at Figure 6.14 (page 365). There is a lot of variation in the humerus lengths of these 5 fossils, from a low of 41 cm to a high of 84 cm. The scatterplot shows that we can explain almost all of this variation by looking at femur length and at the regression line. As femur length increases, it pulls humerus length up with it along the line. There is very little leftover variation in humerus length, which appears in the scatter of points about the line. Because $r = 0.994$ for these data, $r^2 = (0.994)^2 = 0.988$. So the variation "along the line" as femur length pulls humerus length with it accounts for 98.8% of all the variation in humerus length. The scatter of the points about the line accounts for only the remaining 1.2%. Little leftover scatter says that prediction will be accurate.

Contrast the voting data in Figure 6.15 (page 366). There is still a straight-line relationship between the 1980 and 1984 Democratic votes, but there is also much more scatter of points about the regression line. Here, $r = 0.704$ and so $r^2 = 0.496$. Only about half the observed variation in the 1984 Democratic vote is explained by the straight-line pattern. You would still guess a higher 1984 Democratic vote for a state that was 45% Democratic in 1980 than for a state that was only 30% Democratic in 1980. But lots of variation remains in the 1984 votes of states with the same 1980 vote. That is the other half of the total variation among the states in 1984. It is due to other factors, such as differences in the main issues in the two elections and the fact that President Reagan's two Democratic opponents came from different parts of the country.

In reporting a regression, it is usual to give r^2 as a measure of how successful the regression was in explaining the response. When you see a correlation, square it to get a better feel for the strength of the association. Perfect correlation ($r = -1$ or $r = 1$) means the points lie exactly on a line. Then $r^2 = 1$ and all of the variation in one variable is accounted for by the straight-line relationship with the other variable. If $r = -0.7$ or $r = 0.7$, $r^2 = 0.49$ and about half the variation is accounted for by the straight-line relationship. In the scale, correlation ± 0.7 is about halfway between 0 and ± 1.

CALCULATOR CORNER   Regression and prediction

In this Calculator Corner, you will learn to use your calculator to compute the equation of the least-squares regression line. Then you will discover how to use your equation to make predictions. We will use the temperature and natural gas data from Example 6.2 (page 340).

- Reload the temperature data into list L_1 and the natural gas consumed data into list L_2. If you saved the lists in the previous Calculator Corner (page 341), you can find them in the LIST menu.
- Make a scatterplot of the data with temperature as the explanatory variable.
- Turn your diagnostics on. This will ensure that the calculator displays both r and r^2 along with the linear regression equation. Go into the CATALOG by pressing $\boxed{\text{2nd}}$ $\boxed{0}$. Scroll down until your cursor is next to DiagnosticOn. Then press $\boxed{\text{ENTER}}$ twice.
- To calculate the equation of the least-squares regression line: from the home

screen, press $\boxed{\text{STAT}}$, arrow right to CALC, and choose 8:LinReg(a+bx).

```
EDIT CALC TESTS
2↑2-Var Stats
3:Med-Med
4:LinReg(ax+b)
5:QuadReg
6:CubicReg
7:QuartReg
8▮LinReg(a+bx)
```

- Then execute the command LinReg (a+bx) L_1, L_2, Y_1. (To find Y_1, press $\boxed{\text{VARS}}$, then arrow to Y-VARS, choose 1:Function and 1:Y_1.) Compare your result with the screen below.

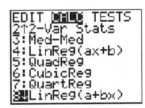

```
LinReg
y=a+bx
a=1425.027417
b=-19.87187775
r²=.9656325261
r=-.9826660298
```

CALCULATOR CORNER   Regression and prediction *(continued)*

- Let's look at the calculator output. The equation of the least-squares regression line is $y = 1425 - 19.9x$, or in plain language,

natural gas consumed
$$= 1425 - (19.9 \times \text{temperature})$$

Note that the value of the correlation, $r = -0.983$, is displayed as part of the regression output.

- Press GRAPH to display the regression line on your scatterplot. Notice how well the line fits the data. The r^2-value displayed earlier, 0.966, tells us that 96.6% of the variation in natural gas consumed is explained by this least-squares regression.

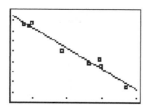

- Let's use the regression line to predict the amount of natural gas Joan will consume in a month with an average temperature of 30° F. From the home screen, execute the command $Y_1(30)$. You should obtain a prediction of 828.9 cubic feet per day. To con-

firm this value, note that

$$y = 1425 - (19.9)(30) = 828$$

which differs slightly from the calculator result due to rounding.

```
Y₁(30)
         828.8710843
```

- Can you interpret the slope and the y intercept of the regression line in this setting? The slope tells us that gas consumption goes down 19.9 cubic feet per day when the average outdoor temperature rises by 1 degree, according to our regression model. The y intercept suggests that if the average outdoor temperature in Joan's area was 0° F, she would use about 1425 cubic feet of natural gas per day.

Caution: Since the data that were used to construct this linear model had *x*-values ranging from about 27 to 57° F, it is risky to use the regression equation to predict gas consumption in a month with an average temperature of 0° F.

The prediction in the Calculator Corner is based on fitting a regression line to the data for eight past months. Joan's house will almost certainly not use exactly 828 cubic feet of gas per day during the next month that has average temperature

30 °F. Because the data points lie close to the line, however, we can be confident that gas consumption in such a month will be quite close to 828 cubic feet per day.

EXERCISES

6.29 Obesity in mothers and daughters A study found that the correlation between the body mass index (BMI) of young girls and their minutes of physical activity in a day was $r = -0.18$.[15] Why might we expect this correlation to be negative? What percent of the variation in BMI among the girls in the study can be explained by the straight-line relationship with minutes of activity?

6.30 Correlation and regression If the correlation between two variables x and y is $r = 0$, there is no straight-line relationship between the variables. It turns out that the correlation is 0 exactly when the slope of the least-squares regression line is 0. Explain why slope 0 means that there is no relationship between x and y. Start by drawing a line with slope 0 and explaining why in this situation x has no value for predicting y.

6.31 r versus r^2 "When $r = 0.7$, this means that y can be predicted from x for 70% of the individuals in the sample." Is this statement true or false? Would it be true if $r^2 = 0.7$? Explain your answers.

6.32 Activity 6.2 follow-up
(a) Use your calculator to find the equation of the least-squares regression line for your data from Activity 6.2 (page 364). Follow the method in the Calculator Corner (page 375). If you saved the lists ARM and HEIT in Activity 6.2, you should use them now.
(b) Interpret the slope and y intercept of the regression line in this setting.
(c) Use your regression equation to predict the height of a student whose measured distance on the arm is 15 inches.

6.33 Beer and BAC How much does drinking beer increase the alcohol content of your blood? This question was addressed in an experiment at Ohio State University. Sixteen students volunteered to take part in an experiment. Before the experiment, each of the students blew into a Breathalyzer to show that their

Did the vote counters cheat?

Republican Bruce Marks was ahead of Democrat William Stinson when the voting machines were tallied in their Pennsylvania election. But Stinson was ahead after absentee ballots were counted by the Democrats who controlled the election board. A court fight followed. The court called in a statistician, who used regression with data from past elections to predict the counts of absentee ballots from the voting-machine results. Marks's lead of 564 votes from the machines predicted that he would get 133 more absentee votes than Stinson. In fact, Stinson got 1025 more absentee votes than Marks. Did the vote counters cheat?

blood alcohol content (BAC) was at the zero mark. The student volunteers then drank a varying number (between 1 and 9) of 12-ounce beers. How much each student drank was assigned by drawing tickets from a bowl. About 30 minutes later, an officer from the OSU Police Department measured their BAC using the Breathalyzer machine. Here are the data:[16]

Student:	1	2	3	4	5	6	7	8
Beers:	5	2	9	8	3	7	3	5
BAC:	0.10	0.03	0.19	0.12	0.04	0.095	0.07	0.06
Student:	9	10	11	12	13	14	15	16
Beers:	3	5	4	6	5	7	1	4
BAC:	0.02	0.05	0.07	0.10	0.085	0.09	0.01	0.05

(a) Suppose you want to estimate how a person's BAC is affected by the number of beers he or she drinks. Make a scatterplot of BAC versus beers using the data in the table above. Which variable did you choose to be the y variable and which did you choose to be the x variable? Explain.

(b) Use your calculator to find the correlation between BAC and beers. Is the correlation an appropriate measure of the strength of the association between BAC and beers? Explain briefly.

(c) Use your calculator to find the least-squares regression line for BAC on beers. Interpret the slope and y intercept of the regression line.

(d) If a student drinks 5 beers, on average what do you predict the student's BAC will be? Show your work.

(e) Would the regression method be as accurate for predicting BAC for a person who drinks 15 beers? Explain.

The question of causation

There is a strong relationship between cigarette smoking and death rate from lung cancer. Does smoking cigarettes *cause* lung cancer? There is a strong association between the availability of handguns in a nation and that nation's homicide rate from guns. Does easy access to handguns *cause* more murders? It says right on the pack that cigarettes cause cancer. Whether more guns cause more murders is hotly debated. Why is the evidence for cigarettes and cancer better than the evidence for guns and homicide?

We already know three big facts about statistical evidence for cause and effect.

Statistics and causation

1. A strong relationship between two variables does not always mean that changes in one variable cause changes in the other.
2. The relationship between two variables is often influenced by other variables lurking in the background.
3. The best evidence for causation comes from randomized comparative experiments.

EXAMPLE 6.9   Does television extend life?

Measure the number of television sets per person x and the life expectancy y for the world's nations. There is a high positive correlation: nations with many TV sets have higher life expectancies.

The basic meaning of causation is that by changing x we can bring about a change in y. Could we lengthen the lives of people in Botswana by shipping them TV sets? No. Rich nations have more TV sets than poor nations. Rich nations also have longer life expectancies because they offer better nutrition, clean water, and better health care. There is no cause-and-effect tie between TV sets and length of life.

"In a new attack on third-world poverty, aid organizations today began delivery of 100,000 television sets."

Example 6.9 illustrates our first two big facts. Correlations such as this are sometimes called "nonsense correlations." The correlation is real. What is nonsense is the conclusion that changing one of the variables causes changes in the other. A lurking variable—such as national wealth in Example 6.9—that influences both x and y can create a high correlation even though there is no direct connection between x and y. We might call this *common response:* both the explanatory and the response variable are responding to some lurking variable.

EXAMPLE 6.10   Obesity in mothers and daughters

What causes obesity in children? Inheritance from parents, overeating, lack of physical activity, and too much television have all been named as explanatory variables.

The results of a study of Mexican American girls aged 9 to 12 years are typical. Measure body mass index (BMI), a measure of weight relative to height, for both the girls and their mothers. People with high BMIs are overweight or obese. Also measure hours of television, minutes of physical activity, and intake of several kinds of food. Result: The girls' BMIs were weakly correlated ($r = -0.18$) with physical activity and also with diet and television. The strongest correlation ($r = 0.506$) was between the BMI of daughters and the BMI of their mothers.[17]

Body type is in part determined by heredity. Daughters inherit half their genes from their mothers. There is therefore a direct causal link between the BMI of mothers and daughters. Of course, the causal link is far from perfect. The mothers' BMIs explain only 25.6% (that's r^2 again) of the variation among the daughters' BMIs. Other factors, some measured in the study and others not measured, also influence BMI. *Even when direct causation is present, it is rarely a complete explanation of an association between two variables.*

Can we use r or r^2 from Example 6.10 to say how much inheritance contributes to the daughters' BMIs? No. Remember *confounding.* It may well be that mothers who are overweight also set an example of little exercise, poor eating habits, and lots of television. Their daughters pick up these habits to some extent, so the influence of heredity is mixed up with influences from the girls' environment. We can't say how much of the correlation between mother and daughter BMIs is due to inheritance.

Figure 6.19 on the facing page shows in outline form how a variety of underlying links between variables can explain association. The dashed line represents an observed association between the variables x and y. Some associations are explained by a *direct cause-and-effect* link between the variables. The first diagram in Figure 6.19 shows "x causes y" by an arrow running from x to y. The second diagram

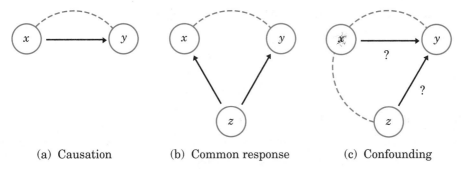

(a) Causation   (b) Common response   (c) Confounding

FIGURE 6.19 Some explanations for an observed association. A dashed line shows an association. An arrow shows a cause-and-effect link. Variable x is explanatory, y is a response variable, and z is a lurking variable.

illustrates *common response.* The observed association between the variables x and y is explained by a lurking variable z. Both x and y change in response to changes in z. This common response creates an association even though there may be no direct causal link between x and y. The third diagram in Figure 6.19 illustrates *confounding.* Both the explanatory variable x and the lurking variable z may influence the response variable y. Variables x and z are themselves associated, so we cannot distinguish the influence of x from the influence of z. We cannot say how strong the direct effect of x on y is. In fact, it can be hard to say if x influences y at all.

Both common response and confounding involve the influence of a lurking variable or variables z on the response variable y. We won't belabor the distinction between the two kinds of relationships. Just remember that "beware the lurking variable" is good advice in thinking about relationships between variables. Here is another example of common response, in a setting where we want to do prediction.

EXAMPLE 6.11   SAT scores and college grades

High scores on the SAT examinations in high school certainly do not cause high grades in college. The moderate association (r^2 is about 27%) is no doubt explained by common response to such lurking variables as academic ability, study habits, and staying sober.

The ability of SAT scores to partly predict college performance doesn't depend on causation. We need only believe that the relationship between SAT scores and college grades that we see in past years will continue to hold for this year's high school graduates. Think once more of our fossils, where femur length predicts humerus length very well. The strong relationship is explained by common response to the overall age and size of the beasts whose fossils we now examine. *Prediction doesn't require causation.*

Discussion of these examples has brought to light two more big facts about causation:

More about statistics and causation

4. The observed relationship between two variables may be due to **direct causation, common response,** or **confounding.** Two or more of these factors may be present together.

5. An observed relationship can, however, be used for prediction without worrying about causation as long as the patterns found in past data continue to hold true.

Evidence for causation

Many of the sharpest disputes in which statistics plays a role involve questions of causation that cannot be settled by experiment. Does taking the drug Bendectin cause birth defects? Does living near power lines cause cancer? Has increased free trade helped to create the growing gap between the incomes of more educated and less educated American workers? All of these questions have become public issues. All concern associations between variables. And all have this in common: they try to pinpoint cause and effect in a setting involving complex relations among many interacting variables. Common response and confounding, along with the number of potential lurking variables, make observed associations misleading. Experiments are not possible for moral or practical reasons. We can't randomly assign people to live near power lines or compare two identical nations with and without free-trade agreements.

EXAMPLE 6.12   Prescription drugs and birth defects

Are children born to women who take a prescription drug during pregnancy more likely to have birth defects? This question has been asked about many drugs. It is not easy to answer. Consider Bendectin, once commonly prescribed for nausea during pregnancy, then accused of causing birth defects and studied intensively in and out of court.[18]

In the absence of an experiment, we must rely on comparative observational studies. That is, we compare children whose mothers took Bendectin with similar children whose mothers did not. Who took Bendectin? Mothers of children with birth defects are more likely to remember taking a drug during pregnancy than those with normal children. This recall bias will overstate the rate of birth defects in the Bendectin group. Perhaps we instead look at prescription records. One study considered that a child had been exposed to Bendectin early in pregnancy if a prescription was filled between 365 and 250 days before birth. There is no way of reliably determining whether the mothers actually took the drug or how regularly they took it.

Constructing a group of "similar children" whose mothers did not take Bendectin is even more difficult than deciding who took the drug. What potential lurking variables should we consider? For example, users of Bendectin also took more other drugs than nonusers. Perhaps the alleged effects of Bendectin are due to confounding with some other drug or to a combination of drugs. Or perhaps birth defects and use of Bendectin and other drugs are common responses to some other medical, economic, or psychological condition. It is hard to say what makes children "similar" to the Bendectin children.

Any comparative observational study must choose some specific rule for forming the two groups. Any specific rule can be criticized. In the case of Bendectin, more than 20 studies mostly found that the drug appeared safe. Most of the lawsuits charging that Bendectin caused birth defects failed in court. The legal expense nonetheless forced the manufacturer to withdraw the drug from the market. Bendectin probably carries little or no risk, but all we can say for sure is "not enough evidence."

Despite the difficulties, it is sometimes possible to build a strong case for causation in the absence of experiments. The evidence that smoking causes lung cancer is about as strong as nonexperimental evidence can be.

Doctors had long observed that most lung cancer patients were smokers. Observational studies comparing smokers and "similar" nonsmokers showed a strong association between smoking and death from lung cancer. Could the association be explained by lurking variables that the studies could not measure? Might there be, for example, a genetic factor that predisposes people both to nicotine addiction and to lung cancer? Smoking and lung cancer would then be positively associated even if smoking had no direct effect on the lungs. How were these objections overcome?

Let's answer this question in general terms: What are the criteria for establishing causation when we cannot do an experiment?

- **The association is strong.** The association between smoking and lung cancer is very strong.

- **The association is consistent.** Many studies of different kinds of people in many countries link smoking to lung cancer. That reduces the chance that a lurking variable specific to one group or one study explains the association.

- **Higher doses are associated with stronger responses.** People who smoke more cigarettes per day or who smoke over a longer period get lung cancer more often. People who stop smoking reduce their risk.

- **The alleged cause precedes the effect in time.** Lung cancer develops after years of smoking. The number of men dying of lung cancer rose as smoking became more common, with a lag of about 30 years. Lung cancer kills more men than any other form of cancer. Lung cancer was rare among women until women began to smoke. Lung cancer in women rose along with smoking, again with a lag of about 30 years, and has now passed breast cancer as the leading cause of cancer death among women.

• **The alleged cause is plausible.** Experiments with animals show that tars from cigarette smoke do cause cancer.

Medical authorities do not hesitate to say that smoking causes lung cancer. The U.S. Surgeon General has long stated that cigarette smoking is "the largest avoidable cause of death and disability in the United States." The evidence for causation is overwhelming—but it is not as strong as the evidence provided by well-designed experiments.

EXERCISES

6.34 Is a little alcohol good for you? A survey of 7000 California men found little correlation between alcohol consumption and chance of dying during the 5 1/2 years of the study. In fact, men who did not drink at all during these years had a slightly higher death rate than did light drinkers. This lack of correlation was somewhat surprising. Explain how common response might account for the higher death rate among men who did not drink at all over a short period.

6.35 Education and income There is a strong positive correlation between years of schooling completed and lifetime earnings for American men. One possible reason for this association is causation: more education leads to higher-paying jobs. Another explanation is confounding: men who complete many years of schooling have other characteristics that would lead to better jobs even without the education. Suggest several possible confounding variables.

6.36 Snow and earthquakes A study measures the average annual snowfall (in inches) for 10 cities over the last decade along with the greatest earth movement (on the Richter scale) over this same time period. The study included 5 cities in California's San Francisco Bay Area and 5 cities from Canada's province of Ontario. The study found a very strong negative correlation between the two variables. Does this mean that a strong snowfall will prevent earthquakes? Explain your answer briefly. You may want to draw a picture like Figure 6.19 (page 381) to illustrate it.

6.37 Is math the key to success in college? Here is the opening of a newspaper account of a College Board study of 15,941 high school graduates:

Minority students who take high school algebra and geometry succeed in college at almost the same rate as whites, a new study says.

The link between high school math and college graduation is "almost magical," says College Board President Donald Stewart, suggesting "math is the gatekeeper for success in college."

"These findings," he says, "justify serious consideration of a national policy that all students take algebra and geometry."[19]

What lurking variables might explain the association between taking several math courses in high school and success in college? Explain why requiring algebra and geometry may have little effect on who succeeds in college.

6.38 Seat belt use In the Australian state of Victoria, a law compelling motorists to wear seat belts went into effect in December 1970. As time passed, an increasing percent of motorists complied. A study found high positive correlation between the percent of motorists wearing belts and the percent reduction in injuries from the 1970 level (Figure 6.20).[20]

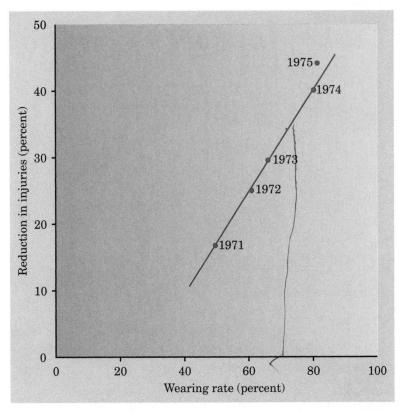

FIGURE 6.20 Reduction in automotive injuries from 1970 level versus percent of motorists wearing seat belts in Victoria, Australia.

(a) Use the regression line in Figure 6.20 to predict the percent reduction from the 1970 injury level that would occur if 75% of motorists wore seat belts.
(b) Is the relationship between these two variables most likely a result of causation, confounding, or common response? Justify your answer.

APPLICATION 6.2 The *Correlation and Regression* Applet

The best way to develop some feeling for how correlation and regression are related is to use an applet that allows you to plot and move data points and to observe how the correlation and least-squares line move as the points change. Go to the *Statistics through Applications* Web site, www.whfreeman.com/sta, and look at the *Correlation and Regression* applet.

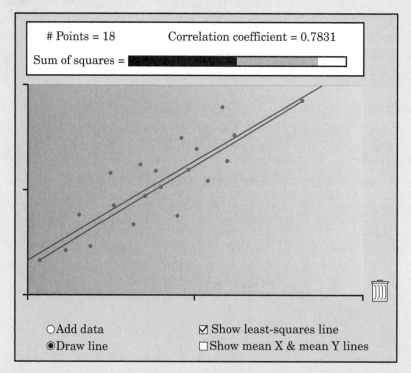

| # Points = 18 | Correlation coefficient = 0.7831 |
| Sum of squares = | |

○Add data   ☑ Show least-squares line
◉Draw line   ☐ Show mean X & mean Y lines

Directions: Use your mouse to click on the blank plot to make red points, creating a scatterplot. Above the plot, the applet shows the correlation. Below the plot are options that can be checked if you wish to draw the regression line on the plot, see the average values of *x* and *y*, or even add your own line to the plot to see how well your line competes with the regression line.

Contemplate:

1. If you create just 2 points, what would the correlation be?

APPLICATION 6.2 The *Correlation and Regression* Applet *(continued)*

2. If you put 3 points in a tight, positively associated linear pattern in the upper left-hand corner of the plot, what will the correlation be? If you then add a fourth point at the lower right-hand corner (near the trash can), what will happen to the correlation? Where would the regression line fall?

3. If you put 10 points in the center of the plot in a U-shaped pattern, what will the correlation be? Where will the regression line fall?

   Verify: Use the applet to test your answers to the three questions above. Did the correlation and regression line behave as you thought they would?

   Experiment:

4. Create a scatterplot of 10 points in a random pattern at the bottom left-hand side of the plot. Move the points around to make the correlation as close to zero as possible. Have the applet show the least-squares regression line. What is its slope? Now add an 11th point at the upper right-hand corner and watch the behavior of the correlation and the regression line.

5. Next, clear the scatterplot by clicking on the trash can, and draw a new group of points in a rough linear pattern with a positive association. Grab one point and drag it around the scatterplot while watching the behavior of the correlation. Where does the placement of the point create the largest correlation? Where does the point create the smallest correlation? Does the correlation ever become negative?

6. Clear the scatterplot again and create a cloud of about 25 points in a weak linear association (e.g., a correlation of about 0.5 or −0.5). Draw a line by hand that goes through the center of the cloud while trying to also capture the slope of your cloud of points. Now have the applet add the regression line to the plot. How does your line compare?

7. Try moving your own line around the plot while watching the green part of the "sum of squares" bar. Can you make the green part disappear without putting your line right on top of the regression line? Explain.

8. Place your line right on top of the regression line, and remove the regression line from the picture. Now switch back to the "add data" mode and put an additional single point on the plot. Grab this point and move it around the scatterplot while watching the "sum of squares" bar. What placement of the point creates the largest green area (indicating your line is no longer close to the regression line)? Can you find several places to put the point where the green part of the bar disappears? Explain.

9. Click on the box to show the "mean *X*" and "mean *Y*" lines. Move some points and observe how these lines relate to the least-squares line.

EXPLORING THE WEB

The *Two Variable Statistical Calculator* applet at the *Statistics through Applications* Web site, www.whfreeman.com/sta, allows you to examine several of the data sets in the text more closely. You can create scatterplots, calculate summary statistics, and graph the least-squares line.

The Web has lots of information about health issues. If you are interested in data and statistical studies, the place to start is the site of the Centers for Disease Control and Prevention, www.cdc.gov. More than you want to know about smoking and tobacco, for example, appears at www.cdc.gov/tobacco/data.htm. Tobacco use, advertising, taxation, surveys of teenagers—it's all here.

STATISTICS IN SUMMARY

Regression is the name for statistical methods that fit some model to data in order to predict a response variable from one or more explanatory variables. The simplest kind of regression fits a straight line on a scatterplot for use in predicting y from x. The most common way to fit a line is the **least-squares** method, which finds the line that makes the sum of the squared vertical distances of the data points from the line as small as possible.

Least-squares regression is closely related to correlation. In particular, the **squared correlation** r^2 tells us what fraction of the variation in the responses is explained by the straight-line relationship between y and x. It is generally true that the success of any statistical prediction depends on the presence of strong patterns in the data. Prediction outside the range of the data is risky because the pattern may be different there.

A strong relationship between two variables is not always evidence that changes in one variable **cause** changes in the other. Lurking variables can create relationships through **common response** or **confounding.** If we cannot do experiments, it is often difficult to get convincing evidence for causation.

SECTION 6.2 EXERCISES

6.39 Heart attacks If you need medical care, should you go to a hospital that handles many cases like yours? Figure 6.21 on the facing page presents some data for heart attacks. The figure plots mortality (the proportion of patients who died) against the number of heart attack patients treated for a large number of hospitals in a recent year. The line on the plot is the least-squares regression line for predicting mortality from number of patients.

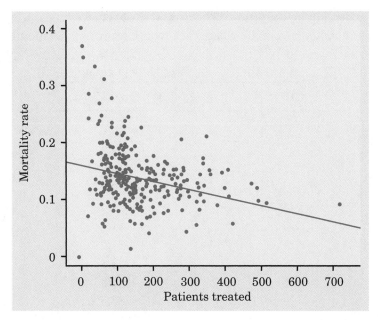

FIGURE 6.21 Mortality of heart attack patients and number of heart attack cases treated for a large group of hospitals.

(a) Do the plot and regression generally support the hypothesis that mortality is lower at hospitals that treat more heart attacks? Is the relationship very strong?
(b) In what way is the pattern of the plot nonlinear? Does the nonlinearity strengthen or weaken the conclusion that heart attack patients should avoid hospitals that treat few heart attacks? Why?

6.40 Beer drinking and cancer A study using data from 41 states found a positive correlation between per capita beer consumption and death rates from some forms of cancer. The states with the highest death rates from these types of cancer were Rhode Island and New York. The beer consumption in those states was 80 quarts per capita per year. People in South Carolina, Alabama, and Arkansas drank only 26 quarts of beer per capita and had cancer rates less than one-third of those in Rhode Island and New York.

Suggest some lurking variables that may be confounded with a state's beer consumption. For a clue, look at the high- and low-consumption states given above.

6.41 How do children grow? The pattern of growth varies from child to child, so we can best understand the general pattern by following the average height of a number of children. The table on the next page presents the mean heights of a group of children in Kalama, an Egyptian village that was the site of a study of nutrition in developing countries.[21] The data were obtained by measuring the heights of 161 children from the village each month from 18 to 29 months of age.

Age x in months	Height y in centimeters
18	76.1
19	77.0
20	78.1
21	78.2
22	78.8
23	79.7
24	79.9
25	81.1
26	81.2
27	81.8
28	82.8
29	83.5

(a) Make a scatterplot that describes the relationship between age and height. Describe what you see.

(b) Use your calculator to obtain the least-squares regression line. Interpret the slope and y intercept in this setting.

(c) Predict the mean height of Kalama children at 32 months of age. How confident are you in the accuracy of this prediction? Explain.

(d) Predict the mean height of Kalama residents at 20 years (240 months). How confident are you in the accuracy of this prediction? Explain.

6.42 Predicting Old Faithful eruptions The calculator screen shots below show both graphical and numerical results from a least-squares regression on data from 222 eruptions of the Old Faithful geyser. Length of eruption is the explanatory variable, and time until the next eruption is the response variable. Both variables were measured in minutes.

(a) Give the equation of the least-squares line. Define any variables you use.

(b) Interpret the slope and the y intercept in this setting.

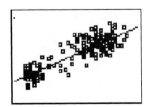

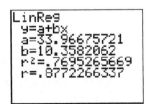

(c) Following a 3-minute eruption, how long would you expect to wait for the next eruption?

(d) What does r^2 tell you?

6.43 Miscarriages among workers A study showed that women who work in the production of computer chips have abnormally high numbers of miscarriages. The union claimed that exposure to chemicals used in production caused the miscarriages. Another possible explanation is that these workers spend most of their work time standing up. Illustrate these relationships in a diagram like one of those in Figure 6.19 (page 381).

6.44 Ranking the states The news media have a weakness for lists. Best place to live, best colleges, healthiest foods, worst-dressed men—a list of best or worst is sure to find a place in the news. When the state-by-state SAT scores come out each year, it's therefore no surprise when we find news articles ranking the states from best (Iowa) to worst (South Carolina) according to the average SAT score achieved by their high school seniors.

The College Board, which sponsors the SAT exams, doesn't like this practice at all. "Comparing or ranking states on the basis of SAT scores alone is invalid and strongly discouraged by the College Board," says the heading on their table of state average SAT scores. To see why, let's look at the data.[22]

(a) Figure 6.22 shows the distribution of average scores on the mathematics part of the SAT exam for the 50 states and the District of Columbia. Describe what you see.

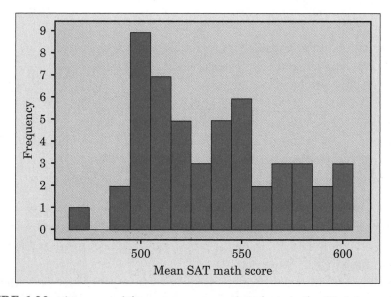

FIGURE 6.22 Histogram of the average scores of students in the 50 states and the District of Columbia on the mathematics part of the SAT exam.

(b) Figure 6.23 plots the average score of each state's high school seniors on the mathematics part of the SAT against the percent of seniors who take the exam. Describe what the scatterplot tells you about this relationship.

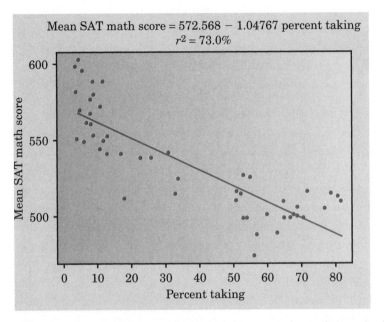

FIGURE 6.23 Scatterplot of average SAT mathematics score for each state against the percent of the state's high school graduates who take the SAT.

(c) The equation of the least-squares regression line relating average SAT score to percent taking is given at the top of Figure 6.23. What does the slope tell us about the relationship between the variables?

(d) In New York State, 77% of high school seniors took the SAT. Predict their average score. (The actual average score in New York was 505.)

(e) How well does percent taking predict average score? (Use r^2 in your answer.)

(f) What is the correlation between percent taking and average SAT math score? What does this tell us?

6.45 Will women outrun men? Table 6.3 (page 361) shows the progress of world record times (in seconds) for the 10,000-meter run for both men and women. You made a scatterplot comparing the progress over time of men's and women's records in Exercise 6.21 (page 360).

(a) The plot of running time in seconds against year is very linear for both men and women. Find the least-squares regression lines for predicting record time from year separately for the two sexes.

(b) Explain in simple language what the slopes of these two lines tell us about the progress of men and women in the 10,000-meter run.

(c) If the linear patterns continue, about when will the women's world record be the same as the men's record? (You can give a rough answer from the scatterplot, or you can use the fitted lines and do some algebra.)

CHAPTER 6 REVIEW

This chapter focused on describing relationships between two quantitative variables. Figure 6.24 retraces the big ideas. We always begin by making graphs of our data. In the case of a scatterplot, we have learned a numerical summary only for data that show a roughly straight-line pattern on the scatterplot. This summary is the means and standard deviations of the two variables and their correlation. A regression line drawn on the plot gives us a compact model of the overall pattern that we can use for prediction. The question marks at the last two stages remind us that correlation and regression describe only straight-line relationships.

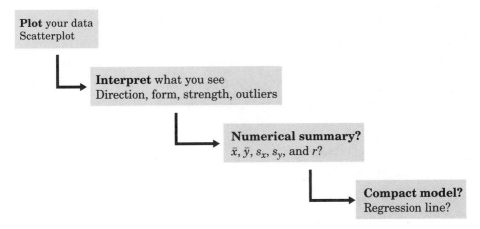

FIGURE 6.24 Flowchart for describing relationships.

Relationships often raise the question of causation. We know that evidence from randomized comparative experiments is the "gold standard" for deciding that one variable causes changes in another variable. Section 6.2 reminded us in more detail how strong associations can appear in data even when there is no direct causation. We must always think about the possible effects of variables lurking in the background.

Here is a review list of the most important skills you should have gained from studying this chapter.

A. SCATTERPLOTS AND CORRELATION

1. Make a scatterplot to display the relationship between two quantitative variables measured on the same subjects. Place the explanatory variable (if any) on the horizontal scale of the plot.

2. Describe the form, direction, and strength of the overall pattern of a scatterplot. In particular, recognize positive or negative associations and straight-line patterns. Recognize outliers in a scatterplot.

3. Judge whether it is appropriate to use correlation to describe the relationship between two quantitative variables. Use a calculator to find the correlation r.

4. Know the basic properties of correlation: r measures the strength and direction of only straight-line relationships; r is always a number between -1 and 1; $r = \pm 1$ only for perfect straight-line relations; r moves away from 0 toward ± 1 as the straight-line relation gets stronger.

B. REGRESSION LINES

1. Explain what the slope b and the intercept a mean in the equation $y = a + bx$ of a straight line.

2. Draw a graph of the straight line when you are given its equation.

3. Use a regression line, given on a graph or as an equation, to predict y for a given x. Recognize the danger of prediction outside the range of the available data.

4. Use r^2, the square of the correlation, to describe how much of the variation in one variable can be accounted for by a straight-line relationship with another variable.

C. STATISTICS AND CAUSATION

1. Give plausible explanations for an observed association between two variables: direct cause and effect, the influence of lurking variables, or both.

2. Assess the strength of statistical evidence for a claim of causation, especially when experiments are not possible.

CHAPTER 6 EXERCISES

6.46 When it rains, it pours Figure 6.25 on the facing page plots the highest *yearly* precipitation ever recorded in each state against the highest *daily* precipitation ever recorded in that state. The points for Alaska, Hawaii, and Texas are marked on the scatterplot.
(a) About what are the highest daily and yearly precipitation values for Alaska?
(b) Alaska and Hawaii have very high yearly maximums relative to their daily maximums. Omit these two states as outliers. Describe the nature of the

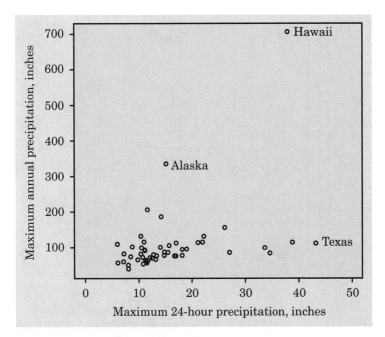

FIGURE 6.25 Record-high yearly precipitation at any weather station in each state plotted against record-high daily precipitation for the state.

relationship for the other states. Would knowing a state's highest daily precipitation be a great help in predicting that state's highest yearly precipitation?

6.47 Measuring crickets For a biology project, you measure the length (centimeters) and weight (grams) of 12 crickets.
(a) Explain why you expect the correlation between length and weight to be positive.
(b) If you measured length in inches, how would the correlation change? (There are 2.54 centimeters in an inch.)

Figure 6.26 (next page) plots the average brain weight in grams versus average body weight in kilograms for 96 species of mammals.[23] *There are many small mammals whose points at the lower left overlap. Exercises 6.48 to 6.53 are based on this scatterplot.*

6.48 Dolphins and hippos, I The points for the dolphin and hippopotamus are labeled in Figure 6.26. Read from the graph the approximate body weight and brain weight for these two species.

6.49 Dolphins and hippos, II One reaction to this scatterplot is "Dolphins are smart, hippos are dumb." What feature of the plot lies behind this reaction?

6.50 Outliers The African elephant is much larger than any other mammal in the data set but lies roughly in the overall straight-line pattern. Dolphins,

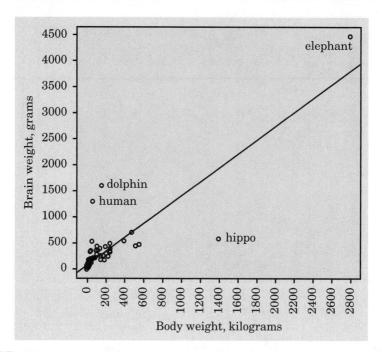

FIGURE 6.26 Scatterplot of the average brain weight (grams) against the average body weight (kilograms) for 96 species of mammals, for Exercises 6.48 to 6.53.

humans, and hippos lie outside the overall pattern. The correlation between body weight for the entire data set is $r = 0.86$.

(a) If we removed elephants, would this correlation increase, decrease, or not change much? Explain your answer.

(b) If we removed dolphins, hippos, and humans, would this correlation increase, decrease, or not change much? Explain your answer.

6.51 Brain and body The correlation between body weight and brain weight is $r = 0.86$. How well does body weight explain brain weight for mammals? Give a number to answer this question, and briefly explain what the number tells us.

6.52 Prediction The line on the scatterplot in Figure 6.26 is the least-squares regression line for predicting brain weight from body weight. Suppose that a new mammal species is discovered hidden in the rainforest with body weight 600 kilograms. Predict the brain weight for this species.

6.53 Slope The line on the scatterplot in Figure 6.26 is the least-squares regression line for predicting brain weight from body weight. The slope of this line is one of the numbers below. Which number is the slope? Why?

(a) $b = 0.5$ (b) $b = 1.3$ (c) $b = 3.2$

6.54 SAT preparation Can intensive preparation significantly improve students' SAT scores? Here's one way to find out. Take a random sample of 100 high school students who took the most recent SAT test and ask them how much time (in hours) they spent preparing and what score they earned on the math section. If we find a correlation of $r = 0.79$ between preparation time and SAT math score, can we conclude that preparation causes higher scores? Why or why not?

6.55 Bad feet Your scatterplot in Exercise 6.23 (page 362) suggests that the severity of the mild foot deformity called MA can help predict the severity of the more serious deformity called HAV.
(a) Use your calculator to find the equation of the least-squares regression line for predicting HAV angle from MA angle. Add this line to the scatterplot you made in Exercise 6.23.
(b) A new patient has MA angle 25 degrees. What do you predict this patient's HAV angle to be?
(c) Does knowing MA angle allow doctors to predict HAV angle accurately? Explain your answer from the scatterplot, then calculate a numerical measure to support your finding.

6.56 Teacher pay and liquor sales A study found a strong positive correlation between average teacher salaries and liquor sales over a 12-month period. Does this suggest that we should not pay teachers more because they would only spend the money on more liquor? Explain your answer clearly.

6.57 Why so small?
(a) Make a scatterplot of the following data:

x	1	2	3	4	10	10
y	1	3	3	5	1	11

(b) Use the method of Example 6.3 (page 348) to show that the correlation is about 0.5.
(c) What feature of the data is responsible for reducing the correlation to this value despite a strong straight-line association between x and y in most of the observations?

Part III

Chance

"What kind of childish nonsense are you working on now?"

"If chance will have me king, why, chance will crown me." So said Macbeth in Shakespeare's great play. Chance does indeed play with us all, and we can do little to manage it. Sometimes, however, chance can be tamed. A roll of dice, a simple random sample, and even the inheritance of eye color or blood type represent chance tied down so that we can understand and work with it. Unlike Macbeth's life or ours, we can roll the dice again. And again, and again. The outcomes are governed by chance, but in many repetitions a pattern emerges. Chance is no longer mysterious, because we can describe its pattern. We humans use mathematics to describe regular patterns, whether those configurations are the circles and triangles of geometry or the movements of the planets. We use mathematics to understand the regular patterns of chance behavior when chance is tamed in a setting where we can repeat the same chance phenomena again and again. The mathematics of chance is called probability. Probability is the topic of this part of the book, though we will go light on the math in favor of experimenting and thinking.

Chance and Probability

7.1 Thinking about Chance
7.2 Probability Models

7.1 THINKING ABOUT CHANCE

What's the chance?

On January 28, 1986, the space shuttle *Challenger* exploded soon after takeoff. A presidential commission investigated. What did those involved estimate the chance of such a failure to be? Some working engineers said roughly 1 in 100. Management said roughly 1 in 100,000. On hearing the latter estimate, physicist Richard Feynman, a member of the commission, asked, "You mean to tell me that if you launched a rocket every day for 300 years, you'd only expect one failure?" Feynman was good at mental arithmetic: 300 years is 109,500 days if we ignore leap years.[1]

Feynman did two important things in this little interchange. Management was clearly just guessing. (OK, they thought they were giving their "informed judgment" of risk, but it comes to the same thing.) That is, they were using the language of chance to express their personal opinion or judgment. Feynman turned this vague personal opinion into the much more concrete image of trying the same thing many times: If we launched a very large number of shuttles, how often would one fail? That should sound familiar, because it's the way we think about sampling: "If we choose many samples from the same population, the truth about the population will be within the margin of error 95% of the time."

Feynman's second tactic was to make the idea of trying something 100,000 times more understandable by bringing it into the real world—do it every day for 300 years. That's still not easy to grasp. Our minds don't deal well with really large numbers or with really small probabilities, like the 1-in-80-million chance of winning the lottery or the 1- in-7-million chance of dying in an airplane crash on your next flight. Chance is a slippery subject. We will follow Feynman in starting with "what would happen if we did this many times" before we try to think about using the language of chance to express personal opinions. We will also start with examples like the 1-in-2 chance of a head in tossing a coin before we try to think about the lottery.

ACTIVITY 7.1A Flipping coins and swatting baseballs

1.   Pretend that you are flipping a fair coin. Without actually flipping a coin, *imagine* the first toss. Write down the result you see in your mind (H or T).

2. Imagine a second coin flip. Write down the result.

3. Keep doing this until you have 50 H's or T's written down. Write your results in blocks of 5 to make them easier to read, like this: HTHTH TTHHT, etc.

4. A **run** is a repetition of the same result. In the example in Step 3, there is a run of two tails followed by a run of two heads in the first 10 coin flips. Read through your 50 imagined coin flips, and count the number of runs of size 2, 3, 4, etc. Record the number of runs of each size in a table.

5. Use your TI-83 to generate a similar list of 50 coin flips. We will let 1 represent a head and 0 represent a tail. Enter this command in the home screen: `randInt(0,1,50)→`L_1. The command `randInt` can be found under `MATH/PRB/5:randInt(`. Record the number of runs of size 2, 3, 4, and so forth.

6. Compare the two results. Did you or your calculator have the longest run? How much longer?

7. Most people are surprised at the occurrence of such long runs in a random sequence. A science writer once said that if you flip a fair coin 250 times, then approximately 32 runs will have at least two heads; about 16 runs will have at least three heads; 8 runs, at least four heads; 4 runs, at least five heads; 2 runs, at least six heads; and 1 run, at least 7 heads.[2] Use your calculator to see if the writer's claim is reasonable.

ACTIVITY 7.1A Flipping coins and swatting baseballs *(continued)*

Economics professor Paul Sommers of Middlebury College, Vermont, addressed the question of whether top home run sluggers hit home runs at random or whether they hit in streaks.[3] One of the records examined by Professor Sommers was Mark McGwire's 1998 season. He analyzed a sequence where 0 represents a game in which no home run was hit and 1 represents a game in which one or more home runs were hit. McGwire hit no home runs in 104 games and one or more home runs in 58 games (games 1, 2, 3, 4, 13, 16, 19, 23, 27, 28, 34, 36, 38, 40, 42, 43, 46, 47, 48, 49, 52, 53, 59, 62, 64, 65, 69, 70, 76, 77, 79, 81, 89, 90, 95, 98, 104, 105, 115, 118, 124, 125, 126, 129, 130, 132, 136, 138, 139, 141, 143, 144, 151, 154, 156, 160, 161, and 162).

8. Make an array of 0s and 1s, in order of games, to represent the games in which McGwire hit no home runs and at least one home run, respectively. Your list should begin 11110 00000.

Sommers reported that "the length of the longest 'run' was nine, which might be regarded as McGwire's longest slump (successive games in which no home run was hit). McGwire's longest 'hot streak' (successive games with one or more home runs per game) was four games and occurred twice during the season."

9. Find the game numbers for McGwire's longest slump and his longest "hot streak."

"The statistical evidence suggests that McGwire belted his 70 homers out of the park at random," Sommers concludes. "An analysis of the data on Sammy Sosa (1998), Roger Maris (1961), and Babe Ruth (1927) suggests that they too punched them out at random."

The idea of probability

Even the rules of football agree that tossing a coin avoids favoritism. Favoritism in choosing subjects for a sample survey or in allotting patients to treatment and placebo groups in a medical experiment is as undesirable as it is in awarding first possession of the ball in football. That's why statisticians recommend random samples and randomized experiments, which are fancy versions of tossing a coin. A big fact emerges when we watch coin tosses or the results of random samples closely: **Chance behavior is unpredictable in the short run but has a regular and predictable pattern in the long run.**

Toss a coin, or choose a simple random sample. The result can't be predicted in advance, because the result will vary when you toss the coin or choose the sample repeatedly. But there is still a regular pattern in the results, a pattern that emerges clearly only after many repetitions. This remarkable fact is the basis for the idea of probability.

EXAMPLE 7.1   Coin tossing

When you toss a coin, there are only two possible outcomes, heads or tails. Figure 7.1 shows the results of tossing a coin 1000 times. For each number of tosses from 1 to 1000, we have plotted the proportion of those tosses that gave a head. The first toss was a head, so the proportion of heads starts at 1. The second toss was a tail, reducing the proportion of heads to 0.5 after two tosses. The next three tosses gave a tail followed by two heads, so the proportion of heads after five tosses is 3/5, or 0.6.

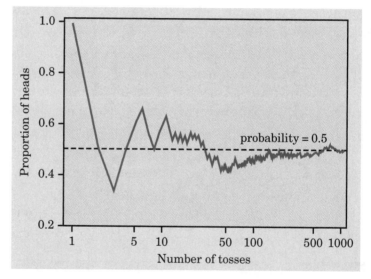

FIGURE 7.1   Toss a coin many times. The proportion of heads changes as we make more tosses but eventually gets very close to 0.5. This is what we mean when we say, "The probability of a head is one-half."

The proportion of tosses that produce heads is quite variable at first, but it settles down as we make more and more tosses. Eventually this proportion gets close to 0.5 and stays there. We say that 0.5 is the *probability* of a head. The probability 0.5 appears as a horizontal line on the graph.

"Random" in statistics is not a synonym for "haphazard" but a description of a kind of order that emerges only in the long run. We encounter the unpredictable side of randomness in our everyday experience, but we rarely see enough repetitions of the same random phenomenon to observe the long-term regularity that probability describes. You can see that regularity emerging in Figure 7.1. In the very long run, the proportion of tosses that give a head is 0.5. This is the intuitive idea of probability. Probability 0.5 means "occurs half the time in a very large number of trials."

We might suspect that a coin has probability 0.5 of coming up heads just because the coin has two sides. But babies must have one of the two sexes, and the probabilities aren't equal—the probability of a boy is about 0.51, not 0.50. The idea of probability is empirical. That is, it is based on data rather than theorizing. Probability describes what happens in very many trials, and we must actually observe many coin tosses or many babies to pin down a probability. In the case of tossing a coin, some diligent people have in fact made thousands of tosses.

EXAMPLE 7.2   Some coin tossers

The French naturalist Count Buffon (1707–1788) tossed a coin 4040 times. Result: 2048 heads, or proportion 2048/4040 = 0.5069 for heads.

Around 1900, the English statistician Karl Pearson heroically tossed a coin 24,000 times. Result: 12,012 heads, a proportion of 0.5005.

While imprisoned by the Germans during World War II, the South African mathematician John Kerrich tossed a coin 10,000 times. Result: 5067 heads, a proportion of 0.5067.

CALCULATOR CORNER   Flipping a coin without getting sore

This investigation uses the TI-83 to flip a fair coin many times. Then the calculator plots the cumulative results. This illustrates the sampling process and what happens after many repetitions. It is an extension of Example 7.1. Begin by clearing lists L_1 through L_4. Then do the following (remember that your results will differ somewhat from those shown):

- Enter the positive integers from 1 to 200 in list L_1. The command seq is 2nd [LIST] OPS/5:seq(.
  seq(X,X,1,200)→L_1.

```
seq(X,X,1,200)→L
1
{1 2 3 4 5 6 7 …
```

CALCULATOR CORNER   Flipping a coin without getting sore *(continued)*

- Generate 200 coin flips and store the values (1 = H and 0 = T) in L_2. The command randInt is $\boxed{\text{MATH}}$ PRB/5:randInt(.
randInt(0,1,200)$\rightarrow L_2$

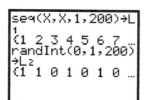

- Calculate the cumulative sum of the observed results, and store these values in list L_3. The command cumSum is $\boxed{2^{nd}}$ [LIST]OPS/6:cumSum(.
cumSum(L_2)$\rightarrow L_3$

- Calculate the proportion of heads and store these values in list L_4.
$L_3/L_1 \rightarrow L_4$

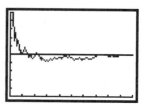

- Specify Plot1 as follows: xyline (second Type icon); Xlist: L_1; Ylist: L_4; Mark: .

- Set the viewing window as follows: X[0,10]$_{10}$. To set the y dimensions, scan the values in list L_4. Or start with Y[0,1]$_1$ and adjust as necessary. Press $\boxed{\text{GRAPH}}$.

- In the WINDOW screen, change Xmax to 100, and press $\boxed{\text{GRAPH}}$ again.

ACTIVITY 7.1B The *Probability* applet

Go to the Web site www.whfreeman.com/sta, and select the *Probability* applet, shown below.

Directions: When you click the Toss button, the applet will toss a coin the number of times you specify. You can set the probability that the coin lands heads by entering a new value and clicking the Reset button.

Contemplate:

1. If you toss the coin just once, what are the possible values for the proportion of heads?

2. If you toss the coin 400 times with a probability of 0.5, which is more likely to happen: (*a*) getting exactly 200 heads or (*b*) getting within a few percent of 200 heads?

3. What will the plot of the proportion of heads look like as you accumulate more and more tosses with a probability of 0.7 for heads?

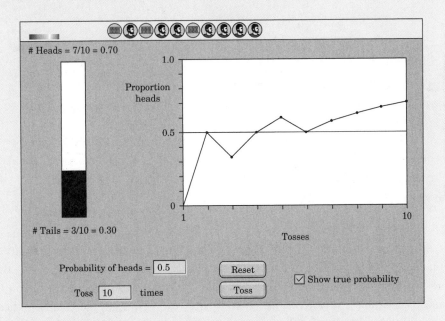

APPLET TIP: You can toss the coin a maximum of 40 times at once. To make longer sequences, toss again without resetting.

Verify: Use the applet to test your answers to the three questions above. Did the proportion of heads behave as you predicted?

ACTIVITY 7.1B The *Probability* applet *(continued)*

Experiment:

1. Set the probability of heads at 0.3 and have the applet repeatedly toss the coin until you accumulate many hundreds of tosses. How does the behavior of the proportion of heads differ at the beginning of the plot compared with the end of the plot?

2. Check the "Show true probability" box to have the applet draw a horizontal line on the plot at 0.3. How does the behavior of the plot of the proportion of heads relate to this line?

3. Change the probability of heads to 0.7 and have the applet toss the coin several hundred times again. How does the plot of the proportion of heads compare to the plot when the probability was 0.3? In what ways is it similar? In what ways does it differ?

Randomness and probability

We call a phenomenon **random** if individual outcomes are uncertain but there is nonetheless a regular distribution of outcomes in a large number of repetitions.

The **probability** of any outcome of a random phenomenon is a number between 0 and 1 that describes the proportion of times the outcome would occur in a very long series of repetitions.

An outcome with probability 0 never occurs. An outcome with probability 1 happens on every repetition. An outcome with probability 1/2 happens half the time in a very long series of trials. Of course, we can never observe a probability exactly. We could always continue tossing the coin, for example. Mathematical probability is an idealization based on imagining what would happen in an indefinitely long series of trials.

We aren't thinking deeply here. That some things are random is simply an observed fact about the world. Probability just gives us a language to describe the long-term regularity of random behavior. The outcome of a coin toss, the time between emissions of particles by a radioactive source, and the sexes of the next litter of lab rats are all random. So is the outcome of a random sample or a randomized experiment. The behavior of large groups of individuals is often as random as the behavior of many coin tosses or many random samples. Life insurance, for example, is based on the fact that deaths occur at random among many individuals.

EXAMPLE 7.3   The probability of dying

We can't predict whether a particular person will die in the next year. But if we observe millions of people, deaths are random. The National Center for Health Statistics says that the proportion of men aged 20 to 24 years who die in any one year is 0.0015. This is the *probability* that a young man will die next year. For women that age, the probability of death is about 0.0005.

If an insurance company sells many policies to people aged 20 to 24, it knows that it will have to pay off next year on about 0.15% of the policies sold to men and on about 0.05% of the policies sold to women. It will charge more to insure a man because the probability of having to pay is higher.

The ancient history of chance

Randomness is most easily noticed in many repetitions of games of chance—rolling dice, dealing shuffled cards, spinning a roulette wheel.[4] Chance devices similar to these have been used from remote antiquity to discover the will of the gods. The most common method of randomization in ancient times was "rolling the bones," tossing several astragali. The astragalus (Figure 7.2) is a solid, quite regular bone from the heel of animals that, when thrown, will come to rest on any of four sides. (The other two sides are rounded.) Cubical dice, made of pottery or bone,

Sheep   Dog

FIGURE 7.2 Animal heel bones (astragali), actual size. (From F. N. David, *Games, Gods, and Gambling,* Charles Griffin & Company, 1962. Reproduced by permission of the publishers.)

Does God play dice?

Few things in the world are truly random in the sense that no amount of information will allow us to predict the outcome. We could in principle apply the laws of physics to a tossed coin, for example, and calculate whether it will land heads or tails. But randomness does rule events inside individual atoms. Albert Einstein didn't like this feature of the new quantum theory. "God does not play dice with the universe," said the great scientist. Almost a century later, it appears that Einstein was wrong.

came later, but even dice existed before 2000 B.C. Gambling on the throw of astragali or dice is, compared with divination, almost a modern development. There is no clear record of this vice before about 300 B.C. Gambling reached flood tide in Roman times, then temporarily receded (along with divination) in the face of Christian displeasure.

Chance devices such as astragali have been used from the beginning of recorded history. Yet none of the great mathematicians of antiquity studied the regular pattern of many throws of bones or dice. Perhaps this is because astragali and most ancient dice were so irregular that each had a different pattern of outcomes. Or perhaps the reasons lie deeper, in the classical reluctance to engage in systematic experimentation.

Professional gamblers, who are not as inhibited as philosophers and mathematicians, did notice the regular pattern of outcomes of dice or cards and tried to adjust their bets to the odds of success. "How should I bet?" is the question that launched mathematical probability. The systematic study of randomness began (we oversimplify, but not too much) when seventeenth-century French gamblers asked French mathematicians for help in figuring out the "fair value" of bets on games of chance. **Probability theory,** the mathematical study of randomness, originated with Pierre de Fermat and Blaise Pascal in the seventeenth century and was well developed by the time statisticians took it over in the twentieth century.

EXERCISES

7.1 Spinning quarter With your forefinger, hold a new quarter (with a state featured on the reverse) upright, on its edge, on a hard surface. Then flick it with your other forefinger so that it spins for some time before it falls and comes to rest. Spin the coin a total of 50 times, and record the results.
(a) Based on 50 spins, estimate the probability of heads.
(b) Estimate the probability of tails.

7.2 How many tosses to get a head? When we toss a penny, experience shows that the probability (long-term proportion) of a head is close to 1/2. Suppose now that we toss the penny repeatedly until we get a head. We want to know the probability that the first head comes up in an odd number of tosses (1, 3, 5, and so on). To find out, perform 50 trials, and keep a record of the number of tosses needed to get a head on each of your 50 trials.

(a) From your experiment, estimate the probability of a head on the first toss. What value should we expect this probability to have?

(b) Use your results to estimate the probability that the first head appears on an odd-numbered toss.

7.3 Random digits The table of random digits (Table A) was produced by a random mechanism that gives each digit probability 0.1 of being a 0. What proportion of the first 200 digits in the table are 0s? This proportion is an estimate, based on 200 repetitions, of the true probability, which in this case is known to be 0.1.

7.4 From words to probabilities Probability is a measure of how likely an event is to occur. Match one of the probabilities that follow with each statement of likelihood given. (The probability is usually a more exact measure of likelihood than is the verbal statement.)

$$0 \qquad 0.01 \qquad 0.3 \qquad 0.6 \qquad 0.99 \qquad 1$$

(a) This event is impossible. It can never occur.

(b) This event is certain. It will occur on every trial.

(c) This event is very unlikely, but it will occur once in a while in a long sequence of trials.

(d) This event will occur more often than not.

7.5 Barry Bonds's home run runs At the end of the current baseball season, it will be possible to inspect the home run performance of the current record holder, Barry Bonds. One place to start finding data on Barry Bonds is the Web site http://baseball.about.com/cs/barrybonds/. See if you can find information on Bonds's home run production by game, similar to the information on Mark McGwire in Activity 7.1A. Did Bonds have any unusual runs where he hit no home runs? Did he have any long runs of home runs in consecutive games? Make a note of any factor such as injury that might help explain a drought in Bonds's home run production.

Myths about chance behavior

The idea of probability seems straightforward. It answers the question "What would happen if we did this many times?" In fact, both the behavior of random phenomena and the idea of probability are a bit subtle. We meet chance behavior constantly, and psychologists tell us that we deal with it poorly.

The myth of short-run regularity The idea of probability is that randomness is regular in the long run. Unfortunately, our intuition about randomness tries to tell us that random phenomena should also be regular in the short run. When they aren't, we look for some explanation other than chance variation.[5]

EXAMPLE 7.4   What looks random?

Toss a coin six times and record heads (H) or tails (T) on each toss. Which of the following outcomes is more probable?

HTHTTH   TTTHHH

Almost everyone says that HTHTTH is more probable, because TTTHHH does not "look random." In fact, both are equally probable. That heads and tails are equally probable says only that about half of a very long sequence of tosses will be heads. It doesn't say that heads and tails must come close to alternating in the short run. The coin has no memory. It doesn't know what past outcomes were, and it can't try to create a balanced sequence.

The outcome TTTHHH in tossing six coins looks unusual because of the runs of 3 straight heads and 3 straight tails. Runs seem "not random" to our intuition but are quite common. Here's an example more striking than tossing coins.

EXAMPLE 7.5   The hot hand in basketball

Belief that runs must result from something other than "just chance" influences behavior. If a basketball player makes several consecutive shots, both the fans and his teammates believe that he has a "hot hand" and is more likely to make the next shot. This is wrong. Careful study has shown that runs of baskets made or missed are no more frequent in basketball than would be expected if each shot is independent of the player's previous shots. Players perform consistently, not in streaks. If a player makes half her shots in the long run, her hits and misses behave just like tosses of a coin—and that means that runs of hits and misses are more common than our intuition expects.[6]

The myth of the surprise meeting Julie is spending the summer in London. One day, on the third floor of the Victoria and Albert Museum, she runs into Jim, a casual friend from college. "How unusual! Maybe we were fated to meet."

Well, maybe not. It is certainly unlikely that Julie would run into *this particular* acquaintance that day, but it is not at all unlikely that Julie would meet *some* acquaintance during her summer in London. After all, a typical adult has about 1500 casual acquaintances. When something unusual happens, we look back and say, "Wasn't that unlikely?" We would have said the same if any of 1500 other unlikely things had happened. Here's an example where we can actually calculate the probabilities.

EXAMPLE 7.6 Winning the lottery twice

In 1986, Evelyn Marie Adams won the New Jersey state lottery for the second time, adding $1.5 million to her previous $3.9 million jackpot. The *New York Times* (February 14, 1986) claimed that the odds of one person winning the big prize twice were about 1 in 17 trillion. Nonsense, said two statistics professors in a letter that appeared in the *Times* two weeks later. The chance that Evelyn Marie Adams would win twice in her lifetime is indeed tiny, but it is almost certain that someone among the millions of regular lottery players in the United States would win two jackpot prizes. The statisticians estimated even odds of another double winner within seven years. Sure enough, Robert Humphries won his second Pennsylvania lottery jackpot ($6.8 million total) in May 1988.

Chance or causation Unusual events—especially distressing events—bring out the human desire to pinpoint a reason, a *cause*. Here's a sequel to our earlier discussion of causation: sometimes it's just the play of chance.

EXAMPLE 7.7 Cancer clusters

In 1984, residents of a neighborhood in Randolph, Massachusetts, counted 67 cancer cases in their 250 residences. This cluster of cancer cases seemed unusual, and the residents expressed concern that runoff from a nearby chemical plant was contaminating their water supply and causing cancer.

In 1979, two of the eight town wells serving Woburn, Massachusetts, were found to be contaminated with organic chemicals. Alarmed citizens began counting cancer cases. Between 1964 and 1983, 20 cases of childhood leukemia were reported in Woburn. This is an unusual number of cases of this rather rare disease. The residents believed that the well water had caused the leukemia and proceeded to sue two companies held responsible for the contamination.[7]

Cancer is a common disease, accounting for more than 23% of all deaths in the United States. That cancer cases sometimes occur in clusters in the same neighborhood is not surprising; there are bound to be clusters *somewhere* simply by chance. But when a cancer cluster occurs in our neighborhood, we tend to suspect the worst and look for someone to blame. State authorities get several thousand

The probability of rain is . . .

You work all week. Then it rains on the weekend. Can there really be a statistical truth behind our perception that the weather is against us? At least on the East Coast of the United States, the answer is "Yes." Going back to 1946, it seems that Sundays receive 22% more precipitation than Mondays. The likely explanation is that the pollution from all those workday cars and trucks forms the seeds for raindrops—with just enough delay to cause rain on the weekend.

calls a year from people worried about "too much cancer" in their area. But, as the National Cancer Institute says, "The majority of cancer clusters are simply the result of chance." Both of the Massachusetts cancer clusters mentioned in Example 7.7 were investigated by statisticians from the Harvard School of Public Health. The investigators tried to obtain complete data on everyone who had lived in the neighborhoods in the periods in question and to estimate their exposure to the suspect drinking water. They also tried to obtain data on other factors that might explain cancer, such as smoking and occupational exposure to toxic substances. The verdict: Chance is the likely explanation of the Randolph cluster, but there is evidence of an association between drinking water from the two Woburn wells and developing childhood leukemia.

The myth of the law of averages Once at a convention in Las Vegas one of the authors roamed the gambling floors, watching money disappear into the drop boxes under the tables. You can see some interesting human behavior in a casino. When the shooter in the dice game craps rolls several winners in a row, some gamblers think she has a "hot hand" and bet that she will keep on winning. Others say that "the law of averages" means that she must now lose so that wins and losses will balance out. Believers in the law of averages think that if you toss a coin six times and get TTTTTT, the next toss must be more likely to give a head. It's true that in the long run heads must appear half the time. What is myth is that future outcomes must make up for an imbalance like six straight tails.

Coins and dice have no memories. A coin doesn't know that the first six outcomes were tails, and it can't try to get a head on the next toss to even things out. Of course, things do even out *in the long run.* After 10,000 tosses, the results of the first six tosses don't matter. They are overwhelmed by the results of the next 9994 tosses, not compensated for.

EXAMPLE 7.8   We want a boy

Belief in this phony "law of averages" can lead to consequences close to disastrous. A few years ago, "Dear Abby" published in her advice column a letter from a distraught mother of eight girls. It seems that she and her husband had planned to limit their family to four children. When all four were girls, they tried again—and again, and again. After seven straight girls, even her doctor had assured her that "the law of averages was in our favor 100 to 1." Unfortunately for this couple, having children is like tossing coins. Eight girls in a row is highly unlikely, but once seven girls have been born, it is not at all unlikely that the next child will be a girl—and it was.

"So the law of averages doesn't guarantee me a girl after seven straight boys, but can't I at least get a group discount on the delivery fee?"

Personal probabilities

Joe sits staring into his beer as his favorite baseball team, the Chicago Cubs, lose another game. The Cubbies have some good young players, so let's ask Joe, "What's the chance that the Cubs will go to the World Series next year?" Joe brightens up. "Oh, about 10%," he says.

Does Joe assign probability 0.10 to the Cubs' appearing in the World Series? The outcome of next year's pennant race is certainly unpredictable, but we can't reasonably ask what would happen in many repetitions. Next year's baseball season will happen only once and will differ from all other seasons in players, weather, and many other ways. The answer to our question seems clear: if probability measures "what would happen if we did this many times," Joe's 0.10 is not a probability. Probability is based on data about many repetitions of the same random phenomenon. Joe is giving us something else, his personal judgment.

Yet we often use "probability" in a way that includes personal judgments of how likely it is that some event will happen. We make decisions based on these judgments—we take the bus downtown because we think the probability of finding a parking spot is low. More serious decisions also take "how likely" judgments into account. A company deciding whether to build a new plant must judge how likely it is that there will be high demand for its products three years from now when the plant is ready. Many companies express "How likely is it?" judgments as numbers—probabilities—and use these numbers in their calculations. High demand in three years, like the Cubs' winning next year's pennant, is a one-time event that doesn't fit the "do it many times" way of thinking. What is more, several company officers may give several different probabilities, reflecting differences in their individual judgment. We need another kind of probability, *personal probability*.

Personal probability

A **personal probability** of an outcome is a number between 0 and 1 that expresses an individual's judgment of how likely the outcome is.

Personal probabilities have the great advantage that they aren't limited to repeatable settings. They are useful because we base decisions on them: "I think the probability that the Vikings will win the Super Bowl is 0.75, so I'm going to bet on the game." Just remember that personal probabilities are different in kind from probabilities as "proportions in many repetitions." Because they express individual opinion, they can't be said to be right or wrong. This is true even in a "many repetitions" setting. If Craig has a gut feeling that the probability of a head on the next toss of this coin is 0.7, that's what Craig thinks and that's all there is to it. Tossing the coin many times may show that the proportion of heads is very close to 0.5, but that's another matter. **There is no reason why a person's degree of confidence in the outcome of one try must agree with the results of many tries.** We stress this because it is common to say that "personal probability" and "what happens in many trials" are somehow two interpretations of the same idea. In fact, they are quite different ideas.

Why do we even use the word "probability" for personal opinions? There are two good reasons. First, we usually do base our personal opinions on data from many trials when we have such data. Data from Buffon, Pearson, and Kerrich (Example 7.2, page 405) and perhaps from our own experience convince us that coins come up heads very close to half the time in many tosses. When we say that a coin has probability 1/2 of coming up heads *on this toss,* we are applying to a single toss a measure of the chance of a head based on what would happen in a long series of tosses. Second, personal probability and probability as long-term proportion both obey the same mathematical rules. To start, both kinds of probabilities are numbers between 0 and 1. This isn't as important to us as it is to mathematicians, but we will look at some of the rules of probability in the next section. These rules apply to both kinds of probability.

Probability and risk

Once we understand that "personal judgment of how likely" and "what happens in many repetitions" are different ideas, we have a good start toward understanding why the public and the experts disagree so strongly about what is risky and what isn't. The experts use probabilities from data to describe the risk of an unpleasant event. Individuals and society, however, seem to ignore data. We worry about some risks that almost never occur while ignoring others that are much more probable.

EXAMPLE 7.9 Asbestos in the schools

High exposures to asbestos are dangerous. Low exposures, such as that experienced by teachers and students in schools where asbestos is present in the insulation around pipes, are not very risky. The probability that a teacher who works for 30 years in a school with typical asbestos levels will get cancer from the asbestos is around 15/1,000,000. The risk of dying in a car accident during a lifetime of driving is about 15,000/1,000,000. That is, driving regularly is 1000 times more risky than teaching in a school where asbestos is present.[8]

Risk does not stop us from driving. Yet the much smaller risk from asbestos launched massive cleanup campaigns and a federal requirement that every school inspect for asbestos and make the findings public.

Why do we take asbestos so much more seriously than driving? Why do we worry about very unlikely threats such as tornadoes and terrorists more than we worry about heart attacks?

- We feel safer when a risk seems under our control than when we cannot control it. We are in control (or so we imagine) when we are driving, but we can't control the risk from asbestos or tornadoes or terrorists.

- It is hard to comprehend very small probabilities. Probabilities of 15 per million and 15,000 per million are both so small that our intuition cannot distinguish between them. Psychologists have shown that we generally overestimate very small risks and underestimate higher risks. Perhaps this is part of the general weakness of our intuition about how probability operates.

What are the odds?

Gamblers often express chance in terms of odds rather than probability. Odds of A to B against an outcome means that the probability of that outcome is $B/(A + B)$. So "odds of 5 to 1" is another way of saying "probability 1/6." A probability is always between 0 and 1, but odds range from 0 to infinity. Although odds are mainly used in gambling, they give us a way to make very small probabilities clearer. "Odds of 999 to 1" may be easier to understand than "probability 0.001."

- The probabilities for risks like asbestos in the schools are not as certain as probabilities for tossing coins. They must be estimated by experts from complicated statistical studies. Perhaps it is safest to suspect that the experts may have underestimated the level of risk.

Our reactions to risk depend on more than probability, even if our personal probabilities are higher than the experts' data-based probabilities. We are influenced by our psychological makeup and by social standards. As one writer noted, "Few of us would leave a baby sleeping alone in a house while we drove off on a 10-minute errand, even though car-crash risks are much greater than home risks."[9]

EXERCISES

7.6 Personal probability versus data Give an example in which you would rely on a probability found as a long-term proportion from data on many trials. Give an example in which you would rely on your own personal probability.

7.7 Personal random numbers? Ask several of your friends (at least 10 people) to choose a four-digit number "at random." How many of the numbers chosen start with 1 or 2? How many start with 8 or 9? (There is strong evidence that people in general tend to choose numbers that start with low digits.)[10]

7.8 Playing "Pick 4" The Pick 4 games in many state lotteries announce a four-digit winning number each day. The winning number is essentially a four-digit group from a table of random digits. You win if your choice matches the winning digits, in any order. The winnings are divided among all players who matched the winning digits. That suggests a way to get an edge.
(a) The winning number might be, for example, either 2873 or 9999. Explain why these two outcomes have exactly the same probability. (It is 1 in 10,000.)
(b) If you asked many people which outcome is more likely to be the randomly chosen winning number, most would favor one of them. Use the information in this chapter to say which one and to explain why. If you choose a number that people think is unlikely, you have the same chance to win, but you will win a larger amount because few other people will choose your number.

7.9 Surprising? You are getting to know your new roommate, assigned to you by the college. In the course of a long conversation, you find that both of you have sisters named Deborah. Should you be surprised? Explain your answer.

7.10 Cold weather coming A meteorologist, predicting a colder-than-normal winter, said, "First, in looking at the past few winters, there has been a lack of really cold weather. Even though we are not supposed to use the law of averages, we are due." Do you think that "due by the law of averages" makes sense in talking about the weather?

APPLICATION 7.1 Estimating the Risk of Lung Cancer

In March 2003, researchers at New York's Memorial Sloan-Kettering Cancer Center published a report of a study of smokers, aged 50 and older, who smoked at least half a pack of cigarettes a day for at least 25 years. It was a large randomized trial of lung cancer prevention. What they found was that how long and how much a person smoked and how long it has been since the last puff were all factors in determining that person's risk of getting lung cancer. Based on a careful analysis of the data they collected in their study, the researchers devised a mathematical model—a formula of sorts—that smokers and ex-smokers with the above characteristics could use to assess their risk of lung cancer. The report stated that "this prediction tool can assess a long-term smoker's risk of developing lung cancer in the next 10 years based on the person's age, sex, smoking history, and asbestos exposure. Knowing about risk can help clinicians and patients make decisions about health care, such as whether to get screening for lung cancer." This new capability is particularly important because there is a new, expensive lung cancer detection device, called a "spiral CT scanner," that is being advertised aggressively but that, at this writing, is still unproven. The National Cancer Institute is studying whether spiral CT scans, which view the lungs at various angles, could improve survival by spotting tumors early. Besides expense, the scans have an additional drawback: up to half detect harmless scar tissue or some other benign lump that requires a risky biopsy or other follow-up testing.

The next several questions involve using the formula to assess the risk of lung cancer for different individuals. Remember that the prediction tool will only work for people who meet these conditions:

* Age: 50 to 75 years old

* Smoking history: 10 to 60 cigarettes a day for 25 to 55 years

* Current status: Current smokers and former smokers who quit 20 or fewer years ago

Go to the Memorial Sloan-Kettering Cancer Center Web site: http://www.mskcc.org/mskcc/html/12463.cfm. To use the formula, click on the picture that says, "Calculate Smokers' Risk of Lung Cancer."

1. Why will the formula not work for people who do not meet all of the conditions above?

2. Specify a 50-year-old woman who smoked a pack a day (20 cigarettes) from age 15 until she stopped 5 years ago. What is her percent risk? Over the next 10 years, if she had not stopped smoking, what would her risk of getting lung cancer be?

3. Click on the [CLEAR] button. Specify a 70-year-old man who has smoked two packs a day since he was 18 and hasn't quit yet. What are his chances of getting lung cancer by his 80th birthday? If he quit smoking today, what would his risk assessment be?

4. Experiment with the prediction tool to see the effects of different situations. Perhaps assess the risk for a family member or an acquaintance who smokes.

5. The Memorial Sloan-Kettering Risk Assessment page states that "Quitting smoking not only reduces risk of lung cancer, but reduces risks of many other smoking-related health problems." Did you find this to be the case for your pretend individuals? If necessary, run the prediction tool several additional times to help you answer this question.

EXPLORING THE WEB

One of the best ways to grasp the idea of probability is to watch the proportion of trials on which an outcome occurs gradually settle down at the outcome's probability. The *Probability* applet that you used in Activity 7.1B, shows this nicely. It can be found at the *Statistics through Applications* Web site, www.whfreeman.com/sta.

The National Cancer Institute has a large amount of information, including maps and statistics, at its information site www.cancer.gov/cancerinfo. Cancer clusters cause so much public concern that the NCI has devoted a separate page to them: cis.nci.nih.gov/fact/3_58.htm.

STATISTICS IN SUMMARY

Some things in the world, both natural and of human design, are **random.** That is, their outcomes have a clear pattern in very many repetitions even though the outcome of any one trial is unpredictable. **Probability** describes the long-term regularity of random phenomena. The probability of an outcome is the proportion of very many repetitions on which that outcome occurs. A probability is a number between 0 (never occurs) and 1 (always occurs). We emphasize this kind of probability because it is based on data.

Probabilities describe only what happens in the long run. Short runs of random phenomena like tossing coins or shooting a basketball often don't look random to us because they do not show the regularity that in fact only emerges in very many repetitions.

Personal probabilities express an individual's personal judgment of how likely outcomes are. Personal probabilities are also numbers between 0 and 1. Different people can have different personal probabilities, and a personal probability need not agree with a proportion based on data about similar cases.

SECTION 7.1 EXERCISES

7.11 Personal probability? When there are few data, we often fall back on personal probability. There had been just 24 space shuttle launches, all successful, before the *Challenger* disaster. The shuttle program management thought the chances of such a failure were only 1 in 100,000. Give some reasons why such an estimate is likely to be too optimistic.

7.12 In the long run Probability works not by compensating for imbalances but by overwhelming them. Suppose that the first six tosses of a coin give six tails and that tosses after that are exactly half heads and half tails. (Exact balance is unlikely, but it illustrates how the first six outcomes are swamped by later outcomes.) What is the proportion of heads after the first six tosses? What is the proportion of heads after 100 tosses if the last 94 produce 47 heads? What is the proportion of heads after 1000 tosses if half of the last 994 produce heads? What is the proportion of heads after 10,000 tosses if half of the last 9994 produce heads?

7.13 An unenlightened gambler
(a) A gambler knows that red and black are equally likely to occur on each spin of a roulette wheel. He observes five consecutive reds occur and bets heavily on black at the next spin. Asked why, he explains that black is "due by the law of averages." Explain to the gambler what is wrong with this reasoning.
(b) After hearing you explain why red and black are still equally likely after five reds on the roulette wheel, the gambler moves to a poker game. He is dealt five straight red cards. He remembers what you said and assumes that the next card dealt in the same hand is equally likely to be red or black. Is the gambler right or wrong, and why?

7.14 Reacting to risks, I The probability of dying if you play high school football is about 10 per million each year you play. The risk of getting cancer from asbestos if you attend a school in which asbestos is present for 10 years is about 5 per million. If we ban asbestos from schools, should we also ban high school football? Briefly explain your position.

7.15 Reacting to risks, II National newspapers such as *USA Today* and the *New York Times* carry many more stories about deaths from airplane crashes than about deaths from automobile crashes. Auto accidents kill about 40,000 people in the United States each year. Crashes of all scheduled air carriers, including commuter carriers, have killed between 0 and 575 people per year in recent years, including the four hijacked planes that crashed on 9/11/01.
(a) Why do the news media give more attention to airplane crashes?
(b) How does news coverage help explain why many people consider flying more dangerous than driving?

7.16 What probability doesn't say The probability of a head in tossing a coin is 1/2. This means that as we make more tosses, the *proportion* of heads will eventually get close to 0.5. It does not mean that the *count* of heads will get close to 1/2 the number of tosses. To see why, imagine that the proportion of heads is 0.51 in 100 tosses, 1000 tosses, 10,000 tosses, and 100,000 tosses of a coin. How many heads came up in each set of tosses? How close is the number of heads to half the number of tosses?

7.17 Flipping out This exercise relates to the Calculator Corner on page 405. In your own words, write a short description of the principle that the Calculator Corner demonstrates.

7.18 A game of chance I have a little bet to offer you. Toss a coin 10 times. If there is no run of three or more straight heads or tails in the 10 outcomes, I'll pay you $2. If there is a run of three or more, you pay me just $1. Surely you will want to take advantage of me and play this game? Is it to your advantage to play?

7.2 PROBABILITY MODELS

Heads, tails, or confusion

Martha: OK, Tom, here's a new quarter featuring the state of Virginia. If I toss the coin, what do you think is the probability of a head?

Tom: Oh, I'd say 60%.

Martha: And what's your probability of a tail?

Tom: Make that 50%.

Martha: 60% chance of a head and 50% chance of a tail add up to 110% for a head or a tail. That doesn't make sense.

Tom: Well, those are my personal probabilities, so they can be anything I want them to be.

Martha: No they can't. They have to make sense when you look at them both together.

Martha says that personal probabilities must obey commonsense rules if they are to make sense. Probabilities that "make sense," however, may not describe what will happen if we toss the coin many times. We know that Tom's personal probabilities may not agree with how often the coin really comes up heads in many tosses. But just as we check to see if data are consistent with each other, we insist that probabilities be consistent with each other. The probability of heads and the probability of tails for the same coin can't add to more than 1. In fact, they must add to exactly 1 unless we allow some other outcome, such as the coin landing on edge and staying upright. That's because all possible outcomes together must have probability exactly 1. And that's one of the "rules of probability" we will explore in this chapter.

"What are the chances? I mean, Dewey makes a hole-in-one, gets struck by lightning and is crushed by a meteor all on the same hole?"

ACTIVITY 7.2 What are the chances?

In the July 27, 1997, issue of *Parade* magazine, a reader posed the following question to Marilyn vos Savant and the "Ask Marilyn" column:

> A woman and a man (unrelated) each have two children. At least one of the woman's children is a boy, and the man's older child is a boy. Do the chances that the woman has two boys equal the chances that the man has two boys?

Courtesy of Marilyn vos Savant

Many people think that the answer to the question is "Yes," and that the probability in both cases is 1/2. But you may be in for a surprise. In this activity you will use simulation methods to answer this question. We will learn more about the art of simulation in Chapter 8.

1. One way to simulate this situation, using a table of random numbers, would be to let an even digit represent a boy and an odd digit represent a girl. Beginning at a random row, read off digits two digits at a time (to represent the two children). If both digits are odd (that is, two girls), discard the pair because that doesn't satisfy the assumption that at least one of the woman's children is a boy. If at least one of the digits is even (a boy), record the results by making a tally mark in a table like this:

1 Boy:
2 Boys:

Continue until you have simulated 50 pairs of children. Each pair of pretend children is called a "trial." The number of tally marks for 2 Boys divided by 50 approximates the probability that the woman has two boys. It would be even better if each student in the class could do 20 replications, and then the results for all members in the class could be accumulated.

2. For the man's family, how would you use the random number table to simulate many trials of a two-child family where the older child is a boy? Devise a strategy, write it down, and then carry it out. What estimate do you obtain for the probability that the man has two boys?

ACTIVITY 7.2 What are the chances? *(continued)*

3. Here is a calculator program that will simulate the woman's situation. Enter it into your calculator or link it from a classmate or your teacher.

```
PROGRAM:ACT7A
:rand→X
:If X<.25:Then:Disp "GG"
:Else:If X<.5:Then:Disp "GB"
:Else:If X<.75:Then:Disp "BG"
:Else:Disp "BB":End
```

Read through the program and make sure you understand what it's doing. Does it assume that the outcomes {GG, GB, BG, BB} are equally likely? The calculator's **rand** function picks a decimal number in the interval $0 \leq X < 1$. The program establishes a correspondence between the four outcomes and certain random numbers. Describe that correspondence.

Run the program and keep pressing the $\boxed{\text{ENTER}}$ key to simulate additional trials. As before, ignore the results "GG" and use tally marks to record the occurrences of 1 Boy and 2 Boys. Simulate 50 trials (not counting GG outcomes), and then determine the proportion of families with two boys.

4. The following program will automate the process. Enter this program into your calculator.

```
PROGRAM:ACT7B
:ClrHome
:Disp "HOW MANY TRIALS"
:Prompt N
:0→C:0→J
:While J≤N
:rand→X
:If X<.25:End
:If X≥.75:C+1→C
:J+1→J
:End
:Disp "REL.FREQ.="
:Disp C/N
```

Again, read through the steps of the program to see what it does. Do you agree that this program will simulate many trials of the experiment? Run the program and specify 50 trials. Do it again several times, and increase the number of trials to 100 or more. Are your results similar to the results you obtained by using the random number table?

ACTIVITY 7.2 What are the chances? *(continued)*

5. Based on these various attempts to simulate the problem, what is your best guess for the probability that both children are boys, given that one child is a boy? Write a short paragraph that summarizes your insight into the probability of this event. Then write a second paragraph about the probability of two boys in a family if we know that the first child is a boy.

Probability models

Choose a woman aged 25 to 29 years old at random and record her marital status. "At random" means that we give every such woman the same chance to be the one we choose. That is, we choose a random sample of size 1. The probability of any marital status is just the proportion of all women aged 25 to 29 who have that status—if we chose many women, this is the proportion we would get. Here is the set of probabilities:

Marital status:	Never married	Married	Widowed	Divorced
Probability:	0.386	0.555	0.004	0.055

This table gives a *probability model* for picking a young woman at random and finding out her marital status. It tells us what are the possible outcomes (there are only four) and it assigns probabilities to these outcomes. The probabilities here are the proportions of all women who are in each marital class. That makes it clear that the probability that a woman is not married is just the sum of the probabilities of the three classes of unmarried women:

$$P(\text{not married}) = P(\text{never married}) + P(\text{widowed}) + P(\text{divorced})$$
$$= 0.386 + 0.004 + 0.055 = 0.445$$

As a shorthand, we often write $P(\text{not married})$ for "the probability that the woman we choose is not married." You see that our model does more than assign a probability to each individual outcome—we can find the probability of any collection of outcomes by adding up individual outcome probabilities.

Probability model

A **probability model** for a random phenomenon describes all the possible outcomes and says how to assign probabilities to any collection of outcomes. We sometimes call a collection of outcomes an **event**.

Probability rules

Because the probabilities in our example are just the proportions of all women who have each marital status, they follow rules that say how proportions behave. Here are some basic rules that any probability model must obey:

A. Any probability is a number between 0 and 1. Any proportion is a number between 0 and 1, so any probability is also a number between 0 and 1. An event with probability 0 never occurs, and an event with probability 1 occurs on every trial. An event with probability 0.5 occurs in half the trials in the long run.

B. All possible outcomes together must have probability 1. Because some outcome must occur on every trial, the sum of the probabilities for all possible outcomes must be exactly 1.

C. The probability that an event does not occur is 1 minus the probability that the event does occur. If an event occurs in (say) 70% of all trials, it fails to occur in the other 30%. The probability that an event occurs and the probability that it does not occur always add to 100%, or 1.

D. If two events have no outcomes in common, the probability that one or the other occurs is the sum of their individual probabilities. If one event occurs in 40% of all trials, a different event occurs in 25% of all trials, and the two can never occur together, then one or the other occurs on 65% of all trials because 40% + 25% = 65%.

Politically correct

In 1950, the Russian mathematician B. V. Gnedenko (1912–1995) wrote a text, *The Theory of Probability*, that was popular around the world. The introduction contains a mystifying paragraph that begins, "We note that the entire development of probability theory shows evidence of how its concepts and ideas were crystallized in a severe struggle between materialistic and idealistic conceptions." It turns out that "materialistic" is jargon for "Marxist-Leninist." It was good for the health of Russian scientists in the Stalin era to add such statements to their books.

EXAMPLE 7.10   Marital status of young women

Look again at the probabilities for the marital status of young women. Each of the four probabilities is a number between 0 and 1. Their sum is

$$0.386 + 0.555 + 0.004 + 0.055 = 1$$

This assignment of probabilities satisfies Rules A and B. **Any assignment of probabilities to all individual outcomes that satisfies Rules A and B is legitimate.** That is, it makes sense as a set of probabilities. Rules C and D are then automatically true.

Here is an example of the use of Rule C. The probability that the woman we draw is not married is, by Rule C,

$$P(\text{not married}) = 1 - P(\text{married})$$
$$= 1 - 0.555 = 0.445$$

That is, if 55.5% are married, then the remaining 44.5% are not married. Rule D says that you can also find the probability that a woman is not married by adding the probabilities of the three distinct ways of being not married, as we did earlier. This gives the same result.

EXAMPLE 7.11   Rolling two dice

Rolling two dice is a common way to lose money in casinos. There are 36 possible outcomes when we roll two dice and record the up-faces in order (first die, second die). Figure 7.3 displays these outcomes. What probabilities should we assign?

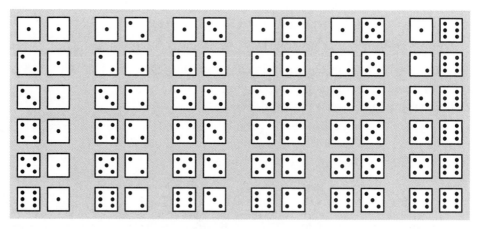

FIGURE 7.3 The 36 possible outcomes from rolling two dice.

Casino dice are carefully made. Their spots are not hollowed out, which would give the faces different weights, but are filled with white plastic of the same density as the red plastic of the body. For casino dice it is reasonable to assign the same probability to each of the 36 outcomes in Figure 7.3. Because these 36 probabilities must have sum 1 (Rule B), each outcome must have probability 1/36.

We are interested in the sum of the spots on the up-faces of the dice. What is the probability that this sum is 5? The event "roll a 5" contains four outcomes, and its probability is the sum of the probabilities of these outcomes:

$$P(\text{roll a 5}) = P\left(\boxed{\cdot} \; \boxed{\vcenter{\hbox{:}}} \right) + P\left(\boxed{\therefore} \; \boxed{\because} \right) + P\left(\boxed{\because} \; \boxed{\therefore} \right) + P\left(\boxed{\vcenter{\hbox{:}}} \; \boxed{\cdot} \right)$$

$$= \frac{1}{36} + \frac{1}{36} + \frac{1}{36} + \frac{1}{36}$$

$$= \frac{4}{36} = 0.111$$

The rules tell us only what probability models *make sense.* They don't tell us whether the probabilities are *correct,* that is, whether they describe what actually happens in the long run. The probabilities in Example 7.11 are correct for casino dice. Inexpensive dice with hollowed-out spots are not balanced, and this probability model does not describe their behavior.

What about personal probabilities? As Tom said, "Well, those are my personal probabilities, so they can be anything I want them to be." We can't say that personal probabilities that don't obey Rules A and B are wrong, but we can say that they are **incoherent.** That is, they don't go together in a way that makes sense. So we usually insist that personal probabilities for all the outcomes of a random phenomenon obey Rules A and B. That is, the same rules govern both kinds of probability.

EXERCISES

7.19 Causes of death Government data assign a single cause for each death that occurs in the United States. The data show that the probability is 0.45 that a randomly chosen death was due to cardiovascular (mainly heart) disease, and 0.23 that it was due to cancer. What is the probability that a death was due either to cardiovascular disease or to cancer? What is the probability that the death was due to some other cause?

7.20 Do husbands do their share? An opinion poll interviewed a random sample of 1025 women. The married women in the sample were asked whether their husbands did their fair share of household chores. Here are the results:

Outcome	Probability
Does more than his fair share	0.12
Does his fair share	0.61
Does less than his fair share	?

These proportions are probabilities for the random phenomenon of choosing a married woman at random and asking her opinion.

(a) What must be the probability that the woman chosen says that her husband does less than his fair share? Why?

(b) The event "I think my husband does at least his fair share" contains the first two outcomes. What is its probability?

7.21 Who gets the interview? Federal Reserve Board Chairman Alan Greenspan is not noted for giving many interviews. As a result, on February 4, 1991, *USA Today* asked Danny Sheridan (author of the *National Sports Newsletter*) to assign probabilities to who would be granted Greenspan's next interview. Here are Sheridan's probabilities:

Washington Post	0.50	*LA Times*	0.46	*TIME* magazine	0.42
Newsweek	0.42	*Business Week*	0.33	*USA Today*	0.25
Chicago Tribune	0.25	*Boston Globe*	0.25	*Fortune* magazine	0.14
Associated Press	0.04	*Wall Street Week*	0.01	All others	0.00001

Does Sheridan give a legitimate assignment of probabilities? Explain why or why not.

7.22 Tetrahedral dice Psychologists sometimes use tetrahedral dice to study our intuition about chance behavior. A tetrahedron (Figure 7.4) is a pyramid with four faces, each a triangle with all sides equal in length. Label the four faces of a tetrahedral die with 1, 2, 3, and 4 spots. Give a probability model for rolling such a die and recording the number of spots on the down-face. Explain why you think your model is at least close to correct.

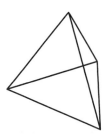

FIGURE 7.4 A tetrahedron. Exercises 7.22 and 7.23 concern dice with this shape.

7.23 More tetrahedral dice Tetrahedral dice are described in Exercise 7.22. Give a probability model for rolling two such dice. That is, write down all possible outcomes and give a probability to each. (Example 7.11 and Figure 7.3 may help you.) What is the probability that the sum of the down-faces is 5?

Probability models for sampling

Choosing a random sample from a population and calculating a statistic such as the sample proportion is certainly a random phenomenon. The *distribution* of the statistic tells us what values it can take and how often it takes those values. That sounds a lot like a probability model.

EXAMPLE 7.12 A sampling distribution

Take a simple random sample (SRS) of 1523 adults. Ask each whether they bought a lottery ticket in the last 12 months. The proportion who say "Yes,"

$$\hat{p} = \frac{\text{number who say "Yes"}}{1523}$$

is the sample proportion $\hat{p}$. Do this 1000 times and collect the 1000 sample proportions from the 1000 samples. The histogram in Figure 7.5 shows the distribution of 1000 sample proportions when the truth about the population is that 60% have bought lottery tickets.

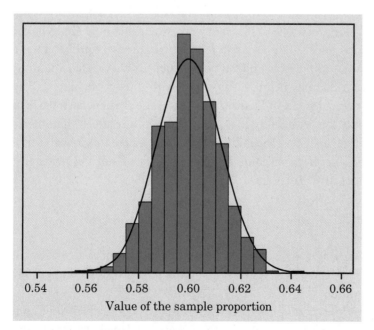

FIGURE 7.5 The sampling distribution of a sample proportion $\hat{p}$ from SRSs of size 1523 drawn from a population in which 60% of the members would give positive answers. The histogram shows the distribution from 1000 samples. The normal curve is the ideal pattern that describes the results of a very large number of samples.

The results of random sampling are of course random: we can't predict the outcome of one sample, but the figure shows that the outcomes of many samples have a regular pattern.

We have seen Figure 7.5 before, in Chapter 5. In fact, we saw the histogram part of this figure earlier, in Chapters 2 and 4. The repetition reminds us that the regular pattern of repeated random samples is one of the big ideas of statistics. The normal curve in the figure is a good approximation to the histogram. The histogram is the result of these particular 1000 SRSs. Think of the normal curve as the idealized pattern we would get if we kept on taking SRSs from this population forever. That's exactly the idea of probability—the pattern we would see in the very long run. *The normal curve assigns probabilities to the outcomes of random sampling.*

This normal curve has mean 0.6 and standard deviation about 0.0125. The "95" part of the 68–95–99.7 rule says that 95% of all samples will give a p falling within two standard deviations of the mean. That's within 0.025 of 0.6, or between 0.575 and 0.625. We now have more concise language for this fact: the *probability* is 0.95 that between 57.5% and 62.5% of the people in a sample will say "Yes." The word "probability" says we are talking about what would happen in the long run, in very many samples.

A statistic from a large sample has a great many possible values. Assigning a probability to each individual outcome worked well for 4 marital classes or 36 outcomes of rolling two dice but is awkward when there are thousands of possible outcomes. Example 7.12 uses a different approach: assign probabilities to intervals of outcomes by using areas under a normal density curve. Density curves have area 1 underneath them, which lines up nicely with total probability 1. The total area under the normal curve in Figure 7.5 is 1, and the area between 0.575 and 0.625 is 0.95, which is the probability that a sample gives a result in that interval. When a normal curve assigns probabilities, you can calculate probabilities from the 68–95–99.7 rule or from Table B of percentiles of normal distributions. These probabilities satisfy Rules A to D.

Sampling distribution

The **sampling distribution** of a statistic tells us what values the statistic takes in repeated samples from the same population and how often it takes those values.

We think of a sampling distribution as assigning probabilities to the values the statistic can take. Because there are usually many possible values, sampling distributions are often described by a density curve such as a normal curve.

EXAMPLE 7.13   Do you approve of gambling?

An opinion poll asks an SRS of 501 teens, "Generally speaking, do you approve of legal gambling or betting?" Suppose that in fact exactly 50% of all teens would say "Yes" if asked. (This is close to what polls show to be true.) The poll's statisticians tell us that the sample proportion who say "Yes" will vary in repeated samples according to a normal distribution with mean 0.5 and standard deviation about 0.022. This is the *sampling distribution* of the sample proportion $\hat{p}$.

The 68–95–99.7 rule says that the probability is 0.16 that the poll gets a sample in which fewer than 47.8% say "Yes." Figure 7.6 shows how to get this result from the normal curve of the sampling distribution.

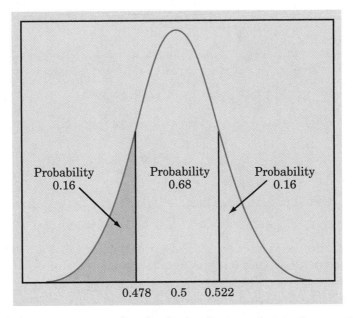

FIGURE 7.6 The normal sampling distribution for Example 7.13. Because 0.478 is one standard deviation below the mean, the area under the curve to the left of 0.478 is 0.16.

EXAMPLE 7.14   Using normal percentiles

What is the probability that the opinion poll in Example 7.13 will get a sample in which 52% or more say "Yes"? Because 0.52 is not 1, 2, or 3 standard deviations away from the mean, we can't use the 68–95–99.7 rule. We will use Table B of percentiles of normal distributions.

To use Table B, first turn the outcome $\hat{p} = 0.52$ into a standard score, z, by subtracting the mean of the distribution and dividing by its standard deviation:

$$z = \frac{0.52 - 0.5}{0.022} = 0.9$$

Now look in Table B. A standard score of 0.9 is the 81.59 percentile of a normal distribution. This means that the probability is 0.8159 that the poll gets a smaller result. By Rule C (or just the fact that the total area under the curve is 1), this leaves probability 0.1841 for outcomes 52% or more "Yes." Figure 7.7 shows the probabilities as areas under the normal curve.

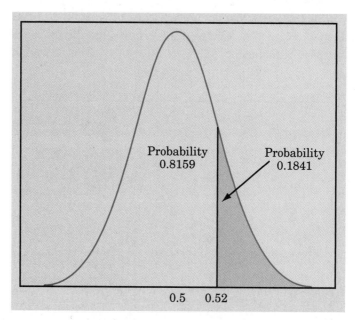

FIGURE 7.7 The normal sampling distribution for Example 7.14. The outcome 0.52 has standard score 0.9, so Table B tells us that the area under the curve to the left of 0.52 is 0.8159.

EXERCISES

7.24 Polling women, I Suppose that 47% of all adult women think they do not get enough time for themselves. An opinion poll interviews 1025 randomly chosen women and records the sample proportion who feel they don't get enough time for themselves. This statistic will vary from sample to sample if the poll is repeated. The sampling distribution is approximately normal with mean 0.47 and standard deviation about 0.016. Sketch this normal curve and use it to answer the following questions.

(a) The truth about the population is 0.47. In what range will the middle 95% of all sample results fall?

(b) What is the probability that the poll gets a sample in which fewer than 45.4% say they do not get enough time for themselves?

7.25 Polling women, II In the setting of Exercise 7.24, what is the probability of getting a sample in which more than 51% of the women think they do not get enough time for themselves? (Use Table B or your calculator.)

7.26 Applying to college You ask an SRS of 1500 college students whether they applied for admission to any other college. Suppose that in fact 35% of all college students applied to colleges besides the one they are attending. (That's close to the truth.) The sampling distribution of the proportion of your sample who say "Yes" is approximately normal with mean 0.35 and standard deviation 0.01. Sketch this normal curve and use it to answer these questions:
(a) Explain in simple language what the sampling distribution tells us about the results of our sample.
(b) What percent of many samples would have a $\hat{p}$ larger than 0.37? (Use the 68–95–99.7 rule.) Explain in simple language why this percent is the probability of an outcome larger than 0.37.
(c) What is the probability that your sample will have a $\hat{p}$ less than 0.33?
(d) Use Rule D: What is the probability that your sample result will be either less than 0.33 or greater than 0.35?

7.27 Do you jog? An opinion poll asks an SRS of 1500 adults, "Do you happen to jog?" Suppose (as is approximately correct) that the population proportion who jog is $p = 0.15$. In a large number of samples, the proportion p who answer "Yes" will be approximately normally distributed with mean 0.15 and standard deviation 0.009. Sketch this normal curve and use it to answer these questions:
(a) What percent of many samples will have a sample proportion who jog that is 0.15 or less? Explain clearly why this percent is the probability that $\hat{p}$ is 0.15 or less.
(b) What is the probability that $\hat{p}$ will take a value between 0.141 and 0.159? (Use the 68–95–99.7 rule.)
(c) Now use Rule C for probability: What is the probability that $\hat{p}$ does not lie between 0.141 and 0.159?

7.28 Credit card purchases Thirty percent of the customers at a hardware store pay using a credit card. Consider the following two probabilities.

Probability I: when 100 customers are selected at random, the probability that between 20 and 40 of them will use a credit card.

Probability II: when 10 customers are selected at random, the probability that between 2 and 4 of them will use a credit card.

Which probability is larger? Explain your reasoning.

APPLICATION 7.2 Zipf's Law

George Kingsley Zipf (1902–1950), an eccentric professor of linguistics (languages) at Harvard University, made a remarkable discovery about the frequency of English words in printed text. He started with the idea that certain familiar words like "the," "of," and "and" occurred very often and that other, not-so-familiar words occurred very rarely. One of the books he examined was *Ulysses,* a formidable tome containing 260,430 words. He made a list of the 29,899 different words and made a tally count of how often each of these words was used. Here is a list of the 20 most common words in *Ulysses,* and the number of times they occurred (the frequency):

Word	Rank, r	Frequency, f
the	1	14,877
of	2	7,786
and	3	7,170
a	4	6,396
to	5	4,907
in	6	4,884
he	7	4,001
his	8	3,326
that	9	3,082
I	10	2,653
with	11	2,506
it	12	2,350
was	13	2,125
on	14	2,095
for	15	1,972
you	16	1,894
her	17	1,775
him	18	1,460
is	19	1,346
all	20	1,311

Actually, Professor Zipf didn't do all of that tedious counting—he was independently wealthy and he hired graduate students to do the work for him.

1. Why are the number of words in *Ulysses* different from the number of different words?

2. Make a new column in the table above, and give it the title "14,877/r." In the new column, calculate the value of this fraction and write the answer in the row for that rank, r. For example, the first three entries would be 14,877, 7,438.5, and 4,959.

APPLICATION 7.2 Zipf's Law *(continued)*

3. Compare the frequency and the values of 14,877/*r*. Are they similar?

4. Make another new column (on the right) and name it "Probability." Calculate the probabilities of each of the 20 words. Enter these 20 probabilities in the extended table.

Zipf noticed that if we let *N* be the number of times the most common word appears (in *Ulysses,* that would be 14,877), then the second most common word will appear about *N*/2 times, the third most common word will appear *N*/3 times, and so forth. In general, the *r*th most common word will appear approximately *N*/*r* times. This is pretty amazing! And this is Zipf's Law, applied to the frequency of English words in text.

Since we've been preaching to always plot your data, let's do that. If you plot a scatterplot of the points in the form (*x*, *y*) = (rank, frequency) on the TI-83, the graph has the shape of one branch of a hyperbola. (See Figure 7.8(a)). The graph can be straightened out by taking the logarithm of both coordinates and plotting the transformed points in the form: log rank, log frequency. (See Figure 7.8(b)). We performed regression on the transformed points, and the least-squares line is also shown. The correlation, incidentally, is very high, −0.9866, indicating a very strong association between the transformed variables.

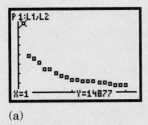

(a)

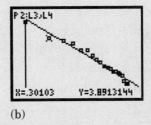

(b)

FIGURE 7.8(a) Plot of frequency of words versus rank.

FIGURE 7.8(b) Plot of log(frequency) versus log(rank). Notice the linear pattern.

Others have attempted to extend Zipf's Law to other relationships. Zipf himself claimed that the population of cities (and communities) as a function of rank obeys Zipf's Law. A very few cities have large populations, and many communities have very small populations. Other examples include a company's revenue as a function of rank and the amount of traffic a Web site experiences. A small number of Web sites are hit many times (www.google.com, for example), and many Web sites are rarely hit. In all of these examples, it helps to picture a curve like the one in Figure 7.8(a).

Zipf's Law works surprisingly well for other novels and for newspapers and magazines.

APPLICATION 7.2 Zipf's Law *(continued)*

5. Find a source of text from the popular culture (not a textbook) or the Internet. Project Gutenberg (www.promo.net/pg/) has lots of "ebooks" that you can download and then read. Use tally marks to determine the frequency of the most common words. Then calculate the probability of the word "the" in your text passage. How does it compare with the probability of "the" in *Ulysses*? (*Note:* You should know that you really need at least a 5000-word passage in order for Zipf's Law to kick in.)

STATISTICS IN SUMMARY

A **probability model** describes a random phenomenon by telling what outcomes are possible and how to assign probabilities to them. There are two simple ways to give a probability model. The first assigns a probability to each individual outcome. These probabilities must be numbers between 0 and 1 (Rule A) and they must add to exactly 1 (Rule B). The probability of an **event** is 1 minus the probability of its complement (Rule C). To find the probability of any event, add the probabilities of the outcomes that make up the event (Rule D).

The second kind of probability model assigns probabilities as areas under a **density curve,** such as a normal curve. The total probability is 1 because the total area under the curve is 1. This kind of probability model is often used to describe the **sampling distribution** of a statistic. This is the pattern of values of the statistic in many samples from the same population.

All legitimate assignments of probability, whether data-based or personal, obey the same **probability rules.** So the mathematics of probability is always the same.

SECTION 7.2 EXERCISES

7.29 Models, legitimate and not A bridge deck contains 52 cards, four of each of the 13 face values ace, king, queen, jack, ten, nine, . . . , two. You deal a single card from such a deck and record the face value of the card dealt. Give an assignment of probabilities to these outcomes that should be correct if the deck is thoroughly shuffled. Give a second assignment of probabilities that is legitimate (that is, obeys the rules of probability) but differs from your first choice. Then give a third assignment of probabilities that is not legitimate, and explain what is wrong with this choice.

7.30 Birth order A couple plan to have three children. There are 8 possible arrangements of girls and boys. For example, GGB means the first two children are girls and the third child is a boy. All 8 arrangements are (approximately) equally likely.

(a) Write down all 8 arrangements of the sexes of three children. What is the probability of any one of these arrangements?

(b) What is the probability that the couple's children are 2 girls and 1 boy?

7.31 16 boys! On December 26, 1991, the Associated Press reported an unusual occurrence at the Canton-Postdam Hospital in Postdam, New York: between December 15 and December 21, all 16 newborns at the hospital were boys.

(a) Would the births of 16 consecutive boys at a different hospital in another small town in the United States be just as newsworthy? Do you think this Associated Press report is very unusual? Explain.

(b) It was also noted that on the two days before the report appeared, all 5 newborns at the Canton-Postdam Hospital had been girls. Does this show the law of averages in action? Explain.

7.32 The girls Suppose that about 48% of all infants are girls. The maternity ward of a large hospital handles 1000 deliveries per year. A smaller hospital in the same city records about 50 deliveries per year. At which hospital is it more likely that between 46% and 50% of the babies born there this year will be girls? Explain briefly.

7.33 How much education? *The Statistical Abstract of the United States* gives this distribution of education for a randomly chosen American over 25 years old:

Education:	Less than high school	High school graduate	College, no bachelor's	Bachelor's degree	Advanced degree
Probability:	0.17	0.34	0.25	0.16	0.08

(a) How do you know that this is a legitimate probability model?

(b) What is the probability that a randomly chosen person over age 25 has at least a high school education?

(c) What is the probability that a randomly chosen person over age 25 has at least a bachelor's degree?

7.34 Spinner probabilities A spinner can land in eight possible positions, numbered 1 through 8; all eight numbers have the same probability of coming up. What can you say about the probabilities of the eight outcomes?

7.35 Selecting a jury A county selects people for jury duty at random from a list of one million registered voters, which happens to be evenly divided in terms of gender (that is, exactly 50% of the county's registered voters are women). Which of the following is true and which is false?

(a) Out of the next 2000 names drawn, it is likely that exactly 1000 will be women.

(b) Out of the next 2000 names drawn, it is likely that the percent of women will be around 50%, to within a percent or two.

7.36 It's not easy being green Assume that 20% of the cars in Columbus, Ohio, are green. An observer stands at the intersection of High Street and Fifth Avenue for an hour and sees 898 cars traveling along High Street and 243 cars traveling along Fifth Avenue.
(a) In which direction is it more likely that more than 25% of the cars seen will be green? Explain.
(b) Suppose that you watch 20 cars in a row traveling along High Street at this intersection and not a single one is green. What is the probability that the next car will be green?
(c) In answering (a) and (b), what assumption have you made about the relationship between the colors of cars?

CHAPTER 7 REVIEW

Some phenomena are random. Although their individual outcomes are unpredictable, there is a regular pattern in the long run. Using gambling devices and taking an SRS are examples of random phenomena. Probability gives us a language to describe randomness. Random phenomena are not haphazard any more than random sampling is haphazard. Randomness is instead a kind of order, a long-run regularity as opposed to either chaos or a determinism that fixes events in advance. Section 7.1 discusses randomness and Section 7.2 presents facts about probability.

When randomness is present, probability answers the question "How often in the long run?" Some probability models assign probabilities to outcomes. Any such model must obey the rules of probability. Another kind of probability model uses a density curve such as a normal curve to assign probabilities as areas under the curve. Personal probabilities express an individual's judgment of how likely some event is. Personal probabilities must follow the rules of probability if they are to be consistent with each other.

Here are the most important skills you should have acquired after studying Chapter 7.

A. RANDOMNESS AND PROBABILITY

1. Recognize that some phenomena are random. Probability describes the long-run regularity of random phenomena.

2. Understand the idea of the probability of an event as the proportion of times the event occurs in very many repetitions of a random phenomenon. Use the idea of probability as long-run proportion to think about probability.

3. Recognize that short runs of random phenomena do not display the regularity described by probability. Accept that randomness is unpredictable in the short run, and avoid seeking causal explanations for random occurrences.

B. PROBABILITY MODELS

1. Use basic probability facts to detect illegitimate assignments of probability: Any probability must be a number between 0 and 1, and the total probability assigned to all possible outcomes must be 1.

2. Use basic probability facts to find the probabilities of events that are formed from other events: The probability that an event does not occur is 1 minus its probability. If two events cannot occur at the same time, the probability that one or the other occurs is the sum of their individual probabilities.

3. When probabilities are assigned to individual outcomes, find the probability of an event by adding the probabilities of the outcomes that make it up.

4. When probabilities are assigned by a normal curve, find the probability of an event by finding an area under the curve.

CHAPTER 7 REVIEW EXERCISES

7.37 Will you have an accident? The probability that a randomly chosen driver will be involved in an accident in the next year is about 0.2. This is based on the proportion of millions of drivers who have accidents. "Accident" includes things like crumpling a fender in your own driveway, not just highway accidents.
(a) What do you think is your own probability of being in an accident in the next year? This is a personal probability.
(b) Give some reasons why your personal probability might be a more accurate prediction of your "true chance" of having an accident than the probability for a random driver.
(c) Almost everyone says their personal probability is lower than the random driver probability. Why do you think this is true?

7.38 Choosing at random Abby, Deborah, Mei-Ling, Sam, and Roberto work in a firm's public relations office. Their employer must choose two of them to attend a conference in Paris. To avoid unfairness, the choice will be made by drawing two names from a hat. (This is an SRS of size 2.)
(a) Write down all possible choices of two of the five names. These are the possible outcomes.
(b) The random drawing makes all outcomes equally likely. What is the probability of each outcome?
(c) What is the probability that Mei-Ling is chosen?

(d) What is the probability that neither of the two men (Sam and Roberto) is chosen?

7.39 We like opinion polls, I Are Americans interested in opinion polls about the major issues of the day? Suppose that 40% of all adults are very interested in such polls. (According to sample surveys that ask this question, 40% is about right.) A polling firm chooses an SRS of 1015 people. If they do this many times, the percent of the sample who say they are very interested will vary from sample to sample following a normal distribution with mean 40% and standard deviation 1.5%. Use the 68–95–99.7 rule to answer these questions.
(a) What is the probability that one such sample gives a result within $\pm 1.5\%$ of the truth about the population?
(b) What is the probability that one such sample gives a result within $\pm 3\%$ of the truth about the population?

7.40 We like opinion polls, II Use the information in Exercise 7.39 and Table B or your calculator to find the probability that one sample misses the truth about the population by 4% or more. (This is the probability that the sample result is either less than 36% or greater than 44%.)

7.41 Generating a sampling distribution Let us illustrate the idea of a sampling distribution in the case of a very small sample from a very small population. The population consists of the numbers 1, 2, 4, and 9. The parameter of interest is the mean of this population. The sample is an SRS of size $n = 3$ drawn from the population.
(a) Find the mean of the four numbers in the population. This is the population mean.
(b) Use your calculator to draw an SRS of size 3 from this population. Enter `randInt(1,4)` and press ENTER. Suppose you get 2. Your sample will then consist of the three numbers except for the second number: {1, 4, 9}. Write the three numbers in *your* sample, and calculate the mean of your sample. This statistic, $\bar{x}$, is an estimate of the population mean.

(c) Repeat step (b) 10 times, pressing ENTER each time to see which number you leave out of your sample. Make a histogram of your 10 values of $\bar{x}$. You are constructing the sampling distribution of $\bar{x}$. Is the center of your histogram close to the population mean you found in (a)?

7.42 What's the probability? Open your local telephone directory to any page in the residential listing. Look at the last four digits of each telephone number, the digits that specify an individual number within an exchange given by the first three digits. Note the first of these four digits in each of the first 100 telephone numbers on the page.

(a) How many of the digits were 1, 2, or 3? What is the approximate probability that the first of the four "individual digits" in a telephone number is 1, 2, or 3?
(b) If all 10 possible digits had the same probability, what would be the probability of getting a 1, 2, or 3? Based on your work in (a), do you think the first of the four "individual digits" in telephone numbers is equally likely to be any of the 10 possible digits?

7.43 Course grades Choose a student at random from all who took this statistics course in recent years. The probabilities for the student's grade are

Grade:	A	B	C	D	F
Probability:	0.2	0.3	0.3	0.1	?

(a) What must be the probability of getting an F?
(b) What's the probability of getting a C or better?
(c) What's the probability of not getting an A or an F?

7.44 An IQ test, I The Wechsler Adult Intelligence Scale (WAIS) is a common "IQ test" for adults. The distribution of WAIS scores for persons over 16 years of age is approximately normal with mean 100 and standard deviation 15. Use the 68–95–99.7 rule to answer these questions.
(a) What is the probability that a randomly chosen individual has a WAIS score of 115 or higher?
(b) In what range do the scores of the middle 95% of the adult population lie?

7.45 An IQ test, II Use the information in Exercise 7.44 and Table B or your calculator to find the probability that a randomly chosen person has a WAIS score of 112 or higher.

7.46 An IQ test, III How high must a person score on the WAIS test to be in the top 10% of all scores? Use the information in Exercise 7.44 and Table B or your calculator to answer this question.

Chapter 8

Simulation and Expected Values

8.1 Simulation
8.2 Expected Values

8.1 SIMULATION

Getting through tollbooths faster

We all know and hate the tollbooth lines at bridges, tunnels, and toll roads.[1] Statistics can't reduce the number of vehicles trying to get through, but it can help them get through faster. Suppose that we must design the toll plaza for a new tunnel. Traffic in both directions will pay at the same location. Drivers may use cash, credit cards, or an electronic pay-while-moving system. How many tollbooths do we need? Should each booth accept all types of payment, or should they specialize? What will happen in tourist season, when fewer drivers use the local electronic payment system and more pay cash? When traffic at a booth doubles, waiting times often quadruple or worse—will backups at peak times block the main road?

Settings like this are too complicated to just "think about." They aren't described by any set of mathematical equations that we can try to solve. What we do is roll out our random number generators and try to imitate the behavior of drivers. The formal term for this imitation is *simulation.* Cars and trucks arrive at random times, but we can see the probability distributions of times and vehicle types from past data. Drivers choose different payment systems, described by more probability distributions (different for local cars, tourist cars, and trucks). Different drivers are more or less aggressive in diving for the shortest line—another probability distribution. The time it takes to serve a driver at a booth is also random. The fellow with a $50 bill who wants a receipt takes longer.

After we have figured out all these distributions, we can watch the tollbooths run on the computer screen. The waiting times and lines of vehicles grow and shrink as the random flow of traffic arrives. Are lines often too long at rush hour in tourist season? What happens if we add a booth? What happens if we make that booth "cash only"? What happens if we make the approach longer so drivers have more time to see the shortest line? What happens if, alas, the amount of traffic doubles? Nothing good will happen, but perhaps we can prevent overflow from the cash lanes from blocking access to the electronic lanes as well. Simulation of complex interconnected systems like our toll plaza is a standard tool. The details are awful, but the ideas are pretty simple. The ideas behind simulation are our topic in this chapter.

ACTIVITY 8.1 Buffon's needle

Materials: 2-inch long matchstick or toothpick, felt-tipped pen, sheet of 1-inch grid paper.

In the early 1700s a French naturalist, Count Buffon (1707–1788), found an interesting way to find an approximate value for π that involved dropping a needle onto a ruled surface. The experiment was described by Buffon in 1733 in the *Proceedings of the Paris Academy of Sciences* and later reproduced with its solution in Buffon's book *Essai d'arithmetique morale,* published in 1777. The experiment is now referred to in the literature as the Buffon Needle Problem.

Here's how to replicate the experiment:

1. Use a felt-tipped pen to draw parallel horizontal lines on a sheet of 1-inch grid paper, marking every fourth line, to obtain lines 4 inches apart.

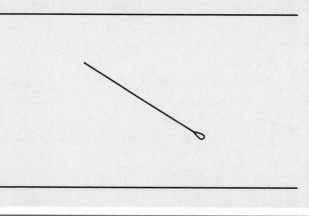

ACTIVITY 8.1 Buffon's needle *(continued)*

2. Cut a matchstick or a toothpick to a length of 2 inches.

3. Drop the stick onto the ruled paper, allowing it to fall anywhere on the sheet. Do this 50 times and record whether the stick touches one of the marked lines.

Hits a line:

Misses a line:

4. Combine your results with those of the other students in your class.

5. Compute the following fraction:

$$\frac{\text{total number of tosses}}{\text{number of times stick touches a line}}$$

6. Compare your answer from Step 5 with the value of $\pi = 3.14159\ldots$

NOTE: It is said that an Italian named Lazzerini tossed a needle 3408 times and got $\pi = 3.141592$, which is six-decimal-place accuracy.

CALCULATOR CORNER   Simulating the Buffon Needle Problem

- Begin by clearing lists 1 through 4.

  $\boxed{\text{STAT}}$ 4:ClrList L$_1$, L$_2$,L$_3$,L$_4$

  Make sure that under MODE, your calculator is set to *radians* instead of degrees.
- Start a counter.

  1→C $\boxed{\text{ENTER}}$
- Type in the following string of commands:

  1.5708rand(100)→L$_1$:2rand
  (100)→L$_2$:
  (L$_2$<cos L$_1$)→L$_3$:(100/sum(L$_3$))→
  L$_4$(C):C+1→C

- Press $\boxed{\text{ENTER}}$. This simulates dropping the needle 100 times on the ruled surface. The number that appears is the value of the counter, that is, the number of times you have simulated 100 repetitions. We will do 10 sets of 100 repetitions for a total of 1000 simulated needle drops.
- Continue to press $\boxed{\text{ENTER}}$. Wait a few moments for the count to appear. Continue to press $\boxed{\text{ENTER}}$ until the count is 11 (you have done 10 repetitions).

- Check your results. You can do this either by entering the command $\boxed{\text{STAT}}$ Calc 1:1-Var Stats(L$_4$) and looking at $\bar{x}$, or by entering the command sum(L$_4$)/10

- Now average all of the results from all of the students doing this simulation. Do you get a better approximation for π than you achieved in Activity 8.1?

Where do probabilities come from?

The probabilities of heads and tails in tossing a coin are very close to 1/2. In principle, these probabilities come from data on many coin tosses. Joe's personal probabilities for the winner of next year's Super Bowl come from Joe's own *individual judgment*. What about the probability that we get a run of three straight heads somewhere in 10 tosses of a coin? We can find this probability by *calculation from a model* that describes tossing coins. That is, once we have used data to give a probability model for the random fall of coins, we don't have to go back to the beginning every time we want the probability of a new event.

The big advantage of probability models is that they allow us to calculate the probabilities of complicated events starting from an assignment of probabilities to simple events like "heads on one toss." This is true whether the model reflects probabilities from data or from personal probabilities. Unfortunately, the math needed to do probability calculations is often tough. Technology rides to the rescue: once we have a probability model, we can use a calculator or computer to *simulate* many repetitions. This is easier than math and much faster than actually running many repetitions in the real world. You might compare finding probabilities by simulation to practicing flying in a computer-controlled flight simulator. Both kinds of simulation are in wide use. Both have similar drawbacks: they are only as good as the model you start with. Flight simulators use a software "model" of how an airplane reacts. Simulations of probabilities use a probability model. We set the model in motion by using our old friends the random digits from Table A.

Simulation

Using random digits from a table, a calculator, or computer software to imitate chance behavior is called **simulation**.

We look at simulation partly because it is how engineers designing a toll plaza, for example, really do find probabilities and partly because simulation forces us to think clearly about probability models. We'll do the hard part—setting up the model—and leave the easy part—telling a computer to do 10,000 repetitions—to those who really need the right probability at the end.

"We've done a computer simulation of your projected performance in five years. You're fired."

Simulation basics

Simulation is an effective tool for finding probabilities of complex events once we have a trustworthy probability model. We can use random digits to simulate many repetitions quickly. The proportion of repetitions on which an event occurs will eventually be close to its probability, so simulation can give good estimates of probabilities. The art of simulation is best learned from a series of examples.

EXAMPLE 8.1   Doing a simulation: Shaquille O'Neal at the free-throw line

Shaquille O'Neal, a "franchise" basketball player for the Los Angeles Lakers, has been one of the NBA all-stars since his freshman year in 1992–93. He has always had impressive statistics, and in 2000, he won all three MVP (Most Valuable Player) awards. At 300+ pounds and 7 feet 4 inches, he is a commanding presence in the paint. Still, he is not yet a "complete" player. Despite much practice and working with special coaches, he has been a 50% free-throw shooter for most of his career. Let's assume that every time Shaq steps up to the free-throw line, the probability that he will make the shot is 0.5. Suppose we want to know how likely he is to make at least 3 free throws in a row out of 10 attempts. Three or more baskets in a row is called a "run."

   Step 1: Give a probability model. Our model has two parts:
- Each free throw has probabilities 0.5 for a basket (a "hit") and 0.5 for a miss.
- Free throws are *independent* of each other. That is, knowing the outcome of one free throw does not change the probabilities for the outcomes of any other free throw. (Studies of consecutive free throws have shown this to be the case.)

Step 2: Assign digits to represent outcomes. Digits in Table A of random digits will stand for the outcomes, in a way that matches the probabilities from Step 1. We know that each digit in Table A has probability 0.1 of being any one of 0, 1, 2, 3, 4, 5, 6, 7, 8, or 9, and that successive digits in the table are independent. Here is one assignment of digits for free throws:

- One digit simulates one free throw.
- Digits 0 to 4 represent a hit; digits 5 to 9 represent a miss.

Successive digits in the table simulate independent tosses.

Step 3: Simulate many repetitions. Ten digits simulate 10 tosses, so looking at 10 consecutive digits in Table A simulates one repetition. Read many groups of 10 digits from the table to simulate many repetitions. Be sure to keep track of whether or not the event we want (a run of 3 hits) occurs on each repetition.

Here are the first three repetitions, starting at line 101 in Table A. We have underlined all runs of 3 or more hits.

	Repetition 1	Repetition 2
Digits	1 9 2 2 3 9 5 0 3 4	0 5 7 5 6 2 8 7 1 3
Hit/miss	HMHHHMMHHH	HMMMMHMMHH
Run of 3?	Yes	No

	Repetition 3
Digits	9 6 4 0 9 1 2 5 3 1
Hit/miss	MMHHMHHMHH
Run of 3?	No

Continuing in Table A, we did 25 repetitions. In 11 of them Shaq made at least 3 consecutive free throws. So we estimate the probability of a run by the proportion

$$\text{estimated probability} = \frac{11}{25} = 0.44$$

Of course, 25 repetitions are not enough to be confident that our estimate is accurate. But now that we understand how to do the simulation, we can tell a computer to do many thousands of repetitions. A long simulation (or hard mathematics) finds that the true probability is about 0.508.

Really random digits

For purists, the RAND Corporation long ago published a book titled *One Million Random Digits*. The book lists 1,000,000 digits that were produced by a very elaborate physical randomization and really are random. An employee of RAND once told me that this is not the most boring book that RAND has ever published.

Once you have gained some experience in simulation, setting up the probability model (Step 1) is usually the hardest part of the process. The model in Example 8.1 is typical of many probability problems because it consists of *independent trials* (the free throws) that all have the *same possible outcomes* with the *same probabilities*. Shooting 10 free throws and observing the number of heads in 10 tosses of a coin have similar models and are simulated in much the same way. Independence simplifies our work because it allows us to simulate each of the 10 free throws in exactly the same way.

Independence

Two random phenomena are **independent** if knowing the outcome of one does not change the probabilities for outcomes of the other.

Independence, like all aspects of probability, can be verified only by observing many repetitions. It is plausible that repeated tosses of a coin are independent (the coin has no memory), and observation shows that they are. It seems less plausible that successive shots by a basketball player are independent, but observation shows that they are at least very close to independent. Step 2 (assigning digits) rests on the properties of the random digit table. Here are some examples of this step.

EXAMPLE 8.2   Assigning digits for simulation

(a) Choose a family at random from a group of which 80% have an Internet connection. One digit simulates one family:

0, 1, 2, 3, 4, 5, 6, 7 = Internet connection

8, 9 = no connection

(b) Choose one family at random from a group of which 83% have an Internet connection. Now two digits simulate one family:

00, 01, 02, . . . , 82 = Internet connection

83, 84, 85, . . . , 99 = no connection

We assigned 83 of the 100 two-digit pairs to "Internet connection" to get probability 0.83. Representing "Internet connection" by 01, 02, . . . , 83 would also be correct.

Was he good or was he lucky?

When a baseball player hits .300, everyone applauds. A .300 hitter gets a hit in 30% of times at bat. Could a .300 year just be luck? Typical major leaguers bat about 500 times a season and hit about .260. A hitter's successive tries seem to be independent. From this model, we can calculate or simulate the probability of hitting .300. It is about 0.025. Out of 100 run-of-the-mill major league hitters, two or three each year will bat .300 because they were lucky.

(c) Choose one family at random from a group of which 50% have a dial-up connection, 20% have no connection, and 30% have a cable or DSL connection. There are now three possible outcomes, but the principle is the same. One digit simulates one family:

$$0, 1, 2, 3, 4 = \text{dial-up}$$
$$5, 6 = \text{no connection}$$
$$7, 8, 9 = \text{cable or DSL}$$

"I've had it! Simulated wood, simulated leather, simulated coffee, and now simulated probabilities!"

EXERCISES

8.1 Which party does it better? An opinion poll selects adult Americans at random and asks them, "Which political party, Democratic or Republican, do you think is more concerned with providing health care for the poor?" Explain carefully how you would assign digits from Table A to simulate the response of one person in each of the following situations.

(a) Of all adult Americans, 50% would choose the Democrats and 50% the Republicans.

(b) Of all adult Americans, 60% would choose the Democrats and 40% the Republicans.

(c) Of all adult Americans, 40% would choose the Democrats, 40% would choose the Republicans, and 20% would be undecided.

(d) Of all adult Americans, 53% would choose the Democrats and 47% the Republicans.

8.2 A small opinion poll Suppose that 80% of a school's students favor abolishing exams. You ask 10 students chosen at random. What is the probability that all 10 favor abolishing exams?

(a) Give a probability model for asking 10 students independently of each other.

(b) Assign digits to represent the answers "Yes" and "No."

(c) Simulate 25 repetitions, starting at line 129 of Table A. What is your estimate of the probability?

8.3 Basic simulation Use Table A to simulate the responses of 10 independently chosen adults in each of the four situations of Exercise 8.1.

(a) For situation (a), use line 110.

(b) For situation (b), use line 111.

(c) For situation (c), use line 112.

(d) For situation (d), use line 113.

8.4 Simulating an opinion poll A recent opinion poll showed that about 75% of Americans would donate their organs upon death. Suppose that this is exactly true. Choosing an American at random then has probability 0.75 of getting one who would donate his or her organs. If we interview Americans separately, we can assume that their responses are independent. We want to know the probability that a simple random sample of 100 Americans will contain at least 80 who would donate their organs. Explain carefully how to do this simulation and simulate *one* repetition of the poll using line 112 of Table A. How many of the 100 would donate their organs? Explain how you would estimate the probability by simulating many repetitions.

8.5 "Ask Marilyn" revisited In Activity 7.2 (page 424), we used the random digit table to simulate two children. How would you use the random number generator on the TI-83 to simulate

(a) two children where the older child is a boy?

(b) two children where one is a boy?

More elaborate simulations

The building and simulation of random models is a powerful tool of contemporary science, yet it can be understood without advanced mathematics. What is more, doing a few simulations will increase your understanding of probability more than many pages of our prose. Having in mind these two goals of understanding simulation for itself and understanding simulation to understand probability, let us look at two more-elaborate examples. The first still has independent trials, but there is no longer a fixed number of trials as there was when we simulated 10 of Shaq's free throws.

EXAMPLE 8.3   First ace

Begin with a standard deck of playing cards. Shuffle the cards thoroughly, and then draw a card. Replace the card in the deck, shuffle the deck, and draw a card again. Continue until you draw an ace, or until you draw 10 cards, whichever comes first. What is the probability of drawing an ace in 10 draws?

Step 1: Give a probability model.

• The probability of drawing an ace on any given draw is $4/52 = 1/13$.

• Draws are independent.

Step 2: Assign digits to represent outcomes.

01 ↔ ace
02 ↔ 2
03 ↔ 3
...
09 ↔ 9
10 ↔ 10
11 ↔ jack
12 ↔ queen
13 ↔ king
14–99 and 00 ↔ discard

Step 3. To simulate one repetition, read pairs of digits from Table A until you draw an ace or you draw 10 cards. The number of pairs needed to simulate one repetition depends on how quickly you draw an ace. Since so many numbers will have to be discarded (14 to 99 and 00), using Table A would be very time-consuming and inefficient. Instead, we will use the random number generator on the TI-83. If we "seed" the random number generator, then every calculator will give the same results. Do this:

$$123 \rightarrow \boxed{\text{MATH}}/\text{PRB}/1:\text{rand} \boxed{\text{ENTER}}$$

Next, generate 10 random numbers between 1 and 13 and store them in list L_1:

$$\boxed{\text{MATH}}/\text{PRB}/5:\text{randInt}(1,13) \boxed{\rightarrow} \boxed{\text{2nd}}/L_1$$

The 10 simulated cards are

$$10 \quad 10 \quad 7 \quad 6 \quad 12 \quad 7 \quad 9 \quad 8 \quad 7 \quad 10$$

none of which is an ace. Repetition 2 produces these cards:

$$\boxed{\text{2nd}} \, [\text{QUIT}] \, \boxed{\text{2nd}} \, [\text{ENTRY}] \, \boxed{\text{ENTER}}$$

$$7 \quad 5 \quad 6 \quad 6 \quad 13 \quad 5 \quad 2 \quad 12 \quad 13 \quad 7$$

Still no ace. Continuing in this fashion, we drew an ace 11 times in 25 repetitions. Our estimate of the probability of drawing an ace is

$$\text{estimated probability} = \frac{11}{25} = 0.44$$

Some mathematics shows that if our probability model is correct, the true probability of obtaining an ace in 10 draws is 0.551.

Now seed your calculator's random number generator with a unique seed. Use the last four digits of your telephone number or Social Security number. For example, $1089 \rightarrow$ rand $\boxed{\text{ENTER}}$.

DILBERT reprinted by permission of United Feature Syndicate, Inc.

Our final example has stages that are *not independent.* That is, the probabilities at one stage depend on the outcome of the preceding stage.

EXAMPLE 8.4   A kidney transplant

Morris's kidneys have failed and he is awaiting a kidney transplant. His doctor gives him this information for patients in his condition: 90% survive the transplant, and 10% die. The transplant succeeds in 60% of those who survive, and the other 40% must return to kidney dialysis. The proportions who survive five years are 70% for those with a new kidney and 50% for those who return to dialysis. Morris wants to know the probability that he will survive for five years.

Step 1. The **tree diagram** in Figure 8.1 (next page) organizes this information to give a probability model in graphical form. The tree shows the three stages and the possible outcomes and probabilities at each stage. Each path through the tree leads to either survival for five years or to death in less than five years. To simulate Morris's fate, we must simulate each of the three stages. The probabilities at Stage 3 depend on the outcome of Stage 2.

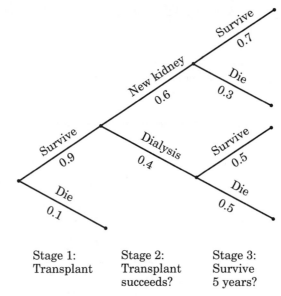

Stage 1:   Stage 2:   Stage 3:
Transplant   Transplant   Survive
                     succeeds?   5 years?

FIGURE 8.1 A tree diagram for the probability model of Example 8.4. Each branch point starts a new stage, with outcomes and their probabilities written on the branches. One repetition of the model follows the tree to one of its endpoints.

Step 2. Here is our assignment of digits to outcomes:

Stage 1:

$$0 = die$$

$$1, 2, 3, 4, 5, 6, 7, 8, 9 = survive$$

Stage 2:

$$0, 1, 2, 3, 4, 5 = transplant\ succeeds$$

$$6, 7, 8, 9 = return\ to\ dialysis$$

Stage 3 with kidney:

$$0, 1, 2, 3, 4, 5, 6 = survive\ five\ years$$

$$7, 8, 9 = fail\ to\ survive$$

Stage 3 with dialysis:

$$0, 1, 2, 3, 4 = survive\ for\ five\ years$$

$$5, 6, 7, 8, 9 = fail\ to\ survive$$

The assignment of digits at Stage 3 depends on the outcome of Stage 2. That's lack of independence.

Step 3. Here are simulations of several repetitions, each arranged vertically. We used random digits from line 110 of Table A.

	Repetition 1	Repetition 2	Repetition 3	Repetition 4
Stage 1	3, survive	4, survive	8, survive	9, survive
Stage 2	8, dialysis	8, dialysis	7, dialysis	1, kidney
Stage 3	4, survive	4, survive	8, die	8, die

Morris survives five years in 2 of our 4 repetitions. Now that we understand how to arrange the simulation, we should turn it over to a computer to do many repetitions. From a long simulation or from mathematics, we find that Morris has probability 0.558 of living for five years.

CALCULATOR CORNER   Simulations with technology

As we've seen, the TI-83 graphing calculator is a terrific device for performing simulations. It can generate random decimals between 0 and 1 using the rand function (rand is found under MATH /PRB/ 1:rand). By multiplying by a constant, you can generate a random real number between 0 and any positive number.

- The command 2rand, for example, generates a random decimal number between 0 and 2 (multiply the endpoints by 2). By adding a positive or negative number, you can slide or translate the interval left or right.
- The command −1+2rand, for example, selects a random number between −1 and 1.

```
rand
            .8158583325
2rand
           1.900888814
-1+2rand
          -.0698249852
```

- The command randInt(1,6,10) (MATH /PRB/5:randInt) generates 10 random *integers* between 1 and 6. So this command can be used to simulate rolling a single die 10 times.
- The command randNorm(100,15,2) generates two random numbers from the normal distribution with mean 100 and standard deviation 15 (randNorm(is found by MATH /PRB/6:randNorm).

CALCULATOR CORNER   Simulations with technology *(continued)*

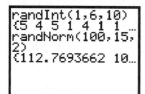

A second nice feature for simulations is that the calculator *lists* provide ample places to store numbers and results of calculator operations. You can simulate rolling a die 100 times, for example, and store the results in a list. Or you can simulate randomly selecting 150 student IQs from a school population having mean IQ 100 and standard deviation 15 and then store these 150 values in a second list.

```
randInt(1,6,100)
→L₁
{4 1 3 4 4 2 6 ...
randNorm(100,15,
150)→L₂
{129.7143582 10...
```

L1	L2	L3	1
4	129.71	------	
1	103.37		
3	102.16		
4	98.546		
4	89.598		
2	104.6		
6	112.29		
L1(1)=4			

A third feature of the calculator is that it knows how to do Boolean algebra (named

after the nineteenth-century British mathematician George Boole). With this capability, the calculator can look at an expression and evaluate it as TRUE or FALSE. If the expression is true, the calculator returns a 1. If the expression is false, it returns a 0. Here are some examples:

```
1+1=2
                      1
1+2=4
                      0
median({1,2,3})=
2
                      1
```

This is an extremely powerful capability and one that we can exploit when we perform simulations on the calculator.

One last technique is essential for what we will do. The calculator allows you to put two or more commands on the same "line" in the home screen by separating the commands with a colon, for example, $\text{rand}(100) \rightarrow L_1 : \text{rand}(100) \rightarrow L_2$. Of course, you could write a calculator program to do these tasks, and you would put these commands on separate lines in the program. We will avoid the program route for now.

EXERCISES

8.6 Generating random decimal numbers Describe how you would use your TI-83 to generate the following random numbers.

(a) A random decimal number between 0 and π.

(b) A random decimal number between -1 and 0.

(c) A random decimal number between $-\pi$ and π.

(d) Here's a way to check your work. For each of your answers in (a) to (c) above, generate 100 such numbers and store them in list L_1 on your calculator. Then calculate 1-variable statistics, and check the mean and five-number summary. Do the mean and five-number summary numbers seem reasonable for your 100 data values?

8.7 Generating random numbers Suppose you want to simulate flipping a fair coin using your calculator. Explain how you would do the following.
(a) Generate either a 1 (representing a head) or a 0 (representing a tail).
(b) Using your answer to (a), simulate flipping a fair coin 100 times, and store these values in a list.
(c) Calculate 1-variable statistics for the data you obtained in (b). What are the mean and the median? Are they close to 0.5? (Note that *none* of the data values are near 0.5.)

8.8 Random selection of subjects
(a) Explain how you would use either Table A or your calculator to randomly select 50 subjects for an experiment from 200 volunteers. The volunteers have been arranged alphabetically by last name and then numbered from 1 to 200.
(b) Use your method from (a) to randomly select the 50 subjects. Store the 50 numbers in a list in your calculator. Suppose that volunteers 1 through 100 are female, and 101 to 200 are male. In a second list, use the Boolean algebra technique described in the Calculator Corner to determine how many of the 50 selected subjects are female.

8.9 Duplicate birthdays There are 30 students in Austin's Pre-calculus class, all unrelated. He wants to bet you $1 that at least two of the students were born on the same day of the year (same month and day, but not necessarily the same year). Should you take the bet? Before you answer, you decide to conduct a simulation to help see how likely that event would be. Here is the probability model:
• The birth date of a randomly chosen person is equally likely to be any of the 365 dates of the year.
• The birth dates of different people in the room are independent.
(a) Describe how you could use either Table A or your calculator to simulate 30 random birthdays from the 365 possible days.
(b) Carry out your simulation. Perform at least 20 repetitions. Based on your results, should you take Austin's bet?

8.10 Blood types Choose a person at random and record his or her blood type. Here are the probabilities for each blood type:

Blood type :	Type O	Type A	Type B	Type AB
Probability :	0.4	0.3	0.2	?

(a) What must be the probability that a randomly chosen person has type AB blood?

(b) To simulate the blood types of randomly chosen people, how would you assign digits to represent the four types?

(c) People with type B blood can receive blood donations from people with either type B or type O blood. Laura has type B blood. What is the probability that 2 or more of Laura's 6 close friends can donate blood to her? Simulate 10 repetitions and estimate this probability. (Your estimate from just 10 repetitions isn't reliable, but you have shown in principal how to find the probability.)

APPLICATION 8.1 The Rock-Paper-Scissors Game

Surely everyone has played the rock-paper-scissors game when they were younger. Two players face each other, and at the count of 3, make a fist (rock), an extended hand, palm side up (paper), or a "V" with the index and middle fingers (scissors). The winner is determined by these rules: rock smashes scissors; paper covers rock; and scissors cut paper. If the two players choose different objects, then there is a winner. But both could choose the same object (for example, rock), in which case there is no winner. One of the questions that we might explore is "What is the probability that a game will result in a winner?"

1. Make a thoughtful guess to answer this question.

2. Play the game many times, and find the long-term relative frequency of a winner. Students form three-member teams and play the game a total of 30 times. Two students play 10 games while the third student records whether there is a winner. The recorder should make a tally sheet like the one below and use tally marks to record the results.

Winner:

No winner:

After each set of 10 games, switch players and recorder, so that every student plays 20 games and records 10 games. Calculate the relative frequency of winners:

$$\frac{\text{number of winners}}{\text{number of games}}$$

Then accumulate your results with the other groups, and obtain a better estimate of the probability.

3. In this step, we will use the calculator to simulate 100 games, count the number of winners, and estimate the probability of a winner. In the home screen, clear lists L_1 through L_4, and then enter the following string of commands:

```
randInt(1,3,100)→L₁:randInt(1,3,100)→L₂:(L₁≠L₂)→L₃:
sum(L₃)/100 ENTER
```

APPLICATION 8.1   The Rock-Paper-Scissors Game *(continued)*

The first command, $\mathtt{randInt(1,3,100) \to L_1}$, randomly selects a choice (rock, paper, scissors) for the first player, records that choice, does it 100 times (games), and assigns those 100 choices to list L_1. The second command does the same thing for the second player. The third command, $\mathtt{(L_1 \neq L_2) \to L_3}$, looks at the first number in L_1 and the first number in L_2 and uses Boolean algebra to determine if the choices are different. If the choices are different, then there is a winner, and a "1" is recorded in the first entry in L_3. If there is no winner ($L_1 = L_2$), then a "0" is recorded. The calculator goes down lists L_1 and L_2 and records a 1 for each winner and a 0 for no winner. The last command, $\mathtt{sum(L_3)/100}$, adds all the 1s in L_3 and divides that sum by the number of games played. Record the results you get for simulating these 100 games.

4. Press the ENTER key to simulate another 100 games. Record the results. Continue to press ENTER until each group has 10 estimates of the probability of a winner. Each group should report their 10 estimates, and the results from all groups should be recorded on the chalkboard or on a transparency.

5. Make a histogram of the results from all of the groups. Estimate the center of the distribution of estimates. Describe the shape, center, and spread of this distribution. From this histogram, what is your best guess for the probability of a winner? How does this final estimate compare with the results you got by hand in Step 2?

EXPLORING THE WEB

Professor George Reese has a nice Web page dedicated to the Buffon Needle Problem:
www.mste.uiuc.edu/reese/buffon/buffon.html.
    The viewer can run the simulation using a Java applet. After specifying the number of tosses, you can see the needles land on the paper and also see the resulting approximation of π. Reese also has a downloadable Macintosh program that does the simulation and a solution by mathematics.

Another Buffon Needle Web site you can visit is the work of Dr. Michael J. Hurben, a physicist who likes to dabble in mathematics and probability:
www.angelfire.com/wa/hurben/buff.html. Hurben has an applet that allows up to 100,000 needle drops. It includes a graph that shows that as the number of drops increases, the estimate get closer and closer to the value of π.

There are lots of sites devoted to Shaquille O'Neal, such as www.nba.com/playerfile/shaquille_oneal/.

The popular Microsoft Flight Simulator, described at www.microsoft.com/games/fs2002/, allows one to experience the sensation of flying an airplane.

STATISTICS IN SUMMARY

We can use random digits to **simulate** random outcomes if we know the probabilities of the outcomes. Use the fact that each random digit has probability 0.1 of taking any one of the 10 possible digits and that all digits in the random number table are **independent** of each other. To simulate more complicated random phenomena, string together simulations of each stage. A common situation is several independent trials with the same possible outcomes and probabilities on each trial. Think of several tosses of a coin or several rolls of a die. Other simulations may require varying numbers of trials or different probabilities at each stage or may have stages that are not independent, so that the probabilities at some stage depend on the outcome of earlier stages. The key to successful simulation is thinking carefully about the **probability model.**

SECTION 8.1 EXERCISES

8.11 Class rank Choose a college student at random and ask his or her class rank in high school. Probabilities for the outcomes are

Class rank:	Top 10%	Top quarter but not top 10%	Top half but not top quarter	Bottom half
Probability:	0.3	0.3	0.3	?

(a) What must be the probability that a randomly chosen student was in the bottom half of his or her high school class?
(b) To simulate the class standing of randomly chosen students, how would you assign digits to represent the four possible outcomes listed?

8.12 Course grades Choose a student at random from all who took Algebra II at Lee-Davis High School in recent years. The probabilities for the student's grade are

Grade:	A	B	C	D or F
Probability:	0.2	0.3	0.3	?

(a) What must be the probability of getting a D or an F?
(b) To simulate the grades of randomly chosen students, how would you assign digits to represent the four possible outcomes listed?
(c) How would you use the TI-83 to randomly select a student's grade?

8.13 More on class rank In Exercise 8.11 you explained how to simulate the high school class rank of a randomly chosen college student. The Random Foundation decides to offer 8 randomly chosen students full college scholarships. What is the probability that no more than 3 of the 8 students chosen are in the bottom half of their high school class? Simulate 10 repetitions of the foundation's choices to estimate this probability.

8.14 More on course grades In Exercise 8.12 you explained how to simulate the grade of a randomly chosen student in an Algebra II course. Five students on the same floor of a dormitory are taking this course. They don't study together, so their grades are independent. Use simulation to estimate the probability that all five get a C or better in the course. (Simulate 20 repetitions.)

8.15 The Asian stochastic beetle We can use simulation to examine the fate of populations of living creatures. Consider the Asian stochastic beetle.[2] Females of this insect have the following pattern of reproduction:

- 20% of females die without female offspring, 30% have 1 female offspring, and 50% have 2 female offspring.

- Different females reproduce independently.

What will happen to the population of Asian stochastic beetles: Will they increase rapidly, barely hold their own, or die out? It's enough to look at only the female beetles, as long as there are some males around.
(a) Assign digits to simulate the offspring of one female beetle.
(b) Make a tree diagram for the female descendants of one beetle through three generations. The second generation, for example, can consist of 0, 1, or 2 females. If it has 0, we stop. Otherwise, we simulate the offspring of each second-generation female. What are the possible numbers of beetles after three generations?
(c) Use line 105 of Table A to simulate the offspring of 5 beetles to the third generation. How many descendants does each have after three generations? Does it appear that the beetle population will grow?

8.16 The rock-paper-scissors game
(a) Use the rules of probability to calculate the probability of a winner in any one game. (*Hint:* Begin by listing all possible outcomes for the game.)
(b) Use your calculator and the methods of Application 8.1 (page 460) to approximate the probability that it takes no more than two games to observe a winner.

8.2 EXPECTED VALUES

The house edge, and the real house edge

If you gamble, you care about how often you will win. The probability of winning tells you what proportion of a large number of bets will be winners. You care even more about how much you will win, because winning a lot is better than winning a little. If you play a game with a 50% chance of winning $1, in the long run you will win half the time and average 50 cents winnings per bet. A 10% chance of winning $100 does better—now you win only one time in 10 in the long run, but you average $10 winnings per bet. Those "average winnings per bet" numbers are *expected values.* Like probabilities, expected values only tell us what will happen on the average in very many bets.

It's no secret that the expected values of games of chance are set so that the house wins on the average in the long run. In roulette, for example, the expected value of a $1 bet is $0.947. The house keeps the other 5.3 cents per dollar bet. Because the house plays hundreds of thousands of times each month, it will keep almost exactly 5.3 cents for every dollar bet.

What is a secret, at least to naive gamblers, is that in the real world a casino does much better than expected values suggest. In fact, casinos keep a bit over 20% of the money gamblers spend on roulette chips. That's because players who win keep on playing. Think of a player who gets back exactly 95% of each dollar bet. After one bet, he has 95 cents; after two bets, he has 95% of that, or 90.25 cents; after three bets, he has 95% of that, or 85.7 cents. The longer he keeps recycling his original dollar, the more of it the casino keeps. Real gamblers don't get a fixed percent back on each bet, but even the luckiest will lose his stake if he plays long enough. The casino keeps 5.3 cents of every dollar bet but 20 cents of every dollar that walks in the door.

That brings us to baccarat, the preferred game of the idle rich. Baccarat is a card game that requires no decisions and no skill, but it looks elegant. Some players bet $100,000 per hand. There are only a few high-stakes players, and they don't play many hands by the standards of more plebeian games like roulette. So "the long run" that the casino profits from is not as long in baccarat. Runs of luck average out over a month or so in roulette but not in baccarat. The quarterly profit statements of some casino companies have suffered from bad luck at the baccarat tables.[3]

ACTIVITY 8.2A Pop a balloon—take a chance!

As a promotion gimmick, a drug store decides to raffle off ice cream bars in the following way. They advertised, "Ice cream bar: 0 to 43 cents. Take a chance." To get one, the customer would pop a balloon and then read the price on a slip of paper that was inside the balloon. Of course, the customer didn't know what his chances were of getting a fairly low price for his ice cream bar. The manager of the drug store had marked prices on slips of paper as shown in the table.

Slips	Price	Probability
30	$0.43	
18	0.42	
20	0.41	
19	0.40	
15	0.34	
10	0.20	
5	0.09	
5	0.02	
5	0.01	
5	0.00	
Total 132		

For example, he made 30 slips with the price of 43 cents. There were 132 slips in all.

1. Suppose that the usual price of an ice cream bar is 40 cents. What are your chances of paying more than 40 cents?

2. What are your chances of getting a free ice cream bar?

3. What are your chances of paying less than 10 cents?

4. Complete the table by finding the probabilities for the remaining prices.

5. How much money will the store take in if all of the balloons are popped? How much would the store have made on these ice cream bars without the promotion? Comment on the difference.

Expected values

Gambling on chance outcomes goes back to ancient times. Both public and private lotteries were common in the early years of the United States. After disappearing for a century or so, government-run gambling reappeared in 1964, when New Hampshire caused a furor by introducing a lottery to raise public revenue without raising taxes. The furor subsided quickly as larger states adopted the idea. Thirty-seven states and all Canadian provinces now sponsor lotteries. State lotteries made gambling acceptable as entertainment. Some form of legal gambling is allowed in 48 of the 50 states. Over half of all adult Americans have gambled legally. They spend more betting than on spectator sports, video games, theme parks, and movie tickets combined. If you are going to bet, you should understand what makes a bet good or bad. As our introductory case study says, we care about how much we win as well as about our probability of winning.

EXAMPLE 8.5   The Tri-State Daily Numbers

Here is a simple lottery wager, the "Straight" from the Pick 3 game of the Tri-State Daily Numbers offered by New Hampshire, Maine, and Vermont. You pay $1 and choose a three-digit number. The state chooses a three-digit winning number at random and pays you $500 if your number is chosen. Because there are 1000 three-digit numbers, you have probability 1/1000 of winning. Here is the probability model for your winnings:

Outcome:	$0	$500
Probability:	0.999	0.001

What are your average winnings? The ordinary average of the two possible outcomes $0 and $500 is $250, but that makes no sense as the average winnings because $500 is much less likely than $0. In the long run you win $500 once in every 1000 bets and $0 on the remaining 999 of 1000 bets. (Of course, if you buy exactly 1000 Pick 3 tickets, there is no guarantee that you will win exactly once. Probabilities are only *long-run* proportions.) Your long-run average winnings from a ticket are

$$\$500\,\frac{1}{1000} + \$0\,\frac{999}{1000} = \$0.50$$

or 50 cents. You see that in the long run the state pays out half the money bet and keeps the other half.

Here is a general definition of the kind of "average outcome" we used to evaluate the bets in Example 8.5.

Expected value

The **expected value** of a random phenomenon that has numerical outcomes is found by multiplying each outcome by its probability and then summing over all possible outcomes.

In symbols, if the possible outcomes are $a_1, a_2, \ldots, a_k$ and their probabilities are $p_1, p_2, \ldots, p_k$, the expected value is

$$\text{expected value} = a_1 p_1 + a_2 p_2 + \ldots + a_k p_k$$

An expected value is an average of the possible outcomes, but it is not an ordinary average in which all outcomes get the same weight. Instead, each outcome is weighted by its probability, so that outcomes that occur more often get higher weights.

EXAMPLE 8.6   The Tri-State Daily Numbers, continued

The Straight wager in Example 8.5 pays off if you match the three-digit winning number exactly. You can choose instead to make a $1 StraightBox wager. You again choose a three-digit number, but you now have two ways to win. You win $292 if you exactly match the winning number, and you win $42 if your number has the same digits as the winning number, in any order. For example, if your number is 123, you win $292 if the winning number is 123 and $42 if the winning number is any of 132, 213, 231, 312, and 321. In the long run, you win $292 once every 1000 bets and $42 five times for every 1000 bets.

The probability model for the amount you win is

Outcome:	$0	$42	$292
Probability:	0.994	0.005	0.001

The expected value is

$$\text{expected value} = (\$0)(0.994) + (\$42)(0.005) + (\$292)(0.001) = \$0.502$$

We see that the StraightBox is a slightly better bet than the Straight bet, because the state pays out slightly more than half the money bet.

The Tri-State Daily Numbers is unusual among state lottery games in that it pays a fixed amount for each type of bet. Most states pay off on the "pari-mutuel" system. New Jersey's Pick 3 game is typical: the state pools the money bet and pays out half of it, equally divided among the winning tickets. You still have probability 1/1000 of winning a Straight bet, but the amount your number 123 wins depends both on how much was bet on Pick 3 that day and on how many other players chose the number 123. Without fixed amounts, we can't find the expected value of today's bet on 123, but one thing is constant: the state keeps half the money bet.

The idea of expected value as an average applies to random outcomes other than games of chance. It is used, for example, to describe the uncertain return from buying stocks or building a new factory. Here is a different example.

EXAMPLE 8.7   How many vehicles per household?

What is the average number of motor vehicles in American households? The Census Bureau tells us that the distribution of vehicles per household (as of 1997) is as follows:

Number of vehicles:	0	1	2	3	4	5
Proportion of households:	0.04	0.25	0.45	0.18	0.06	0.02

This is a probability model for choosing a household at random and counting its vehicles. (We ignored the very few households with more than 5 vehicles.) The expected value for this model is the average number of vehicles per household. This average is

$$\text{expected value} = (0)(0.04) + (1)(0.25) + (2)(0.45) + (3)(0.18) +$$
$$(4)(0.06) + (5)(0.02) = 2.03$$

EXERCISES

8.17 Pop the balloon, I In Activity 8.2A (page 465) what is the *expected price* of an ice cream bar when you pop a balloon?

8.18 Pop the balloon, II Make a table of numbers of slips and prices so that the store neither makes money nor loses money. Have at least 100 slips.

8.19 The numbers racket Pick 3 lotteries (Example 8.5) copy the numbers racket, an illegal gambling operation common in the poorer areas of large cities. One version works as follows. You choose any one of the 1000 three-digit numbers 000 to 999 and pay your local numbers runner $1 to enter your bet. Each day, one three-digit number is chosen at random and pays off $600. What is the expected value of a bet on the numbers? Is the numbers racket more or less favorable to gamblers than the Pick 3 game in Example 8.5?

8.20 Pick 4 The Tri-State Daily Numbers Pick 4 is much like the Pick 3 game of Example 8.5. You pay $1 and pick a four-digit number. The state chooses a four-digit number at random and pays you $5000 if your number is chosen. What are the expected winnings from a $1 Pick 4 wager?

8.21 More Pick 4 Just as with Pick 3 (Example 8.6), you can make more elaborate bets in Pick 4. In the $1 StraightBox bet, if you choose 1234 you win $2604 if the randomly chosen winning number is 1234, and you win $104 if the winning number has the digits 1, 2, 3, and 4 in any other order. What is the expected amount you win?

The law of large numbers

The definition of "expected value" says that it is an average of the possible outcomes, but an average in which outcomes with higher probability count more. We argued that the expected value is also the average outcome in another sense: it represents the long-run average we will actually see if we repeat a bet many times or choose many households at random. This is more than intuition. Mathematicians can prove, starting from just the basic rules of probability, that the expected value calculated from a probability model really is the "long-run average." This famous fact is called the *law of large numbers.*

Rigging the lottery

We have all seen the televised live lottery drawings, in which numbered balls bubble about and are randomly popped out by air pressure. How might we rig such a drawing? In 1980, the Pennsylvania lottery was in fact rigged by the smiling host and several stagehands. They injected paint into all balls bearing 8 of the 10 digits. This weighed them down and guaranteed that all three balls for the winning number would have the remaining 2 digits. The perps then bet on all combinations of these digits. When 6-6-6 popped out, they won $1.2 million. Yes, they were caught.

> ### The law of large numbers
>
> If a random phenomenon with numerical outcomes is repeated many times independently, the mean of the actually observed outcomes approaches the expected value.

The law of large numbers is closely related to the idea of probability. In many independent repetitions, the proportion of each possible outcome will be close to its probability, and the average outcome obtained will be close to the expected value. These facts express the long-run regularity of chance events. They are the true version of the "law of averages."

The law of large numbers explains why gambling, which is a recreation or an addiction for individuals, is a business for a casino. The "house" in a gambling operation is not gambling at all. The average winnings of a large number of customers will be quite close to the expected value. The house has calculated the expected value ahead of time and knows what its take will be in the long run. There is no need to load the dice or stack the cards to guarantee a profit. Casinos concentrate on inexpensive entertainment and cheap bus trips to keep the customers flowing in. If enough bets are placed, the law of large numbers guarantees the house a profit. Life insurance companies operate much like casinos—they bet that the people who buy insurance will not die. Some do die, of course, but the insurance company knows the probabilities and relies on the law of large numbers to predict the average amount it will have to pay out. Then the company sets its premiums high enough to guarantee a profit.

The following applet lets you explore the expected value when you roll different numbers of dice. Go to the *Statistics through Applications* Web site, at www.whfreeman.com/sta, and look at the *Expected Value* applet.

ACTIVITY 8.2B The *Expected Value* applet

Directions: Use the "Fewer dice" and "More dice" buttons to fix the number of dice to be rolled with each trial. Next, enter the number

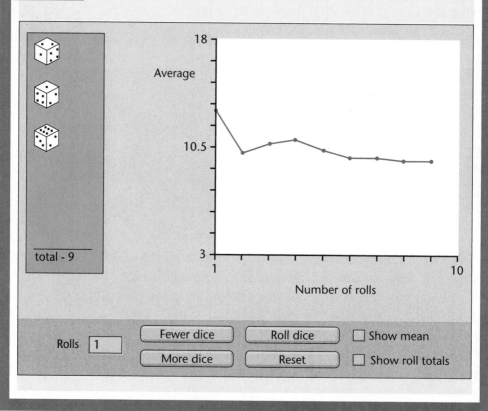

ACTIVITY 8.2B The *Expected Value* applet *(continued)*

of rolls (100 maximum) and then click the "Roll dice" button to have the applet roll the dice and plot the accumulating average of the roll totals.

Contemplate:

1. When you roll just 1 die, what values can you get? Are they all equally likely?

2. When you roll 10 dice, what values can you get for the roll total? Are they all equally likely?

3. Which process is likely to get close to its mean value in fewer trials: (a) the accumulating averages of rolling 1 die at a time or (b) the accumulating averages of rolling 10 dice at a time?

Verify: Test your answers to the three questions above through repeated use of the applet. Did the individual roll totals and the accumulating averages behave as you thought they would?

APPLET TIP: Click the "Show roll totals" button to see the results of each individual roll (shown as blue dots).

Experiment:

1. Have the applet roll 1 die 100 times and observe the behavior of the accumulating-average plot. How far from the long-run mean was the average after 10 rolls? After 20 rolls? After 50 rolls? After 100 rolls? Explain the pattern.

2. Reset the applet and roll 10 dice 100 times. Describe the pattern you see in the accumulating average. How is this pattern similar to the pattern for rolling 1 die? How does it differ?

3. The average of the numbers from 1 to 6 is $(1 + 2 + 3 + 4 + 5 + 6)/6 = 3.5$. Set the applet to roll 1 die and check the "Show mean" box to see that the expected value for rolling a single die is 3.5. Now alternately click the "More dice" button and the "Roll dice" button while you watch the behavior of the expected value for the sum of the rolls as more dice are added. What is the pattern?

4. What is the smallest total you can get with one roll and what is the largest total you can get? (*Hint:* The answer depends on how many dice you are rolling.) Regardless of the number of dice rolled, the vertical axis on the accumulating-average plot is marked off in tenths of the whole range of possible values. Have the applet roll 1 die 20 times and count how many times the individual roll total falls within the first tick mark to either side of the mean. Have the applet make 20 replications of rolling 2 dice at a time, then

ACTIVITY 8.2B The *Expected Value* applet *(continued)*

3 dice at a time, then 4, and so on—each time checking how many of the individual roll totals fall within one tick mark of the mean value. Describe and explain the pattern you observe.

Thinking about expected values

As with probability, it is worth exploring a few fine points about expected values and the law of large numbers.

How large is a large number? The law of large numbers says that the actual average outcome of many trials gets closer to the expected value as more trials are made. It doesn't say how many trials are needed to guarantee an average outcome close to the expected value. That depends on the *variability* of the random outcomes.

The more variable the outcomes, the more trials are needed to ensure that the mean outcome is close to the expected value. In Activity 8.1 (Buffon Needle Problem), you discovered that the estimates of π were quite variable. Many trials (tens of thousands or more) are needed to ensure that the estimate of π is correct to two decimal places. Games of chance must be quite variable if they are to hold the interest of gamblers. Even a long evening in a casino has an unpredictable outcome. Gambles with extremely variable outcomes, like state lottos with their very large but very improbable jackpots, require impossibly large numbers of trials to ensure that the average outcome is close to the expected value. (The state doesn't rely on the law of large numbers—lotto payoffs, unlike casino games, use the pari-mutuel system.)

Though most forms of gambling are less variable than lotto, the practical answer to the applicability of the law of large numbers is usually that the house plays often enough to rely on it, but you don't. Much of the psychological allure of gambling is its unpredictability for the player. The business of gambling rests on the fact that the result is not unpredictable for the house.

Is there a winning system? Serious gamblers often follow a system of betting in which the amount bet on each play depends on the outcome of previous plays. You might, for example, double your bet on each spin of the roulette wheel until you win—or, of course, until your fortune is exhausted. Such a system tries to take advantage of the fact that you have a memory even though the roulette wheel does not. Can you beat the odds with a system? No. Mathematicians have established a stronger version of the law of large numbers that says that if you do not have an infinite fortune to gamble with, your average winnings (the expected value) remain the same as long as successive trials of the game (such as spins of the roulette wheel) are independent. Sorry.

Finding expected values by simulation

How can we calculate expected values in practice? You know the mathematical recipe, but that requires that you start with the probability of each outcome. Expected values too difficult to compute in this way can be found by simulation. The procedure is as before: give a probability model, use random digits to imitate it, and simulate many repetitions. By the law of large numbers, the average outcome of these repetitions will be close to the expected value.

High-tech gambling

There are more than 450,000 slot machines in the United States. Once upon a time, you put in a coin and pulled the lever to spin three wheels, each with 20 symbols. No longer. Now the machines are video games with flashy graphics and outcomes produced by random number generators. Machines can accept many coins at once, can pay off on a bewildering variety of outcomes, and can be networked to allow common jackpots. Gamblers still search for systems, but in the long run the random number generator guarantees the house its 5% profit.

EXAMPLE 8.8	**Drawing an ace**

You draw cards, one after the other, from a standard deck of 52 playing cards until you draw an ace. We count the number of cards drawn until an ace is drawn. On average, how many cards can you expect to turn over until you get an ace? That is, what is the expected value for the number of cards drawn? This is not an easy question to answer analytically. But with simulation methods, we can obtain a fairly good estimate, although it is a bit tedious.

Clearly, the cards drawn are not independent, since the first card drawn is no longer in the deck and so cannot be drawn again. If we use the random digit table, we would let the numbers 01 to 52 represent the 52 cards and arbitrarily let numbers 01, 02, 03, and 04 represent the four aces. Read pairs of digits from the table, discarding numbers in the range 53 to 99 and 00. Because about half of the pairs of numbers would be thrown out, we will use the calculator rather than Table A. So that you can check our work, seed the calculator's random digit generator, and then randomly generate numbers in the range 1 to 52. (That way, we will all get the same results and the process will still be random.) As soon as a 1, 2, 3, or 4 is observed, stop and record the number of cards drawn.

1234→rand ENTER and then randInt(1,52) ENTER

If the calculator generates a duplicate number, it is ignored, and we draw again.

Here are the cards drawn in the first 6 repetitions and the counts of number of cards drawn until an ace is observed:

Rep. no.	Cards drawn	Count
1	2	1
2	10, 5, 2	10
3	20, 42, 23, 52, 31, 19, 1	
4	9, 8, 45, 46, 33, 43, 17, 47, 22, 14, 11, 5, 16, 32, 20, 31, 30, 34, 38, 51, 49, 13, 3	23
5	8, 1	2
6	41, 18, 33, 28, 44, 24, 27, 47, 14, 31, 29, 34, 32, 16, 17, 7, 42, 3	18

The average number of cards drawn for these 6 repetitions is 9. But notice that there appears to be a lot of variability. We got lucky on the first repetition and drew an ace on the first card. But it took 23 cards in the fourth repetition to draw an ace. Because of the large variability, it will take a large number of repetitions until we have some confidence in our estimate. Math or a long simulation shows that the actual expected value is 10.6. We were lucky to get a result as close to 10.6 as we did after only 6 repetitions.

Before you continue, seed your calculator's random number generator with a unique number, such as the last four digits of your telephone number or Social Security number. For example, $1089 \rightarrow \text{rand}$.

EXERCISES

8.22 Birthdays revisited On average, how many different people can you expect to check before you find two with the same birthday? To simulate birthdays, let each three-digit group in Table A stand for one person's birth date. That is, 001 is January 1, and 365 is December 31. Ignore leap years and skip groups that don't label birth dates. Use your calculator to randomly select a line (101 to 150) in Table A to simulate birthdays of randomly chosen people until you hit the same date a second time.
(a) How many people did you look at to find two with the same birthday?
(b) Combine your results with those from other members of the class and average the results. This is an estimate of the expected value.

8.23 It pays to eat cereal In the mid-1980s, Cheerios cereal boxes displayed a dollar bill on the front of the box and a cartoon character who said, "Free $1 bill in every 20th box." Mae Lee wants to know how many boxes of Cheerios she can expect to buy in order to get one of the "free" dollar bills.
(a) Describe how you would establish a correspondence between boxes of cereal that do and do not contain a dollar bill and numbers from Table A.
(b) Describe how you could use your calculator to simulate buying boxes of Cheerios and checking for dollar bills.

(c) Carry out a simulation to answer Mae Lee's question. Perform enough repetitions so that when you pool your results with your classmates, you will have at least 25 repetitions (don't forget that the more repetitions, the better the results).

8.24 A risky investment Picard Partners is planning a major investment. The amount of profit X is uncertain, but a probabilistic estimate gives the following distribution (in millions of dollars):

Profit:	1	1.5	2	4	10
Probability:	0.1	0.2	0.4	0.2	0.1

What is the expected value of the profit?

8.25 One die What is the expected number of spots observed in rolling a carefully balanced die once?

8.26 Poker Deal a five-card poker hand from a shuffled deck. The probabilities of several types of hand are approximately as follows:

Hand:	Worthless	One pair	Two pairs	Better hands
Probability:	0.50	0.42	0.05	?

(a) What must be the probability of getting a hand better than two pairs?
(b) What is the expected number of hands a player is dealt before the first hand better than one pair appears? Explain how you would use simulation to answer this question, then simulate just two repetitions.

APPLICATION 8.2 Benford's Law

1. Using the newspaper, an almanac, the *Statistical Abstract of the United States,* or similar text material, locate a compact collection of numbers such as closing stock prices or sports statistics. Select a block of at least 100 separate numbers. Copy the table below onto your paper. Then use tally marks to record the frequency of each first digit, from 1 to 9. (Note: For decimals, like 0.00305, you would take the first nonzero digit.) Calculate the relative frequency for each digit.

Digit	Tally marks	Counts	Relative frequency
1			
2			
3			
4			
5			
6			
7			
8			
9			

You might think that each of the digits from 1 to 9 would occur equally often in the distribution of first digits. Were your results uniformly distributed? Scientist Simon Newcomb (1835–1909) discovered in 1881 that the digits are not uniformly distributed. In an article that year in the *American Journal of Mathematics,* Newcomb conjectured that with naturally occurring data, you are most likely to observe a 1 as the leading digit, and that each digit from 2 to 9 is less likely than its predecessor to occur.[4] In fact, he conjectured that many (but not all) such data have the following distribution:

Digit	Formula	Probability
1	$\log(1 + 1/1) = \log(2)$	0.301
2	$\log(1 + 1/2) = \log(1.5)$	0.176
3	$\log(1 + 1/3) = \log(1.333333)$	0.125
4	$\log(1 + 1/4) = \log(1.25)$	0.097
5	$\log(1 + 1/5) = \log(1.2)$	0.079
6	$\log(1 + 1/6) = \log(1.166666)$	0.067
7	$\log(1 + 1/7) = \log(1.142857)$	0.058
8	$\log(1 + 1/8) = \log(1.125)$	0.051
9	$\log(1 + 1/9) = \log(1.111111)$	0.046

APPLICATION 8.2 Benford's Law *(continued)*

Figure 8.2 is a histogram of the probabilities for each first digit.

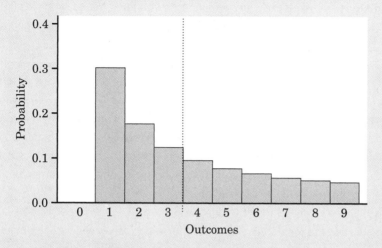

FIGURE 8.2 Digits between 1 and 9 chosen from records that obey Benford's Law.

Notice that we can't locate the mean of the right-skewed distribution by eye—calculation is needed. The mean is shown in Figure 8.2 as a dotted line.

2. Compare your results with the theoretical probability for each digit. Construct a bar graph that shows two bars for each digit: one bar for the probability and a bar to the right of it for the relative frequency from your data. Use color or shading to distinguish the two bars. This is a good way to visually compare how well your results agree with the conjectured relative frequencies.

Newcomb got scant notice for a result that is truly quite remarkable. Almost 60 years later, in 1938, Frank Benford, a physicist working for General Electric, noticed that the early pages in a table of logarithms were dirty and torn, while pages in the back of the book were cleaner and showed little evidence of use. Apparently unaware of Newcomb's paper, Benford examined a large collection of data that were not produced randomly, like lottery results, or that were not constrained by certain factors. He published his findings and his explanation, which he called Benford's Law, in another journal.[5] Here is a portion of Benford's first-digit table showing relative frequencies:

	1	2	3	4	5	6	7	8	9
Reader's Digest	33.4	18.5	12.4	7.5	7.1	6.5	5.5	4.9	4.2
Population	33.9	20.4	14.2	8.1	7.2	6.2	4.1	3.7	2.2
American League	32.7	17.6	12.6	9.8	7.4	6.4	4.9	5.6	3.0
Rivers, area	31.0	16.4	10.7	11.3	7.2	8.6	5.5	4.2	5.1

APPLICATION 8.2 Benford's Law *(continued)*

3. Calculate the average first digit (expected value) for data that obey Benford's Law.

In 1992, a mathematically inclined accountant named Mark Nigrini published his PhD dissertation, in which he showed that Benford's Law applies to accounting summaries and balance sheets of companies. He observed that data such as sales figures, expense claims, insurance claims, and tax returns follow Benford's Law. Nigrini has become somewhat of a celebrity as an expert in applying Benford's Law to ferret out fraud and detect "cooked books," and he has a substantial side business serving as a forensic accountant.[6] Benford's Law is almost certainly being applied by auditors looking for financial misdeeds in the sudden demise of several large corporations, such as Enron Energy Corporation and Arthur Anderson Consulting. In a relatively short time, Benford's Law has become an important tool for uncovering fraud committed by top managers in publicly held companies.[7]

In 1998, Professor Ted Hill of Georgia Institute of Technology published a comprehensive article in the journal *American Scientist* in which he reviewed the mathematical foundations and applications of the first-digit phenomenon.[8] In addition to its value as an accounting tool, other applications of Benford's Law include tasks as diverse as determining whether data from drug trials have been faked and methods to efficiently organize storage space on computer hard drives.

4. Explain clearly why the distribution of first digits in the blocks of 5 digits in Table A would *not* obey Benford's Law.

EXPLORING THE WEB

The debate over legalized gambling continues. For the case against, visit the National Coalition against Legalized Gambling, www.ncalg.org. For the defense by the casino industry, visit the American Gaming Association, www.americangaming.org. State lotteries make their case via the National Association of State and Provincial Lotteries, www.naspl.org. The National Indian Gaming Association, www.indiangaming.org, is more assertive; click on "Myths," for example. The report of a commission established by Congress to study the impact of gambling is at www.ngisc.gov. You'll find lots of facts and figures at all of these sites.

A search on Benford's Law will provide numerous hits. Look for an article by *SCIENCE News* math columnist Ivars Peterson and another by popular math writer and Temple University professor John Allen Paulos.

STATISTICS IN SUMMARY

Probability tells us how often (in the very long run) a random phenomenon takes each of its possible outcomes. When the outcomes are numbers, as in games of chance, we are also interested in the long-run average outcome. The **law of large numbers** says that the mean outcome in many repetitions eventually gets close to the **expected value.** The expected value is found as an average of all the possible outcomes, each weighted by its probability. If you don't know the outcome probabilities, you can estimate the expected value (along with the probabilities) by simulation.

Cooking the books

Professor Nigrini, now at Southern Methodist University, once gave his accounting students an assignment to go through some real sales figures from a real company. One of the students used the figures from his brother-in-law's hardware store. The sales figures were astonishing: 93% began with the digit 1, none of them began with the digits between 2 and 7, four began with the digit 8, and 21 began with the digit 9. According to Benford's Law, the brother-in-law was cooking the books—and Benford's Law was right!

SECTION 8.2 EXERCISES

8.27 Gambling in ancient Rome Tossing four astragali was the most popular game of chance in Roman times. Many throws of a present-day sheep's astragalus show that the approximate probability distribution for the four sides of the bone that can land uppermost are

Outcome	Probability
Narrow flat side of bone	1/10
Broad concave side of bone	4/10
Broad convex side of bone	4/10
Narrow hollow side of bone	1/10

The best throw of four astragali was the "Venus," when all the four uppermost sides were different.[9]

(a) Explain how to simulate the throw of a single astragalus. Then explain how to simulate throwing four astragali independently of each other.

(b) Simulate 25 throws of four astragali. Estimate the probability of throwing a "Venus." Be sure to say what part of Table A you used.

8.28 Course grades, I The distribution of grades in a large statistics course is as follows:

Grade:	A	B	C	D	F
Probability:	0.2	0.3	0.3	0.1	0.1

To calculate student grade point averages, grades are expressed in a numerical scale with A = 4, B = 3, and so on down to F = 0.

(a) Find the expected value. This is the average grade in this course.

(b) Explain how to simulate choosing students at random and recording their grades. Simulate 50 students and find the mean of their 50 grades. Compare this estimate of the expected value with the exact expected value from (a). (The law of large numbers says that the estimate will be very accurate if we simulate a very large number of students.)

8.29 Course grades, II If you choose 5 students at random from all those who have taken the course described in Exercise 8.28, what is the probability that all the students chosen got a C or better? Simulate 10 repetitions of this random choosing and use your results to estimate the probability. (Your estimate from only 10 repetitions isn't reliable, but if you can do 10, you could do 10,000.)

8.30 Course grades, III Choose a student at random from the course of Exercise 8.28 and observe what grade that student earns (A = 4, B = 3, C = 2, D = 1, F = 0). The expected grade is not one of the 5 grades possible for one student. Explain why your result nevertheless makes sense as an expected value.

8.31 Benford's Law, I There are lots of ways Benford's Law could be used to detect fraud in financial papers. Investigators could count the proportion of times the first digit is 5 or more, for example, and compare that result with what Benford's Law says it should be. Define the events A = {first digit is 1} and B = {first digit is 5 or greater}.

(a) Use probability rule C (page 427) to find the probability that event A does not happen (that is, that the first digit is not 1).

(b) Find the probability that event B happens, using Benford's distribution. (See Application 8.2 on page 476.)

(c) Find the probability that the first digit is odd.

8.32 Benford's Law, II This exercise extends Exercise 8.31. Define the event C = {the first digit is odd}.

(a) Find the probability that event B or event C happens.

(b) Is the probability that event B or event C happens equal to the sum of their individual probabilities? Explain.

CHAPTER 8 REVIEW

To calculate the probability of a complicated event without using complicated math, we use random digits to simulate many repetitions. You can also find expected values by simulation. Section 8.1 shows how to do simulations. First give a probability model for the outcomes, then assign random digits to imitate the assignment of probabilities. The table of random digits now imitates repetitions. Keep track of the proportion of repetitions on which an event occurs to estimate its probability. Keep track of the mean outcome to estimate an expected value.

Here are the most important skills you should have after studying Chapter 8.

8.1 SIMULATION

1. Specify simple probability models that assign probabilities to each of several stages and take the stages to be independent of each other.

2. Assign random digits to simulate such models.

3. Estimate either a probability or an expected value by repeating a simulation many times.

8.2 EXPECTED VALUES

1. Understand the idea of expected value as the average of numerical outcomes in very many repetitions of a random phenomenon.

2. Find the expected value from a probability model that lists all outcomes and their probabilities (when outcomes are numerical).

CHAPTER 8 REVIEW EXERCISES

8.33 Application 8.1 follow-up You want to determine the expected number of games of rock-paper-scissors until a winner is observed.
(a) Describe the steps involved in designing a simulation to answer this question, using either Table A or your calculator.
(b) Carry out a sufficient number of repetitions to obtain a reliable estimate for this expected value.

8.34 Two warning systems An airliner has two independent automatic systems that sound a warning if there is terrain ahead (which means that the airplane is about to fly into a mountain). Neither system is perfect. System A signals in time with probability 0.9. System B does so with probability 0.8. The pilots are alerted if either system works.
(a) Explain how to simulate the response of System A to terrain.
(b) Explain how to simulate the response of System B.
(c) Both systems are in operation simultaneously. Draw a tree diagram with System A as the first stage and System B as the second stage. Simulate 100 trials of the reaction to terrain ahead. Estimate the probability that a warning will sound. The probability for the combined system is higher than the probability for either A or B alone.

8.35 Playing craps The game of craps is played with two dice. The player rolls both dice and wins immediately if the outcome (the sum of the faces) is 7 or 11.

If the outcome is 2, 3, or 12, the player loses immediately. If he rolls any other outcome, he continues to throw the dice until he either wins by repeating his first outcome or loses by rolling a 7.

(a) Explain how to simulate the roll of a single fair die. (*Hint:* Just use digits 1 to 6 and ignore the others.) Then explain how to simulate a roll of two fair dice.

(b) Draw a tree diagram for one play of craps. In principle, a player could continue forever, but stop your diagram after four rolls of the dice. Use Table A, beginning at line 114, to simulate plays and estimate the probability that the player wins.

8.36 The airport van Your company operates a van service from the airport to downtown hotels. Each van carries 7 passengers. Many passengers who reserve seats don't show up—in fact, the probability is 0.25 that a randomly chosen passenger will fail to appear. Passengers are independent. If you allow 9 reservations for each van, what is the probability that more than 7 passengers will appear? Do a simulation to estimate this probability.

8.37 More on the airport van Let's continue the simulation of Exercise 8.36. You have a backup van, but it serves several stations. The probability that it is available to go to the airport at any one time is 0.6. You want to know the probability that some passengers with reservations will be left stranded because the first van is full and the backup van is not available. Draw a tree diagram with the first van (full or not) as the first stage and the backup (available or not) as the second stage. In Exercise 8.36 you simulated a number of repetitions of the first stage. Add simulations of the second stage whenever the first van is full. What is your estimate of the probability of stranded passengers?

8.38 Selling cars Bill sells new cars for a living. On a weekday afternoon, he will deal with 1 customer with probability 0.2, 2 customers with probability 0.4, and 3 customers with probability 0.4. Each customer has probability 0.2 of buying a car. Customers buy independently of each other.

(a) Describe how you would simulate the number of cars Bill sells in an afternoon. You must first simulate the number of customers, then simulate the buying decisions of 1, 2, or 3 customers.

(b) Simulate one afternoon to demonstrate your procedure.

8.39 Life insurance You might sell insurance to a 21-year-old friend. The probability that a man aged 21 will die in the next year is about 0.0015. You decide to charge $200 for a policy that will pay $100,000 if your friend dies.

(a) What is your expected profit on this policy?

(b) Although you expect to make a good profit, you would be foolish to sell your friend the policy. Why?

(c) A life insurance company that sells thousands of policies, on the other hand, would do very well selling policies on exactly these same terms. Explain why.

8.40 A multiple-choice exam Charlene takes a quiz with 10 multiple-choice questions, each with five answer choices. If she just guesses independently at each question, she has probability 0.2 of guessing right on each. Use simulation to estimate Charlene's expected number of correct answers. (Simulate 20 repetitions.)

8.41 A common expected value Here is a common setting that we simulated in Section 8.1: there are a fixed number of independent trials with the same two outcomes and the same probabilities on each trial. Tossing a coin, shooting basketball free throws, and drawing a card (as long as the card is replaced in the deck each time) are all examples of this setting. Call the outcomes "hit" and "miss." We can see what the expected number of hits should be. If Shaquille O'Neal shoots 10 free throws and has probability 1/2 of hitting each one, the expected number of hits is 1/2 of 10, or 5. By the same reasoning, if we have n trials with probability p of a hit on each trial, the expected number of hits is np. This fact can be proved mathematically. Can we verify it by simulation? Simulate 10 tosses of a fair coin 100 times. (To do this quickly, use the first 10 digits and the last 10 digits in each of the 50 rows of Table A, with odd digits meaning a head and even digits a tail.) What is the expected number of heads by the np formula? What is the mean number of heads in your 100 repetitions?

8.42 Your state's lottery Most states have a lotto game that offers large prizes for choosing (say) 6 out of 51 numbers. If your state has a lotto game, find out what percent of the money bet is returned to the bettors in the form of prizes. What percent of the money bet is used by the state to pay lottery expenses? What percent is net revenue to the state? For what purposes does the state use lottery revenue?

8.43 An expected rip-off? A "psychic" runs the following ad in a magazine:
Expecting a baby? Renowned psychic will tell you the sex of the unborn child from any photograph of the mother. Cost $10. Moneyback guarantee.
This may be a profitable con game. Suppose that the psychic simply replies "boy" to all inquiries. In the worst case, everyone who has a girl will ask for her money back. Find the expected value of the psychic's profit by filling in the table below.

Sex of child	Probability	The psychic's profit
Boy	0.51	
Girl	0.49	

8.44 Collecting cereal box prizes Several years ago, every box of Frosted Mini-Wheats contained a NASCAR driver's decal. There was a set of 6 decals, and consumers were encouraged to "collect all 6." The probability model is

• Every box of Frosted Mini-Wheats cereal contains one NASCAR driver's decal.

• A single decal is randomly selected and placed in the cereal, one per box.

• There are 6 decals in the set.

The question we want to answer is "How many boxes of Frosted Mini-Wheats cereal can we expect to buy in order to collect the complete set of 6 NASCAR decals?"
(a) Begin with a simpler problem. Suppose there are only 2 decals. How would you use the table of random digits to simulate this problem? How would you use a single die to simulate the problem? How would you use your calculator to simulate the problem? Choose one of these methods, and carry out 10 repetitions. Then average the 10 results to estimate the expected number of boxes needed to collect a set of 2 decals.
(b) Answer the questions in (a) if you assume a set of 3 decals, and then 4 decals.
(c) Now answer the original question for a set of 6 decals. Is there a pattern? If so, describe it.

Part IV

Inference

HOWARD HUGE*

"Howard's been like this ever since he heard that having a dog lowers your blood pressure."

© 2003 WM. Hoest Enterprises, Inc.

To *infer* is to draw a conclusion from evidence. Statistical *inference* draws a conclusion about a population from evidence provided by a sample. Drawing conclusions in mathematics is a matter of starting from a hypothesis and using logical argument to prove without doubt that the conclusion follows. Statistics isn't like that. Statistical conclusions are uncertain, because the sample isn't the entire population. So statistical inference has to state conclusions and also say how uncertain they are. We use the language of probability to express uncertainty. Because inference must both give conclusions and say how uncertain they are, it is the most technical part of statistics. Texts and courses intended to *train people to do* statistics spend most of their time on inference. Our aim in this book is to *help you understand* statistics, which takes less technique but often more thought. We will look only at a few basic techniques of inference. The techniques are simple, but the ideas are subtle, so prepare to think. To start, think about what you already know and don't be too impressed by elaborate statistical techniques: even the fanciest inference cannot remedy fundamental flaws such as voluntary response samples or uncontrolled experiments.

Chapter 9

Introduction to Inference

9.1 WHAT IS A CONFIDENCE INTERVAL?

Don't get mad

Know someone who is prone to anger? Nature has a way to calm such people: they get heart disease more often. Several observational studies have discovered a link between anger and heart disease. The best study looked at 12,986 people, both black and white, chosen at random from four communities.[1] When first examined, all subjects were between the ages of 45 and 64 and were free of heart disease. Let's focus on the 8474 people in this sample who had normal blood pressure.

A short psychological test, the Spielberger Trait Anger Scale, measured how easily each person became angry. There were 633 people in the high range of the anger scale, 4731 in the middle, and 3110 in the low range. Now follow these people forward in time for almost six years and compare the rate of heart disease in the high and low groups. There are some lurking variables: people in the high-anger group are somewhat more likely to be male, to have less than a high school education, and to be smokers and drinkers. After adjusting for these differences, high-anger people were 2.2 times as likely to get heart disease and 2.7 times as likely to have an acute heart attack as low-anger people.

That makes anger sound serious. But only 53 people in the low group and 27 in the high group got heart disease during the period of the study. We know that the numbers 2.2 and 2.7 won't be exactly right for the population of all people aged 45 to 64 years with normal blood pressure. News reports of the study cited those numbers, but the full report in the medical journal *Circulation* gave confidence intervals. With 95% confidence, high-anger people are between 1.36 and 3.55 times as likely to get heart disease as low-anger people. They are between 1.48 and 4.90 times as likely to have an acute heart attack. The intervals remind us that any statement we make about the population is uncertain because we have data only from a sample. For the sample, we can say "exactly 2.2 times as likely." For the whole population, the sample data allow us to say only "between 1.36 and 3.55 times as likely," and only with 95% confidence. To go behind the news to the real thing, in medicine and in other areas, we must speak the language of confidence intervals.

ACTIVITY 9.1A Estimating the proportion of brown M&M candies

Materials: For each student: 1.69-ounce bag of M&M's milk chocolate candies. For each group: quart plastic reclosable bag; standard ground-coffee measure or 1-ounce plastic medicine cup or 3-ounce paper bathroom cup.

The Mars candy company makes M&M's milk chocolate candies in brown, red, yellow, orange, green, and blue and randomly mixes the colors according to a specific percent distribution. In this Activity, we'll investigate the proportion of M&M's that are brown. To get an estimate of this proportion, we need to take a sample. But the sample proportion is subject to chance variation. One sample might include more than its fair share of browns, while other samples might not have very many browns at all. We can express this variability by constructing a confidence interval.

1. Your teacher will supply each student with a 1.69-ounce bag of M&M candies and will divide the class into small groups of 3 to 5 students for convenience. Combine the M&M's from all members of your group in a quart plastic reclosable bag.

2. Your group members should decide on the procedure you will use to draw a sample of 30 M&M's from the plastic bag. Describe the precautions you will take to avoid bias.

3. Have each student in your group take a sample of 30 M&M's using the agreed-upon procedure. Be sure to put the M&M's from each sample back in the plastic bag before the next sample is chosen. Were there any problems with your procedure that were difficult to resolve?

ACTIVITY 9.1A Estimating the proportion of brown M&M candies *(continued)*

4. Define the population of interest and the parameter p being investigated.

5. Each student: Count the number of brown M&M's that you obtained. Divide this number by the total number of M&M's in your sample to obtain the proportion, $\hat{p}$, of brown M&M's in your sample.

   The variability in your sample proportion of brown M&M's can be presented as

$$\text{sample proportion} = \text{population proportion} + \text{random variation}$$

where the random variation from sample to sample is approximately normally distributed. The estimated standard deviation for the proportion of brown M&M's is

$$\sqrt{\frac{\hat{p}\,(1-\hat{p})}{n}}$$

   Remember the question we started with: What is the true proportion of brown candies among all M&M's? The 68–95–99.7 rule tells us that the size of the random variation will be less than 1 standard deviation about 68% of the time, so the interval

$$\hat{p} \pm \text{estimated standard deviation of } \hat{p}$$

should contain the population proportion for about 68% of the samples. The expression

$$\hat{p} \pm \sqrt{\frac{\hat{p}\,(1-\hat{p})}{n}}$$

is an example of a *confidence interval*.

6. Use the expression above to calculate the left and right endpoints of your 68% confidence interval.

7. The teacher will draw a horizontal axis at the top of the chalkboard with a scale that goes from 0 to 0.6. Each student, in turn, should draw his or her confidence interval below the axis. According to the Department of Consumer Information of Mars, Inc., M&M's milk chocolate candies are randomly mixed to contain 30% brown candies.[2]

> **ACTIVITY 9.1A Estimating the proportion of brown M&M candies *(continued)***
>
> **8.** Did the 68% confidence interval that you constructed contain the $p = 0.30$ figure given by the company? What proportion of the confidence intervals that your classmates drew contained 0.30?
>
> **9.** The theory tells us that about 68% of the intervals should contain p. Compare the proportion of your classmates' confidence intervals that contained the true population proportion, $p = 0.30$, with the 68% figure that we would expect.

Estimating

Statistical inference draws conclusions about a population on the basis of data about a sample. One kind of conclusion answers questions like "What percent of employed women have a college degree?" or "What is the mean survival time for patients with this type of cancer?" These questions ask about a number (a percent, a mean) that describes a population. Numbers that describe a population are **parameters.** To estimate a population parameter, choose a sample from the population and use a **statistic,** a number calculated from the sample, as your estimate. Here's an example.

EXAMPLE 9.1 Teen smoking

How common is behavior that puts people at risk of lung cancer? The 2001 Youth Risk Behavioral Survey questioned a nationally representative sample of 12,960 students in grades 9–12. Of these, 3340 said they had smoked cigarettes at least one day in the past month. That's 25.77% of the sample.[3] This result may be biased by reluctance to tell the truth about any kind of risky behavior. For now, assume that the people in the sample told the truth. Based on these data, what can we say about the percent of all high schoolers who smoked?

Our population is high school students in the United States. The parameter is the proportion that have smoked cigarettes in the past month. Call this unknown parameter p, for "proportion." The statistic that estimates the parameter p is the **sample proportion**

$$\hat{p} = \frac{\text{count in the sample}}{\text{size of the sample}}$$

$$= \frac{3340}{12{,}960} = 0.2577$$

A basic move in statistical inference is to use a sample statistic to estimate a population parameter. Once we have the sample in hand, we estimate that the

proportion of all high school students who have smoked cigarettes in the past month, p is "about 25.77%" because the proportion in the sample was exactly 25.77%. We can only estimate that the truth about the population is "about" 25.77% because we know that the sample result is unlikely to be exactly the same as the true population proportion. A confidence interval makes that "about" precise.

95% confidence interval

A **95% confidence interval** is an interval calculated from sample data that is guaranteed to capture the true population parameter in 95% of all samples.

We will first march straight through to the interval for a population proportion, then reflect on what we have done and generalize a bit.

Estimating with confidence

We want to estimate the proportion p of the individuals in a population who have some characteristic—they are employed, or they approve of the president's performance, for example. Let's call the characteristic we are looking for a "success." We use the proportion $\hat{p}$ of successes in a simple random sample (SRS) to estimate the proportion p of successes in the population. How good is the statistic $\hat{p}$ as an estimate of the parameter p? To find out, we ask, "What would happen if we took many samples?" Well, we know that $\hat{p}$ would vary from sample to sample. We also know that this sampling variability isn't haphazard. It has a clear pattern in the long run, a pattern that is pretty well described by a normal curve. Here are the facts.

Sampling distribution of a sample proportion

The **sampling distribution** of a statistic is the distribution of values taken by the statistic in all possible samples of the same size from the same population.

Sampling distribution of a sample proportion *(continued)*

Take an SRS of size n from a large population that contains proportion p of successes. Let $\hat{p}$ be the **sample proportion** of successes,

$$\hat{p} = \frac{\text{count of successes in the sample}}{n}$$

Then, if the sample is large enough:

- The sampling distribution of $\hat{p}$ is **approximately normal.**
- The **mean** of the sampling distribution is p.
- The **standard deviation** of the sampling distribution is

$$\sqrt{\frac{p\,(1-p)}{n}}$$

These facts can be proved by mathematics, so they are a solid starting point. Figure 9.1 summarizes them in a form that also reminds us that a sampling distribution describes the results of lots of samples from the same population.

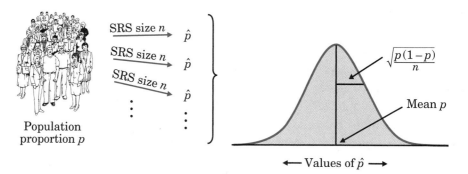

FIGURE 9.1 Repeat many times the process of selecting an SRS of size n from a population of which the proportion p are successes. The values of the sample proportion of successes have this normal sampling distribution.

EXAMPLE 9.2   Risky behavior

Suppose, for example, that the truth is that 25% of high school students smoke. Then in the setting of Example 9.1, $p = 0.25$. The Youth Risk Behavioral Survey sample of size $n = 12{,}960$ would, if repeated many times, produce sample proportions that closely follow the normal distribution with

$$\text{mean} = p = 0.25$$

and

$$\text{standard deviation} = \sqrt{\frac{p\,(1-p)}{n}}$$

$$= \sqrt{\frac{0.25\,(0.75)}{12{,}960}}$$

$$= \sqrt{0.0000144676}$$

$$= 0.0038$$

The center of this normal distribution is at the truth about the population. That's the absence of bias in random sampling once again. The standard deviation is small because the sample is quite large. So almost all samples will produce a statistic that is close to the true p. In fact, the 95 part of the 68–95–99.7 rule says that 95% of all sample outcomes will fall between

$$\text{mean} - 2 \text{ standard deviations} = 0.25 - 0.0076 = 0.2424$$

and

$$\text{mean} + 2 \text{ standard deviations} = 0.25 + 0.0076 = 0.2576$$

Figure 9.2 displays these facts.

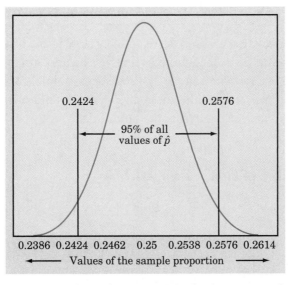

FIGURE 9.2 Repeat many times the process of selecting an SRS of size 12,960 from a population of which the proportion $p = 0.25$ are successes. The middle 95% of the values of the sample proportion will lie between 0.2424 and 0.2576.

So far, we have just put numbers on what we already knew. We can trust the results of large random samples because almost all such samples give results that are close to the truth about the population. The numbers say that in 95% of all samples of size 12,960, the statistic $\hat{p}$ and the parameter p are within 0.0076 of each other. We can put this another way: 95% of all samples give an outcome $\hat{p}$ such that the population truth p is captured by the interval from $\hat{p} - 0.0076$ to $\hat{p} + 0.0076$. The 0.0076 came from substituting $p = 0.25$ into the formula for the standard deviation of $\hat{p}$. For any value of p, the general fact is: **When the population proportion has the value p, 95% of all samples catch p in the interval extending 2 standard deviations on either side of $\hat{p}$.** That's the interval

$$\hat{p} \pm 2\sqrt{\frac{p(1-p)}{n}}$$

Is this the 95% confidence interval we want? Not quite. The interval can't be found just from the data because the standard deviation involves the population proportion p, and in practice we don't know p. In Example 9.2, we applied the formula for $p = 0.25$, but this may not be the true p.

What to do? Well, the standard deviation of the statistic does depend on the parameter p, but it doesn't change a lot when p changes. Go back to Example 9.2 and redo the calculation for other values of p. Here's the result:

Value of p:	0.23	0.24	0.25	0.26	0.27
Standard deviation:	0.0037	0.00375	0.0038	0.00385	0.0039

We see that if we guess a value of p reasonably close to the true value, the standard deviation found from the guessed value will be about right. We know that when we take a large random sample, the statistic $\hat{p}$ is almost always close to the parameter p. So we will use $\hat{p}$ as the guessed value of the unknown p. Now we have an interval that we can calculate from the sample data.

95% confidence interval for a proportion

Choose an SRS of size n from a large population that contains an unknown proportion p of successes. Call the proportion of successes in this sample $\hat{p}$. An approximate 95% confidence interval for the parameter p is

$$\hat{p} \pm 2\sqrt{\frac{\hat{p}(1-\hat{p})}{n}}$$

EXAMPLE 9.3 A confidence interval for youthful smokers

The Youth Risk Behavioral Survey random sample of 12,960 high school students found that 3340 had smoked cigarettes at least one day in the past month, a sample proportion $\hat{p} = 0.2577$. The 95% confidence interval for the proportion of all high school students who had smoked in the past month is

$$\hat{p} \pm 2 \sqrt{\frac{\hat{p}(1-\hat{p})}{n}} = 0.2577 \pm 2 \sqrt{\frac{(0.2577)(0.7423)}{12,960}}$$

$$= 0.2577 \pm 2 \, (0.0038)$$

$$= 0.2577 \pm 0.0076$$

or the interval $(0.2501, 0.2653)$

Interpret this result as follows: we got this interval by using a recipe that catches the true unknown population proportion in 95% of all samples. The shorthand is: we are **95% confident** that the true proportion of high school students who smoked in the past month lies between 25.01% and 26.53%.

EXERCISES

9.1 A student survey Tonya wants to estimate what proportion of the seniors in her school plan to attend the senior prom. She interviews an SRS of 50 of the 175 seniors in her school. She finds that 36 plan to attend the prom.
(a) What population does Tonya want to draw conclusions about?
(b) What is the population proportion p in this setting?
(c) What is the value of the sample proportion $\hat{p}$ from Tonya's sample?

9.2 Our dirty little secret, I More than 95% of people surveyed say they always wash their hands after using public restrooms. But observers posted in public facilities in five major cities in August 2000 found that only 5250 of 7836 observed individuals actually did.[4]
(a) What population does inference concern here?
(b) Explain clearly what the population proportion p is in this setting.
(c) What is the numerical value of the sample proportion $\hat{p}$?

9.3 Teen ecstasy use A 2001–2002 survey of teens conducted by the Partnership for a Drug-Free America found that 12% of 6937 U.S. teenagers surveyed reported use of ecstasy, the so-called love drug. The report states, "With 95% confidence, we can say that between 10.2% and 13.8% of all American teenagers have used ecstasy."[5] Explain to someone who knows no statistics what the phrase "95% confidence" means in this report.

9.4 Gun violence The Harris Poll asked a sample of 1009 adults which causes of death they thought would become more common in the future. Gun violence topped the list: 706 members of the sample thought deaths from guns would increase. Although the samples in national polls are not SRSs, they are close enough that our method gives approximately correct confidence intervals.
(a) Say in words what the population proportion p is for this poll.
(b) Find a 95% confidence interval for p.
(c) Harris announced a margin of error of plus or minus three percentage points for this poll result. How well does your work in (b) agree with this margin of error?

9.5 Polling women A *New York Times* Poll on women's issues interviewed 1025 women randomly selected from the United States, excluding Alaska and Hawaii. Of the women in the sample, 482 said they do not get enough time for themselves. Although the samples in national polls are not SRSs, they are close enough that our method gives approximately correct confidence intervals.
(a) Explain in words what the parameter p is in this setting.
(b) Use the poll results to give a 95% confidence interval for p.
(c) Write a short explanation of your findings in (b) for someone who knows no statistics.

Understanding confidence intervals

Our 95% confidence interval for a population proportion has the familiar form

$$\text{estimate} \pm \text{margin of error}$$

We know that news reports of sample surveys, for example, usually give the estimate and the margin of error separately: "A new Gallup Poll shows that 66% of women favor new laws restricting guns. The margin of error is plus or minus four percentage points." We also know that news reports usually leave out the level of confidence.

Not all confidence intervals have this form. Here's a complete description of a *confidence interval.*

Confidence interval

A **level *C* confidence interval** for a parameter has two parts:

• An **interval** calculated from the data.

• A **confidence level** *C*, which gives the probability that the interval will capture the true parameter value in repeated samples.

ACTIVITY 9.1B The *Confidence Intervals* applet

Go to the *Statistics Through Applications* Web site www.whfreeman.com/sta and find the *Confidence Intervals* applet.

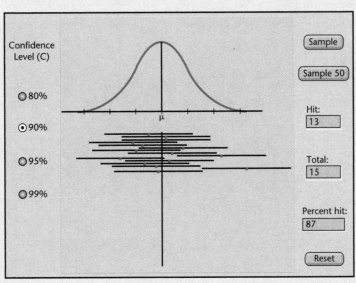

Directions: Click the "Sample" button to have the applet take a random sample and construct a confidence interval. Click the "Sample 50" button to have the applet take 50 different random samples and make 50 different confidence intervals from the resulting data.

1. If you make just one 90% confidence interval, will it necessarily capture the parameter value?

2. If you make 10 different 90% confidence intervals, how many of them do you think will capture the parameter?

3. If you make a thousand 90% confidence intervals, how many of them do you think will capture the parameter?

APPLET TIP: The cumulative totals are cleared when you click the "Reset" button. When you change the confidence level, these totals pertain only to the displayed intervals.

4. Test your answers to the three questions above through repeated use of the applet. How far did the results vary from what you expected in questions 2 and 3? How far did the results vary from what you expected on a percentage basis?

ACTIVITY 9.1B The *Confidence Intervals* applet *(continued)*

5. Set the confidence level to 80%, and sample 50 confidence intervals. How many of these intervals fail to capture the parameter (they are drawn in red on the computer screen)?

6. Increase the confidence level in succession to 90%, then to 95%, and then to 99% while watching the changing behavior of the 50 intervals. What happens to the lengths of the intervals? What happens to the number of intervals that fail to capture the parameter?

7. Have the applet make 50 new confidence intervals. The dots in the center of the intervals mark the sample mean. Are the intervals produced by the applet symmetric around the sample mean?

   The tick marks on the horizontal axis below the normal curve are each separated by one standard deviation of the sample mean. According to the 68–95–99.7 rule, how many of the 50 sample means would you expect to fall within one standard deviation of the parameter? How many of the 50 samples did fall within one standard deviation?

8. How is the margin of error related to the length of the interval? Explain. Examine the 50 intervals at the 95% confidence level. What is the typical margin of error for these intervals measured in units of the standard deviation of the mean? Explain how this is related to the normal curve.

There are many recipes for statistical confidence intervals for use in many situations. Be sure you understand how to interpret a confidence interval. The interpretation is the same for any recipe, and you can't use a calculator or a computer to do the interpretation for you. Confidence intervals use the central idea of probability: ask what would happen if we repeated the sampling many times. The 95% in a 95% confidence interval is a probability, the probability that the method produces an interval that does capture the true parameter.

EXAMPLE 9.4 How confidence intervals behave

The Youth Risk Behavioral Survey sample of 12,960 high school seniors found 3340 smokers, so the sample proportion was

$$\hat{p} = \frac{3340}{12{,}960} = 0.2577$$

and the 95% confidence interval was

$$\hat{p} \pm 2\sqrt{\frac{\hat{p}\,(1-\hat{p})}{n}} = 0.2577 \pm 0.0038$$

Draw a second sample from the same population. It finds 3127 of its 12,960 respondents who smoke. For this sample,

$$\hat{p} = \frac{3127}{12{,}960} = 0.2413$$

$$\hat{p} \pm 2\sqrt{\frac{\hat{p}\,(1-\hat{p})}{n}} = 0.2413 \pm 0.0038$$

Draw another sample. Now the count is 3479 and the sample proportion and confidence interval are

$$\hat{p} = \frac{3479}{12{,}960} = 0.2684$$

$$\hat{p} \pm 2\sqrt{\frac{\hat{p}\,(1-\hat{p})}{n}} = 0.2684 \pm 0.0039$$

National Baseball Hall of Fame and Museum, Inc., Cooperstown, N.Y.

A baseball poll

Polls can sometimes yield surprising and entertaining results. A *Sports Illustrated* poll conducted in summer 2003 asked 550 major league baseball players to name the greatest living player. Barry Bonds was named most often (39.9%). Next was Alex Rodriguez (12.8%), followed by Willie Mays (12.1%), Nolan Ryan (7.1%), Hank Aaron (6.7%), and Pete Rose (3.6%). Babe Ruth was named by 8 of the players (1.5%). Babe Ruth died in 1948.

Keep sampling. Each sample yields a new estimate and a new confidence interval. *If we sample forever, 95% of these intervals capture the true parameter.* This is true no matter what the true value is. Figure 9.3 summarizes the behavior of the confidence interval in graphical form.

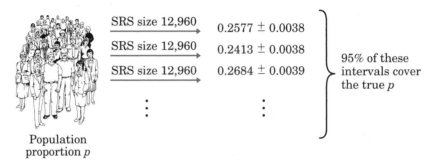

FIGURE 9.3 Repeated samples from the same population give different 95% confidence intervals, but 95% of these intervals capture the true population proportion *p*.

Figures 9.3 and 9.4 (on the facing page) show how confidence intervals behave. Example 9.4 and Figure 9.3 remind us that repeated samples give different results and that we are guaranteed only that 95% of the samples give a correct result.

Figure 9.4 goes behind the scenes. The vertical line is the true value of the population proportion p. The normal curve at the top of the figure is the sampling distribution of the sample statistic, which is centered at the true p. We are behind the scenes because in real-world statistics we don't know p. The 95% confidence intervals from 25 SRSs appear one after the other. The central dots are the values of $\hat{p}$, the centers of the intervals. The arrows on either side span the confidence interval. In the long run, 95% of the intervals will cover the true p and 5% will miss. Of the 25 intervals in Figure 9.4, 24 hit and 1 misses. (Remember that probability describes only what happens in the long run—we don't expect exactly 95% of 25 intervals to capture the true parameter.)

Don't forget that our interval is only *approximately* a 95% confidence interval. It isn't exact for two reasons. The sampling distribution of the sample proportion $\hat{p}$ isn't exactly normal. And we don't get the standard deviation of $\hat{p}$ exactly right, because we used $\hat{p}$ in place of the unknown p. Both of these difficulties become less serious as the sample size n gets larger. So our recipe is good only for large samples. What is more, the recipe assumes that the population is really big—at least 10 times the size of the sample. Professional statisticians use

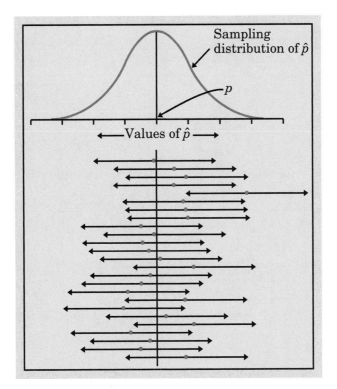

FIGURE 9.4 Twenty-five samples from the same population give these 95% confidence intervals. In the long run, 95% of all such intervals cover the true population proportion, marked by the vertical line.

more elaborate methods that take the size of the population into account and work even for small samples.[6] But our method works well enough for many practical uses. More important, it shows how we get a confidence interval from the sampling distribution of a statistic. That's the reasoning behind any confidence interval.

Confidence intervals for a population proportion

We used the 95 part of the 68–95–99.7 rule to get a 95% confidence interval for the population proportion. Perhaps you think that a method that works 95% of the time isn't good enough. You want to be 99% confident. For that, we need to mark off the central 99% of a normal distribution. For any probability C between 0 and 1, there is a number z^* such that any normal distribution has probability C within z^* standard deviations of the mean. Figure 9.5 (next page) shows how the probability C and the number z^* are related.

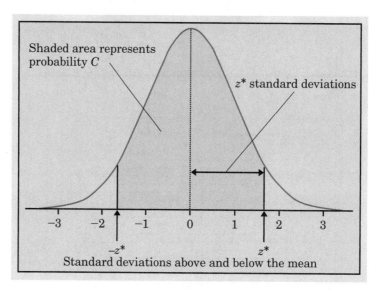

Shaded area represents probability C

z^* standard deviations

−3   −2   −1   0   1   2   3

$-z^*$ z^*
Standard deviations above and below the mean

FIGURE 9.5 Critical values z^* of the normal distributions. In any normal distribution, there is area (probability) C under the curve between $-z^*$ and z^* standard deviations away from the mean.

Table 9.1 gives the numbers z^* for various choices of C. For convenience, the table gives C as a confidence level in percent. The numbers z^* are called **critical values** of the normal distributions. Table 9.1 shows that any normal distribution has probability 99% within ±2.58 standard deviations of its mean. The table also shows that any normal distribution has probability 95% within ±1.96 standard deviations of its mean. The 68–95–99.7 rule uses 2 in place of the critical value $z^* = 1.96$. That is good enough for practical purposes, but the table gives the more exact value.

TABLE 9.1 Critical Values of the Normal Distributions

Confidence level C	Critical value z^*	Confidence level C	Critical value z^*
50%	0.67	90%	1.64
60%	0.84	95%	1.96
70%	1.04	99%	2.58
80%	1.28	99.9%	3.29

From Figure 9.5 we see that with probability C, the sample proportion takes a value within z^* standard deviations of p. That is just to say that with probability C, the interval extending z^* standard deviations on either side of the observed $\hat{p}$ captures the unknown p. Using the estimated standard deviation of $\hat{p}$ produces the following recipe.

Confidence interval for a population proportion

Choose an SRS of size n from a population of individuals of which proportion p are successes. The proportion of successes in the sample is $\hat{p}$. When n is large, an approximate level C confidence interval for p is

$$\hat{p} \pm z^* \sqrt{\frac{\hat{p}(1 - \hat{p})}{n}}$$

where z^* is the critical value for probability C from Table 9.1.

EXAMPLE 9.5 A 99% confidence interval

The Youth Risk Behavioral Survey random sample of 12,960 high school students found that 3340 were smokers. We want a 99% confidence interval for the proportion p of all high school students who smoked cigarettes in the past month. Table 9.1 says that for 99% confidence, we must go out $z^* = 2.58$ standard deviations. Here are our calculations:

$$\hat{p} = \frac{3340}{12,960} = 0.2577$$

$$\hat{p} \pm z^* \sqrt{\frac{\hat{p}(1 - \hat{p})}{n}} = 0.2577 \pm 2.58 \sqrt{\frac{0.2577(1 - 0.2577)}{12,960}}$$

$$= 0.2577 \pm 2.58(0.0038)$$

$$= 0.2577 \pm 0.0098$$

So our 99% confidence interval is (0.2479, 0.2675).

We are 99% confident that the true population proportion is between 24.79% and 26.75%. That is, we got this range of percents by using a method that gives a correct answer 99% of the time.

Compare Example 9.5 with the calculation of the 95% confidence interval in Example 9.3. The only difference is the use of the critical value 2.58 for 99% confidence in place of 2 for 95% confidence. That makes the margin of error for 99% confidence larger and the confidence interval wider. Higher confidence isn't free—we pay for it with a wider interval. Figure 9.5 reminds us why this is true. To cover a higher percent of the area under a normal curve, we must go farther out from the center.

EXERCISES

9.6 Random digits, I We know that the proportion of 0s among a large set of random digits is $p = 0.1$ because all 10 possible digits are equally probable. The entries in a table of random digits are a sample from the population of all random digits. To get an SRS of 200 random digits, look at the first digit in each of the 200 five-digit groups in lines 101 to 125 of Table A in the back of the book. How many of these 200 digits are 0s? Give a 95% confidence interval for the proportion of 0s in the population from which these digits are a sample. Does your interval cover the true parameter value, $p = 0.1$?

Who is a smoker?

When estimating a proportion p, be sure you know what counts as a "success." The news says that 25% of adolescents smoke. Shocking. It turns out that this is the percent who smoked at least once in the past month. If we say that a smoker is someone who smoked on at least 20 of the past 30 days and smoked at least half a pack on those days, fewer than 4% of adolescents qualify.

9.7 Random digits, II Use the methods of Section 8.1 to simulate Exercise 9.6 with your TI-83.
(a) What command generates 200 single digits? Store these digits in list L_1.
(b) Next, have the calculator put a 1 in list L_2 for each 0 in list L_1, and put a 0 otherwise. What command does this?
(c) Write a command to sum the 1s and divide by 200.
(d) Construct a 95% confidence interval for the proportion of 0s in the population.

9.8 Random digits, III Suppose your simulation in Exercise 9.7 produced a value of $\hat{p} = 0.095$. The critical value for a 98% confidence level is $z^* = 2.326$.
(a) Use this information to find the 98% confidence interval for the proportion of 0s in the population. Use the value $\hat{p} = 0.095$.
(b) Write a sentence to interpret this interval.

9.9 Going to the prom The prom committee wants to know how many seniors they can expect so that they can develop a budget.

(a) Use the information from Exercise 9.1 (page 495) to construct a 95% confidence interval for the true proportion of seniors who will attend the prom.

(b) The committee asks if you can provide a "more precise" interval. To oblige them, construct a 90% confidence interval.

(c) Interpret the 90% confidence interval for them in simple language.

9.10 Teens and their TV sets The *New York Times* and CBS News conducted a nationwide poll of 1048 randomly selected 13- to 17-year-olds. Of these teenagers, 692 had a television in their room and 189 named Fox as their favorite television network.[7] We can consider the sample to be an SRS and still get results that are approximately correct.

(a) Give 95% confidence intervals for the proportion of all people in this age-group who had a TV in their room at the time of the poll and the proportion that would choose Fox as their favorite network.

(b) The news article says, "In theory, in 19 cases out of 20, the poll results will differ by no more than three percentage points in either direction from what would have been obtained by seeking out all American teenagers." Explain how your results agree with this statement.

CALCULATOR CORNER Confidence Intervals

You can use your TI-83 to construct confidence intervals for population proportions. We will use the teen smoking data from Example 9.3 (page 495) to illustrate.

```
1-PropZInt
 x:3340
 n:12960
 C-Level:.95
 Calculate
```

- Press STAT , choose TESTS, and then choose A:1-PropZInt.

- Enter the number of teen smokers as the count *x* and the number of teens in the sample *n*. Specify the confidence level as shown.

- Highlight "Calculate" and press ENTER .

```
1-PropZInt
 (.25019,.26525)
 p̂=.2577160494
 n=12960
```

APPLICATION 9.1　　1068

Example 1.5 (page 10) stated that polls like the weekly Gallup Poll typically sample between 1000 and 1500 American adults who are interviewed by telephone. The magic number is 1068. This Application shows why. First, national polls like to use the 95% confidence level because it has become a standard and people are used to it. Second, pollsters like to report results with a ±3 percentage points margin of error. Newspaper and news magazine reports of polls frequently have information about the mechanics of how the poll was conducted and how to interpret the results. How many people do the pollsters need to interview in order to give a ±3 percentage points margin of error at the 95% confidence level? The details require some algebra with inequalities. The form for the confidence interval is

$$\hat{p} \pm z^* \sqrt{\frac{\hat{p}\,(1 - \hat{p})}{n}}$$

where $\hat{p}$ is the proportion of people who respond in a certain way and the ± term is the margin of error. For the 95% confidence interval, $z^* = 1.960$, correct to three decimal places (remember the middle part of the 68–95–99.7 rule). For technical reasons, we use $\hat{p} = 0.5$ under the radical in this case. Since we want the margin of error to be *at most* 3 percentage points, we have to solve the inequality

$$z^* \sqrt{\frac{\hat{p}\,(1 - \hat{p})}{n}} \le 0.03$$

Notice that we use the decimal form for 3%. Substituting, we have

$$1.960 \sqrt{\frac{0.5\,(1 - 0.5)}{n}} \le 0.03$$

Cross-multiplying and simplifying,

$$1.960 \left(\frac{0.5}{0.03} \right) \le \sqrt{n}$$

$$\sqrt{n} \ge \frac{(1.960)\,(0.5)}{0.03} = 32.667$$

Squaring both sides,　$n \ge (32.667)^2 = 1067.111$

This says that you have to interview *at least* 1067.111 people. Since you can't interview 0.111 person, you have to round up to the next whole number of people: 1068. That's the magic number.

APPLICATION 9.1 1068 *(continued)*

1. Why is $z^* = 1.960$? (*Hint:* Recall the 68–95–99.7 rule, and explain why we use 1.960 instead of 2.)

2. Go to the Web and find information on five recent polls. For each of the five polls, record the name of the polling organization, the topic of the poll, the main finding, the confidence level, and the margin of error. Do any of these polls tell you how many people they originally selected to interview?

3. Many polls nowadays randomly select 2000 or more individuals to contact. If they need only 1068, why do you think they have so many people in their pool?

EXPLORING THE WEB

Go to the Gallup Web site, www.gallup.com, and find the article "How Polls Are Conducted," by Frank Newport, editor-in-chief, the Gallup Poll. It's a classic and easy to understand.

There are lots of interesting poll results on the Web. Perform a search on a topic (preferably controversial) that interests you. You can do a Google search on "poll president job performance," for example, and get multiple hits. For other good starting points, see Exploring the Web in Chapter 2 (pages 101 and 122).

STATISTICS IN SUMMARY

Statistical inference draws conclusions about a population on the basis of data from a sample. Because we don't have data for the entire population, our conclusions are uncertain. A **confidence interval** estimates an unknown parameter in a way that tells us how uncertain the estimate is. The interval itself says how closely we can pin down the unknown parameter. The **confidence level** is a probability that says how often in many samples the method produces an interval that does contain the parameter. We find confidence intervals starting from the **sampling distribution** of a statistic, which shows how the statistic varies in repeated sampling.

In this chapter we found one specific confidence interval, for the proportion p of "successes" in a population, based on an SRS from the population. You will find more advice on interpreting confidence intervals in Section 9.3.

SECTION 9.1 EXERCISES

9.11 Activity 9.1A follow-up, I
(a) What is the margin of error for your 68% confidence interval in Activity 9.1A?
(b) What would your 95% confidence interval be?
(c) What would your 99.7% confidence interval be?

9.12 Wildlife management Wildlife biologists know that males of most big-game species can be harvested without affecting the total population because they tend to mate with several females a season. It is the female that requires the most protection. In 2002, Virginia hunters killed 928 black bears during the fall hunting season. Of these, 328 were females.[8]

Consider the 928 bears a sample of the population of all bears killed over many seasons.
(a) What is the sample proportion of females harvested?
(b) Find a 95% confidence interval for the true proportion of female bears killed in Virginia over many seasons.
(c) What is the margin of error for your confidence interval in (b)? What two factors determine the margin of error?

9.13 Activity 9.1A follow-up, II There is a trade-off between confidence and precision. You can have more confidence in an interval estimate if you are willing to be less precise.
(a) Calculate the width of the 95% confidence interval for the true proportion of brown M&M's that you calculated in Activity 9.1A. Record this width in the table below.
(b) Find the 99% and 90% confidence intervals for your value of $\hat{p}$. Record the widths in the table.
(c) Using your 68% confidence interval from Activity 9.1A, complete the table.

Confidence level (%)	Width of interval
68	
90	
95	
99	

(d) Write a sentence or two that describes your findings in (c).

9.14 How many polled? Suppose a local poll wanted to report the results at the 95% confidence level but with a margin of error of $\pm$ 4 percentage points. Use the method described in Application 9.1 (page 506) to find the number of subjects they would need to interview.

9.15 The good life, I In February 2003, Roper ASW, a market research and consulting firm, conducted in-person interviews with 2004 adults. When asked if they felt they had achieved the "good life," only 180 of the respondents answered in the affirmative. This was despite the fact that a majority of respondents had the things they said constituted the good life: a house, good health, a car, and children. What is the margin of error for this poll?

9.16 Count Buffon's coin The eighteenth-century French naturalist Count Buffon tossed a coin 4040 times. He got 2048 heads. Give a 95% confidence interval for the probability that Buffon's coin lands heads up. Are you confident that this probability is not 1/2? Why?

9.17 Harley motorcycles Harley-Davidson motorcycles make up 14% of all the motorcycles registered in the United States. You plan to interview an SRS of 600 motorcycle owners.
(a) What is the sampling distribution of the proportion of your sample who own Harleys?
(b) How likely is your sample to contain 18.2% or more who own Harleys? How likely is it to contain at least 11.2% Harley owners? Use the 68–95–99.7 rule and your answer to (a).

9.18 The quick method The quick method of Section 2.2 uses $\hat{p} \pm 1/\sqrt{n}$ as a rough recipe for a 95% confidence interval for a population proportion. The margin of error from the quick method is a bit larger than needed. It differs most from the more accurate method of this chapter when $\hat{p}$ is close to 0 or 1. An SRS of 500 motorcycle registrations finds that 68 of the motorcycles are Harley-Davidsons. Give a 95% confidence interval for the proportion of all motorcycles that are Harleys by the quick method and then by the method of this chapter. How much larger is the quick-method margin of error?

9.2 WHAT IS A TEST OF SIGNIFICANCE?

There goes the neighborhood?

Despite the Fair Housing Act and other laws, most black and white Americans still live in segregated neighborhoods. Whites tend to move out when a neighborhood becomes too heavily black. Have white attitudes changed over time, so that more whites are willing to stay as black families move in?

The Detroit Area Study interviewed a random sample of 1104 adults in the Detroit metropolitan area in 1976 and another random sample of 1543 adults in 1992.[9] Detroit is the most segregated large city in the United States, so the study looked in detail at attitudes toward mixed-race neighborhoods. One question asked whites to imagine that they lived in an all-white neighborhood, illustrated by the first card in Figure 9.6. Most of the respondents did in fact live in all-white neighborhoods. Then they were shown the second card in Figure 9.6, in which 3 of the 15 houses are occupied by blacks. This is the actual proportion of blacks in the entire Detroit area. Would they try to move away from such a neighborhood?

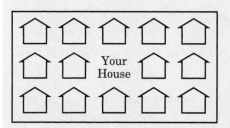 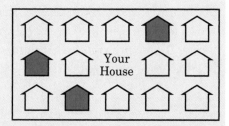

FIGURE 9.6 The Detroit Area Study showed white residents these cards. It asked them to imagine that their neighborhood looked like the one on the left. Now some black families move in and the neighborhood looks like the one on the right. Would the whites try to move out?

In 1976, 24% of whites would try to leave. By 1992, this percent had dropped to 15%. It appears that changing attitudes were making "white flight" less likely. (This assumes that the respondents told the truth. It is possible that between 1976 and 1992 it just became less acceptable to give segregationist answers.) These findings come from two samples of modest size. Could it be that the difference between the two samples is just due to the luck of the draw in randomly choosing the respondents? The study authors answered that challenge by finding the probability that two random samples would differ by as much as 24% versus 15% just by chance. This probability is small, less than 0.01. So the difference is "statistically significant." That idea, and using probabilities such as that 0.01 to express it, will be our focus in this section.

ACTIVITY 9.2 Roll the dice!

Materials: pair of dice (one red, one white)
Students pair off, Player 1 with the red die and Player 2 with the white die. Each player rolls the die, and the higher number wins that round. A tie is ignored and the dice are rolled again. The purpose is to look at the distribution of wins by the red die over the long term.

ACTIVITY 9.2 Roll the dice! *(continued)*

1. Roll the dice 10 times and use tally marks to record the winner.

Red wins:
White wins:

To make sure that no one has an unfair advantage, exchange the dice and roll another 10 times. Change dice again and roll 10 more times, and then exchange dice and roll 10 more times. Continue to record your results. You should have a total of 40 rolls.

2. Combine your results with the other teams to obtain a grand total of red wins and white wins.

3. We will state an assumption about the dice that says, "Each die should win half the time." We call this our *null hypothesis*, and we denote this null hypothesis by the symbol H_0. Are your class results consistent with H_0? Or do your results provide evidence that H_0 is not true? Are you surprised at these results?

4. Keep these data handy. We will use them later in this chapter to perform some formal statistical analyses.

The reasoning of statistical tests

The local hot-shot playground basketball player claims to make 80% of his free throws. "Show me," you say. He shoots 20 free throws and makes 8 of them. "Aha," you conclude, "if he makes 80%, he would almost never make as few as 8 of 20. So I don't believe his claim." That's the reasoning of statistical tests at the playground level: *An outcome that is very unlikely if a claim is true is good evidence that the claim is not true.*

Statistical inference uses data from a sample to draw conclusions about a population. So once we leave the playground, statistical tests deal with claims about a population. Tests ask if sample data give good evidence *against* a claim. A test says, "If we took many samples and the claim were true, we would rarely get a result like this." To get a numerical measure of how strong the sample evidence is, replace the vague term "rarely" by a probability. Here is an example of this reasoning at work.

EXAMPLE 9.6   Is the coffee fresh?

People of taste are supposed to prefer fresh-brewed coffee to the instant variety. On the other hand, perhaps many coffee drinkers just want their caffeine fix. A skeptic claims that only half of all coffee drinkers prefer fresh coffee. Let's do an experiment to test this claim.

Each of 50 subjects tastes two unmarked cups of coffee and says which he or she prefers. One cup in each pair contains instant coffee; the other, fresh-brewed coffee. The statistic that records the result of our experiment is the proportion of the sample who say they like the fresh-brewed coffee better. We find that 36 of our 50 subjects choose the fresh coffee. That is,

$$\hat{p} = \frac{36}{50} = 0.72 = 72\%$$

To make a point, let's compare our outcome $\hat{p} = 0.72$ with another possible result. If only 28 of the 50 subjects like the fresh coffee better than instant coffee, the sample proportion is

$$\hat{p} = \frac{28}{50} = 0.56 = 56\%$$

Surely 72% is stronger evidence against the skeptic's claim than 56%. But how much stronger? Is even 72% in favor in a sample convincing evidence that a majority of the *population* prefer fresh coffee? Statistical tests answer these questions.

Here's the answer in outline form:

• **The claim.** The skeptic claims that only half of all coffee drinkers prefer fresh-brewed coffee. That is, he claims that the population proportion p is only 0.5. *Suppose for the sake of argument that this claim is true.*

• **The sampling distribution (from page 492).** If the claim $p = 0.5$ were true and we tested many random samples of 50 coffee drinkers, the sample proportion would vary from sample to sample according to (approximately) the normal distribution with

$$\text{mean} = p = 0.5$$

and

$$\text{standard deviation} = \sqrt{\frac{p(1-p)}{n}}$$

$$= \sqrt{\frac{(0.5)(0.5)}{50}}$$

$$= 0.0707$$

Figure 9.7 on the facing page displays this normal curve.

• **The data.** Place the sample proportion on the sampling distribution. You see in Figure 9.7 that $p = 0.56$ isn't an unusual value, but that $p = 0.72$ is unusual. We would rarely get 72% of a sample of 50 coffee drinkers preferring fresh-brewed coffee if only 50% of all coffee drinkers felt that way. So the sample data do give evidence against the claim.

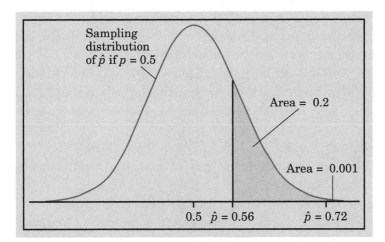

FIGURE 9.7 The sampling distribution of the proportion of 50 coffee drinkers who prefer fresh-brewed coffee. This distribution would hold if the truth about all coffee drinkers is that 50% prefer fresh coffee. The shaded area is the probability that the sample proportion is 56% or greater.

• **The probability.** We can measure the strength of the evidence against the claim by a probability. What is the probability that a sample gives a $\hat{p}$ this large or larger if the truth about the population is that $p = 0.5$? If $\hat{p} = 0.56$, this probability is the shaded area under the normal curve in Figure 9.7. This area is 0.20. Our sample actually gave $\hat{p} = 0.72$. The probability of getting a sample outcome this large is only 0.001, an area too small to see in Figure 9.7. An outcome that would occur just by chance in 20% of all samples is *not* strong evidence against the claim. But an outcome that would happen only 1 in 1000 times *is* good evidence.

Be sure you understand why this evidence is convincing. There are two possible explanations of the fact that 72% of our subjects prefer fresh to instant coffee:

1. The skeptic is correct ($p = 0.5$) and by bad luck a very unlikely outcome occurred.

2. In fact, the population proportion favoring fresh coffee is greater than 0.5, so the sample outcome is about what would be expected.

We cannot be certain that explanation 1 is untrue. Our taste test results *could* be due to chance alone. But the probability that such a result would occur by chance is so small (0.001) that we are quite confident that explanation 2 is right.

EXERCISES

9.19 The good life, II Refer to the good life poll in Exercise 9.15 (page 509).
(a) What is the population proportion of interest? What is the sample proportion?
(b) Calculate the standard deviation of the sampling distribution of the sample proportion.

9.20 Putting the baby at risk A 1998 study by a researcher at the University of Minnesota Cancer Center analyzed the first urine samples collected from 48 babies of smokers and nonsmokers in Germany. The researcher looked for NNK, an especially powerful carcinogen found only in tobacco products, and for by-products of NNK after it had been processed by the body. NNK can cause adeno-carcinoma, a kind of lung cancer found largely in smokers. He found that positive samples came only from the newborns of mothers who smoked. Of the 31 samples from babies of mothers who smoked, 22 contained NNK and its by-products. Babies of nonsmokers had none of those substances in their urine.[10]
(a) Was this an experiment?
(b) How would you define the population proportion of interest?
(c) How many smoking mothers were included in the study? How many non-smoking mothers?
(d) Describe the sample proportions for the two groups of mothers, and calculate the two values.
(e) Explain why we need not perform a test of significance in this case.
(f) What do you think the researcher's conclusion was from this study? Write a brief statement that you think sums up his findings.

9.21 Smoking by teens in decline The American Legacy Foundation, an anti-smoking organization, began a "truth campaign" in the form of television ads in 2000. These ads focus on the disease and death that smoking causes. In 2002, the National Youth Tobacco Survey involved 6853 middle school students in 69 schools in 27 states. A student was considered a smoker if he or she had smoked a cigarette within 30 days. The study found that middle school smoking had declined from 11.2% of students in 2000, before the truth campaign began, to 10.6% in 2002. The report said this represented a 5.4% decline among the middle school students.
(a) Was this an experiment? Explain.
(b) Is it clear how the 5.4% decline was calculated?
(c) The foundation president said, "The study shows that the truth campaign is the anti-smoking vaccine for kids we've been waiting for."[11] Can you conclude that the truth campaign caused the decline in smoking among middle school students? Explain.

9.22 The scourge of high heels Does wearing high heels cause knee osteoarthritis in women? Osteoarthritis is caused when the surface covering the knee joints degenerates. Over time, the cartilage breaks down and bones rub together, causing severe pain. In a 2003 study, Dr. Casey Kerrigan, a University of Virginia professor of physical medicine and rehabilitation, used motion analysis and force sensors in the ground to calculate stress on the knees. She had 20 women walk in 2- to 3-inch heels and then had the same group walk barefoot. She reported that "the strain on the parts of the knee that are the most vulnerable to osteoarthritis increased by 23 percent when women wore heels."[12]

(a) Is this an experiment? Explain.

(b) The sample proportion is the percent increase in the strain on the knee. Find the standard deviation of the sample proportion.

9.23 Who uses tanning booths? Are teens, particularly girls, ignoring the dangers of skin cancer for the sake of sporting a good tan? Dermatology experts point out that there is evidence that indoor tanning contributes to the risk of malignant melanoma, the most serious kind of skin cancer. A Case Western Reserve University study investigated tanning-booth use. A national survey involving 6903 white teens found that 28% of teenage girls and 7% of teenage boys reported using tanning booths three or more times. For girls 18 and 19 years old, the figure was 47%.[13]

(a) Is this study subject to bias? If so, what kind?

(b) It is not necessary to perform formal inference on these data in order to conclude that significantly more teenage girls use tanning beds than teen boys. Why is that?

Hypotheses and *P*-values

Tests of significance refine (and perhaps hide) basic reasoning. In most studies, we hope to show that some definite effect is present in the population. In Example 9.6, we suspect that a majority of coffee drinkers prefer fresh-brewed coffee. A statistical test begins by supposing for the sake of argument that the effect we seek is not present. We then look for evidence against this supposition and in favor of the effect we hope to find. The first step in a test of significance is to state a claim that we will try to find evidence *against*.

Null hypothesis H_0

The claim being tested in a statistical test is called the **null hypothesis.** The test is designed to assess the strength of the evidence against the null hypothesis. Usually the null hypothesis is a statement of "no effect" or "no difference."

The term "null hypothesis" is abbreviated H_0, read as "H-naught." It is a statement about the population and so must be stated in terms of a population parameter. In Example 9.6, the parameter is the proportion p of all coffee drinkers who prefer fresh to instant coffee. The null hypothesis is

$$H_0: p = 0.5$$

The statement we hope or suspect is true instead of H_0 is called the **alternative hypothesis** and is abbreviated H_a. In Example 9.6, the alternative hypothesis is that a majority of the population favor fresh coffee. In terms of the population parameter, this is

$$H_a : p > 0.5$$

A significance test looks for evidence against the null hypothesis and in favor of the alternative hypothesis. The evidence is strong if the outcome we observe would rarely happen if the null hypothesis is true but is more probable if the alternative hypothesis is true. For example, it would be surprising to find 36 of 50 subjects favoring fresh coffee if in fact only half of the population feel this way. How surprising? A significance test answers this question by giving a probability: the probability of getting an outcome at least as far as the actually observed outcome is from what we would expect when H_0 is true. What counts as "far from what we would expect" depends on H_a as well as H_0. In the taste test, the probability we want is the probability that 36 or more of 50 subjects favor fresh coffee. If the null hypothesis $p = 0.5$ is true, this probability is very small (0.001). That's good evidence that the null hypothesis is not true.

P-value

The probability, computed assuming that H_0 is true, that the sample outcome would be as extreme or more extreme than the actually observed outcome is called the **P-value** of the test. The smaller the P-value is, the stronger is the evidence against H_0 provided by the data.

In practice, most statistical tests are carried out by computer software that calculates the P-value for us. It is usual to report the P-value in describing the results of studies in many fields. You should therefore understand what P-values say even if you don't do statistical tests yourself, just as you should understand what "95% confidence" means even if you don't calculate your own confidence intervals.

EXAMPLE 9.7   Count Buffon's coin

The French naturalist Count Buffon (1707–1788) tossed a coin 4040 times. He got 2048 heads. The sample proportion of heads is

$$\hat{p} = \frac{2048}{4040} = 0.507$$

That's a bit more than one-half. Is this evidence that Buffon's coin was not balanced? This is a job for a significance test.

The truth about Count Buffon's coin-tossing experiment.

The hypotheses. The null hypothesis says that the coin is balanced ($p = 0.5$). We did not suspect a bias in a specific direction before we saw the data, so the alternative hypothesis is just "the coin is not balanced." The two hypotheses are

$$H_0 : p = 0.5$$
$$H_a : p \neq 0.5$$

The sampling distribution. If the null hypothesis is true, the sample proportion of heads has approximately the normal distribution with

$$\text{mean} = p = 0.5$$

$$\text{standard deviation} = \sqrt{\frac{p(1-p)}{n}}$$

$$= \sqrt{\frac{(0.5)(0.5)}{4040}}$$

$$= 0.00787$$

Figure 9.8 shows this sampling distribution with Buffon's sample outcome $\hat{p} = 0.507$ marked. The picture already suggests that this is not an unlikely outcome that would give strong evidence against the claim that $p = 0.5$.

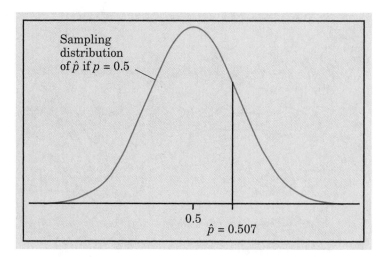

FIGURE 9.8 The sampling distribution of the proportion of heads in 4040 tosses of a balanced coin. Count Buffon's result, proportion 0.507 heads, is marked.

The P-value. How unlikely is an outcome as far from 0.5 as Buffon's $\hat{p} = 0.507$? Because the alternative hypothesis allows p to lie on either side of 0.5, values of $\hat{p}$ far from 0.5 in either direction provide evidence against H_0 and in favor of H_a. The P-value is therefore the probability that the observed $\hat{p}$ lies as far from 0.5 *in either direction* as the

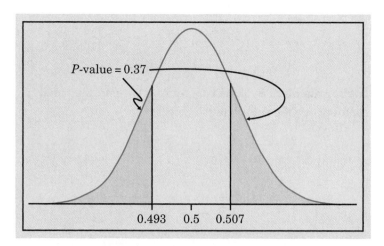

FIGURE 9.9 The P-value for testing whether Count Buffon's coin was balanced. This is the probability, calculated assuming a balanced coin, of a sample proportion as far or farther from 0.5 as Buffon's result of 0.507.

observed $\hat{p} = 0.507$. Figure 9.9 on the facing page shows this probability as area under the normal curve. It is $P = 0.37$.

Conclusion. A truly balanced coin would give a result this far or farther from 0.5 in 37% of all repetitions of Buffon's trial. His result gives no reason to think that his coin was not balanced.

The alternative $H_a : p > 0.5$ in Example 9.6 is a **one-sided alternative** because the effect we seek evidence for says that the population proportion is greater than one-half. The alternative $H_a : p \neq 0.5$ in Example 9.7 is a **two-sided alternative** because we ask only whether or not the coin is balanced. Whether the alternative is one-sided or two-sided determines whether sample results extreme in one or in both directions count as evidence against H_0 and in favor of H_a.

Statistical significance

We can decide in advance how much evidence against H_0 we will insist on. The way to do this is to say how small a P-value we require. The decisive value of P is called the **significance level.** It is usual to write it as α, the Greek letter alpha. If we choose $\alpha = 0.05$, we are requiring that the data give evidence against H_0 so strong that it would happen no more than 5% of the time (1 time in 20) when H_0 is true. If we choose $\alpha = 0.01$, we are insisting on stronger evidence against H_0, evidence so strong that it would appear only 1% of the time (1 time in 100) if H_0 is in fact true.

> ## Statistical significance
>
> If the P-value is as small or smaller than α, we say that the data are **statistically significant at level α.**

"Significant" in the statistical sense does not mean "important." It means simply "not likely to happen just by chance." We used these words in Section 3.1 (page 145). Now we have attached a number to statistical significance to say what "not likely" means. You will often see significance at level 0.01 expressed by the statement "The results were significant ($P < 0.01$)." Here P stands for the P-value.

We don't need to make use of traditional levels of significance such as 5% and 1%. The P-value is more informative because it allows us to assess significance at any level we choose. For example, a result with $P = 0.03$ is significant at the $\alpha = 0.05$ level but not significant at the $\alpha = 0.01$ level. Nonetheless, the traditional significance levels are widely accepted guidelines for "how much evidence is enough." We might say that $P < 0.10$ indicates "some evidence" against the null hypothesis, $P < 0.05$ is "moderate evidence," and $P < 0.01$ is "strong evidence." Don't take these guidelines too literally, however. We will say more about interpreting tests in the next section.

Calculating *P*-values

Finding the *P*-values we gave in Examples 9.6 and 9.7 requires doing normal distribution calculations using Table B of normal percentiles. That was addressed in Section 5.1. In practice, software or a calculator does the calculation for us, but here is an example that shows how to use Table B.

EXAMPLE 9.8   Tasting coffee

The hypotheses. In Example 9.6, we want to test the hypotheses

$$H_0: p = 0.5$$
$$H_a: p > 0.5$$

Here p is the proportion of the population of all coffee drinkers who prefer fresh coffee to instant coffee.

The sampling distribution. If the null hypothesis is true, so that $\hat{p} = 0.5$, then $\hat{p}$ follows a normal distribution with mean 0.5 and standard deviation 0.0707.

The data. A sample of 50 people found that 36 preferred fresh coffee. The sample proportion is $\hat{p} = 0.72$.

The P-value. The alternative hypothesis is one-sided on the high side. So the *P*-value is the probability of getting an outcome at least as large as 0.72. Figure 9.7 (page 513) displays this probability as an area under the normal sampling distribution curve. To find any normal curve probability, move to the standard scale. The standard score for the outcome $\hat{p} = 0.72$ is

$$\text{standard score} = \frac{\text{observation} - \text{mean}}{\text{standard deviation}}$$

$$= \frac{0.72 - 0.5}{0.0707} = 3.1$$

Table B says that standard score 3.1 is the 99.9 percentile of a normal distribution. That is, the area under a normal curve to the left of 3.1 (in the standard scale) is 0.999. The area to the right is therefore 0.001, and that is our *P*-value.

Conclusion. The small *P*-value means that the data provide very strong evidence that a majority of the population prefer fresh coffee.

EXERCISES

9.24 Affirmative action A 2003 comprehensive survey of American attitudes toward colleges and universities commissioned by the *Chronicle of Higher Education* found that 640 of the 1000 American adults aged 25 to 64 surveyed said that schools should not admit minorities who have lower grades than other qualified candidates.[14] Is this sufficient evidence to conclude that a majority of adults have this view of affirmative action?

(a) State null and alternative hypotheses.

(b) Describe the sampling distribution.

(c) How extreme is the sample outcome? Find the *P*-value.

(d) Explain your conclusions in nontechnical language.

9.25 Bullies in middle school If one study is any indication, many middle school students are not shy about admitting aggressive behavior in school. A 1999 University of Illinois study on aggressive behavior surveyed 558 students in a midwestern middle school. When asked to describe their behavior in the last 30 days, 445 students said their behavior included physical aggression, social ridicule, teasing, name-calling, and issuing threats. This behavior was not defined as bullying in the questionnaire.[15]

(a) Is this evidence that more than three-quarters of the students at that middle school engage in bullying behavior? Carry out an inference procedure to answer this question. State hypotheses, describe the sampling distribution, and find the *P*-value. State your conclusions.

(b) Does this study provide strong evidence of pervasive bullying in America's middle schools? Explain.

9.26 Allergy medicine After conducting clinical trials on a new allergy drug, the pharmaceutical company reports, "In our sample, perennial allergic rhinitis symptoms were significantly reduced for patients taking [our product] when compared with a placebo ($P < 0.01$)." Explain to someone who knows no statistics what this means.

9.27 Where's the blunder? Write a sentence or two to explain the blunder in each of the following statements.

(a) Austin says that a result that is significant at the $\alpha = 0.05$ level must also be significant at the 0.01 level.

(b) Asked to explain the meaning of "statistically significant at the $\alpha = 0.05$ level," Carla says: "This means that the probability that the null hypothesis is true is less than 0.05."

9.28 Diet and diabetes Does eating more fiber reduce the blood cholesterol levels of patients with diabetes? A randomized clinical trial compared normal and high-fiber diets. Here is part of the researchers' conclusion: "The high-fiber diet reduced plasma total cholesterol concentrations by 6.7 percent ($P = 0.02$), triglyceride concentrations by 10.2 percent ($P = 0.02$), and very-low-density lipoprotein cholesterol concentrations by 12.5 percent ($P = 0.01$)."[16] A doctor who knows no statistics says that a drop of 6.7% in cholesterol isn't a lot—maybe it's just an accident due to the chance assignment of patients to the two diets. Explain in simple language how "$P = 0.02$" answers this objection.

APPLICATION 9.2 Clinical Trials for Drugs

Schering Corporation, maker of Clarinex, publishes a "physician's prescribing information sheet" with magazine advertisements for this drug. Clarinex is an allergy medication that contains an antihistamine. It is designed to combat a condition known as perennial allergic rhinitis. The information sheet includes several tables. One table shows the reduction in total nasal symptom score for the treatment group and the control group, and indicates the P-value for the comparison between the Clarinex group and the placebo group. Another table shows the incidence of adverse events (side effects) reported by at least 2% of the subjects in the clinical trials. We will look at one of these "adverse experiences." The condition "dry mouth" was experienced by 3.0% of the 1655 subjects taking Clarinex, while 1.9% of the 1652 subjects receiving the placebo experienced dry mouth. (It is customary for pharmaceutical companies to report side effects in terms of percents instead of actual counts.) Is this significant evidence that Clarinex causes more dry mouth than the placebo?

Converting from percents, let's say that 49 of the 1655 subjects in the Clarinex/treatment group experienced dry mouth, and that 31 of the 1652 placebo subjects reported dry mouth.

Our null and alternative hypotheses are

$H_0: p_1 = p_2$ (There is no difference between the proportion of Clarinex users experiencing dry mouth and the proportion of placebo recipients experiencing dry mouth.)

$H_a: p_1 \neq p_2$ (The proportions are different.)

Notice that we have elected to perform a two-sided test, so we will be looking at areas in the left tail and the right tail of the normal distribution. Software or your graphing calculator tells us that the probability of observing a result this extreme by chance alone is 0.04. (The Calculator Corner on page 524 shows the details.) Most would agree that this is sufficient evidence that Clarinex users will experience significantly more dry mouth than those taking a placebo.

APPLICATION 9.2 Clinical Trials for Drugs *(continued)*

1. The side effect of dry mouth for the Clarinex group was significant at the $\alpha = 0.05$ level. Is it significant at the 0.10 level? At the 0.01 level? Explain.

2. The information sheet lists some other side effects. The condition of pharyngitis (inflammation of the pharynx/throat) was experienced by 4.1% of the treatment group versus 2.0% of the control (placebo) group. For this difference, using a count of 68 for the treatment group and 33 for the control group, software tells us the standard score is 3.528 and the *P*-value is 0.0004. Write a sentence or two that interprets these results.

3. The information sheet also states that fatigue was experienced by 2.1% of the Clarinex group, while only 1.2% of the placebo group experienced fatigue. Estimate the counts and then use your calculator to perform inference on these data. Write a conclusion for whether the effect is significant.

CALCULATOR CORNER Inference for proportions with the TI-83

The TI-83 can be used to test a claim about a population proportion. Consider Example 9.7, Buffon's coin-tossing activity. In $n = 4040$ tosses, Buffon observed $X = 2048$ heads. Recall that our hypotheses were

$$H_0: p = 0.5$$

$$H_a: p \neq 0.5$$

```
1-PropZTest
 p0:.5
 x:2048
 n:4040
 prop≠p0 <p0 >p0
 Calculate Draw
```

- If you select the "Calculate" choice and press ENTER, you will see that the z statistic is 0.88 and the *P*-value is 0.3783.

To perform a significance test:
- Press STAT, then chose TESTS and 5:1-PropZTest.
- On the 1-PropZTest screen, enter these values: $p_0 = 0.5$, $x = 2048$, and $n = 4040$. Specify the alternative hypothesis as prop≠p_0.

```
1-PropZTest
 prop≠.5
 z=.8810434857
 p=.3782942021
 p̂=.5069306931
 n=4040
```

CALCULATOR CORNER   Inference for proportions with the T1-83 *(continued)*

- If you select the "Draw" option, you will see the screen shown here.

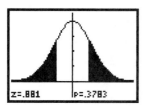

Compare these results with those in Example 9.7.

The TI-83 can also be used to compare two population proportions using significance tests. Here, we use the data from Application 9.2. In the treatment (Clarinex) group of 1655 subjects, 49 experienced the side effect of dry mouth. In the control (placebo) group, 31 of the 1652 subjects experienced dry mouth. The hypotheses were

$$H_0: p_1 = p_2$$
$$H_a: p_1 \neq p_2$$

The alternative hypothesis says that the proportions of users experiencing this side effect are different in the two groups. To perform the test of the null hypothesis:

- Press STAT , then choose TESTS and 6:2-PropZTest.
- On the 2-PropZTest screen, enter these values: $x_1 = 49$, $n_1 = 1655$, $x_2 = 31$, and $n_2 = 1652$. Specify the alternative hypothesis as $p_1 \neq p_2$.

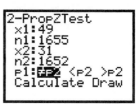

- If you select the "Calculate" choice and press ENTER , you will see that the z statistic is 2.03, and the P-value is 0.0425, as shown here.

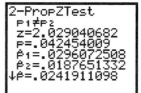

- If you select the "Draw" option, you will see the screen shown here.

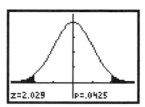

Compare these results with those in Application 9.2.

EXPLORING THE WEB

Confidence intervals and statistical significance appear constantly in reports of studies in many fields. Most scientific journals now have Web sites that display at least summaries of articles appearing in the journals. For example, the case study that opens Section 9.1 comes from *Circulation* (circ.ahajournals.org). The opening case study in this section is from the *American Journal of Sociology* (www.journals.uchicago.edu/AJS/home.html).

Choose a major journal in an area of interest. Use a Web search engine to find its Web site—just search on the journal's name. Look at the summaries of recent papers. In fields where studies are done that produce data, you will surely find phrases like "95% confidence" and "significant ($P = 0.01$)."

Many of the well-known drugs have their own Web sites. We found information on Clarinex, for example, at www.clarinex.com. To find results of clinical trials of new drugs, try a similar URL. To get comparative data from these trials, look for wording like "physician's prescribing information."

STATISTICS IN SUMMARY

A confidence interval estimates an unknown parameter. A **test of significance** assesses the evidence for some claim about the value of an unknown parameter. In practice, the purpose of a statistical test is to answer the question "Could the effect we see in the sample just be an accident due to chance, or is it good evidence that the effect is really there in the population?"

Significance tests answer this question by giving the probability that a sample effect as large as the one we see in this sample would arise just by chance. This probability is the *P*-value. A small *P*-value says that our outcome is unlikely to happen just by chance. To set up a test, state a **null hypothesis** that says the effect you seek is not present in the population. The **alternative hypothesis** says that the effect is present. The *P*-value is the probability, calculated taking the null hypothesis to be true, of an outcome as extreme in the direction specified by the alternative hypothesis as the actually observed outcome. A sample result is **statistically significant at the 5% level** if it would occur just by chance no more than 5% of the time in repeated samples.

This section concerns the basic reasoning of tests and the details of tests for hypotheses about a population proportion. There is more discussion of the practical interpretation of statistical tests in the next section.

SECTION 9.2 EXERCISES

9.29 Body temperature We have all heard that 98.6 degrees Fahrenheit (or 37 degrees Celsius) is "normal body temperature." In fact, there is evidence that most people have a slightly lower body temperature. You plan to measure the body temperature of a random sample of people very accurately. You hope to show that a majority have temperatures lower than 98.6 degrees.
(a) Say clearly what the population proportion p stands for in this setting.
(b) In terms of p, what are your null and alternative hypotheses?

9.30 Activity 9.2 follow-up Use the data from Activity 9.2 (on page 510) to perform a test of significance.
(a) Give null and alternative hypotheses in terms of the red die and the proportion of times the red die should win.
(b) Determine the sampling distribution and calculate the P-value. Is there sufficient evidence to reject H_0?
(c) State your conclusion.
(d) Carefully inspect the red die. Can you explain the results that you observed?

9.31 Hispanics and discrimination A *New York Times*/CBS News poll of 1078 Hispanics conducted in mid-2003 found that "most Hispanics in the United States are optimistic about prospects for themselves and their children." Almost two-thirds say they have experienced no discrimination in this country. The Associated Press article that reported on the story did not include counts, so we will take some liberties and estimate them. Interpret "almost two-thirds" of the 1078 Hispanic respondents to be 700.
(a) What is the proportion of Hispanics who say they have experienced no discrimination in this country?
(b) Can you conclude that a majority of Hispanics have experienced no discrimination? Carry out inference to answer this question.

9.32 Job satisfaction In July 2003, the Conference Board's Consumer Research Center surveyed 5000 employed American workers to determine their job satisfaction. Of those surveyed, 2445 said that they were satisfied with their jobs. The figure was nearly 59% in 1995. Can we conclude that the level of job satisfaction has changed in those eight years?
(a) Write hypotheses in preparation for performing inference.
(b) Carry out a test of significance by determining the sampling distribution and calculating a P-value, and then state your conclusions.

9.33 Middle school smokers Exercise 9.21 (page 514) described a survey of smoking among middle school students. In 2000, 11.2% of middle school students smoked. A follow-up study conducted two years later found that 726 out of 6853 students were smokers. Is there enough evidence that the proportion of middle school smokers has changed in those two years? Use an appropriate inference procedure to answer this question.

9.34 Home run king In 1998, baseball fans witnessed an exciting personal competition between long-ball hitters Mark McGwire and Sammy Sosa. By hitting 70 home runs, McGwire broke Roger Maris's record of 61 home runs set in 1961. Sosa was not far behind with 66. But if McGwire had many more at-bats, then he would have had an advantage, so perhaps it would be interesting to compare the number of home runs relative to the number of opportunities to hit home runs. A quick Web check reveals that McGwire had 509 at-bats that year, while Sosa had 643 at-bats. Consider the number of home runs to be a sample of the population of number of home runs each player could have hit in a season. Using the technique shown in the Calculator Corner for comparing two proportions (page 524), compare McGwire's home run success rate (number of home runs divided by number of at-bats, as a proportion) to that of Sosa. Was either hitter's proportion of home runs significantly better than that of the other batter? Perform your analyses, and report your findings.

9.35 Treatment for depression Doctors and other health-care professionals know that 1 in 4 women and 1 in 10 men can expect to develop depression during their lifetime. The good news? Up to 80% of people who receive proper treatment improve. Celexa is a relatively new drug to treat depression. In clinical trials the most frequent adverse events reported with Celexa versus a placebo were nausea (21% versus 14%), dry mouth (20% versus 14%), somnolence (18% versus 10%), insomnia (15% versus 14%), increased sweating (11% versus 9%), tremor (8% versus 6%), and diarrhea (8% versus 5%). The number of subjects receiving Celexa was 1063. The number receiving a placebo was 446.[17] Select one of the side effects above and perform a test of significance like we did in Application 9.2 (page 522) and the Calculator Corner (page 524). Report your findings and conclusions.

9.36 Are cell phones distracting? The American Automobile Association (AAA) commissioned a 2003 study by researchers at the University of North Carolina on sources of distraction for drivers. The study tracked 70 drivers aged 18 to 80 for a week. Miniature cameras were placed in the drivers' cars, with their knowledge, and their driving behavior was randomly viewed for a week. The first three hours of each tape were discarded in the hope that drivers would act more naturally later in the week. The researchers considered a wide range of activities to be distracting, such as leaning over to reach for something, fiddling with radio controls, attending to babies, talking to passengers, applying makeup, and opening and reading their mail. The study found that 21 of the subjects used cell phones while their vehicles were moving.

(a) Is there significant evidence that fewer than half of all drivers use cell phones while their vehicles are moving? Carry out appropriate inference, and report your results and conclusions.

(b) Do you believe that this study is subject to bias? In what regard?

9.3 USE AND ABUSE OF STATISTICAL INFERENCE

I have perfect hindsight

Inference is subtle, so mistakes in inference are also often a bit subtle. Subtle mistakes are less important than glaring, in-your-face blunders, so don't forget that most really big statistical mistakes involve things like voluntary response samples, ignoring lurking variables, and using invalid measures. Nonetheless, inference also allows room for mistakes and misunderstandings. This is particularly true of statistical significance.

Let's look at the 8000 mutual funds on sale to the investing public. Any Internet investment site worth clicking on will tell you which fund produced the highest return over the past (say) three years. As the year 2000 opened, one site claimed that the winner was the Kinetics Internet Fund. If you had bought this fund three years earlier, you would have gained 112% per year. This return is significantly higher than the average for all funds.

It should be clear that "look back and take the best" isn't a suitable foundation for a significance test. Significance tests work when we form a hypothesis such as "Kinetics Internet Fund will have higher-than-average returns" and then wait for data. It makes no sense to look at past data, take the fund that happened to be the best out of 8000 funds, and ask if this fund was above average. It does make sense to ask if funds that do better than average in one period tend to stay better than average. If they do, it would make sense to buy funds that have done well in the past. Professors of finance have devoted lots of computer time and lots of significance tests to this question. The answer seems to be that the worst funds do tend to stay bad (until they disappear), but that there is no statistically significant evidence for persistent good performance. In fact, Kinetics Internet Fund was in the *bottom* 25 out of 6700 mutual funds at midyear, 2000.[18]

ACTIVITY 9.3A The *Test of Significance* applet

Go to www.whfreeman.com/sta and select the *Test of Significance* applet. Enter the number of shots (100 maximum) and click the "Shoot" button to gather data on the free-throw ability of the shooter.

ACTIVITY 9.3A The *Test of Significance* applet *(continued)*

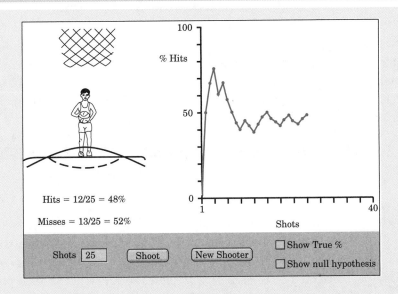

Click the "New Shooter" button to change the true value of $p =$ the probability that the shooter makes a basket. The plot shows the accumulating percent of shots made ($\hat{p}$) and can include horizontal lines at the null hypothesis (0.8) and at the true value of p for the current shooter if those boxes are checked.

1. Draw a picture that shows what you think the plot of the accumulating percent of shots made will look like when the null hypothesis is true.

2. Draw a picture that shows what you think the plot of the accumulating percent of shots made will look like when the null hypothesis is not true.

APPLET TIP: The null hypothesis is always $p = 0.8$, and the alternative hypothesis is $p < 0.8$ since the applet never fixes p at a value higher than 0.8.

3. Use the applet to test your two answers above. Did the plots look similar to your drawings? What key aspect(s) of the plots did you capture? What aspect(s) differed?

4. Use the applet to have the shooter make five shots. Based on these data, can you decide whether the shooter can hit 80% of his shots overall? Explain why or why not.

5. Add another five shots to the total. Can you decide at this point?

6. Continue to add five shots at a time until you are confident that you know whether the null hypothesis is true or false. How many shots did it take?

7. Check the "Show true %" box to see the true value of p.

ACTIVITY 9.3A The *Test of Significance* applet *(continued)*

8. Remove the true value of p from the plot and pick a new shooter. Repeat the process of using the applet to add five shots at a time until you feel confident that you know whether the null hypothesis is true or false. Did it take more shots or fewer shots this time? View the true value of p. Did it take longer for you to determine whether the null hypothesis was true or false when p was closer to 0.8 or when it was farther from 0.8? Explain why this happened.

Using inference wisely

We have met the two major types of statistical inference: confidence intervals and significance tests. We have, however, seen only one inference method of each type, designed for inference about a population proportion p. There are libraries of both books and software filled with methods for inference about various parameters in various settings. The reasoning of confidence intervals and significance tests remains the same, but the details can seem overwhelming. The first step in using inference wisely is to understand your data and the questions you want to answer and to fit the method to its setting. Here are some tips on inference, adapted to the one setting we are familiar with.

The design of the data production matters "Where do the data come from?" remains the first question to ask in any statistical study. Any inference method is intended for use in a specific setting. For our confidence interval and test for a proportion p:

Dropping out

An experiment found that weight loss is significantly more effective than exercise for reducing high cholesterol and high blood pressure. The 170 subjects were randomly assigned to a weight-loss program, an exercise program, or a control group. Only 111 of the 170 subjects completed their assigned treatment, and the analysis used data from these 111. Did the dropouts create bias? Always ask about details of the data before trusting inference.

- The data must be a simple random sample (SRS) from the population of interest. When you use these methods, you are acting as if the data are an SRS. In practice, it is often not possible to actually choose an SRS from the population. Your conclusions may then be open to challenge.

- These methods are not correct for sample designs more complex than an SRS, such as stratified samples. There are other methods that fit these settings.

- There is no correct method for inference from data haphazardly collected with bias of unknown size. Fancy formulas cannot rescue badly produced data.

• Other sources of error, such as dropouts and nonresponse, are important.
Remember that confidence intervals and tests use the data you give them and
ignore these practical difficulties.

ACTIVITY 9.3B How to get a shorter confidence interval

Materials: TI-83 calculator

1. Make up five numbers like five grade point averages or the
number of boyfriends or girlfriends five of your classmates have ever had. Then
make up a condition, such as "GPA $\geq$ 3" or "number of boyfriends $\geq$ 3." Record
a 1 for those data that satisfy your condition and 0 for those that fail your
condition. Calculate your sample proportion $\hat{p}$ of successes. For example, here
are some imaginary GPAs, with the 1s and 0s for success/failure (GPA $\geq$ 3):

4	3.7	2.6	2.8	3
1	1	0	0	1

so $\hat{p} = 3/5 = 0.6$.

2. Calculate the 95% confidence interval for your sample proportion. Find the
length of the interval.

3. Make up 15 more data values and append these values to your list from Step
1 to make a total of 20 data values. Code the rest of these values as 1 (success)
or 0 (failure), according to your condition.

4. Find the length of the confidence interval for this larger data set. Compare it
to the length of the confidence interval in Step 2, and look for a relationship
between the two lengths.

5. Write a rule for a way to cut a confidence interval in half. Does your rule
work for the data of other members of your class?

6. Why do you think the confidence interval in Step 4 is not *exactly* half as wide
as the interval in Step 2?

Know how confidence intervals behave A confidence interval estimates the unknown value of a parameter and also tells us how uncertain the estimate is. All confidence intervals share these behaviors:

- The confidence level says how often the *method* catches the true parameter in very many uses. We never know whether a specific data set gives us an interval that contains the true parameter. All we can say is that "we got this result from a method that works 95% of the time." Our data set might be one of the 5% that produce an interval that misses the parameter. If that risk bothers you, use a 99% confidence interval.

- High confidence is not free. A 99% confidence interval will be wider than a 95% confidence interval based on the same data. There is a trade-off between how closely we can pin down the parameter and how confident we are that we have caught the parameter.

- Larger samples give shorter intervals. If we want high confidence and a short interval, we must take a larger sample. The length of our confidence interval for p goes down in proportion to the square root of the sample size. To cut the interval in half, we must take four times as many observations (see Activity 9.3B). This is typical of many types of confidence intervals.

Know what statistical significance says Many statistical studies hope to show that some claim is true. A clinical trial compares a new drug with a standard drug because the doctors hope that patients given the new drug will do better. A psychologist studying gender differences suspects that women will do better than men (on the average) on a test that measures social-networking skills. The purpose of significance tests is to weigh the evidence that the data give in favor of such claims. That is, a test helps us know if we have found what we were looking for.

To do this, we ask what would happen if the claim were not true. That's the null hypothesis—no difference between the two drugs, no difference between women and men. A significance test answers only one question: "How strong is the evidence that the null hypothesis is not true?" A test answers this question by giving a P-value. The P-value tells us how unlikely our data would be if the null hypothesis were true. Data that are very unlikely are good evidence that the null hypothesis is not true. We never know whether the hypothesis is true for this specific population. All we can say is, "Data like these would occur only 5% of the time if the hypothesis were true."

This kind of indirect evidence against the null hypothesis (and for the effect we hope to find) is less straightforward than a confidence interval.

Know what your methods require Our test and confidence interval for a proportion p require that the population be much larger than the sample. They also require that the sample itself be reasonably large so that the sampling distribution of the sample proportion is close to normal. We have said little about the specifics of these requirements because the reasoning of inference is more important. Just as there are inference methods that fit stratified samples, there are methods that fit small samples and small populations. If you plan to use statistical inference in practice, you will need help from a statistician (or need to learn lots more statistics) to manage the details.

Most of us read about statistical studies more often than we actually work with data ourselves. Concentrate on the big issues, not on the details of whether the authors used exactly the right inference methods. Does the study ask the right questions? Where did the data come from? Do the results make sense? Does the study report confidence intervals so you can see both the estimated values of important parameters and how uncertain the estimates are? Does it report P-values to help convince you that findings are not just good luck?

EXERCISES

9.37 Cholesterol and breast cancer The October 2003 issue of the *Journal of Women's Health* included a report of a study of 7528 women aged 65 and older who were participants in a study of osteoporotic fractures. At one clinic visit during the study, researchers asked the women whether they took any cholesterol-lowering drugs (such as the popular statins). None of the women had breast cancer at the time the researchers collected information about their use of cholesterol-lowering medication. The scientists then followed the women, for nearly seven years on average, to see who developed breast cancer. After accounting for age and weight (two factors related to breast cancer risk), the researchers found that statin users were about 75% less likely to develop breast cancer than women who weren't taking any cholesterol-lowering medication.[19] A news report said the study suggested that cholesterol-lowering medications might reduce women's risk of breast cancer. Can we be confident in this conclusion? Explain.

9.38 How big is the moon? If 1000 80% confidence intervals are independently calculated for the equatorial diameter of the moon, would you expect about 200 (20%) of these intervals to fail to include the moon's true equatorial diameter? Explain.

9.39 True or false? Because 95/68 ≈ 1.4, 95% confidence intervals are usually about 1.4 times as long as 68% intervals.

9.40 Margin of error Which of the following would decrease the margin of error in a confidence interval, and which would have little effect on the margin of error?

(a) If you had a random sample that was only one-fourth as large as before.

(b) If you had twice as many observations.

(c) If you created a 68% confidence interval instead of an 80% confidence interval.

9.41 More true/false Which of the following statements are true and which are false? For those statements that are false, briefly explain why or give a counter-example.

(a) If the P-value is small, then the null hypothesis does not provide a plausible explanation of the data.

(b) If the P-value is small, then the sample statistic is a poor estimate of the population parameter.

(c) A P-value of 0.999 says that the null hypothesis is a reasonable explanation of the data.

(d) When the P-value is 0.5, there is a 50-50 chance that the alternative hypothesis is true.

(e) A test statistic is used to measure the difference between the data and what is expected under the null hypothesis.

The woes of significance tests

The purpose of a significance test is usually to give evidence for the presence of some effect in the population. The effect might be a probability of heads different from one-half for a coin or a longer mean survival time for patients given a new cancer treatment. If the effect is large, it will show up in most samples—the proportion of heads among our tosses will be far from one-half, or the patients who get the new treatment will live much longer than those in the control group. Small effects, such as a probability of heads only slightly different from one-half, will often be hidden behind the chance variation in a sample. This is as it should be: big effects are easier to detect. That is, the P-value will usually be small when the population truth is far from the null hypothesis.

The "woes" of testing start with the fact that a test just measures the strength of evidence against the null hypothesis. It says nothing about how big or how important the effect we seek in the population really is. For example, our hypothesis might be "This coin is balanced." We express this hypothesis in terms of the probability p of getting a head as $H_0 : p = 1/2$. No real coin is exactly balanced, so we know that this hypothesis is not exactly true. If this coin has probability $p = 0.502$ of a head, we might say that for practical purposes it is balanced. A statistical test doesn't think about "practical purposes." It just asks if there is evidence that p is not exactly equal to 0.5. The focus of tests on the strength of the evidence against an exact null hypothesis is the source of much confusion in using tests.

Pay particular attention to the size of the sample when you read the result of a significance test. Here's why:

- Larger samples make tests of significance more sensitive. If we toss a coin hundreds of thousands of times, a test of $H_0 : p = 0.5$ will often give a very low P-value when the truth for this coin is $p = 0.502$. The test is right—it found good evidence that p really is not exactly equal to 0.5—but it has picked up a difference so small that it is of no practical interest. **A finding can be statistically significant without being practically important.**

- On the other hand, tests of significance based on small samples are often not sensitive. If you toss a coin only 10 times, a test of $H_0 : p = 0.5$ will often give a large P-value even if the truth for this coin is $p = 0.7$. Again the test is right—10 tosses are not enough to give good evidence against the null hypothesis. **Lack of significance does not mean that there is no effect, only that we do not have good evidence for an effect. Small samples often miss effects that are really present in the population.**

Whatever the truth about the population—whether $p = 0.7$ or $p = 0.502$—more observations allow us to pin down p more closely. If p is not 0.5, more observations will give more evidence of this, that is, a smaller P-value. Because significance depends strongly on the sample size as well as on the truth about the population, statistical significance tells us nothing about how large or how practically important an effect is. Large effects (like $p = 0.7$ when the null hypothesis is $p = 0.5$) often give data that are insignificant if we take only a small sample. Small effects (like $p = 0.502$) often give data that are highly significant if we take a large sample. Let's return to a favorite example to see how significance changes with sample size.

EXAMPLE 9.9   Count Buffon's coin again

Count Buffon tossed a coin 4040 times and got 2048 heads. His sample proportion of heads was

$$\hat{p} = \frac{2048}{4040} = 0.507$$

Is Buffon's coin balanced? The hypotheses are

$$H_0 : p = 0.5$$
$$H_a : p \neq 0.5$$

The test of significance works by locating the sample outcome $\hat{p} = 0.507$ on the sampling distribution that describes how $\hat{p}$ would vary if the null hypothesis were true. Figure 9.10

repeats Figure 9.8. It shows that the observed $\hat{p} = 0.507$ is not surprisingly far from 0.5 and therefore is not good evidence against the hypothesis that the true p is 0.5. The P-value, which is 0.37, just makes this precise.

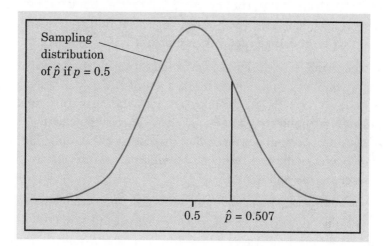

FIGURE 9.10 The sampling distribution of the proportion of heads in 4040 tosses of a coin if in fact the coin is balanced. Sample proportion 0.507 is not an unusual outcome.

Suppose that Count Buffon got the same result, $\hat{p} = 0.507$, from tossing a coin 1000 times and also from tossing a coin 100,000 times. The sampling distribution of $\hat{p}$ when the null hypothesis is true always has mean 0.5, but its standard deviation gets smaller as the sample size n gets larger. Figure 9.11 displays the three sampling distributions, for $n = 1000$, $n = 4040$, and $n = 100,000$. The middle curve in this figure is the same normal curve

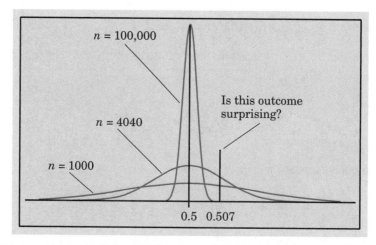

FIGURE 9.11 The three sampling distributions of the proportion of heads in 1000, 4040, and 100,000 tosses of a balanced coin. Sample proportion 0.507 is not unusual in 1000 or 4040 tosses but is very unusual in 100,000 tosses.

as in Figure 9.10, but drawn on a scale that allows us to show the very tall and narrow curve for $n = 100,000$. Locating the sample outcome $\hat{p} = 0.507$ on the three curves, you see that the same outcome is more or less surprising depending on the size of the sample.

The P-values are $P = 0.66$ for $n = 1000$, $P = 0.37$ for $n = 4040$, and $P = 0.000009$ for $n = 100,000$. Imagine tossing a balanced coin 1000 times repeatedly. You will get a proportion of heads at least as far from one-half as Buffon's 0.507 in about two-thirds of your repetitions. If you toss a balanced coin 100,000 times, however, you will almost never (9 times in a million repeats) get an outcome this unbalanced.

The outcome $\hat{p} = 0.507$ is not evidence against the hypothesis that the coin is balanced if it comes up in 1000 tosses or in 4040 tosses. It is completely convincing evidence if it comes up in 100,000 tosses.

Beware the naked *P*-value

The *P*-value of a significance test depends strongly on the size of the sample, as well as on the truth about the population.

It is bad practice to report a *P*-value without also giving the sample size and a statistic or statistics that describe the sample outcome.

The advantages of confidence intervals

Example 9.9 suggests that we not rely on significance alone in understanding a statistical study. Just knowing that the sample proportion was $\hat{p} = 0.507$ helps a lot. You can decide whether this deviation from one-half is large enough to interest you. Of course, $\hat{p} = 0.507$ isn't the exact truth about the coin, just the chance result of Count Buffon's tosses. So a confidence interval, whose width shows how closely we can pin down the truth about the coin, is even more helpful. Here are the 95% confidence intervals for the true probability of a head p, based on the three sample sizes in Example 9.9. You can check that the method of Section 9.1 gives these answers.

Number of tosses	95% confidence interval
$n = 1000$	0.507 ± 0.031, or 0.476 to 0.538
$n = 4040$	0.507 ± 0.015, or 0.492 to 0.522
$n = 100,000$	0.507 ± 0.003, or 0.504 to 0.510

The confidence intervals make clear what we know (with 95% confidence) about the true p. The intervals for 1000 and 4040 tosses include 0.5, so we are not confident the coin is unbalanced. For 100,000 tosses, however, we are confident that the true p lies between 0.504 and 0.510. In particular, we are confident that it is not 0.5.

Give a confidence interval

Confidence intervals are more informative than tests because they actually estimate a population parameter. They are also easier to interpret. It is good practice to give confidence intervals whenever possible.

Significance at the 5% level isn't magical

The purpose of a test of significance is to describe the degree of evidence provided by the sample against the null hypothesis. The P-value does this. But how small a P-value is convincing evidence against the null hypothesis? This depends mainly on two circumstances:

- *How plausible is H_0?* If H_0 represents an assumption that the people you must convince have believed for years, strong evidence (small P) will be needed to persuade them.

- *What are the consequences of rejecting H_0?* If rejecting H_0 in favor of H_a means making an expensive changeover from one type of product packaging to another, you need strong evidence that the new packaging will boost sales.

These criteria are a bit subjective. Different people will often insist on different levels of significance. Giving the P-value allows each of us to decide individually if the evidence is sufficiently strong.

Users of statistics have often emphasized standard levels of significance such as 10%, 5%, and 1%. For example, courts have tended to accept 5% as a standard in discrimination cases. This emphasis reflects the time when tables of critical values rather than computer software dominated statistical practice. The 5% level ($\alpha = 0.05$) is particularly common. **There is no sharp border between "significant" and "insignificant," only increasingly strong evidence as the P-value decreases.** There is no practical distinction between the P-values 0.049 and 0.051. It makes no sense to treat $P \leq 0.05$ as a universal rule for what is significant.

Beware of searching for significance

Statistical significance ought to mean that you have found an effect that you were looking for. The reasoning behind statistical significance works well if you decide what effect you are seeking, design a study to search for it, and use a test of significance to weigh the evidence you get. In other settings, significance may have little meaning. We saw one example at the beginning of this section: it makes no sense to look at 8000 mutual funds, find the one that happened to have the highest return, then ask if this fund "significantly" outperformed the average. Here is another example.

EXAMPLE 9.10   Predicting success of trainees

You want to learn what distinguishes managerial trainees who eventually become executives from those who, after expensive training, don't succeed and leave the company. You have abundant data on past trainees—data on their personalities and goals, their college preparation and performance, even their family backgrounds and their hobbies. Statistical software makes it easy to perform dozens of significance tests on these dozens of variables to see which ones best predict later success. Aha! You find that future executives are significantly more likely than washouts to have an urban or suburban upbringing and an undergraduate degree in a technical field.

Before you base future recruiting on these findings, recall that results significant at the 5% level occur 5 times in 100 in the long run even when H_0 is true. When you make dozens of tests at the 5% level, you expect a few of them to be significant by chance alone. Running one test and reaching the $\alpha = 0.05$ level is reasonably good evidence that you have found something. Running several dozen tests and reaching that level once or twice is not.

In the mutual-fund example, we looked for the best, then tested it as if we had not first sought it out. In Example 9.10, we just tested everything and took the most significant. Both bad practices confuse the roles of exploratory analysis of data and formal statistical inference.

Searching data for suggestive patterns is certainly legitimate. Exploratory data analysis is an important part of statistics. But the reasoning of formal inference does not apply when your search for a striking effect in the data is successful. The remedy is clear. Once you have a hypothesis, design a study to search specifically for the effect you now think is there. If the result of this study is statistically significant, you have real evidence.

EXERCISES

9.42 A television poll   A television news program conducts a call-in poll about a proposed city ban on handgun ownership. Of the 2372 calls, 1921 oppose the ban. The station, following recommended practice, makes a confidence statement: "81% of the Channel 13 Pulse Poll sample opposed the ban. We can be 95% confident that the true proportion of citizens opposing a handgun ban is within 1.6% of the sample result." The confidence interval calculation is correct, but the conclusion is not justified. Why?

9.43 Ages of presidents Joe is writing a report on the backgrounds of American presidents. He looks up the ages of all 43 presidents when they entered office. Because Joe took a statistics course, he uses these 43 numbers to get a 95% confidence interval for the mean age of all men who have been president. This makes no sense. Why not?

9.44 Who will win? A poll taken shortly before an election finds that 52% of the voters favor candidate Shrub over candidate Snort. The poll has a margin of sampling error of plus or minus three percentage points at 95% confidence. The poll press release says the election is too close to call. Why?

9.45 Why we seek significance Asked why statistical significance appears so often in research reports, a student says, "Because saying that results are significant tells us that they cannot easily be explained by chance variation alone." Do you think that this statement is essentially correct? Explain your answer.

9.46 What is significance good for? Which of the following questions does a test of significance answer?
(a) Is the sample or experiment properly designed?
(b) Is the observed effect due to chance?
(c) Is the observed effect important?

Putting significance tests in perspective

Research studies in many fields rely on tests of significance. Often habit leads to overreliance. Robert Rosenthal of Harvard, an eminent psychologist well known for statistical work, says, "Many of us were trained that we're not supposed to look too carefully at the data. You come up with a hypothesis, decide on a statistical test, do the test, and if your results are significant at .05, you've supported your hypothesis. If not, you stick it all in a drawer and never look at your data."

That should shock you. "Always plot your data" has been one of our mottoes, along with "Always ask where the data come from." Psychologists generally think carefully about how their data are produced. How can it be that many psychologists hardly glance at the data? The reason, according to some psychologists, is the tyranny of significance tests and of 0.05 as the magical sign that a result is important. In particular, custom dictates that results should be significant at the 5% level in order to be published, so researchers fall into the bad habits that Rosenthal describes. Significant at 5%, success. Not significant at 5%, failure. The limitations of tests are so severe, the risks of misinterpretation so high, and bad habits so ingrained, say these critics, that significance tests should be banned from professional journals in psychology.

In response, the American Psychological Association appointed a Task Force on Statistical Inference. Robert Rosenthal was one of the cochairs of this group.

The task force did not want to ban tests. Its report was in fact a summary of good statistical practice. Define your population clearly. Describe your data production and prefer randomized methods whenever possible. Describe your variables and how they were measured. Give your sample size and explain how you decided on the sample size. If there were dropouts or other practical problems, mention them. "As soon as you have collected your data, before you compute any statistics, *look at your data.*" Ask whether the results of computations make sense to you. Recognize that "inferring causality from nonrandomized designs is a risky enterprise."

There is more, but here is the punch line about statistical tests: "It is hard to imagine a situation in which a dichotomous accept-reject decision is better than reporting an actual *p* value or, better still, a confidence interval.... Always provide some effect-size estimate when reporting a *p* value." Ban tests? "Although this might eliminate some abuses, the committee thought there were enough counterexamples to justify forbearance."[20]

APPLICATION 9.3   Test Results and the "No Child Left Behind" Act

In 2002, the Bush administration sponsored legislation known as the "No Child Left Behind" act, which attempted to improve public education by mandating a testing program and minimal standards. The act left it up to the states to develop a statewide

APPLICATION 9.3 Test Results and the "No Child Left Behind" Act *(continued)*

testing program in reading and math and a minimum proficiency level. Students and schools that didn't meet the minimum performance level would be classified as failing schools. Students at those school would be able to transfer to other, supposedly better, schools. Many states established 40% as the minimum passing rate. To many people, this minimum passing rate is cast in stone. If a state mandates a 40% pass rate, then a school that has 40% or more of its students pass the test would be considered a passing school. A school that has less than 40% of students pass the test would be designated a failing school.

Principals and school system administrators don't want their schools to be declared failing schools, so many states have adopted a controversial procedure to buy some "wiggle room" on test results. Here is their position. Instead of considering test results for one year to be the entire population of interest, they argue that test results should be viewed as a *sample* of school performance from a population of multiple years. Each year's results come from a sample of students gathered at one point in time, they say. The act requires states to examine performance of subgroups (for example, based on gender, racial and ethnic backgrounds, special needs), and so sample sizes become a factor. Consequently, it makes sense to construct confidence intervals for the population passing rate. At this writing, 35 states use the confidence interval approach or a variation of it to interpret test scores.

On the other side, critics are aghast that despite Oregon's requirement that 40% of the students pass a reading test, a small high school in that state that had only 28% of the students pass the test was considered to have met the 40% pass requirement. And in Maryland, an elementary school met that state's standards even though only 31% of its pupils passed a state math test when a 41.5% pass rate was required.[21]

Let's look at the numbers. In the case of the Oregon high school, if Oregon adopted the 95% confidence level, then the (sample) proportion of $\hat{p} = 0.28$ passing the test has to be in the 95% confidence interval centered at 0.40.

$$(\text{------}|\text{------})$$
$$0.28 \quad\; 0.40 \quad\; 0.52$$

The margin of error is half the width of the interval, or 0.12. The formula for the margin of error for a proportion is

$$z^* \sqrt{\frac{\hat{p}\,(1 - \hat{p})}{n}}$$

where $z^* = 1.960$ for a 95% confidence interval. Set this equal to or greater than the margin of error, 0.12, and solve for n.

APPLICATION 9.3 Test Results and the "No Child Left Behind" Act *(continued)*

$$1.960 \sqrt{\frac{(0.28)\,(0.72)}{n}} \geq 0.12$$

$$1.960\,\frac{\sqrt{0.2016}}{\sqrt{n}} \geq 0.12$$

$$1.960\,\frac{\sqrt{0.2016}}{\sqrt{0.12}} \geq \sqrt{n}$$

$$7.33365 \geq \sqrt{n}$$

$$53.78 \geq n$$

If this small Oregon high school tested up to 53 students and 28% of them passed the test, then 0.28 would be inside the 95% confidence interval, and that school would have been labeled passing. Fourteen states use a 95% confidence interval.

1. Check our work. Substitute the appropriate numbers into the expression for the margin of error and show that you get a number less than the margin of error, 0.12.

2. Write a sentence to interpret this 95% confidence interval. Be sure to mention the true (population) proportion.

Now the story takes an even more interesting turn. If there were more than 53 students tested, then the school might still be declared passing if it adopted the 99% confidence level. Here is the analysis. For the 99% confidence level, $z^* = 2.576$, $\hat{p}$ is still 0.28, and the margin of error is still 0.12.

$$z^* \sqrt{\frac{\hat{p}\,(1 - \hat{p})}{n}} = 2.576 \sqrt{\frac{(0.28)\,(0.72)}{n}} \geq 0.12$$

$$\frac{1.15662}{0.12} \geq \sqrt{n}$$

$$\sqrt{n} \leq 9.6385$$

$$n \leq 92.9$$

So if they had up to 92 students take the test and 28% passed, then the school's passing rate would be within the 99% confidence interval centered at 40%, and the school would be considered passing. Life is good. Thirteen states use a 99% confidence interval.

APPLICATION 9.3 Test results and the "No Child Left Behind" act *(continued)*

3. Why does the 99% confidence level produce a wider interval? (*Hint:* Look at the formula for the margin of error.)

4. Use the same procedure to determine how many pupils in the Maryland elementary school could have been tested in order for the sample proportion passing (0.31) to be inside a 95% confidence interval centered at 0.415.

5. Answer the same question as in (4), but this time use a 99% confidence interval.

6. Do you think a confidence interval approach to interpreting test scores is appropriate? What do you think is the most compelling reason for or against this practice?

EXPLORING THE WEB

The report of the American Psychological Association's Task Force on Statistical Inference is an excellent brief introduction to wise use of inference. The report appeared in the journal *American Psychologist* in 1999. You can find it on the Web at www.apa.org/journals/amp/amp548594.html.

Use a search engine to find more information on the "No Child Left Behind" act. You might also want to find out what your state's pass/fail policy is under this act.

STATISTICS IN SUMMARY

Statistical inference is less widely applicable than exploratory analysis of data. Any inference method requires the right setting, in particular the right design for a random sample or randomized experiment. Understanding the meaning of confidence levels and statistical significance helps avoid improper conclusions. Increasing the number of observations has a straightforward effect on confidence intervals: the interval gets shorter for the same level of confidence. More observations usually drive down the *P*-value of a test when the truth about the population stays the same, making tests harder to interpret than confidence intervals. A finding with a small *P*-value may not be practically interesting if the sample is large, and an important truth about the population may fail to be significant if the sample is small. Avoid depending on fixed significance levels such as 5% to make decisions.

SECTION 9.3 EXERCISES

9.47 Why are larger samples better? Statisticians prefer large samples. Describe briefly the effect of increasing the size of a sample (or the number of subjects in an experiment) on each of the following:
(a) The margin of error of a 95% confidence interval.
(b) The P-value of a test when H_0 is false and all facts about the population remain unchanged as n increases.

9.48 Is this convincing? You are planning to test a new vaccine for a virus that now has no vaccine. Since the disease is usually not serious, you will expose 100 volunteers to the virus. After some time, you will record whether or not each volunteer has been infected.
(a) Explain how you would use these 100 volunteers in a designed experiment to test the vaccine. Include all important details of designing the experiment (but don't actually do any random allocation).
(b) You hope to show that the vaccine is more effective than a placebo. State H_0 and H_a. (Notice that this test compares two population proportions.)
(c) The experiment gave a P-value of 0.25. Explain carefully what this means.
(d) Your fellow researchers do not consider this evidence strong enough to recommend regular use of the vaccine. Do you agree?

9.49 Searching for ESP A researcher looking for evidence of extrasensory perception (ESP) tests 500 subjects. Four of these subjects do significantly better $(P < 0.01)$ than random guessing.
(a) Is it proper to conclude that these four people have ESP? Explain your answer.
(b) What should the researcher now do to see whether any of these four subjects have ESP?

9.50 Comparing package designs A company compares two package designs for a laundry detergent by placing bottles with both designs on the shelves of several markets. Checkout scanner data on more than 5000 bottles bought show that more shoppers bought Design A than Design B. The difference is statistically significant $(P = 0.02)$. Can we conclude that consumers strongly prefer Design A? Explain your answer.

9.51 Color blindness in Africa An anthropologist suspects that color blindness is less common in societies that live by hunting and gathering than in settled agricultural societies. He tests a number of adults in two populations in Africa, one of each type. The proportion of color-blind people is significantly lower $(P < 0.05)$ in the hunter-gatherer population. What additional information would you want to help you decide whether you accept the claim about color blindness?

9.52 Blood types in Southeast Asia One way to assess whether two human groups should be considered separate populations is to compare their distributions of blood types. An anthropologist finds significantly different ($P = 0.01$) proportions of the main human blood types (A, B, AB, O) in different tribes in central Malaysia. What other information would you want before you agree that these tribes are separate populations?

CHAPTER 9 REVIEW

Statistical inference draws conclusions about a population on the basis of sample data and uses probability to indicate how reliable the conclusions are. A confidence interval estimates an unknown parameter. A significance test shows how strong the evidence is for some claim about a parameter. Sections 9.1 and 9.2 present the reasoning of confidence intervals and tests and give details for inference about a population proportion p.

The probabilities in both confidence intervals and tests tell us what would happen if we used the formula for the interval or test very many times. A confidence level is the probability that the formula for a confidence interval actually produces an interval that contains the unknown parameter. A 95% confidence interval gives a correct result 95% of the time when we use it repeatedly. Figure 9.12 illustrates the reasoning using the approximate 95% confidence interval for a population proportion p.

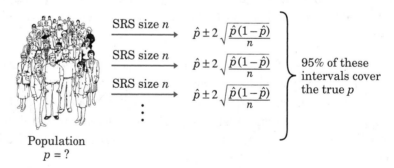

FIGURE 9.12 The idea of a confidence interval.

A P-value is the probability that the test would produce a result at least as extreme as the observed result if the null hypothesis really were true. Figure 9.13 on the facing page illustrates the reasoning, placing the sample proportion $\hat{p}$ from our one sample on the normal curve that shows how $\hat{p}$ would vary in all possible samples if the null hypothesis were true. A P-value tells us how surprising the observed outcome is. Very surprising outcomes (small *P*-values) are good evidence that the null hypothesis is not true.

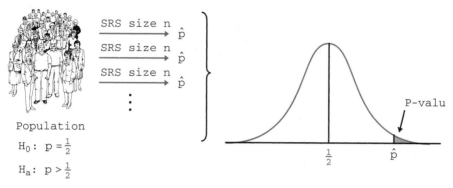

$H_0: p = \frac{1}{2}$

$H_a: p > \frac{1}{2}$

FIGURE 9.13 The idea of a significance test.

To detect sound and unsound uses of inference, you must know the basic reasoning and also be aware of some fine points and pitfalls. Section 9.3 will help.

Here are the most important skills you should have acquired after studying Chapter 9.

A. SAMPLING DISTRIBUTIONS

1. Explain the idea of a sampling distribution. See Figure 9.1 (page 492).

2. Use the normal sampling distribution of a sample proportion and the 68–95–99.7 rule to find probabilities involving $\hat{p}$.

B. CONFIDENCE INTERVALS

1. Explain the idea of a confidence interval. See Figure 9.12.

2. Explain in nontechnical language what is meant by "95% confidence" and other statements of confidence in statistical reports.

3. Use the basic formula $\hat{p} \pm 2\sqrt{\dfrac{\hat{p}(1 - \hat{p})}{n}}$ to obtain an approximate 95% confidence interval for a population proportion p.

4. Understand how the margin of error of a confidence interval changes with the sample size and the level of confidence.

5. Detect major mistakes in applying inference, such as improper data production, selecting the best of many outcomes, ignoring high nonresponse and outliers.

C. SIGNIFICANCE TESTS

1. Explain the idea of a significance test. See Figure 9.13.

2. State the null and alternative hypotheses in a testing situation when the parameter in question is a population proportion p.

3. Explain in nontechnical language the meaning of the P-value when you are given the numerical value of P for a test.

4. Explain the meaning of "statistically significant at the 5% level" and other statements of significance. Explain why significance at a specific level such as 5% is less informative than a P-value.

5. Recognize that significance testing does not measure the size or importance of an effect.

6. Recognize and explain the effect of small and large samples on the significance of an outcome.

CHAPTER 9 REVIEW EXERCISES

9.53 Big events, I An SRS of 489 adults found that 347 chose World War II from a list of events as the most important event of the twentieth century. Give a 95% confidence interval for the proportion of all adults who think World War II is the century's most important event.

9.54 Drinking problems, I An SRS of 1039 adults found that 374 said that drinking had been a problem in their families. Give a 95% confidence interval for the proportion of all adults who have had a drinking-related problem in their families.

9.55 Big events, II Exercise 9.53 concerns an SRS of 489 adults. Suppose that (unknown to the pollsters) exactly 70% of all adults would choose World War II as the most important event of the twentieth century. Imagine that we take very many SRSs of size 489 from this population and record the percent in each sample who choose World War II. Where would the middle 95% of all values of this percent lie?

9.56 Drinking problems, II Exercise 9.54 concerns an SRS of 1039 adults. Suppose that in the population of all adults, exactly 35% would say that drinking had been a problem in their families. Imagine that we take a very large number of SRSs of size 1039. For each sample, record the proportion of the sample who have had a drinking problem in their families.
(a) What is the sampling distribution that describes the values $\hat{p}$ would take in our samples?
(b) Use this distribution and the 68–95–99.7 rule to find the approximate percent of all samples in which more than 36.5% of the respondents have had a drinking problem in their families.

9.57 Our dirty little secret, II In August 2000, researchers from the American Society for Microbiology surveyed a random sample of 1021 American adults. More than 95% of the respondents said they washed their hands after using public restrooms. But observers stationed in public facilities in Chicago, New Orleans, San Francisco, Atlanta, and New York reported that 67% of 7836 people did. How could you explain the huge difference in these two results?

9.58 Our dirty little secret, III Researchers in the study described in Exercise 9.57 found that men are less likely than women to wash their hands after using public bathrooms. About 58% of men and 75% of women were seen washing up. The study did not provide a breakdown of the sample by gender, so assume that half of the total sample of 7836 people were men.
(a) Is there significant evidence that more women wash up after using public restrooms than men? Use the 1% significance level for the test.
(b) Find a 95% confidence interval for the true proportion of men who wash their hands after using public facilities.

9.59 Senior citizens Table 4.4 (page 225) records the percent of residents aged 65 or older in each of the 50 states. You can check that this percent is 14% or higher in 9 of the states. So the sample proportion of states with at least 14% of elderly residents is $\hat{p} = 9/50 = 0.18$. Explain why it does not make sense to go on to obtain a 95% confidence interval for the population proportion p.

9.60 Roulette A roulette wheel has 18 red slots among its 38 slots. You observe many spins and record the number of times that red occurs. Now you want to use these data to test whether the probability p of a red has the value that is correct for a fair roulette wheel. State the hypotheses H_0 and H_a that you will test.

9.61 When shall we call you? As you might guess, telephone sample surveys get better response rates during the evening than during the weekday daytime. One study called 2304 randomly chosen telephone numbers on weekday mornings. Of these, 1313 calls were answered and only 207 resulted in interviews. Of 2454 calls on weekday evenings, 1840 were answered and 712 interviews resulted. Give two 95% confidence intervals, for the proportions of all calls that are answered on weekday mornings and on weekday evenings. Are you confident that the proportion is higher in the evening?

9.62 Alternative medicine A nationwide random survey of 1500 adults asked about attitudes toward "alternative medicine" such as acupuncture, massage therapy, and herbal therapy. Among the respondents, 660 said they would use alternative medicine if traditional medicine was not producing the results they wanted. Do these data provide good evidence that more than 1/3 of all adults would use alternative medicine if traditional medicine did not produce the results they wanted?

(a) State the hypotheses to be tested.

(b) If your null hypothesis is true, what is the sampling distribution of the sample proportion $\hat{p}$? Sketch this distribution.

(c) Mark the actual value of $\hat{p}$ on the curve. Does it appear surprising enough to give good evidence against the null hypothesis?

(d) Carry out the significance test called for in detail. Show the five steps of the inference procedure (hypotheses, sampling distribution, data, P-value, conclusion) clearly.

Chapter 10

Inference for Tables and Means

10.1   Two-Way Tables and the Chi-Square Test
10.2   Inference about a Population Mean

10.1 TWO-WAY TABLES AND THE CHI-SQUARE TEST

Female college professors

Purdue University is a Big Ten university that emphasizes engineering, scientific, and technical fields. In the 1998–1999 academic year, Purdue had 1621 professors, of whom 335 were women. That's just over 20%, or 1 out of every 5 professors. These numbers don't tell us much about the place of women on the faculty. As usual, we must look at relationships among several variables, not just at gender alone. For example, female faculty are more common in the humanities than in agriculture.

Let's look at the relationship between gender and a variable particularly important to faculty members—academic rank. Professors typically start as assistant professors, are promoted to associate professor (and gain tenure then), and finally reach the rank of full professor. Here is a *two-way table* that breaks down Purdue's 1621 faculty members by both gender and academic rank:

	Female	Male	Total
Assistant professors	126	213	339
Associate professors	149	411	560
Full professors	60	662	722
Total	335	1286	1621

The table makes the place of women on the faculty much clearer. The number of men goes up as we climb the ladder of ranks, and the number of women goes down. Rates speak more clearly than counts, so let's calculate some percents. More than 37% of the assistant professors are women, but women make up only about 27% of the associate professors and only about 8% of the full professors. Women are strikingly underrepresented in the highest rank.

The table reports the facts but does not explain them. It typically takes about 6 years as an assistant professor before promotion to associate professor and still more time to become a full professor. It may be that many women joined the faculty only in the past decade and are still in the lower ranks because they are young. Or, more seriously, it may be that women have a harder time gaining promotion. More data are needed to say which explanation fits.

ACTIVITY 10.1 Sports preference

Materials: 3 × 5 cards or slips of paper

Is there a relationship between gender and sports preference? Do girls and boys like the same sports, or does sports preference depend on one's gender?

1. Do we need to clarify "preference"? Some might prefer to play one sport and watch another. Talk it over as a class and decide how you want to handle it.

2. Do you think that sports preference depends on whether you are male or female? If you think boys and girls prefer different sports, do you think this difference would be statistically significant for your class? Or do you think that any difference, if there is one, is insignificant? Write down your opinion and why you think that way.

3. On a slip of paper or a 3 × 5 card, write your gender (male or female) and your sports preference (basketball, football, soccer, or other). The teacher will collect your data. What kind of data are these?

4. Copy the table below onto your paper. As the teacher reveals the votes, use tally marks to record the results in this two-way table.

	Basketball	Football	Soccer	Other
Girls				
Boys				

When you finish your tally, make a new table and record the counts in each category.

5. Calculate the row and column totals. Calculate the percent of boys who prefer basketball. Then calculate the percent of girls who prefer basketball. Do the same for the other categories.

Keep your results handy; you will need them for a later exercise.

Two-way tables

The rank and gender of college faculty are both categorical variables. That is, they place individuals into categories but do not have numerical values that allow us to describe relationships by scatterplots, correlation, or regression lines. To display relationships between two categorical variables, use a **two-way table** like the table of rank and gender of Purdue faculty. Rank is the **row variable** because each row in the table describes faculty in one rank. Gender is the **column variable** because each column describes one gender. The entries in the table are the counts of faculty in each rank-by-gender class. Although both rank and gender are categorical variables, rank has a natural order from lowest to highest. The order of the rows in the table reflects the order of the categories.

How can we best grasp the information contained in this table? First, *look at the distribution of each variable separately.* The distribution of a categorical variable says how often each outcome occurred. The "Total" column at the right of the table contains the totals for each of the rows. These row totals give the distribution of rank for all faculty, men and women combined. The "Total" row at the bottom of the table gives the distribution of gender for faculty, all ranks combined. It is often clearer to present these distributions using percents. We might report the distribution of gender as

$$\text{percent female} = \frac{335}{1621} = 0.207 = 20.7\%$$

$$\text{percent male} = \frac{1286}{1621} = 0.793 = 79.3\%$$

The two-way table contains more information than the two distributions of rank alone and gender alone. The nature of the relationship between rank and gender cannot be deduced from the separate distributions but requires the full table. **To describe relationships among categorical variables, calculate appropriate percents from the counts given.**

EXAMPLE 10.1 Rank and gender of the faculty

Because there are only two genders, we can see the relationship between gender and academic rank by comparing the percents of women in the three ranks:

Assistant professors

$$\frac{126}{339} = 37.2\%$$

Associate professors

$$\frac{149}{560} = 26.6\%$$

Full professors

$$\frac{60}{722} = 8.3\%$$

The percents make it clear that women are less common in the higher ranks. That's the nature of the association between gender and rank.

In working with two-way tables, you must calculate lots of percents. Here's a tip to help decide what fraction gives the percent you want. Ask, "What group represents the total that I want a percent of?" The count for that group is the denominator of the fraction that leads to the percent. In Example 10.1, we wanted the percent *of each rank* who are women, so the counts in the ranks form the denominators.

Simpson's paradox

As is the case with quantitative variables, the effects of lurking variables can change or even reverse relationships between two categorical variables. Let's continue the theme of gender and higher education in looking at an example. The numbers here are artificial for simplicity, but they illustrate a phenomenon that often appears in real data.

"Yes, on the surface it would appear to be sex-bias but let us ask the following questions..."

EXAMPLE 10.2   Discrimination in admissions?

A university offers only two degree programs, one in electrical engineering and one in English. Admission to these programs is competitive, and the women's caucus suspects discrimination against women in the admissions process.[1] The caucus obtains the following data from the university, a two-way table of all applicants by gender and admission decision:

	Male	Female
Admit	35	20
Deny	45	40
Total	80	60

These data do show an association between the gender of applicants and their success in obtaining admission. To describe this association more precisely, we compute some percents from the data.

$$\text{percent of male applicants admitted} = \frac{35}{80} = 44\%$$

$$\text{percent of female applicants admitted} = \frac{20}{60} = 33\%$$

Aha! Almost half of the males but only one-third of the females who applied were admitted.

The university replies that although the observed association is correct, it is not due to discrimination. In its defense, the university produces a **three-way table** that classifies applicants by sex, admission decision, and the program to which they applied. We present a three-way table as several two-way tables side by side, one for each value of the third variable. In this case there are two two-way tables, one for each program:

	Engineering				English	
	Male	Female			Male	Female
Admit	30	10	Admit		5	10
Deny	30	10	Deny		15	30
Total	60	20	Total		20	40

Check that these entries add to the entries in the two-way table. The university has simply broken down that table by department. We now see that engineering admitted exactly half of all applicants, both male and female, and that English admitted one-fourth of both males and females. There is *no association* between sex and admission decision in either program.

How can no association in either program produce strong association when the two are combined? Look at the data: English is hard to get into, and mainly females apply to that program. Electrical engineering is easier to get into and attracts mainly male applicants. English had 40 female and 20 male applicants, while engineering had 60 male and only 20 female applicants. The original two-way table, which did not take account of the difference between programs, was misleading. This is an example of *Simpson's paradox*.

Simpson's paradox

An association or comparison that holds for all of several groups can disappear or even reverse direction when the data are combined to form a single group.

Simpson's paradox is just an extreme form of the fact that observed associations can be misleading when there are lurking variables. Remember the caution from Section 6.2: *Beware the lurking variable.*

The Inventor of Simpson's Paradox at work.

"No, I am **NOT** Homer Simpson."

EXAMPLE 10.3   Discrimination in mortgage lending?

Studies of applications for home mortgage loans from banks show a strong racial pattern: banks reject a higher percent of black applicants than of white applicants.[2] One lawsuit in the Washington, D.C., area contends that a bank rejected 17.5% of blacks but only 3.3% of whites.

The bank replies that lurking variables explain the difference in rejection rates. Blacks have (on the average) lower incomes, poorer credit records, and less secure jobs than whites. Unlike race, these are legitimate reasons to turn down a mortgage application. It is because these lurking variables are confounded with race, the bank says, that it rejects a higher percent of black applicants. It is even possible, thinking of Simpson's paradox, that the bank accepts a *higher* percent of black applicants than of white applicants if we look at people with the same income and credit record.

Who is right? Both sides will hire statisticians to examine the effects of the lurking variables. The court will eventually decide.

EXERCISES

10.1 Extracurricular activities and grades, I North Carolina State University studied student performance in a course required by its chemical engineering major. One question of interest was the relationship between time spent in extracurricular activities and whether a student earned a C or better in the course. Here are the data for the 119 students who answered a question about extracurricular activities:[3]

	Extracurricular activities (hours per week)		
	<2	2 to 12	>12
C or better	11	68	3
D or F	9	23	5

Calculate percents that describe the nature of the relationship between time spent on extracurricular activities and performance in the course. Give a brief summary in words.

10.2 Smoking by students and their parents, I How are the smoking habits of students related to their parents' smoking? Here is a two-way table from a survey of students in eight Arizona high schools:[4]

	Student smokes	Student does not smoke
Both parents smoke	400	1380
One parent smokes	416	1823
Neither parent smokes	188	1168

Write a brief answer to the question posed, including a comparison of selected percents.

10.3 Python eggs, I How is the hatching of water python eggs influenced by the temperature of the snake's nest? Researchers assigned newly laid eggs to one of three water temperatures: hot, neutral, or cold. Hot duplicates the extra warmth provided by the mother python, and cold duplicates the absence of the mother. Here are the data on the number of eggs and the number that hatched:[5]

	Eggs	Hatched
Cold	27	16
Neutral	56	38
Hot	104	75

(a) Make a two-way table of temperature by outcome (hatched or not).
(b) Calculate the percent of eggs in each group that hatched. The researchers anticipated that eggs would not hatch in cold water. Do the data support that anticipation?

10.4 Trying to quit, I A 1982 study of 177 people who were trying to quit smoking looked at whether or not alcohol consumption was a factor in relapse (resuming smoking).[6] Here are the results:

Alcohol consumption	Smoked	Did not smoke
Yes	20	13
No	48	96

The entry 20 means that 20 of the study participants consumed alcohol and smoked during the cessation period.
(a) Find the row and column totals.
(b) Among the drinkers, what percent smoked? Among the teetotalers, what percent smoked? What is your preliminary conclusion?
(c) Draw a bar graph that shows these percents graphically. Be sure to label your graph appropriately.

10.5 Airline flight delays Here are the numbers of flights on time and delayed for two airlines at five airports in one month. Overall on-time percents for each airline are often reported in the news. The airport that flights serve is a lurking variable that can make such reports misleading.[7]

	Alaska Airlines		America West	
	On time	Delayed	On time	Delayed
Los Angeles	497	62	694	117
Phoenix	221	12	4840	415
San Diego	212	20	383	65
San Francisco	503	102	320	129
Seattle	1841	305	201	61

(a) What percent of all Alaska Airlines flights were delayed? What percent of all America West flights were delayed? These are the numbers usually reported.

(b) Now find the percent of delayed flights for Alaska Airlines at each of the five airports. Do the same for America West.

(c) America West does worse at *every one* of the five airports, yet does better overall. That sounds impossible. Explain carefully, referring to the data, how this can happen. (The weather in Phoenix and Seattle lies behind this example of Simpson's paradox.)

Inference for a two-way table

We often gather data and arrange them in a two-way table to see if two categorical variables are related to each other. The sample data are easy to investigate: turn them into percents and look for an association between the row and column variables. Is the association in the sample evidence of an association between these variables in the entire population? Or could the sample association easily arise just from the luck of random sampling? This is a question for a significance test.

EXAMPLE 10.4   Treating cocaine addiction

Cocaine addicts need the drug to feel pleasure. Perhaps giving them a medication that fights depression will help them stay off cocaine. A three-year study compared an antidepressant called desipramine with lithium (a standard treatment for cocaine addiction) and a placebo. The subjects were 72 chronic users of cocaine who wanted to break their drug habit. Twenty-four of the subjects were randomly assigned to each treatment. Here are the counts and percents of the subjects who succeeded in staying off cocaine during the study:

Group	Treatment	Subjects	Successes	Percent
1	Desipramine	24	14	58.3%
2	Lithium	24	6	25.0%
3	Placebo	24	4	16.7%

The sample proportions of subjects who stayed off cocaine are quite different. In particular, desipramine was much more successful than lithium or a placebo. The bar graph in Figure 10.1 compares the results visually. Are these data good evidence that there is a relationship between treatment and outcome in the population of all cocaine addicts?[8]

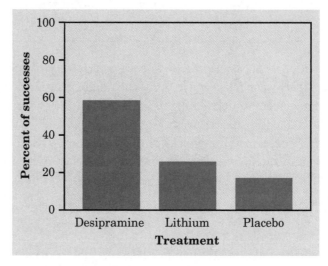

FIGURE 10.1 Bar graph comparing the success rates of three treatments for cocaine addiction.

The test that answers this question starts with a two-way table. Here's the table for the data of Example 10.4:

	Success	Failure	Total
Desipramine	14	10	24
Lithium	6	18	24
Placebo	4	20	24
Total	24	48	72

Our null hypothesis, as usual, says that the treatments have no effect. That is, addicts do equally well on any of the three treatments. The differences in the sample are just the play of chance. Our null hypothesis is

H_0: There is no association between treatment and success in the population of all cocaine addicts.

Expressing this hypothesis in terms of population parameters can be a bit messy, so we will be content with the verbal statement. The alternative hypothesis just says, "Yes, there is some association between the treatment an addict receives and whether or not he or she succeeds in staying off cocaine." The alternative doesn't specify the nature of the relationship. It doesn't say, for example, "Addicts who take desipramine are more likely to succeed than addicts given lithium or a placebo."

To test H_0, we compare the observed counts in a two-way table with the *expected counts*, the counts we would expect—except for random variation—if H_0 were true. If the observed counts are far from the expected counts, that is evidence against H_0. We can guess the expected counts for the cocaine study. In all, 24 of the 72 subjects succeeded. That's an overall success rate of one-third, because 24/72 is one-third. If the null hypothesis is true, there is no difference among the treatments. So we expect one-third of the subjects in each group to succeed. There were 24 subjects in each group, so we expect 8 successes and 16 failures in each group. If the treatment groups differ in size, the expected counts will differ also, even though we still expect the same proportion in each group to succeed. Fortunately, there is a rule that makes it easy to find expected counts. Here it is.

More tests

There are also tests for hypotheses more specific than "no relationship." Place people in classes by social status, wait 10 years, then classify the same people again. The row and column variables are the classes at the two times. We might test the hypothesis that there has been no change in the overall distribution of social status. Or we might ask if moves up in status are balanced by matching moves down. There are statistical tests for these and other hypotheses.

Expected counts

The **expected count** in any cell of a two-way table when H_0 is true is

$$\text{expected count} = \frac{\text{row total} \times \text{column total}}{\text{table total}}$$

Try it. For example, the expected count of successes in the desipramine group is

$$\text{expected count} = \frac{\text{row 1 total} \times \text{column 1 total}}{\text{table total}}$$

$$= \frac{(24)(24)}{72} = 8$$

If the null hypothesis of no treatment differences is true, we expect 8 of the 24 desipramine subjects to succeed. That's just what we guessed.

EXERCISES

10.6 Extracurricular activities and grades, II Refer to Exercise 10.1 (page 557).
(a) Write the null and alternative hypotheses for the question of interest.
(b) Calculate the expected number of C's or better for those who spend 2 to 12 hours per week in extracurricular activities.

10.7 Smoking by students and their parents, II Refer to Exercise 10.2 (page 557).
(a) Write the null and alternative hypotheses for the question of interest.
(b) Calculate the expected number of students who do not smoke even though both parents smoke.

10.8 Python eggs, II Refer to Exercise 10.3 (page 558).
(a) Write the null and alternative hypotheses for the question of interest.
(b) What is the expected count of hatched pythons in the hot group?

10.9 Trying to quit, II Refer to Exercise 10.4 (page 558).
(a) Write the null and alternative hypotheses for the question of interest.
(b) Find the expected counts prior to conducting inference.

10.10 Firearm deaths, I Firearms are second to motor vehicles as a cause of nondisease deaths in the United States. Here are counts from a study of all firearm-related deaths in Milwaukee, Wisconsin, between 1990 and 1994. We want to compare the types of firearms used in homicides and in suicides.[9]

	Handgun	Shotgun	Rifle	Unknown	Total
Homicides	468	28	15	13	524
Suicides	124	22	24	5	175

(a) Write the null and alternative hypotheses for the question of interest.
(b) Find the expected counts prior to conducting inference.

The chi-square test

To see if the data give evidence against the null hypothesis of "no relationship," compare the counts in the two-way table with the counts we would expect if there really were no relationship. If the observed counts are far from the expected counts, that's the evidence we were seeking. The test uses a statistic that measures how far apart the observed and expected counts are.

Chi-square statistic

The chi-square statistic is a measure of how far the observed counts in a two-way table are from the expected counts. The formula for the statistic is

$$X^2 = \sum \frac{(\text{observed count} - \text{expected count})^2}{\text{expected count}}$$

The symbol $\sum$ means "sum over all cells in the table."

The chi-square statistic is a sum of terms, one for each cell in the table. In the cocaine example, 14 of the desipramine group succeeded. The expected count for this cell is 8. So the term in the chi-square statistic from this cell is

$$\frac{(\text{observed count} - \text{expected count})^2}{\text{expected count}} = \frac{(14 - 8)^2}{8}$$

$$= \frac{36}{8} = 4.5$$

EXAMPLE 10.5 The cocaine study

Here are the observed and expected counts for the cocaine study side by side:

	Observed		Expected	
	Success	Failure	Success	Failure
Desipramine	14	10	8	16
Lithium	6	18	8	16
Placebo	4	20	8	16

We can now find the chi-square statistic, adding six terms for the six cells in the two-way table:

$$X^2 = \frac{(14 - 8)^2}{8} + \frac{(10 - 16)^2}{16} + \frac{(6 - 8)^2}{8}$$

$$+ \frac{(18 - 16)^2}{16} + \frac{(4 - 8)^2}{8} + \frac{(20 - 16)^2}{16}$$

$$= 4.50 + 2.25 + 0.50 + 0.25 + 2.00 + 1.00 = 10.50$$

Because X^2 measures how far the observed counts are from what would be expected if H_0 were true, large values are evidence against H_0. Is $X^2 = 10.5$ a large value? You know the drill: compare the observed value 10.5 against the sampling distribution that shows how X^2 would vary if the null hypothesis were true. This sampling distribution is not a normal distribution. It is a right-skewed distribution that allows only positive values because X^2 can never be negative. Moreover, the sampling distribution is different for two-way tables of different sizes. Here are the facts.

The chi-square distributions

The sampling distribution of the chi-square statistic X^2 when the null hypothesis of no association is true is called a **chi-square distribution.**

The chi-square distributions are a family of distributions that take only positive values and are skewed to the right. A specific chi-square distribution is specified by giving its **degrees of freedom.**

The chi-square test for a two-way table with r rows and c columns uses critical values from the chi-square distribution with $(r-1)(c-1)$ degrees of freedom.

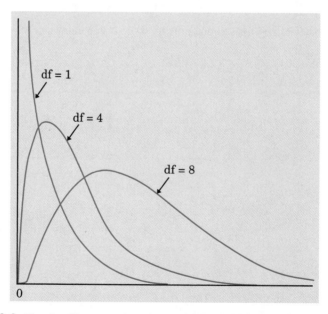

FIGURE 10.2 The density curves for three members of the chi-square family of distributions. The sampling distributions of chi-square statistics belong to this family.

Figure 10.2 on the facing page shows the density curves for three members of the chi-square family of distributions. As the degrees of freedom (df) increase, the density curves become less skewed and larger values become more probable. We can't find P-values as areas under a chi-square curve by hand, though software can do it for us.

Table 10.1 is a shortcut. It shows how large the chi-square statistic X^2 must be in order to be significant at various levels. This isn't as good as an actual P-value, but it is often good enough. Each number of degrees of freedom has a separate row in the table. We see, for example, that a chi-square statistic with 3 degrees of freedom is significant at the 5% level if it is greater than 7.81 and is significant at the 1% level if it is greater than 11.34.

TABLE 10.1 To Be Significant at Level α, a Chi-Square Statistic Must Be Larger Than the Table Entry for α

	Significance level α						
df	0.25	0.20	0.15	0.10	0.05	0.01	0.001
1	1.32	1.64	2.07	2.71	3.84	6.63	10.83
2	2.77	3.22	3.79	4.61	5.99	9.21	13.82
3	4.11	4.64	5.32	6.25	7.81	11.34	16.27
4	5.39	5.99	6.74	7.78	9.49	13.28	18.47
5	6.63	7.29	8.12	9.24	11.07	15.09	20.51
6	7.84	8.56	9.45	10.64	12.59	16.81	22.46
7	9.04	9.80	10.75	12.02	14.07	18.48	24.32
8	10.22	11.03	12.03	13.36	15.51	20.09	26.12
9	11.39	12.24	13.29	14.68	16.92	21.67	27.88

EXAMPLE 10.6   The cocaine study, conclusion

We have seen that desipramine produced markedly more successes and fewer failures than lithium or a placebo. Comparing observed and expected counts gave the chi-square statistic $X^2 = 10.5$. The last step is to assess significance.

The two-way table of 3 treatments by 2 outcomes for the cocaine study has 3 rows and 2 columns. That is, $r = 3$ and $c = 2$. The chi-square statistic therefore has degrees of freedom

$$(r - 1)(c - 1) = (3 - 1)(2 - 1) = (2)(1) = 2$$

Look in the df $= 2$ row of Table 10.1. We see that $X^2 = 10.5$ is larger than the critical value 9.21 required for significance at the $\alpha = 0.01$ level but smaller than the critical value 13.82

for $\alpha = 0.001$. The cocaine study shows a significant relationship ($P < 0.01$) between treatment and success.

The significance test only says we have strong evidence of some association between treatment and success. We must look at the two-way table to see the nature of the relationship: desipramine works better than the other treatments.

EXERCISES

10.11 Extracurricular activities and grades, III In Exercise 10.1 (page 557), you described the relationship between extracurricular activities and success in a required course. Is the observed association between these variables statistically significant? To find out, proceed as follows.
(a) Find the chi-square statistic. Which cells contribute most to this statistic?
(b) What are the degrees of freedom? Use Table 10.1 on the previous page to say how significant the chi-square test is.
(c) Write a brief conclusion for your study.

10.12 Smoking by students and their parents, III In Exercise 10.2 (page 557), you saw that there is an association between smoking by parents and smoking by their children. Children are more likely to smoke if their parents smoke. We want to know whether this association is statistically significant.
(a) What do you think the population is?
(b) Find the expected cell counts. Write a sentence that explains in simple language what "expected counts" are.
(c) Find the chi-square statistic, its degrees of freedom, and the P-value.
(d) What is your conclusion about significance?

10.13 Python eggs, III Exercise 10.3 (page 558) presents data on the hatching of python eggs in water at three different temperatures. Does temperature have a significant effect on hatching? Write a clear summary of your work and your conclusion.

10.14 Trying to quit, III In Exercise 10.4 (page 558), you saw an apparent association between alcohol consumption and smoking. In Exercise 10.9 (page 562) you wrote hypotheses and calculated expected counts.
(a) Calculate a X^2 statistic and determine the P-value.
(b) Write your conclusion from this study in plain language.

10.15 Firearm deaths, II Refer to Exercise 10.10 (page 562).
(a) Calculate a X^2 statistic and determine the P-value.
(b) Write your conclusion from this study in plain language.

Using the chi-square test

Like our test for a population proportion, the chi-square test uses some approximations that become more accurate as we take more observations. Here is a rough rule for when it is safe to use this test.[10]

> ## Cell counts required for the chi-square test
>
> You can safely use the chi-square test when no more than 20% of the expected counts are less than 5 and all individual expected counts are 1 or greater.

The cocaine study easily passes this test: all the expected cell counts are either 8 or 16. Here is a concluding example that outlines the examination of a two-way table.

EXAMPLE 10.7   Do angry people have more heart disease?

People who get angry easily tend to have more heart disease. That's the conclusion of a study that followed a random sample of 12,986 people from three locations for about four years. All subjects were free of heart disease at the beginning of the study. The subjects took the Spielberger Trait Anger Scale, which measures how prone a person is to sudden anger. Here are data for the 8474 people in the sample who had normal blood pressure. CHD stands for "coronary heart disease." This includes people who had heart attacks and those who needed medical treatment for heart disease.

	Anger score		
	Low	Moderate	High
Sample size	3110	4731	633
CHD count	53	110	27
CHD percent	1.7%	2.3%	4.3%

There is a clear trend: as the anger score increases, so does the percent who suffer heart disease. Is this relationship between anger and heart disease statistically significant?[11]

The first step is to write the data as a two-way table by adding the counts of subjects who did not suffer from heart disease. We also add the row and column totals, which we need to find the expected counts.

	Low anger	Moderate anger	High anger	Total
CHD	53	110	27	190
No CHD	3057	4621	606	8284
Total	3110	4731	633	8474

We can now follow the steps for a significance test, familiar from Section 9.2.

The hypotheses. The chi-square method tests these hypotheses:

$$H_0: \text{no relationship between anger and CHD}$$

$$H_a: \text{some relationship between anger and CHD}$$

The sampling distribution. We will see that all the expected cell counts are larger than 5, so we can safely apply the chi-square test. The two-way table of anger versus CHD has 2 rows and 3 columns. We will use critical values from the chi-square distribution with degrees of freedom df $= (2 - 1)(3 - 1) = 2$.

The data. First find the expected cell counts. For example, the expected count of high-anger people with CHD is

$$\text{expected count} = \frac{\text{row 1 total} \times \text{column 3 total}}{\text{table total}}$$

$$= \frac{(190)(633)}{8474} = 14.19$$

Here is the complete table of observed and expected counts side by side:

	Observed			Expected		
	Low	Moderate	High	Low	Moderate	High
CHD	53	110	27	69.73	106.08	14.19
No CHD	3057	4621	606	3040.27	4624.92	618.81

Looking at these counts, we see that the high-anger group has more CHD than expected and the low-anger group has less CHD than expected. This is consistent with what the percents in Example 10.7 show. The chi-square statistic is

$$X^2 = \frac{(53 - 69.73)^2}{69.73} + \frac{(110 - 106.08)^2}{106.08} + \frac{(27 - 14.19)^2}{14.19}$$

$$+ \frac{(3057 - 3040.27)^2}{3040.27} + \frac{(4621 - 4624.92)^2}{4624.92} + \frac{(606 - 618.81)^2}{618.81}$$

$$= 4.014 + 0.145 + 11.557 + 0.092 + 0.003 + 0.265 = 16.077$$

In practice, statistical software can do all this arithmetic for you. Look at the six terms that we sum to get X^2. Most of the total comes from just one cell: high-anger people have more CHD than expected.

Significance? Look at the df = 2 line of Table 10.1 (page 565). The observed chi-square $X^2 = 16.077$ is larger than the critical value 13.82 for $\alpha = 0.001$. We have highly significant evidence ($P < 0.001$) that anger and heart disease are related. Statistical software can give the actual P-value. It is $P = 0.0003$.

Can we conclude that proneness to anger *causes* heart disease? The anger and CHD study is an observational study, not an experiment. It isn't surprising that some lurking variables are confounded with anger. For example, people prone to anger are more likely than others to be men who drink and smoke. The study report used advanced statistics to adjust for many differences among the three anger groups. The adjustments raised the P-value from $P = 0.0003$ to $P = 0.002$ because the lurking variables explain some of the heart disease. But this is still good evidence for a relationship. Because the study started with a random sample of people who had no CHD and followed them forward in time, and because many lurking variables were measured and accounted for, it does give some evidence for causation. The next step might be an experiment that shows anger-prone people how to change. Will this reduce their risk of heart disease?

CALCULATOR CORNER   Chi-square tests on the TI-83

To perform a chi-square test of the anger and heart disease data in Example 10.7 on the TI-83, use matrix [A] to store the observed counts. Here are the keystrokes, along with several calculator screens for you to check your progress.

- Press 2nd x^{-1} (MATRIX), cursor right to EDIT, and choose 1: [A].

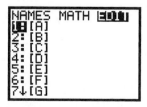

- Specify a 2 × 3 matrix. Enter the observed counts from the two-way table in the matrix in the same locations.

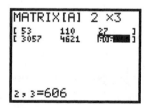

- Specify the chi-square test, the matrix where the observed counts are found, and the matrix where the expected counts will be stored.

CALCULATOR CORNER   Chi-square tests on the TI-83 *(continued)*

- Press STAT , cursor right to TESTS, and choose C: X^2 Test...

```
EDIT CALC TESTS
0↑2-SampTInt…
A:1-PropZInt…
B:2-PropZInt…
█C:X²-Test…
D:2-SampFTest…
E:LinRegTTest…
F:ANOVA(
```

```
X²-Test
 Observed:[A]
 Expected:[B]
 Calculate Draw
```

- Choose "Calculate" or "Draw" to carry out the test. If you choose "Calculate," you should get these results.

```
X²-Test
 X²=16.07676213
 p=3.2283117ε-4
 df=2
```

If you choose "Draw," the chi-square curve with 2 degrees of freedom will be drawn, the

critical area in the tail will be shaded, and the *P*-value will be displayed.

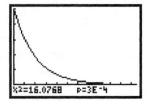

```
X²=16.0768    p=3ε-4
```

If you want to see the expected counts, simply ask for a display of the matrix [B].

- Press 2ⁿᵈ x⁻¹ (MATRIX), choose 2:[B], and then press ENTER .

```
df=2

[B]
[[69.7309417  1…
 [3040.269058  4…
```

```
df=2

[B]
…7   106.0762332…
…58  4624.923767…
```

Verify that these calculator results agree with the results on page 568.

EXERCISES

For Exercises 10.16 through 10.20, do the following:

(a) Use your calculator and the techniques from the Calculator Corner to determine the table of expected counts. Are the conditions for performing chi-square analysis satisfied? If the answer is "yes," then proceed to (b).

(b) Calculate the chi-square test statistic and the *P*-value. Is there sufficient evidence to reject H_0?

(c) Write your conclusion(s) in plain language.

10.16 Extracurricular activities and grades, IV Refer to Exercises 10.1, 10.6, and 10.11.

10.17 Smoking by students and their parents, IV Refer to Exercises 10.2, 10.7, and 10.12.

10.18 Python eggs, IV Refer to Exercises 10.3, 10.8, and 10.13.

10.19 Trying to quit, IV Refer to Exercises 10.4, 10.9, and 10.14.

10.20 Firearm deaths, III Refer to Exercises 10.10 and 10.15.

APPLICATION 10.1 **Matching Products with Markets: A Study in Economics**

A company that makes detergents wants to investigate consumer preferences. Knowing which consumers would be interested in different types of detergent products will help the company determine how to market their detergent products. Everyone uses detergents to wash clothes, dishes, cars, and virtually everything that gets dirty, so the market is huge and the competition for consumer dollars is intense. Specifically, the company wants to know if income level determines which detergent a consumer will purchase. The company decides to study four leading brands of detergent. Detergents A and B are premium brands, and Detergents C and D are economy brands. They decide to classify income into four income levels: Lower, Middle, Upper Middle, and Upper. Incidentally, this was a real study conducted by a marketing consultant for a real company that makes detergents. Using a stratified sampling procedure, researchers selected a representative sample of 600 consumers for the study. The number of consumers who fell into each of the income level and detergent choice categories was recorded in a 4 × 4 table. The actual frequencies observed in the study appear below.[12]

| | Detergent | | | | |
	A	B	C	D	Total
Lower	25	15	55	65	160
Middle	30	25	35	30	120
Upper Middle	50	55	20	22	147
Upper	60	80	15	18	173
Total	165	175	125	135	600

1. Hypotheses. Write null and alternative hypotheses for this study.

APPLICATION 10.1 Matching Products with Markets: A Study in Economics *(continued)*

2. **Conditions.** State and verify the conditions that need to be satisfied in order to perform a chi-square analysis of these data.

3. **Mechanics.** This is too much arithmetic to do by hand. Use your calculator to calculate the X^2 statistic and the *P*-value.

4. **Interpretation and conclusion.** Explain what the *P*-value tells you in terms of rejecting or failing to reject the null hypothesis. Then state your conclusion for this study, as if your job is to analyze the data and make a report and recommendation to company management.

5. **Significance level.** Suppose that as the study was being planned the decision was made to adopt a 5% significance level. Are the data significant at the 5% level? At the 1% level? Explain.

STATISTICS IN SUMMARY

Categorical variables group individuals into classes. To display the relationship between two categorical variables, make a **two-way table** of counts in the classes. We describe the nature of an association between categorical variables by comparing selected percents. As always, lurking variables can make an observed association misleading. In some cases, an association that holds for every level of a lurking variable disappears or changes direction when we lump all levels together. This is **Simpson's paradox.**

The **chi-square test** tells us whether an observed association in a two-way table is statistically significant. The **chi-square statistic** compares the counts in the table with the counts we would expect if there were no association between the row and column variables. The sampling distribution is not normal. It is a new distribution, the **chi-square distribution.**

SECTION 10.1 EXERCISES

10.21 Who earns academic degrees? How do women and men compare in the pursuit of academic degrees? The following table presents counts (in thousands) of degrees earned in the year 2000 categorized by the level of the degree and the gender of the recipient.[13]

	Bachelor's	Master's	Professional	Doctorate
Female	708	265	28	20
Male	530	192	36	25
Total	1238	457	64	45

(a) How many people earned bachelor's degrees?

(b) What percent of each level of degree is earned by women?

(c) Write a brief description of what the data show about the relationship between gender and degree level.

10.22 Majors for men and women in business, I A study of the career plans of young women and men sent questionnaires to all 722 members of the senior class in the College of Business Administration at the University of Illinois. One question asked which major within the business program the student had chosen. Here are the data from the students who responded:[14]

	Female	Male
Accounting	68	56
Administration	91	40
Economics	5	6
Finance	61	59

Describe the differences between the distributions of majors for women and men with percents, with a graph, and in words.

10.23 Firearm deaths, IV Refer to Exercises 10.10, 10.15, and 10.20. We suspect that long guns (shotguns and rifles) will more often be used in suicides because many people keep them at home for hunting. Make a careful comparison of homicides and suicides with a bar graph. What did you find about long guns versus handguns?

10.24 Activity 10.1 follow-up Perform a chi-square analysis on the sports preference data you collected in Activity 10.1, providing the conditions are met for this analysis.

(a) State hypotheses, check conditions, and if conditions are satisfied, calculate a chi-square statistic and a P-value.

(b) How would you describe the population of interest?

(c) What do you conclude?

(d) Could your conclusion extend to everyone in your school? Could it extend to high school students in other schools?

10.25 Conditions One of the conditions for performing a chi-square analysis is that all expected counts be at least 1. Why do you suppose you can't have an expected count of 0? (*Hint:* Look at the formula for calculating a chi-square statistic.)

10.26 Majors for men and women in business, II Exercise 10.22 gives the responses to questionnaires sent to all 722 members of a business school graduating class.
(a) Two of the observed cell counts are small. Do these data satisfy our guidelines for safe use of the chi-square test?
(b) Is there a statistically significant relationship between the gender and major of business students?
(c) What percent of the students did not respond to the questionnaire? The non-response weakens conclusions drawn from these data.

10.27 Race and the death penalty Whether a convicted murderer gets the death penalty seems to be influenced by the race of the victim. Here are data on 326 cases in which the defendant was convicted of murder:[15]

	White defendant				Black defendant	
	White victim	Black victim			White victim	Black victim
Death	19	0		Death	11	6
Not	132	9		Not	52	97

(a) Use these data to make a two-way table of defendant's race (white or black) versus death penalty (yes or no).
(b) Show that Simpson's paradox holds: a higher percent of white defendants are sentenced to death overall, but for both black and white victims a higher percent of black defendants are sentenced to death.
(c) Use the data to explain the paradox in language that a judge could understand.

10.28 Totals aren't enough Here are the row and column totals for a two-way table with two rows and two columns:

a	*b*	50
c	*d*	50
60	40	100

Find *two different* sets of counts *a*, *b*, *c*, and *d* for the body of the table that give these same totals. This shows that the relationship between two variables cannot be obtained from the two individual distributions of the variables.

10.2 INFERENCE ABOUT A POPULATION MEAN

Is 98.6 degrees normal?

We have all heard that 98.6 degrees Fahrenheit (that's 37.0 degrees Celsius) is "normal body temperature" as measured by a thermometer stuck under the tongue. Of course, your actual body temperature varies during the day. Body temperature is usually highest around 6 A.M. and lowest around 4 to 6 P.M. Body temperature also varies among people and is a bit higher in children. So that 98.6 degrees really refers to a *mean* temperature. Presumably the claim is that if we measured all healthy adults at many times during the day, the mean temperature would be 98.6 degrees.

Is this claim correct? Some enterprising doctors measured 148 healthy adults (aged 18 to 40 years) four times a day for three days. They claimed that the traditional value is wrong and that the correct mean is 98.2 degrees rather than 98.6 degrees. The old value really is old—it goes back to 1861. We might ask if new data, from just 148 subjects but with more accurate thermometers, justifies giving a new value. That's a question of inference about the mean of a population. A confidence interval can answer the question "What can we say about the true value of the mean?" A significance test can answer the question "Is the true mean different from 98.6 degrees?"[16]

A calculator or computer will do the inference if we give it the data. We should also ask some questions that the computer won't ask for us. Why are we interested in the mean—don't we really want to know how high or low body temperature must be to signal that something is wrong? Is 98.2 versus 98.6 too small a difference to be medically important? Was the 1861 result rounded to the nearest degree (37 degrees Celsius), so that translating it to 98.6 degrees Fahrenheit claims more accuracy than is justified? These are good questions. They remind us even more than the new study that "98.6 degrees is normal" is too simple a claim to be much help to doctors or to patients.

ACTIVITY 10.2 Exploring sampling distributions

Professor David Lane of Rice University has developed a wonderful applet that probably ranks #1 among high school and college students studying statistics. It provides a dynamic way for students to understand a fundamental concept in statistics: sampling distributions. And it is just plain fun to manipulate it. Let's begin.

ACTIVITY 10.2 Exploring sampling distributions *(continued)*

1. Go to the Web site www.ruf.rice.edu/~lane/stat_sim/sampling_dist/. When the BEGIN button appears on the left side of the screen, click on it. You will see a yellow page entitled "Sampling Distributions."

2. There are choices of populations: normal, uniform, skewed, and custom. Select uniform. Specify "mean" and sample size $n = 2$. Then click on "animated samples." Click again. Continue. What do the black boxes represent? What is the blue square that drops down onto the plot below? What does the red horizontal band under the population histogram tell us? Notice the left panel. Important numbers are displayed there. Did you notice that the colors of the numbers match up with the objects to the right? As you make things happen, the numbers change accordingly, like an automatic scorekeeper.

3. Click on "Clear lower 3" to start clean. Click on the 5 samples button repeatedly until you have simulated 100 repetitions (look for "Reps = 100" on the left panel in black letters). Does the sampling distribution (blue bars) have a recognizable shape? Click in the box next to "Fit normal." Compare the center (mean) of the population with the center of the sampling distribution.

4. Click "Clear lower 3." Click on 1000 samples. Click four more times until you have 5000 samples. What is the range of the data? How well does the normal curve fit your histogram now?

5. Click "Clear lower 3." Select sample size $n = 16$. Repeat Steps 3 and 4, but with the bigger sample size. Then "Clear lower 3" and do the same for sample size $n = 25$. Complete the statements: *The center of the sampling distribution is* (how related to) *the center of the population distribution. As the sample size increases, the spread of the sampling distribution* _____."

6. Clear the page, and select "Skewed" population. Repeat the steps above for the skewed distribution as you increase the sample size.

7. Clear the page, and select "Custom" distribution. Click on a point on the population histogram to insert a bar of that height. Or click on a point on the horizontal axis, and drag up to define a bar. Make a distribution as strange as you can. (Note: You can shorten a bar or get rid of it completely by clicking on the top of the bar and dragging down to the axis.) Then repeat the steps above for your custom distribution as you increase the sample size. Cool, huh?

8. Now go to your textbook's Web site: www.whfreeman.com/sta and click on APPLETS. Select the *Central Limit Theorem* applet. Click and drag the slider to

ACTIVITY 10.2 Exploring sampling distributions *(continued)*

change the sample size, and watch how the idealized curve for the sampling distribution changes with it.

9. What did you discover about the shape and spread of the sampling distribution? Does something special happen as you gradually increase the sample size? Fill in the blank to describe this phenomenon: *As the sample size (increases) (decreases), the sampling distribution of the sample mean gets closer and closer to*

_____ .

There is a name for this important result: the **central limit theorem**. We will learn more about the central limit theorem in this section. In particular, there is a very explicit relationship between the sample size and the spread of the sampling distribution of the sample mean. We will learn what that relationship is.

Although the reasoning of confidence intervals and significance tests is always the same, specific recipes vary greatly. The form of an inference procedure depends first on the parameter you want information about—a population proportion or mean or median or whatever. The second influence is the design of the sample or experiment. Estimating a population proportion from a stratified sample requires a different recipe than if the data come from an SRS. This chapter presents a confidence interval and a significance test for inference about a population mean when the data are an SRS from the population.

The sampling distribution of a sample mean

What is the mean number of hours your school's seniors study each week? What is their mean grade point average? We often want to estimate the mean of a population. To distinguish the population mean (a parameter) from the sample mean $\bar{x}$, we write the population mean as μ (the Greek letter mu). We use the mean $\bar{x}$ of an SRS to estimate the unknown mean μ of the population.

Like the sample proportion $\hat{p}$, the sample mean $\bar{x}$ from a large SRS has a sampling distribution that is close to normal. Because the sample mean of an SRS is an unbiased estimator of μ, the sampling distribution of $\bar{x}$ has μ as its mean. The standard deviation of $\bar{x}$ depends on the standard deviation of the population, which is usually written as σ (the Greek letter sigma). By mathematics we can discover the following facts.

> ## Sampling distribution of a sample mean
>
> Choose an SRS of size n from a population in which individuals have mean μ and standard deviation σ. Let $\bar{x}$ be the mean of the sample. Then:
>
> • The sampling distribution of $\bar{x}$ is **approximately normal** when the sample size n is large.
>
> • The **mean** of the sampling distribution is equal to μ.
>
> • The **standard deviation** of the sampling distribution is $\sigma/\sqrt{n}$.

It isn't surprising that the values that $\bar{x}$ takes in many samples are centered at the true mean μ of the population. That's the lack of bias in random sampling once again. The other two facts about the sampling distribution make precise two very important properties of the sample mean $\bar{x}$:

• The mean of a number of observations is less variable than individual observations.

• The distribution of a mean of a number of observations is more normal than the distribution of individual observations.

Figure 10.3 illustrates the first of these properties. It compares the distribution of a single observation with the distribution of the mean $\bar{x}$ of 10 observations.

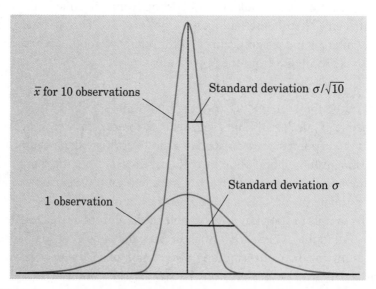

FIGURE 10.3 The sampling distribution of the sample mean $\bar{x}$ of 10 observations compared with the distribution of individual observations.

Both have the same center, but the distribution of $\bar{x}$ is less spread out. In Figure 10.3, the distribution of individual observations is normal. If that is true, then the sampling distribution of $\bar{x}$ is exactly normal for any size sample, not just approximately normal for large samples. A remarkable statistical fact, called the **central limit theorem**, says that as we take more and more observations at random from any population, the distribution of the mean of these observations eventually gets close to a normal distribution. (There are some technical qualifications to this big fact, but in practice we can ignore them.) The central limit theorem lies behind the use of normal sampling distributions for sample means.

EXAMPLE 10.8 The central limit theorem in action

Figure 10.4 shows the central limit theorem in action. The top left density curve describes individual observations from a population. It is strongly right-skewed. Distributions like this describe the time it takes to repair a household appliance, for example. Most repairs are quickly done, but some are lengthy.

The other three density curves in Figure 10.4 show the sampling distributions of the sample means of 2, 10, and 25 observations from this population. As the sample size n increases, the shape becomes more normal. The mean remains fixed and the standard

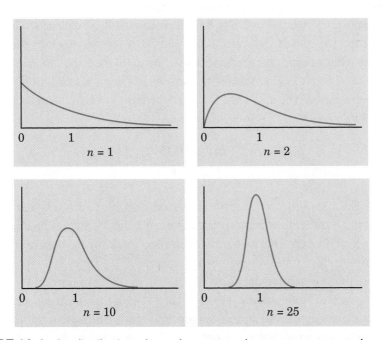

FIGURE 10.4 The distribution of sample means $\bar{x}$ becomes more normal as the size of the sample increases. The distribution of individual observations ($n = 1$) is far from normal. The distributions of means of 2, 10, and finally 25 observations move closer to the normal shape.

deviation decreases, following the pattern $\sigma/\sqrt{n}$. The distribution for 10 observations is still somewhat skewed to the right but already resembles a normal curve. The density curve for $n = 25$ is yet more normal. The contrast between the shapes of the population distribution and of the distribution of the mean of 10 or 25 observations is striking.

EXERCISES

10.29 The idea of a sampling distribution Figure 9.1 (page 492) shows the idea of the sampling distribution of a sample proportion in picture form. Draw a similar picture that shows the idea of the sampling distribution of a sample mean $\bar{x}$.

10.30 Averages versus individuals Scores on the American College Testing (ACT) college admissions examination vary normally with mean $\mu = 18$ and standard deviation $\sigma = 6$. The range of reported scores is 1 to 36.
(a) What range contains the middle 95% of all individual scores?
(b) If the ACT scores of 25 randomly selected students are averaged, what range contains the middle 95% of the averages $\bar{x}$?

10.31 Student attitudes The Survey of Study Habits and Attitudes (SSHA) is a psychological test that measures students' study habits and attitude toward school. Scores range from 0 to 200. The mean score for U.S. college students is about 115, and the standard deviation is about 30. A teacher suspects that older students have better attitudes toward school. She gives the SSHA to 25 students who are at least 30 years old. Assume that scores in the population of older students are normally distributed with standard deviation $\sigma = 6$. The teacher wants to test the hypotheses

$$H_0: \mu = 115$$
$$H_a: \mu > 115$$

(a) What is the sampling distribution of the mean score $\bar{x}$ of a sample of 25 older students if the null hypothesis is true? Sketch the density curve of this distribution. (*Hint:* Sketch a normal curve first, then mark the axis using what you know about locating μ and σ on a normal curve.)
(b) Suppose that the sample data give $\bar{x} = 118.6$. Mark this point on the axis of your sketch. In fact, the outcome was $\bar{x} = 125.7$. Mark this point on your sketch. Using your sketch, explain in simple language why one outcome is good evidence that the mean score of all older students is greater than 115 and why the other outcome is not.
(c) Shade the area under the curve that is the *P*-value for the sample result $\bar{x} = 118.6$.

10.32 A sampling distribution, I We got the TI-83 to generate samples of 100 random numbers repeatedly. Here are the sample means $\bar{x}$ for 50 samples of size 100:

0.532	0.450	0.481	0.508	0.510	0.530	0.499	0.461	0.543	0.490
0.497	0.552	0.473	0.425	0.449	0.507	0.472	0.438	0.527	0.536
0.492	0.484	0.498	0.536	0.492	0.483	0.529	0.490	0.548	0.439
0.473	0.516	0.534	0.540	0.525	0.540	0.464	0.507	0.483	0.436
0.497	0.493	0.458	0.527	0.458	0.510	0.498	0.480	0.479	0.499

The sampling distribution of $\bar{x}$ is the distribution of the means from all possible samples. We actually have the means from 50 samples.

(a) Make a histogram of these 50 observations. Does the distribution appear to be roughly normal, as the central limit theorem says will happen for large enough samples?

(b) What is the value of the population mean?

10.33 A sampling distribution, II Exercise 10.32 presents 50 sample means $\bar{x}$ from 50 random samples of size 100. Find the mean and standard deviation of these 50 values. Then answer these questions.

(a) The mean of the population from which the 50 samples were drawn is $\mu = 0.5$ if the random number generator is accurate. What do you expect the mean of the distribution of $\bar{x}$'s from all possible samples to be? Is the mean of these 50 samples close to this value?

(b) The standard deviation of the distribution of $\bar{x}$ from samples of size $n = 100$ is supposed to be $\sigma/10$, where σ is the standard deviation of individuals in the population. Use this fact with the standard deviation you calculated for the 50 $\bar{x}$'s to estimate σ.

Confidence intervals for a population mean

The standard deviation of $\bar{x}$ depends on both the sample size n and the standard deviation σ of individuals in the population. We know n but not σ. When n is large, the sample standard deviation s is close to σ and can be used to estimate it, just as we use the sample mean $\bar{x}$ to estimate the population mean μ. The estimated standard deviation of $\bar{x}$ is therefore $s/\sqrt{n}$. Now we can find *confidence intervals for μ* following the same reasoning that led us to confidence intervals for a proportion p in Section 9.1. The big idea is that to cover the central area C under a normal curve, we must go out a distance z^* on either side of the mean. Look again at Figure 9.5 (page 502) to see how C and z^* are related.

Confidence interval for a population mean

Choose an SRS of size n from a large population of individuals having mean μ. The mean of the sample observations is $\bar{x}$. When n is large, an approximate level C confidence interval for μ is

$$\bar{x} \pm z^* \frac{s}{\sqrt{n}}$$

where z^* is the critical value for confidence level C from Table 9.1 (page 502).

The cautions we noted in estimating p apply here as well. The recipe is valid only when an SRS is drawn and the sample size n is reasonably large. The margin of error again decreases only at a rate proportional to $\sqrt{n}$ as the sample size n increases. One additional caution: Remember that $\bar{x}$ and s are strongly influenced by outliers. Inference using $\bar{x}$ and s is suspect when outliers are present. Always look at your data.

EXAMPLE 10.9   NAEP quantitative scores

The National Assessment of Educational Progress (NAEP) includes a short test of quantitative skills, covering mainly basic arithmetic and the ability to apply it to realistic problems. Scores on the test range from 0 to 500. For example, a person who scores 233 can add the amounts of two checks appearing on a bank deposit slip; someone scoring 325 can determine the price of a meal from a menu; a person scoring 375 can transform a price in cents per ounce into dollars per pound.

In a recent year, 840 men 21 to 25 years of age were in the NAEP sample. Their mean quantitative score was $\bar{x} = 272$, and the standard deviation of their scores was $s = 59$. These 840 men are a simple random sample from the population of all young men. On the basis of this sample, what can we say about the mean score in the population of all 9.5 million young men of these ages?

The 95% confidence interval for μ uses the critical value from Table 9.1 (page 502). The interval is

$$\bar{x} \pm z^* \frac{s}{\sqrt{n}} = 272 \pm 1.96 \frac{59}{\sqrt{840}}$$

$$= 272 \pm (1.96)(2.036) = 272 \pm 4.0$$

We are 95% confident that the mean for all young men lies between 268 and 276.[17]

Tests for a population mean

As with confidence intervals, the reasoning that leads to significance tests for hypotheses about a population mean μ follows the reasoning that led to tests about a population proportion p. The big idea is to use the sampling distribution that the sample mean $\bar{x}$ would have if the null hypothesis were true. Locate the $\bar{x}$ from your data on this distribution and see if it is unlikely. A value of $\bar{x}$ that would rarely appear if H_0 were true is evidence that H_0 is not true. The four steps are also similar to those in tests for a proportion. Here are two examples, the first one-sided and the second two-sided.

EXAMPLE 10.10   Can you balance your checkbook?

In a discussion of the education level of the American workforce, a pessimist says, "The average young person can't even balance a checkbook." The NAEP survey says that a score of 275 or higher on its quantitative test reflects the skill needed to balance a checkbook. The NAEP random sample of 840 young men had mean score $\bar{x} = 272$, a bit below the checkbook-balancing level. Is this sample result good evidence that the mean for *all* young men is less than 275? The standard deviation of the scores in the sample was $s = 59$.

The hypotheses. The pessimist's claim is that the mean NAEP score is less than 275. That's our alternative hypothesis, the statement we seek evidence *for*. The hypotheses are

$$H_0: \mu = 275$$
$$H_a: \mu < 275$$

The sampling distribution. *If the null hypothesis is true,* the sample mean $\bar{x}$ has approximately the normal distribution with mean μ and standard deviation

$$\frac{s}{\sqrt{n}} = \frac{59}{\sqrt{840}} = 2.036$$

We once again use the sample standard deviation s in place of the unknown population standard deviation σ.

The data. The NAEP sample gave $\bar{x} = 272$. The standard score for this outcome is

$$\text{standard score} = \frac{\text{observation} - \text{mean}}{\text{standard deviation}}$$

$$= \frac{272 - 275}{2.036} = -1.47$$

That is, the sample result is about 1.47 standard deviations below the mean we would expect if on the average a young man had just enough skill to balance a checkbook.

The P-value. Figure 10.5 locates the sample outcome -1.47 (in the standard scale) on the normal curve that represents the sampling distribution if H_0 is true. This curve has mean 0 and standard deviation 1 because we are using the standard scale. The P-value for our one-sided test is the shaded area to the left of -1.47. To use Table B, round the standard score to -1.5. Table B says that -1.5 is the 6.68 percentile, so the area to its left is 0.0668. That is our P-value. (This result is approximate because we rounded the standard score to use Table B. Software gives $P = 0.071$.)

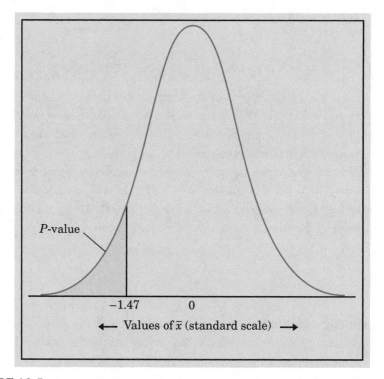

FIGURE 10.5 The P-value for a one-sided test when the standard score for the sample mean is -1.47.

Conclusion. A P-value of about $P = 0.07$ suggests that the mean score for all young men is below the checkbook-balancing level, but it is not convincing evidence.

EXAMPLE 10.11 Executives' blood pressures

The National Center for Health Statistics reports that the mean systolic blood pressure for males 35 to 44 years of age is 128. The medical director of a large company looks at the medical records of 72 executives in this age-group and finds that the mean systolic blood pressure in this sample is $\bar{x} = 126.1$ and that the standard deviation is $s = 15.2$. Is this evidence that the company's executives have a different mean blood pressure from the general population?

The hypotheses. The null hypothesis is "no difference" from the national mean. The alternative is two-sided, because the medical director did not have a particular direction in mind before examining the data. So the hypotheses about the unknown mean μ of the executive population are

$$H_0: \mu = 128$$
$$H_a: \mu \neq 128$$

Catching cheaters

Lots of students take a long multiple-choice exam. Can the computer that scores the exam also screen for papers that are suspiciously similar? Clever people have created measures that take into account not just identical answers but the popularity of those answers and the total score on the similar papers. The measure has close to a normal distribution, and the computer flags pairs of papers with a measure outside ±4 standard deviations as significant.

The sampling distribution. *If the null hypothesis is true,* the sample mean $\bar{x}$ has approximately the normal distribution with mean μ and standard deviation

$$\frac{s}{\sqrt{n}} = \frac{15.2}{\sqrt{72}} = 1.79$$

The data. The sample mean is $\bar{x} = 126.1$. The standard score for this outcome is

$$\text{standard score} = \frac{\text{observation} - \text{mean}}{\text{standard deviation}}$$

$$= \frac{126.1 - 128}{1.79} = -1.06$$

We know that an outcome a little more than 1 standard deviation away from the mean of a normal distribution is not very surprising. The last step is to make this formal.

The P-value. Figure 10.6 (next page) locates the sample outcome -1.06 (in the standard scale) on the normal curve that represents the sampling distribution if H_0 is true. The two-sided P-value is the probability of an outcome at least this far out in either direction. This is the shaded area under the curve.

To use Table B, round the standard score to -1.1. This is the 13.57 percentile of a normal distribution. So the area to the left of -1.1 is 0.1357. The area to the left of -1.1 and to the right of 1.1 is double this, or about 0.27. This is our approximate P-value. (The exact P-value, from software, is $P = 0.289$.)

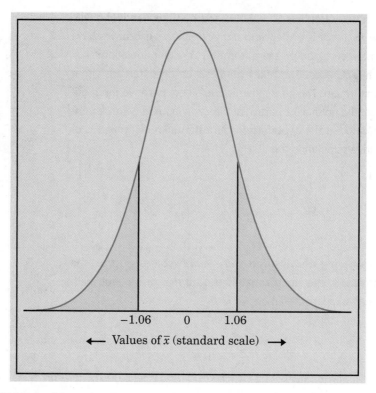

FIGURE 10.6 The *P*-value for a two-sided test when the standard score for the sample mean is −1.06.

Conclusion. The large *P*-value gives us no reason to think that the mean blood pressure of the executive population differs from the national average.

The test assumes that the 72 executives in the sample are an SRS from the population of all middle-aged male executives in the company. We should check this assumption by asking how the data were produced. If medical records are available only for executives with recent medical problems, for example, the data are of little value for our purpose. It turns out that all executives are given a free annual medical exam, and that the medical director selected 72 exam results at random.

The data in Example 10.11 do *not* establish that the mean blood pressure μ for this company's executives is 128. We sought evidence that μ differed from 128 and failed to find convincing evidence. That is all we can say. No doubt the mean blood pressure of the entire executive population is not exactly equal to 128. A large enough sample would give evidence of the difference, even if it were very small.

CALCULATOR CORNER Inference for means

Confidence intervals and tests of significance can be performed on the TI-83, thus avoiding table lookups. Here is a brief summary of the techniques using the 50 means of random numbers generated in Exercise 10.32 (page 581). If you haven't entered these 50 means into list L_1, do so now.

• Remember that it's always a good idea to plot your data first to inspect the shape, center, and spread and to look for possible outliers. A histogram is shown here.

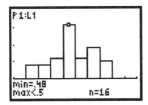

There are no outliers, so we will proceed with a test of significance. Our hypotheses are

$$H_0: \mu = 0.5$$

$$H_a: \mu \neq 0.5$$

• Go to STAT/TESTS. Choose 2:T-Test. Adjust your settings as shown.

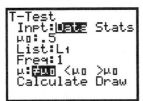

• If you select "Calculate," the following screen appears:

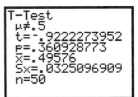

The test statistic is −0.92 and the P-value is 0.36. Our conclusion is: If the null hypothesis is true (the population mean of random numbers generated by the calculator is really 0.5), then we would observe a sample mean like the one we actually saw, $\bar{x} = 0.49576$, about a third of the time. There is insufficient evidence to reject H_0. We can't conclude that the population mean is anything other than 0.5. That's a relief. If we had been able to reject H_0, that would say that the calculator's random number generator was defective. That would be bad for TI's business.

To determine a confidence interval for these data:

• Choose 8:Tinterval. Choose "Data" (and not "Stats") because your data are in list L_1. Select the 99% confidence level (0.01 significance level).

CALCULATOR CORNER   Inference for means (continued)

- Select "Calculate" and press [ENTER].

```
TInterval
 (.48344,.50808)
 x̄=.49576
 Sx=.0325096909
 n=50
```

The result tells us that the 99% confidence interval for the true population mean random number is between 0.48344 and 0.50808. Notice that this interval contains the value 0.5, which tells us that 0.5 is a plausible value for the population mean.

CALCULATOR CORNER   Comparing two means

Constructing confidence intervals and performing tests of significance when you want to compare two means is very similar to inference for a single mean. To illustrate, consider the following question: Do males or females have better social insight into other people? The Chapin Social Insight Test is a psychological test designed to measure how accurately a person appraises other people. The possible scores on the test range from 0 to 41. Here are the results for male and female college students:

Group	Gender	n	$\bar{x}$	s
1	Male	133	25.34	5.05
2	Female	162	24.94	5.44

Do these data support the contention that female and male students differ in average social insight?

Our hypotheses are

H_0: $\mu_1 = \mu_2$ (the means are the same)

H_a: $\mu_1 \neq \mu_2$ (the means are different)

- Go to STAT/TESTS. Choose 4:2-SampTTest. Adjust your settings as shown.

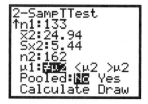

CALCULATOR CORNER   Comparing two means *(continued)*

Notice that we specified "Stats" because we had summary statistics instead of raw data. Note also that we select the default "Pooled: No." Pooling is beyond the scope of this course. Suffice it to say that beginners should always select "No" for pooling.

- If you select "Calculate," the first of two screens appears:

```
2-SampTTest
 μ1≠μ2
 t=.6536989197
 P=.5138263391
 df=288.5762887
 x̄1=25.34
↓x̄2=24.94
```

```
2-SampTTest
 μ1≠μ2
↑x̄2=24.94
 Sx1=5.05
 Sx2=5.44
 n1=133
 n2=162
```

The test statistic is 0.65 and the *P*-value is 0.51. There is insufficient evidence to reject H_0. We can't conclude that there is any difference in social insight between male and female students.

To determine a confidence interval for these data:

- Choose 0:2-SampTInt. Choose "Stats." The calculator remembers the numbers you entered for the test of significance. Select the 95% confidence level (0.05 significance level).

```
2-SampTInt
 Inpt:Data Stats
 x̄1:25.34
 Sx1:5.05
 n1:133
 x̄2:24.94
 Sx2:5.44
↓n2:162
```

```
2-SampTInt
↑n1:133
 x̄2:24.94
 Sx2:5.44
 n2:162
 C-Level:.95
 Pooled:No Yes
 Calculate
```

- Highlight "Calculate" and press ENTER.

```
2-SampTInt
 (-.8044,1.6044)
 df=288.5762887
 x̄1=25.34
 x̄2=24.94
 Sx1=5.05
↓Sx2=5.44
```

The result tells us that the 95% confidence interval for the true difference of population means $\mu_1 - \mu_2$ is between -0.8044 and 1.6044. We are 95% confident that this interval captures the true difference in social insight for male and female college students.

EXERCISES

10.34 Executives' blood pressure Example 10.11 (page 584) found that the mean blood pressure of a sample of executives did not differ significantly at the 10% level from the national mean, which is 128. Use the data in that example to give a 90% confidence interval for the mean of the executive population. Why do you expect 128 to be inside this interval?

10.35 IQ test scores, I Here are the IQ test scores of 31 seventh-grade girls in a Midwest school district:[18]

114	100	104	89	102	91	114	114	103	105	
108	130	120	132	111	128	118	119	86	72	
111	103	74	112	107	103	98	96	112	112	93

(a) We expect the distribution of IQ scores to be close to normal. Make a histogram of the distribution of these 31 scores. Does your plot show outliers, clear skewness, or other nonnormal features? Using a calculator, find the mean and standard deviation of these scores.

(b) Treat the 31 girls as an SRS of all middle school girls in the school district. Give a 95% confidence interval for the mean score in the population.

(c) In fact, the scores are those of all seventh-grade girls in one of the several schools in the district. Explain carefully why we cannot trust the confidence interval from (b).

10.36 Blood pressure A randomized comparative experiment studied the effect of diet on blood pressure. Researchers divided 54 healthy white males at random into two groups. One group received a calcium supplement; the other, a placebo. At the beginning of the study, the researchers measured many variables on the subjects. The paper reporting the study gives $\bar{x} = 114.9$ and $s = 9.3$ for the seated systolic blood pressure of the 27 members of the placebo group.

(a) Give a 95% confidence interval for the mean blood pressure of the population from which the subjects were recruited.

(b) The recipe you used in part (a) requires an important assumption about the 27 men who provided the data. What is this assumption?

10.37 IQ test scores, II The mean IQ for the entire population in any age-group is supposed to be 100. Treat the IQ scores in Exercise 10.35 as if they were an SRS from all middle school girls in this district. Do the scores provide good evidence that the mean IQ of this population is not 100?

(a) State the hypotheses to be tested.

(b) If your hypothesis is true, what is the sampling distribution of the sample mean $\bar{x}$? Sketch this distribution.

(c) Mark the actual value of $\bar{x}$ on the curve. Does it appear surprising enough to give good evidence against the null hypothesis?

(d) Carry out the rest of the significance test called for. Use your calculator to find the test statistic, the P-value, and then state your conclusion clearly.

10.38 Testing a random number generator, I Your TI-83 has a "random number generator" that is supposed to produce numbers scattered at random between 0 and 1. If this is true, the numbers generated come from a population with $\mu = 0.5$. Generate 100 random numbers on your TI-83 and store them in list L_1 (`MATH/PRB/1:rand,` and complete the command `rand(100)` $\rightarrow L_1$). Then find the mean $\bar{x}$ and standard deviation s_x. Use the T-test function on your calculator to carry out all the steps of the significance test: hypotheses, sampling distribution, data, and P-value. State your conclusion.

APPLICATION 10.2 Can Listening to Classical Music Make You Smarter?

In Section 3.2, we described the "Mozart Effect" and the 1993 study by researchers at the University of California at Irvine that coined the term. They recruited students at the university and divided them into three groups. After ten minutes spent listening to either Mozart's Sonata in D Major for Two Pianos or a "relaxation tape," or simply sitting in silence, the students were given a paper folding and cutting test. (A piece of paper is folded over several times and then cut. You have to mentally unfold it and choose the right shape from five examples.) The diagram on the next page illustrates a sample question. The top row shows how a piece of paper is folded and cut. The dotted line shows where the paper is folded, and the solid, heavy line shows where the paper is cut. From the bottom row the student has to select what the paper will look like after it is unfolded. (The correct answer is C.)

1. Sketch a diagram that shows the design of this experiment.

2. Three groups received three different treatments. Write a sentence that you think would be an appropriate null hypothesis. (Use your common sense here.) Now write an alternative hypothesis.

The result of the study was that the group who listened to the Mozart sonata increased their mean spatial-reasoning scores the equivalent of 8 to 9 points on portions

APPLICATION 10.2 Can Listening to Classical Music Make You Smarter?
(continued)

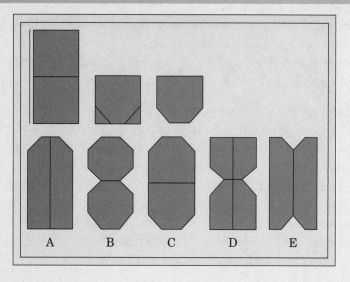

of the Stanford-Binet Intelligence Scale, commonly known as an IQ test. This increase was statistically significant. The researchers also reported that this "Mozart effect," as it was called, disappeared within 10 to 15 minutes.[19]

Some people reasoned that if classical music could enhance college students' intelligence, then babies and in fact all of us might benefit as well. Other psychologists familiar with the study thought that the media and laymen were giving the music too much credit, so there were numerous follow-up studies trying to duplicate the results. Among them, Kenneth Steele, a psychology professor at Appalachian State University in Boone, North Carolina, and two of his colleagues replicated the original study. They made a great effort to duplicate the conditions and the procedure of the earlier experiment as accurately as possible. Their conclusions were similar to those of other independent investigators: the students who listened to the sonata did no better than control groups who listened to other types of music or simply relaxed before taking the test. In the report of their study, Steele and colleagues wrote, "There was no significant effect of treatment, $F(2, 122) = 0.11, p = 0.89$."[20] If they had compared the change in scores from pretreatment to posttreatment in, for example, the Mozart group and the silence group, they could have used a significance test like the one we studied in this section. Because they were comparing increases in scores for *three* groups, they had to use a more powerful procedure, called analysis of variance.

3. The value of the test statistic was 0.11 and the *P*-value was 0.89. Explain in simple terms what this *P*-value tells you.

Since the mid-1990s a number of studies by different researchers have tried to replicate the original 1993 results, but so far none have found significance.

FRAZZ   **BY JEF MALLETT**

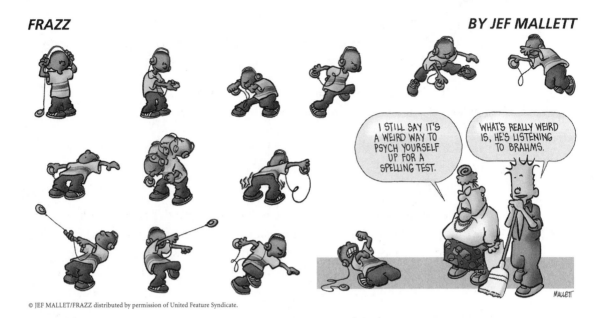

© JEF MALLET/FRAZZ distributed by permission of United Feature Syndicate.

EXPLORING THE WEB

Use a search engine and specify "Mozart effect" to find a large number of sites, both technical and popular, that discuss whether there is such an effect.

STATISTICS IN SUMMARY

We estimate a population mean μ using the sample mean $\bar{x}$ of an SRS from the population. Confidence intervals and significance tests for μ are based on the **sampling distribution** of $\bar{x}$. When the sample size n is large, the **central limit theorem** says that this distribution is approximately normal. Although the details of the methods differ, inference about μ is quite similar to inference about a population proportion p because both are based on normal sampling distributions.

SECTION 10.2 EXERCISES

10.39 Confidence level and margin of error The NAEP test (Example 10.9, page 582) was also given to a sample of 1077 women aged 21 to 25 years. Their mean quantitative score was 275, and the standard deviation was 58.
(a) Give a 95% confidence interval for the mean score μ in the population of all young women.

(b) Give the 90% and 99% confidence intervals for μ.

(c) What are the margins of error for 90%, 95%, and 99% confidence? How does increasing the confidence level affect the margin of error of a confidence interval?

10.40 Mice in a maze Experiments on learning in animals sometimes measure how long it takes mice to find their way through a maze. The mean time is 18 seconds for one particular maze. A researcher thinks that a loud noise will cause the mice to complete the maze faster. She measures how long each of several mice takes with a noise as stimulus. What are the null hypothesis H_0 and alternative hypothesis H_a?

10.41 Response time Last year, your company's service technicians took an average of 2.6 hours to respond to trouble calls from business customers who had purchased service contracts. Do this year's data show a significantly different average response time? What null and alternative hypotheses should you test to answer this question?

10.42 Will they charge more? A bank wonders whether omitting the annual credit card fee for customers who charge at least $2400 in a year will increase the amount charged on its credit cards. The bank makes this offer to an SRS of 200 of its credit card customers. It then compares how much these customers charge this year with the amount that they charged last year. The mean increase in the sample is $332, and the standard deviation is $108. Is there significant evidence at the 1% level that the mean amount charged increases under the no-fee offer? State H_0 and H_a and carry out a test.

10.43 Testing a random number generator, II Give a 90% confidence interval for the mean of all numbers produced by the calculator in Exercise 10.38 (page 591).

10.44 Will they charge more? Exercise 10.42 reports a bank's trial of eliminating credit card fees for clients who charge at least $2400. Give a 99% confidence interval for the mean amount charges would have increased if this benefit had been extended to all such customers.

10.45 Water quality. An environmentalist group collects a liter of water from each of 45 random locations along a stream and measures the amount of dissolved oxygen in each specimen. The mean is 4.62 milligrams (mg), and the standard deviation is 0.92 mg. Is this strong evidence that the stream has a mean oxygen content of less than 5 mg per liter?

10.46 Will they charge more? In Exercises 10.42 and 10.44, you carried out the calculations for a test and confidence interval based on a bank's experiment in changing the rules for its credit cards. You ought to ask some questions about this study.

(a) The distribution of the amount charged is skewed to the right, but outliers are prevented by the credit limit that the bank enforces on each card. Why can we use a test and confidence interval based on a normal sampling distribution for the sample mean $\bar{x}$?

(b) The bank's experiment was not comparative. The increase in amount charged over last year may be explained by lurking variables rather than by the rule change. What are some plausible reasons why charges might go up? Outline the design of a comparative randomized experiment to answer the bank's question.

CHAPTER 10 REVIEW

This chapter discusses a few more specific inference procedures. Section 10.1 deals with two-way tables, both for description and for inference. The descriptive part of this chapter completes Part II's discussion of relationships among variables by describing relationships among categorical variables. Section 10.2 concerns inference about the mean of a population.

Here are the most important skills you should have acquired after reading Chapter 10.

A. TWO-WAY TABLES

1. Arrange data on two categorical variables for a number of individuals into a two-way table of counts.

2. Use percents to describe the relationship between any two categorical variables starting from the counts in a two-way table.

3. Explain what null hypothesis the chi-square statistic tests in a specific two-way table.

4. Calculate expected cell counts, the chi-square statistic, and its degrees of freedom from a two-way table.

5. Use Table 10.1 for chi-square distributions to assess significance. Interpret the test result in the setting of a specific two-way table.

B. SAMPLING DISTRIBUTIONS
Use the normal sampling distribution of a sample mean $\bar{x}$ to find probabilities involving $\bar{x}$.

C. CONFIDENCE INTERVALS
Use the formula $\bar{x} \pm z* \, s/\sqrt{n}$ to obtain confidence intervals for a population mean μ.

D. SIGNIFICANCE TESTS

Carry out a one-sided or a two-sided test about a mean μ using the sample mean $\bar{x}$ and Table B.

CHAPTER 10 REVIEW EXERCISES

10.47 Simpson's paradox If we compare average NAEP quantitative scores, we find that eighth-grade students in Nebraska do better than eighth-grade students in New Jersey. But if we look only at white students, New Jersey does better. If we look only at minority students, New Jersey again does better. That's Simpson's paradox: the comparison reverses when we lump all students together. Explain carefully why this makes sense, using the fact that a much higher percent of Nebraska eighth-graders are white.[21]

10.48 When shall we call you? In Exercise 9.61 (page 549), we learned of a study that dialed telephone numbers at random during two periods of the day. Of 2304 calls on weekday mornings, 1313 were answered. Of 2454 calls on weekday evenings, 1840 were answered.
(a) Make a two-way table of time of day versus answered or not. What percent of calls were answered in each time period?
(b) It should be obvious that there is a highly significant relationship between time of day and answering. Why?
(c) Nonetheless, carry out the chi-square test. What do you conclude?

10.49 Unhappy HMO patients A study of complaints by HMO members compared those who filed complaints about medical treatment and those who filed nonmedical complaints with an SRS of members who did not complain that year. Here are the data on the total number in each group and the number who voluntarily left the HMO:[22]

	No complaint	Medical complaint	Nonmedical complaint
Total	743	199	440
Left	22	26	28

(a) Find the percent of each group who left.
(b) Make a two-way table of complaint status by left or not.
(c) Find the expected counts and check that you can safely use the chi-square test.
(d) The chi-square statistic for this table is $X^2 = 31.765$. What null and alternative hypotheses does this statistic test? What are its degrees of freedom? How significant is it? What do you conclude about the relationship between complaints and leaving the HMO?

10.50 Favorite car Do males and females like the same types of vehicles? First, decide on a few categories (for example, domestic car, foreign car, truck, SUV, other). Then, as in Activity 10.1, have students in the class vote by secret ballot (slip of paper). Write your gender and which type of vehicle you prefer. Then perform a complete chi-square analysis and answer the gender/vehicle question. If you have too many cells with expected counts less than 5, you may want to combine vehicle types into the "other" category. Don't forget to state hypotheses and check conditions.

10.51 Good news for chocolate lovers? A German study concluded that dark chocolate might reduce blood pressure.[23] The subjects in the study were patients with untreated mild hypertension. Six subjects ate a three-ounce white-chocolate bar each day for two weeks, while seven subjects ate a three-ounce dark-chocolate bar. Dark chocolate contains plant substances called polyphenols, which scientists think are responsible for the heart-healthy attributes of red wine. Polyphenols also have been shown to lower blood pressure in animals. Prior to the study, the participants had average systolic and diastolic blood pressure readings of about 153 over 84. The systolic blood pressure readings for the dark-chocolate group dropped an average of 5 points to 148. Assume a standard deviation of 6 points for the dark-chocolate group.
(a) Is there sufficient evidence to conclude that eating a three-ounce dark-chocolate bar every day will lower one's systolic blood pressure? Begin to answer this question by writing null and alternative hypotheses for this experiment.
(b) Calculate the test statistic and the P-value. Is there sufficient evidence to reject your null hypothesis?
(c) Write your conclusion in plain language.
(d) Do you have any concerns about the design of this experiment? If you were replicating this experiment, would you do anything differently? Explain.

10.52 What's the average height? One hundred and fifty students attend the lecture of a college statistics course on the first day of class. The course instructor wants to estimate the average height of the students in the room and asks the nine students in the front row to write down their heights. The average height of these nine students turns out to be 64 inches with a standard deviation of 3 inches.
(a) What is the parameter in this problem? What is the sample statistic?
(b) Would it be appropriate to use the normal distribution and the formula $\bar{x} \pm z^* s/\sqrt{n}$ to make a confidence statement about the average height of the students in the class, based on the data given in the problem? If yes, construct the appropriate confidence interval. If no, explain what is wrong.

10.53 Sharks, I Great white sharks are big and hungry. Here are the lengths in feet of 44 great whites:

18.7	12.3	18.6	16.4	15.7	18.3	14.6	15.8	14.9	17.6	12.1
16.4	16.7	17.8	16.2	12.6	17.8	13.8	12.2	15.2	14.7	12.4
13.2	15.8	14.3	16.6	9.4	18.2	13.2	13.6	15.3	16.1	13.5
19.1	16.2	22.8	16.8	13.6	13.2	15.7	19.7	18.7	13.2	16.8

Source : Data provided by Chris Olsen, who found the information in scuba-diving magazines.

(a) Make a stemplot with feet as the stems and tenths of feet as the leaves. There are two outliers, one in each direction. These won't change $\bar{x}$ much but will pull up the standard deviation s.
(b) Give a 90% confidence interval for the mean length of all great white sharks. (The interval may be too wide to be useful due to the influence of the outliers on s.)
(c) What do we need to know about these sharks in order to interpret your result in (b)?

10.54 Sharks, II Return to the data in Exercise 10.53. Is there good evidence that the mean length of sharks in the population these sharks represent is greater than 15 feet?

10.55 Pleasant smells, I Do pleasant odors help work go faster? Twenty-one subjects worked a paper-and-pencil maze wearing a mask that was either unscented or carried the smell of flowers. Each subject worked the maze three times with each mask, in random order. (This is a matched pairs design.) Here are the differences in their average times (in seconds), unscented minus scented. If the floral smell speeds work, the difference will be positive because the time with the scent will be lower.[24]

−7.37	−3.14	4.10	−4.40	19.47	−10.80	−0.87
8.70	2.94	−17.24	14.30	−24.57	16.17	−7.84
8.60	−10.77	24.97	−4.47	11.90	−6.26	6.67

(a) We hope to show that work is faster on the average with the scented mask. State null and alternative hypotheses in terms of the mean difference in times μ for the population of all adults.
(b) Using a calculator, find the mean and standard deviation of the 21 observations. Did the subjects work faster with the scented mask? Is the mean improvement big enough to be important?

(c) Make a stemplot of the data (round to the nearest whole second). Are there outliers or other problems that might hinder inference?

(d) Test the hypotheses you stated in (a). Is the improvement statistically significant?

10.56 Pleasant smells, II Return to the data in Exercise 10.55. Give a 95% confidence interval for the mean improvement in time to solve a maze when wearing a mask with a floral scent. Are you confident that the scent does improve mean working time?

PREFACE

1. From the National Council of Teachers of Mathematics (NCTM) Standards Web site, http://standards.nctm.org/.
2. The quotations are from a summary of the committee's report that was unanimously endorsed by the Board of Directors of the American Statistical Association. The full report is George Cobb, "Teaching statistics," in L. A. Steen (ed.), *Heeding the Call for Change: Suggestions for Curricular Action*, Mathematical Association of America, 1990, pp. 3–43.

PRELUDE

1. Arthur Nielsen quotation from "Making statistics more effective in schools of business: the Chicago meeting, 1986," available online at www.weatherhead.cwru.edu/msmesb/1986Chicago/nielsenpanel.
2. The information about mammograms comes from H. C. Sox, "Editorial: benefit and harm associated with screening for breast cancer," *New England Journal of Medicine*, 338 (1998), p. 1145.
3. On suicide, see Howard I. Kushner, *Self-Destruction in the Promised Land*, Rutgers University Press, 1989.

CHAPTER 1

1. We found these year-end sales data at the Recording Industry Association of America's Web site, www.riaa.com.
2. Steve Sternberg, "Researchers on the trail of a better smallpox vaccine," *USA Today*, December 30, 2002.
3. Robert J. Blendon et al., "The public and the smallpox threat," *New England Journal of Medicine*, 348 (January 30, 2003), pp. 381–382.
4. "Traffic safety facts 2000: young drivers," published by the National Highway Traffic Safety Administration on its Web site, www.nhtsa.dot.gov.
5. Nanci Hellmich, "Extra weight shaves years off lives," *USA Today*, January 7, 2003.
6. The Chicago Homeless Study is described in P. H. Rossi, J. D. Wright, G. A. Fisher, and G. Willis, "The urban homeless: estimating composition and size," *Science*, 235 (1987), pp. 1336–1339.
7. Example 1.2 is suggested by Maxine Pfannkuch and Chris J. Wild, "Statistical thinking and statistical practice: themes gleaned from professional statisticians," unpublished manuscript, 1998.
8. The study is by J. David Cassidy et al., "Effect of eliminating compensation for pain and suffering on the outcome of insurance claims for whiplash injury," *New England Journal of Medicine*, 342 (2000), pp. 1179–1186.
9. The estimates of the census undercount come from Howard Hogan, "The 1990 post-enumeration survey: operations and results," *Journal of the American Statistical Association*, 88 (1993), pp. 1047–1060.
10. Read "Is vitamin C really good for colds?" in *Consumer Reports*, February 1976, pp. 68–70, for a discussion of this conclusion. The article also reviews the need for controlled experiments and the Toronto study.
11. J. E. Muscat et al., "Handheld cellular telephone use and risk of brain cancer," *Journal of the American Medical Association*, 284 (2000), pp. 3001–3007.
12. S. G. Stolberg, "Link found between behavioral problems and time in child care," *New York Times*, April 19, 2001. The study is the National Institute of Child Health and Human Development Study of Early Child Care.
13. Janny Scott, "Working hard, more or less: studies of leisure time point both up and down," *New York Times*, July 10, 1999. We learned of this article at the *Chance News* Web site, www.dartmouth.edu/~chance/chance_news.
14. "Trial and error," *Economist*, October 31, 1998, pp. 87–88.
15. Quotation from a FairTest press release dated August 31, 1999, and appearing on the organization's Web page, www.fairtest.org. Most other information in this example and in Application 1.2A comes from chapter 2 of the National Science Foundation report *Women, Minorities, and Persons with Disabilities in Science and Engineering: 1998*, NSF99-338, 1999.
16. The table reports r^2-values, calculated from correlations given on the College Board Web site, www.collegeboard.org.
17. The deviations of NIST time from BIPM time are from the Web site of the NIST Time and Frequency Division, www.boulder.nist.gov/timefreq.
18. From the Electronic Encyclopedia of Statistical Examples and Exercises (EESEE) story "Blinded Knee Doctors."
19. The "missing vans" case is based on news articles in the *New York Times*, April 18, 1992, and September 10, 1992.
20. R. J. Newan, "Snow job on the slopes," *US News and World Report*, December 17, 1994, pp. 62–65.
21. Robert Niles of the *Los Angeles Times*, at nilesonline.com.
22. "Colleges inflate SATs and graduation rates in popular guidebooks," *Wall Street Journal*, April 5, 1995.
23. Robyn Meredith, "Oops, Cadillac says, Lincoln won after all," *New York Times*, May 6, 1999.

24. B. Yuncker, "The strange case of the painted mice," *Saturday Review/World*, November 30, 1974, p. 53.

25. Letter by J. L. Hoffman, *Science*, 255 (1992), p. 665. The original article appeared on December 6, 1991.

26. *Organic Gardening*, March 1983.

27. *Fine Gardening*, September/October 1989, p. 76.

28. E. Marshall, "San Diego's tough stand on research fraud," *Science*, 234 (1986), pp. 534–535.

29. Both items are from *Chance News*, www.dartmouth. edu/~chance/chance_news.

30. Darryl Nester spotted the ad featured in this example.

31. *Science*, 192 (1976), p. 1081.

32. Quotation from the American Heart Association statement "Cardiovascular disease in women," found at www. americanheart.org.

33. Edwin S. Rubenstein, "Inequality," *Forbes*, November 1, 1999, pp. 158–160; and Bureau of the Census, *Money Income in the United States, 1998*.

34. *Providence (R. I.) Journal*, December 24, 1999. We found it in *Chance News* 9.02.

35. *New York Times*, April 21, 1986.

36. *Science*, 189 (1975), p. 373.

37. *Lafayette (Ind.) Journal and Courier*, October 23, 1988.

38. Letter by L. Jarvik in the *New York Times*, May 4, 1993. The editorial, "Muggings in the kitchen," appeared on April 23, 1993.

39. Data are from the NBA Web site, www.nba.com.

40. *Condé Nast Traveler* magazine, June 1992.

41. *Purdue* (West Lafayette, Ind.) *Exponent*, September 21, 1977.

CHAPTER 2

1. "Acadian ambulance officials, workers flood call-in poll," *Baton Rouge Advocate*, January 22, 1999.

2. "Majority of smokers want to quit, consider themselves addicted," Gallup Poll press release by Mark Gillespie, November 18, 1999. Read online at www.gallup.com/poll/releases.

3. From the *New York Times*, January 4, 1970. Quoted with extensive discussion by Stephen E. Fienberg, "Randomization and social affairs: the 1970 draft lottery," *Science*, 171 (1971), pp. 255–261.

4. It's even a little more complicated than we have described. For all the details, see Joan R. Rosenblatt and James J. Filliben, "Randomization and the draft lottery," *Science*, 171 (1971), pp. 306–308.

5. D. Horvitz in his contribution to "Pseudo-opinion polls: SLOP or useful data?" *Chance*, 8, No. 2 (1995), pp. 16–25.

6. From "Working women count," a report released by the U.S. Department of Labor on October 15, 1997. For more information, visit its Web site, www. dol.gov.

7. The Gallup Organization, "Gambling in America— 1999: a comparison of adults and teenagers." Topline and Trends detailed report; read online at www.gallup.com/poll.

8. Warren McIsaac and Vivek Goel, "Is access to physician services in Ontario equitable?" Institute for Clinical Evaluative Sciences in Ontario, October 18, 1993.

9. *New York Times*, August 21, 1989.

10. Gallup Poll described on the Gallup Web site, www. gallup.com.

11. Information about restrictions on releasing election forecasts from Richard Morin, "Crackdown on pollsters," *Washington Post*, January 19, 1998.

12. Gallup Poll described on the Gallup Web site, www. gallup.com.

13. Based on a report at the Gallup Web site, www. gallup.com.

14. Gregory Flemming and Kimberly Parker, "Race and reluctant respondents: possible consequences of non-response for pre-election surveys," Pew Research Center for the People and the Press, 1997, found at www.people-press.org.

15. For more detail on nonsampling errors, along with references, see P. E. Converse and M. W. Traugott, "Assessing the accuracy of polls and surveys," *Science*, 234 (1986), pp. 1094–1098.

16. For more detail on the limits of memory in surveys, see N. M. Bradburn, L. J. Rips, and S. K. Shevell, "Answering autobiographical questions: the impact of memory and inference on surveys," *Science*, 236 (1987), pp. 157–161.

17. The nonresponse rate for the CPS comes from "Technical notes to household survey data published in *Employment and Earnings*," found on the Bureau of Labor Statistics Web site at http:// stats.bls.gov/cpshome.htm. The General Social Survey reports its response rate on its Web site, www.norc.uchicago.edu/gss/homepage.htm. The claim that "pollsters say response rates have fallen as low as 20 percent in some recent polls" is made in Don Van Natta, Jr., "Polling's 'dirty little secret': no response," *New York Times*, November 11, 1999.

18. The quotation is typical of Gallup Polls, from the Gallup Web site, www.gallup.com.

19. P. H. Lewis, "Technology" column, *New York Times*, May 29, 1995.

20. The responses on welfare are from a *New York Times*/CBS News Poll reported in the *New York Times*, July 5, 1992. Those for Scotland are from "All set for independence?" *Economist*, September 12, 1998.

21. D. Goleman, "Pollsters enlist psychologists in quest for unbiased results," *New York Times*, September 7, 1993.

22. The quotation on weighting is from Adam Clymer and Janet Elder, "Poll finds greater confidence in Democrats," *New York Times*, November 10, 1999.

23. Richard Morin, "It depends on what your definition of 'do' is," *Washington Post*, December 21, 1998.

24. From the *New York Times,* August 18, 1980.

25. The most recent account of the design of the CPS is Bureau of Labor Statistics, *Design and Methodology,* Current Population Survey Technical Paper 63, March 2000. Available in print or online at www.bls.census.gov/cps/tp/tp63.htm. The account in Example 2.18 omits many complications, such as the need to separately sample "group quarters" like college dormitories.

26. From the online "Supplementary Material" for G. Gaskell et al., "Worlds apart? The reception of genetically modified foods in Europe and the U.S.," *Science,* 285 (1999), pp. 384–387.

27. John Simons, "For risk takers, system is no longer sacred," *Wall Street Journal,* March 11, 1999.

28. See Note 21.

29. Adam Clymer and Janet Elder, "Poll finds greater confidence in Democrats," *New York Times,* November 10, 1999.

30. See www.louisharris.com.

CHAPTER 3

1. Allan H. Schulman and Randi L. Sims, "Learning in an online format versus an in-class format: an experimental study," *T.H.E. Journal,* June 1999, pp. 54–56.

2. L. L. Miao, "Gastric freezing: an example of the evaluation of medical therapy by randomized clinical trials," in J. P. Bunker, B. A. Barnes, and F. Mosteller (eds.), *Costs, Risks and Benefits of Surgery,* Oxford University Press, 1977, pp. 198–211.

3. Details of the Carolina Abecedarian Project, including references to published work, can be found online at www.fpg.unc.edu/overview/abc/abc-ov.htm.

4. Linda Rosa et al., "A close look at therapeutic touch," *Journal of the American Medical Association,* 279 (1998), pp. 1005–1010.

5. See Note 2.

6. Samuel Charache et al., "Effects of hydroxyurea on the frequency of painful crises in sickle cell anemia," *New England Journal of Medicine,* 332 (1995), pp. 1317–1322.

7. Marilyn Ellis, "Attending church found factor in longer life," *USA Today,* August 9, 1999.

8. Dr. Daniel B. Mark, in Associated Press, "Age, not bias, may explain differences in treatment," *New York Times,* April 26, 1994. Dr. Mark was commenting on Daniel B. Mark et al., "Absence of sex bias in the referral of patients for cardiac catheterization," *New England Journal of Medicine,* 330 (1994), pp. 1101–1106. See the correspondence from D. Douglas Miller and Leslee Shaw, "Sex bias in the care of patients with cardiovascular disease," *New England Journal of Medicine,* 331 (1994), p. 883, for comments on an opposed study.

9. C. S. Fuchs et al., "Alcohol consumption and mortality among women," *New England Journal of Medicine,* 332 (1995), pp. 1245–1250.

10. Martin Enserink, "Fickle mice highlight test problems," *Science,* 284 (1999), pp. 1599–1600. There is a full report of the study in the same issue.

11. The original study on the Mozart Effect is described in F. H. Rauscher, G. L. Shaw, and K. N. Ky, "Music and spatial task performance," *Nature,* 365 (1993), p. 611. For an opposing perspective, see Kenneth Steele, "The 'Mozart Effect': an example of the scientific method in operation," *Psychology Teacher Network,* November–December 2001.

12. The fact about rats comes from E. Street and M. B. Carroll, "Preliminary evaluation of a new food product," in J. M. Tanur et al. (eds.), *Statistics: A Guide to the Unknown,* 3rd edition, Wadsworth, 1989, pp. 161–169.

13. The placebo effect examples are from Sandra Blakeslee, "Placebos prove so powerful even experts are surprised," *New York Times,* October 13, 1998.

14. The "three-quarters" estimate is cited by Martin Enserink, "Can the placebo be the cure?" *Science,* 284 (1999), pp. 238–240. An extended treatment is Anne Harrington (ed.), *The Placebo Effect: An Interdisciplinary Exploration,* Harvard University Press, 1997.

15. The flu trial quotation is from Kristin L. Nichol et al., "Effectiveness of live, attenuated intranasal influenza virus vaccine in healthy, working adults," *Journal of the American Medical Association,* 282 (1999), pp. 137–144.

16. "Cancer clinical trials: barriers to African American participation," *Closing the Gap,* newsletter of the Office of Minority Health, December 1997–January 1998.

17. Michael H. Davidson et al., "Weight control and risk factor reduction in obese subjects treated for 2 years with Orlistat: a randomized controlled trial," *Journal of the American Medical Association,* 281 (1999), pp. 235–242.

18. Edward P. Sloan et al., "Diaspirin cross-linked hemoglobin (DCLHb) in the treatment of severe traumatic hemorrhagic shock," *Journal of the American Medical Association,* 282 (1999), 1857–1864.

19. "Advertising: the cola war," *Newsweek,* August 30, 1976, p. 67.

20. Mary O. Mundinger et al., "Primary care outcomes in patients treated by nurse practitioners or physicians," *Journal of the American Medical Association,* 238 (2000), pp. 59–68.

21. Carol A. Warfield, "Controlled-release morphine tablets in patients with chronic cancer pain," *Cancer,* 82 (1998), pp. 2299–2306.

22. From the Electronic Encyclopedia of Statistical Examples and Exercises (EESEE) case study "Is Caffeine Dependence Real?"

23. The quotation is from Thomas B. Freeman et al., "Use of placebo surgery in controlled trials of a cellular-based therapy for Parkinson's disease," *New England Journal of Medicine,* 341 (1999), pp. 988–992. Freeman supports the Parkinson's disease trial. The opposition is represented by Ruth Macklin, "The ethical problems with sham surgery in clinical research," *New England Journal of Medicine,* 341 (1999), pp. 992–996.

24. From the EESEE case study "Stepping Up Your Heart Rate."

25. John C. Bailar III, "The real threats to the integrity of science," *Chronicle of Higher Education,* April 21, 1995, pp. B1–B2.

26. The difficulties of interpreting guidelines for informed consent and for the work of institutional review boards in medical research are a main theme of Beverly Woodward, "Challenges to human subject protections in U.S. medical research," *Journal of the American Medical Association,* 282 (1999), pp. 1947–1952. The references in this paper point to other discussions.

27. Quotation from the *Report of the Tuskegee Syphilis Study Legacy Committee,* May 20, 1996. A detailed history is James H. Jones, *Bad Blood: The Tuskegee Syphilis Experiment,* Free Press, 1993.

28. Dr. Hennekens's words are from an interview in the Annenberg/Corporation for Public Broadcasting video series *Against All Odds: Inside Statistics.*

29. Dr. C. Warren Olanow, chief of neurology, Mt. Sinai School of Medicine, quoted in Margaret Talbot, "The placebo prescription," *New York Times Magazine,* January 8, 2000, pp. 34–39, 44, 58–60.

30. For extensive background, see Jon Cohen, "AIDS trials ethics questioned," *Science,* 276 (1997), pp. 520–523. Search the archives at www.sciencemag.org for recent episodes in the continuing controversies of Exercises 3.50, 3.63, and 3.68.

31. See Paul Meier, "The biggest public health experiment ever," in Judith M. Tanur et al. (eds.), *Statistics: A Guide to the Unknown,* Holden-Day, 1972, pp. 2–13, for a discussion of the experiment that established the effectiveness of the Salk polio vaccine, a case that fits this exercise.

32. Stanley Milgram, "Group pressure and action against a person," *Journal of Abnormal and Social Psychology,* 69 (1964), pp. 137–143.

33. Postexperimental attitudes and the final quotation are in Stanley Milgram, "Liberating effects of group pressure," *Journal of Personality and Social Psychology,* 1 (1965), pp. 127–134.

34. Quotations from Gina Kolata and Kurt Eichenwald, "Business thrives on unproven care leaving science behind," *New York Times,* October 3, 1999. Background and details about the first clinical trials appear in a National Cancer Institute press release "Questions and answers: high-dose chemotherapy with bone marrow or stem cell transplants for breast cancer," April 15, 1999. That one of the studies reported there involved falsified data is reported by Denise Grady, "Breast cancer researcher admits falsifying data," *New York Times,* February 5, 2000.

35. Exercise 3.70 is based on Christopher Anderson, "Measuring what works in health care," *Science,* 263 (1994), pp. 1080–1082.

36. See note 15.

CHAPTER 4

1. Data from the *Statistical Abstract of the United States,* 2002.

2. Data provided by the College Board.

3. The prices in Figure 4.7 combine the consumer price index component for unleaded regular gasoline (an index number) with the Energy Information Administration report of the actual price in January 1999. This information is available online at www.bls.gov/cpi/.

4. Data from Jupiter Communications.

5. We found the college student credit card data at the Web site www.nelliemae.com.

6. *1997 Statistical Yearbook of the Immigration and Naturalization Service,* U.S. Department of Justice, 1999.

7. Organ transplant data from the *Statistical Abstract of the United States,* 2002.

8. 1999–2000 tuition and fees from the College Board's *Annual Survey of Colleges,* 1999–2000.

9. Cereal data provided by Professor Dennis Pearl, Ohio State University, Columbus.

10. Percent low-birth-weight data from the *Statistical Abstract of the United States,* 2002.

11. Figure 4.20 is based on an episode in the Annenberg/Corporation for Public Broadcasting telecourse *Against All Odds: Inside Statistics.*

12. Data from the government's Survey of Earned Doctorates, in the online supplement to Jeffery Mervis, "Wanted: a better way to boost numbers of minority PhDs," *Science,* 281 (1998), pp. 1268–1270.

13. *Consumer Reports,* June 1986, pp. 366–367. A more recent study of hot dogs appears in *Consumer Reports,* July 1993, pp. 415–419. The newer data cover few brands of poultry hot dogs and take calorie counts mainly from the package labels, resulting in suspiciously round numbers.

14. Data on income and education from the March 1999 CPS supplement, downloaded from the Census Bureau Web site using the government's FERRET system. These are raw survey data, including many people with no income or negative income. To simplify comparisons, we report only four education classes.

15. The SAT data behind the table of state SAT scores come from College Board Online, www.collegeboard.com.

16. AP data from the *Statistical Abstract of the United States,* 2002.

17. See Note 15.

CHAPTER 5

1. The IQ scores in Figure 5.11 were collected by Darlene Gordon, Purdue University School of Education.

2. Ulric Neisser, "Rising scores on intelligence tests," *American Scientist,* September–October 1997, online edition.

3. Bureau of Labor Statistics, "Updated response to the recommendations of the advisory commission to study the consumer price index," June 1998. The entire December 1996 issue of the BLS *Monthly Labor Review* is devoted to the major revision of the CPI that became effective in January 1998. The effect of changes in the CPI is discussed by Kenneth J. Stewart and Stephen B. Reed, "CPI research series using current methods, 1978–98," *Monthly Labor Review,* 122 (1999), pp. 29–38. All of these are available on the BLS Web site, www.bls.gov/data/.

4. Exercises 5.30 to 5.33 use information from the BLS Web site, www.bls.gov/data/.

5. The government statistics survey is reported in "The good statistics guide," *Economist,* September 13, 1993, p. 65.

CHAPTER 6

1. Data from the Electronic Encyclopedia of Statistical Examples and Exercises (EESEE) story "Is Old Faithful Faithful?"

2. Data from the World Bank's *1999 World Development Indicators.* Life expectancy is estimated for 1997, and GDP per capita (purchasing power parity basis) for 1998.

3. Christer G. Wiklund, "Food as a mechanism of density-dependent regulation of breeding numbers in the merlin *Falco columbarius,*" *Ecology,* 82 (2001), pp. 860–867.

4. Environmental Protection Agency, *2001 Fuel Economy Guide,* www.fueleconomy.gov.

5. From the EESEE story "Is It Tough to Crawl in March?"

6. Figure 6.7 is based on data from the Web site of Professor Kenneth French of Dartmouth, mba.tuck.dartmouth.edu/pages/faculty/ken.french/data_library.html.

7. Data from Daren Starnes's Advanced Placement™ Statistics class, 1998–1999.

8. National Center for Education Statistics, nces.ed.gov/nationsreportcard/.

9. M. A. Houck et al., "Allometric scaling in the earliest fossil bird, *Archaeopteryx lithographica,*" *Science,* 247 (1990), pp. 195–198. The authors conclude from a variety of evidence that all specimens represent the same species.

10. Data from T. J. Lorenzen, "Determining statistical characteristics of a vehicle emissions audit procedure," *Technometrics,* 22 (1980), pp. 483–493.

11. From a study of 4511 white, non-Hispanic children reported by Judith Blake, "Number of siblings and educational attainment," *Science,* 245 (1989), pp. 32–36.

12. From the *Runner's World* Web site, www.runnersworld.com.

13. Alan S. Banks et al., "Juvenile hallux abducto valgus association with metatarsus adductus," *Journal of the American Podiatric Medical Association,* 84 (1994), pp. 219–224.

14. The quotation is from the Galateia Corporation Web site, at www.voicenet.com/~mitochon.

15. Laura L. Calderon et al., "Risk factors for obesity in Mexican-American girls: dietary factors, anthropometric factors, physical activity, and hours of television viewing," *Journal of the American Dietetic Association,* 96 (1996), pp. 1177–1179.

16. From the EESEE story "Blood Alcohol Content."

17. See Note 15.

18. See L. Lasagna and S. R. Schulman, "Bendectin and the language of causation," in K. R. Foster, D. E. Bernstein, and P. W. Huber (eds.), *Phantom Risk: Scientific Inference and the Law,* MIT Press, 1994, pp. 101–122.

19. Quotation from a Gannett News Service article appearing in the *Lafayette (Ind.) Journal and Courier,* April 23, 1994.

20. From G. Grime, "A review of research on the protection afforded to occupants of cars by seat belts which provide upper-torso restraint," *Accident Analysis and Prevention,* 11 (1979), p. 299.

21. Data provided by Linda McCabe and the Agency for International Development.

22. Data obtained from the College Board Web site, www.collegeboard.com.

23. The data plotted in Figure 6.26 come from G. A. Sacher and E. F. Staffelt, "Relation of gestation time to brain weight for placental mammals: implications for the theory of vertebrate growth," *American Naturalist,* 108 (1974), pp. 593–613. We found them in Fred L. Ramsey and Daniel W. Schafer, *The Statistical Sleuth: A Course in Methods of Data Analysis,* Duxbury, 1997, p. 228.

CHAPTER 7

1. Richard Feynman's comments on the probability of a shuttle failure can be found in his appendix to the Presidential Commission report, "Personal observations on the reliability of the shuttle by R. P. Feynman." See the Kennedy Space Center Web site, at www.ksc.nasa.gov/shuttle/missions/51-l/docs/rogers-commission/Appendix-F.txt.

2. Ivars Peterson, "Random home runs," *Science News,* 159, No. 26 (June 30, 2001). Available at www. sciencenews.org.

3. Paul M. Sommers, "Sultans of swat and the runs test," *Journal of Recreational Mathematics,* 30, No. 2 (2001), p. 118.

4. More historical detail can be found in the opening chapters of F. N. David, *Games, Gods and Gambling,* Charles Griffin and Co., 1962. The historical information given here comes from this excellent and entertaining book.

5. For a discussion and amusing examples, see A. E. Watkins, "The law of averages," *Chance,* 8, No. 2 (1995), pp. 28–32.

6. R. Vallone and A. Tversky, "The hot hand in basketball: on the misperception of random sequences," *Cognitive Psychology,* 17 (1985), pp. 295–314.

7. For Woburn, see S. W. Lagakos, B. J. Wessen, and M. Zelen, "An analysis of contaminated well water and health effects in Woburn, Massachusetts," *Journal of the American Statistical Association,* 81 (1986), pp. 583–596, and the following discussion. For Randolph, see R. Day, J. H. Ware, D. Wartenberg, and M. Zelen, "An investigation of a reported cancer cluster in Randolph, Ma.," *Harvard School of Public Health Technical Report,* June 27, 1988.

8. Estimated probabilities from R. D'Agostino, Jr., and R. Wilson, "Asbestos: the hazard, the risk, and public policy," in K. R. Foster, D. E. Bernstein, and P. W. Huber (eds.), *Phantom Risk: Scientific Inference and the Law,* MIT Press, 1994, pp. 183–210. See also the similar conclusions in B. T. Mossman et al., "Asbestos: scientific developments and implications for public policy," *Science,* 247 (1990), pp. 294–301.

9. The quotation is from R. J. Zeckhauser and W. K. Viscusi, "Risk within reason," *Science,* 248 (1990), pp. 559–564.

10. T. Hill, "Random-number guessing and the first digit phenomenon," *Psychological Reports,* 62 (1988), pp. 967–971.

CHAPTER 8

1. The introductory case study is based in part on Nico M. van Dijk et al., "Designing the Westerscheldetunnel toll plaza using a combination of queueing and simulation," in P. A. Farrington et al. (eds.), *Proceedings of the 1999 Winter Simulation Conference,* INFORMS, 1999, pp. 1272–1279.

2. Stochastic beetles are well known in the folklore of simulation, if not in entomology. They are said to be the invention of Arthur Engle of the School Mathematics Study Group.

3. Andrew Pollack, "In the gaming industry, the house can have bad luck, too," *New York Times,* July 25, 1999, reports that Mirage Resorts issued a quarterly profit warning due in part to its bad luck at baccarat.

4. S. Newcomb, "Note on the frequency of the use of digits in natural numbers," *American Journal of Mathematics,* 4 (1881), p. 39.

5. F. Benford, "The law of anomalous numbers," *Proceedings of the American Philosophical Society,* 78 (1938), pp. 551–572.

6. Professor Nigrini has an interesting Web site at www.nigrini.com/benford's_law.htm.

7. Examples of how the first-digit phenomenon is being used to detect fraud can be found at www.fm.co.za/97/0926/infotech/audit.htm and www.bham.ac.uk/EAA/eaa97/abstracts/BUSTA.HTM.

8. T. P. Hill, "The first digit phenomenon," *American Scientist,* 86 (July–August 1998), p. 358.

9. F. N. David, *Games, Gods and Gambling,* Charles Griffin and Co., 1962.

CHAPTER 9

1. Janice E. Williams et al., "Anger proneness predicts coronary heart disease risk," *Circulation,* 101 (2000), pp. 2034–2039.

2. M&M plain candy color proportions can be found at the Mars Web site, www.mms.com/factory/history/faq1.html.

3. The 2001 national school-based Youth Risk Behavioral Survey (YRBS) was carried out by the Centers for Disease Control (CDC) to monitor the prevalence of youth behaviors that most influence health.

4. Report of the American Society for Microbiology. We found notice of it at www.usatoday.com/life/health/general/lhgen083.htm.

5. Data from the Web site www.drugfreeamerica.org.

6. For the state of the art in confidence intervals for p, see Alan Agresti and Brent Coull, "Approximate is better than 'exact' for interval estimation of binomial proportions," *American Statistician,* 52 (1998), pp. 119–126. The authors note that the accuracy of our confidence interval for p can be greatly improved by simply "adding 2 successes and 2 failures." That is, replace by (count of successes + 2) / $(n + 4)$. Texts on sample surveys give confidence intervals that take into account the fact that the population has finite size and also give intervals for sample designs more complex than an SRS.

7. Laurie Goodstein and Marjorie Connelly, "Teenage poll finds support for tradition," *New York Times,* April 30, 1998.

8. Hunting data provided by the Virginia Department of Game and Inland Fisheries, Richmond, Virginia.

9. This is a somewhat simplified account of part of the study described by Reynolds Farley et al., "Stereotypes and segregation: neighborhoods in the Detroit area," *American Journal of Sociology,* 100 (1994), pp. 750–780. Figure 9.6 is based on part of a figure on p. 754 of this article.

10. Research by a team led by Stephen S. Hecht of the University of Minnesota Cancer Center, as reported by the *Los Angeles Times/Washington Post* News Service, August 24, 1998.

11. Quoting Cheryl Healton, American Legacy Foundation president.

12. Study conducted by Dr. D. Casey Kerrigan, University of Virginia, professor and chairwoman of the Department of Physical Medicine and Rehabilitation. Reported by Claudia Pinto, Media General News Service, March 6, 2003.

13. Catherine A. Demko et al., "Use of indoor tanning facilities by white adolescents in the United States," *Archives of Pediatrics and Adolescent Medicine,* 157 (September 8, 2003), pp. 854–860.

14. Jeffrey Selingo, "What Americans think about higher education," *Chronicle of Higher Education,* 49, No. 34 (May 2, 2003), pp. A10–A15.

15. Dorothy Espelage et al., "Factors associated with bullying behavior in middle school students," *Journal of Early Adolescence,* 19, No. 3 (August 1999), pp. 341–362.

16. Manisha Chandalia et al., "Beneficial effects of high dietary fiber intake in patients with type 2 diabetes mellitus," *New England Journal of Medicine,* 342 (2000), pp. 1392–1398.

17. We found these data at the www.celexa.com Web site.

18. We looked at the list of top funds at the Smart Money site, www. smartmoney.com, in early May 2000. Here's part of the abstract of a typical recent study, Mark Carhart, "On persistence in mutual fund performance," *Journal of Finance,* 52 (1997), pp. 57–82: "Using a sample free of survivor bias, I demonstrate that common factors in stock returns and investment expenses almost completely explain persistence in equity mutual funds' mean and risk-adjusted returns.... The only significant persistence not explained is concentrated in strong underperformance by the worst-return mutual funds. The results do not support the existence of skilled or informed mutual fund portfolio managers."

19. Jane A. Cauley et al., "Lipid-lowering drug use and breast cancer in older women: a prospective study," *Journal of Women's Health,* 12, No. 8 (October 2003), pp. 749–756.

20. Robert Rosenthal is quoted in B. Azar, "APA statistics task force prepares to release recommendations for public comment," *APA Monitor Online,* 30 (May 1999), at www.apa.org/monitor. The task force report, Leland Wilkinson et al., "Statistical methods in psychology journals: guidelines and explanations," *American Psychologist,* 54 (August 1999), pp. 594–604, offers a summary of the elements of good statistical practice.

21. Application 9.3 uses data in the article "Schools toying with test results," by *Chicago Tribune* staff reporters Diane Rado and Darnelle Little, September 29, 2003, available online at www. chicagotribune.com.

CHAPTER 10

1. An example of real data similar in spirit is P. J. Bickel and J. W. O'Connell, "Is there a sex bias in graduate admissions?" *Science,* 187 (1975), pp. 398–404.

2. Data for Nationsbank, from S. A. Holmes, "All a matter of perspective," *New York Times,* October 11, 1995.

3. Richard M. Felder et al., "Who gets it and who doesn't: a study of student performance in an introductory chemical engineering course," *1992 ASEE Annual Conference Proceedings,* American Society for Engineering Education, 1992, pp. 1516–1519.

4. S. V. Zagona (ed.), *Studies and Issues in Smoking Behavior,* University of Arizona Press, 1967, pp. 157–180.

5. R. Shine, T. R. L. Madsen, M. J. Elphick, and P. S. Harlow, "The influence of nest temperatures and maternal brooding on hatchling phenotypes in water pythons," *Ecology,* 78 (1997), pp. 1713–1721.

6. We found these data at Patrick Meirmans's Web site, where he attributed the study to a paper by H. R. Schiffman in a 1982 issue of the *Journal of Counseling and Psychology.*

7. From reports submitted by airlines to the Department of Transportation, found in A. Barnett, "How numbers can trick you," *Technology Review,* October 1994, pp. 38–45.

8. D. M. Barnes, "Breaking the cycle of addiction," *Science,* 241 (1988), pp. 1029–1030.

9. S. W. Hargarten et al., "Characteristics of firearms involved in fatalities," *Journal of the American Medical Association,* 275 (1996), pp. 42–45.

10. There are many computer studies of the accuracy of chi-square critical values for X^2. For a brief discussion and some references, see Section 3.2.5 of David S. Moore, "Tests of chi-squared type," in Ralph B. D'Agostino and Michael A. Stephens (eds.), *Goodness-of-Fit Techniques,* Marcel Dekker, 1986, pp. 63–95.

11. Janice E. Williams et al., "Anger proneness predicts coronary heart disease risk," *Circulation,* 101 (2000), pp. 2034–2039.

12. Data reported by P. K. Viswanathan, Adjunct Professor and Management Consultant, Chennai, India, in a paper, "Glimpses into application of chi-square tests in marketing," found at David Lane's Hyperstat Web site, http://davidmlane.com/hyperstat/chi_square.html.

13. *Digest of Education Statistics 1997,* accessed on the National Center for Education Statistics Web site, http://www.ed.gov/NCES.

14. Francine D. Blau and Marianne A. Ferber, "Career plans and expectations of young women and

men," *Journal of Human Resources,* 26 (1991), pp. 581–607.

15. M. Radelet, "Racial characteristics and imposition of the death penalty," *American Sociological Review,* 46 (1981), pp. 918–927.

16. The study is P. A. Mackowiak, S. S. Wasserman, and M. M. Levine, "A critical appraisal of 98.6 degrees F, the upper limit of normal body temperature, and other legacies of Carl Reinhold August Wunderlich," *Journal of the American Medical Association,* 268 (1992), pp. 1578–1580. We owe the reference, and data based on plots in the original article, to Allen L. Shoemaker, "What's normal? Temperature, gender, and heart rate," *Journal of Statistics Education,* 1996. (This electronic journal is found at www.amstat. org/publications/jse.)

17. Information about the NAEP test is from Francisco L. Rivera-Batiz, "Quantitative literacy and the likelihood of employment among young adults," *Journal of Human Resources,* 27 (1992), pp. 313–328.

18. Data provided by Darlene Gordon, Purdue University.

19. F. H. Rauscher, G. L. Shaw, and K. N. Ky, "Listening to Mozart enhances spatial-temporal reasoning: towards a neuralpsychological basis," *Neuroscience Letters* 185 (1993), pp. 44–47.

20. Kenneth M. Steele, Karen E. Bass, and Melissa D. Crooke, "The mystery of the Mozart Effect," *Psychological Science* 10, No. 4 (1999), pp. 366–369.

21. This example is in Howard Wainer's "Visual revelations" column, *Chance,* 12, No. 2 (1999), pp. 43–44.

22. Sara J. Solnick and David Hemenway, "Complaints and disenrollment at a health maintenance organization," *Journal of Consumer Affairs,* 26 (1992), pp. 90–103.

23. Dirk Taubert, Reinhard Berkels, Renate Roesen, and Wolfgang Klaus, "Chocolate and blood pressure in elderly individuals with isolated systolic hypertension," *Journal of the American Medical Association,* 290 (2003) pp. 1029–1030.

24. A. R. Hirsch and L. H. Johnston, "Odors and learning," *Journal of Neurological and Orthopedic Medicine and Surgery,* 17 (1996), pp. 119–126. We found the data in a case study, "Floral Scents and Learning," in the Electronic Encyclopedia of Statistical Examples and Exercises (EESEE), W. H. Freeman, 2000.

Solutions to Odd-Numbered Exercises

CHAPTER 1

1.1 (a) Students in a statistics class. (b) Sex, home-room teacher, grade level, calculator number, score on Test 1; grade level and score on Test 1 take numerical values.

1.3 For example, whether a household uses its recycling bin.

1.5 (a) *Statistics Through Applications* textbooks. (b) Length. (c) Because students are estimating where the edge of the book falls along the special ruler. Also, books are not perfectly identical in length.

1.7 (a) All pieces of hardwood in the lot; the 5 pieces of wood that are selected. (b) All auto insurance claims filed in a month at an insurance company; the sample of claims selected.

1.9 Population: all apples in the truckload. Sample: three large buckets of apples selected. Individuals: apples. Variable(s): weight, color, number of blemishes.

1.11 (a) Television shows. (b) Viewers (in millions), household rating (in points; 1 point = 1,067,000 households), household share (% of TV sets in use). (c) From journals kept by viewers and electronic monitoring devices used in a sample of households.

1.13 An experiment—because a treatment (eating two brands of muffins) is imposed on the subjects.

1.15 (a) What is the most popular lunch entrée served in the school cafeteria? (b) Which make of car is most popular at your school? (c) Does listening to relaxing music for ten minutes immediately prior to a test improve students' scores?

1.17 (a) Answers will vary. (b) An observational study using police records.

1.19 A sample survey based on interviews of individuals living in public housing.

1.21 (a) No; subjects are assigned to groups based on fitness level. (b) Population: for example, "business executives" or "men in leadership positions" or "adults." Variables: fitness level and leadership ability.

1.23 (a) An experiment, since each student will be assigned to one of two treatments: wear red clothes or wear white clothes. (b) Clothing color and number of bee stings. (c) No; there may be reasons other than clothing color that cause bees to sting (or not). (d) There are ethical concerns associated with conducting this experiment. There are many other variables, such as smell or size, that might perhaps affect number of bee stings.

1.25 Answers will vary.

1.27 (a) Instrument: hand. Units: seconds. Variable: temperature of fire. (b) The method seems valid: the hotter the fire, the less time you can hold your hand over it. (c) This method is probably not very accurate. Individuals will vary considerably in their tolerance for heat.

1.29 A count does not allow for the growing size of the labor force.

1.31 Compare death rates (deaths in buses and in cars divided by the number of passengers in each).

1.33 (a) Cancer deaths increase as the population becomes older. (b) Cancer death rates rise as the health of the population improves and fewer people die of other causes. Also, perhaps the environment is becoming more hazardous. (c) Survival times could increase if the disease is being discovered earlier; that is, if diagnosis (not treatment) becomes more effective.

1.35 Local agencies may deliberately underreport crimes to the FBI or may simply be careless in keeping records. Survey respondents may remember dates inaccurately, lie, or not understand what constitutes a crime.

1.37 Since IQ tests do not relate directly to job performance, the issue of concern is lack of validity.

1.39 (a) The average of several repeated measurements will yield greater accuracy. (b) Since they are using the same process to make the measurements, the two investigators should have about the same bias in their reported results.

1.41 Answers will vary.

1.43 Texas (12.28 executions per million) is the highest; Florida (3.19) is the lowest.

1.45 (a) Bias. (b) For example, randomly assign workers to a training program rather than letting them choose whether or not to participate.

1.47 Statement I addresses measurement validity; Statement II addresses measurement reliability.

1.49 (a) This depends on your class's measurements. (b) This would not be likely to affect the reliability of the measurements (that is, the measurements would still bounce around by about the same amount) but would create a bias (since the distance from the $\frac{1}{0}$ mark to the end of the page would be added to every value).

1.51 (a) The folks at Snapple probably surveyed either consumers or pizza makers to obtain this estimate. (b) That's a lot of pizza!

1.53 More than twice as many unmarried men as women is far too high.

1.55 Friday through Sunday is about 42.8% of the week.

1.57 57% of 20 equals 11.4 studies, and 42% would be 8.4 studies.

1.59 (a) It is 2.1 million beatings, not 21 million. (b) The survey counts cases, not incidents: a woman in an abusive relationship (one case) likely is beaten more than once a year (several incidents).

1.61 Both Anacin and Bufferin are brands of aspirin. Bufferin also contains ingredients to prevent upset stomach and is likely specified for that reason.

1.63 This would mean no bags were lost.

1.65 For example, over 20 years, 150,000 suicides means an average of 20 suicides per day, which would be hard to ignore.

1.67 About a 1 in 10,000 chance.

1.69 Answers will vary.

1.71 Answers will vary.

1.73 (a) Population: all multi-airline tickets sold. Sample: 12% of the multi-airline tickets sold. (b) They will probably not result in exactly the correct amount of money for each airline. With a good sample, however, the sample results should be close to the truth about the population.

1.75 (a) Method 3 would increase reliability by taking the average of several measurements. (b) Method 2 would have better validity. (c) Method 1 would reduce bias.

1.77 (a) Method 2 is more reliable because you are measuring the pulse for a longer time interval. (b) Neither method is clearly more biased, since either could result in an overestimate or underestimate of an individual's true pulse rate. (c) With Method 1 and Method 2, you are likely to get caught "between beats." This is not accounted for in your estimate of the pulse rate. Counting beats avoids this problem.

1.79 (a) Deaths are almost certain to be reported, so the listed number of deaths is quite accurate. (b) The increase in injuries from 1970 to 1990 is likely due to increased reporting, not to an increase in actual injuries. (c) 1980 to 1990: 36.4%; 1990 to 1997: 5.3%; 1980 to 1997: 39.8%.

CHAPTER 2

2.1 (a) 29,777. (b) It's a voluntary response sample. (c) If more men responded, there may be more "No" responses in the sample than there would be in the population.

2.3 Possible answers: (a) A call-in poll. (b) Interviewing students as they enter the student center.

2.5 (a) Put two elements in the population. Sample one element at a time. Let 1 represent "heads" and 2 represent "tails." (b) Put 52 elements in the population. Sample all 52 elements. (c) Put 500 elements in the population and sample 12 elements.

2.7 Apartment complexes 16, 32, 18: Fairington, Waterford Court, and Fowler.

2.9 (a) Use labels 001 to 440. (b) 381, 262, 183, 322, 341, 185, 414, 273, 190, 325, 330, 029, 079, 078, 118, 209, 354, 239, 421, 426, 435, 437, 193, 099, 224.

2.11 Using line 122: 13, 15, 05—Lee, Milhalko, and Brockman.

2.13 (a) All black adult residents of Miami; 300 adults from the SRS. (b) Because a black police officer is asking the questions.

2.15 Answers will vary.

2.17 (a) Label from 0001 to 3478. (b) 2940, 0769, 1481, 2975, 1315.

2.19 It's a convenience sample.

2.21 Supporters of an individual's right to bear arms have tremendous political influence. Opinion polls on such subjects do not always result in legislative action.

2.23 Yes, since the library surveyed card holders.

2.25 All smokers who signed a card on November 20, 1995; the percent (or proportion) who had not smoked in the past six months; 1000 people who were surveyed; 21%.

2.27 Statistic, parameter.

2.29 This is a comparative experiment because the researchers assigned the ducks to the groups being compared (inside boxes versus outside boxes).

2.31 (a) High bias, high variability. (b) Low bias, low variability. (c) Low bias, high variability. (d) High bias, low variability.

2.33 Answers will vary.

2.35 The margin of error is half as big with the larger sample.

2.37 0.031; yes.

2.39 Quadruple the sample size, to $n = 4036$.

2.41 Smaller sample size means larger margin of error.

2.43 Larger.

2.45 Greater, since the variability in the estimate depends on the sample size.

2.47 Parameter, statistic.

2.49 4.5%. The announced margin of error (5%) is slightly higher.

2.51 (a) 0.59; the proportion of all adults who prefer balancing the budget over cutting taxes. (b) We are 95% confident that between 56% and 62% of all adults prefer balancing the budget over cutting taxes.

2.53 (a) All adults. (b) We are 95% confident that between 33% and 39% of adults feel that the government should provide money to low-income families who want to send their children to non-public schools.

2.55 We can be 95% confident that between 66.5% and 71.5% of U.S. adults are satisfied with the way things were going in the United States in January 2000.

2.57   For example, undercoverage or nonresponse.

2.59   (a) 11.3%. (b) 2.6%; yes. (c) No, since a voluntary response sample was used.

2.61   Answers will vary.

2.63   "Adding to the Constitution" sounds easier to support.

2.65   Closed questions allow for easier tabulation of results but may omit response choices that individuals would prefer to select. Open questions give individuals more freedom to respond, but the tabulation of results could then be more difficult.

2.67   (a) The sample was picked by a procedure involving several steps, perhaps similar to how the Current Population Survey sample is chosen. (b) The strata are the 17 countries. (c) At least one of the stages in choosing the sample involved random selection.

2.69   (a) 40. (b) Stratified random samples, to allow for separate conclusions about males and females.

2.71   (a) 69, 169, 269, 369, and 469. (b) 1/100. (c) Not every sample of five addresses is equally likely to be chosen.

2.73   (a) We are told that each region was proportionately represented in the sample. (b) An SRS could result in heavy representation from some areas and light (or no) representation from other areas. This could lead to a biased estimate.

2.75   Some people probably said that they voted when they actually didn't.

2.77   (a) All students at your school; the proportion (or percent) of students who rate teachers as "sufficiently available." (b) to (d) Answers will vary.

2.79   The margin of error depends on the sample size, not the population size.

2.81   Wording of questions can greatly affect individual responses.

2.83   (a) Your estimate of amount spent is likely to be low since you are not including those sitting in expensive seats in the sample. (b) Undercoverage is a sampling error.

2.85   (a) All adult women in the country; the percent who would say "highly desirable." (b) Statistic. (c) We are 95% confident that between 71% and 81% of adult women in the country would say that an "elegant" car is "highly desirable."

2.87   (a) Assign labels 0001 through 3500 and use random digits. (b) Randomly select one of the first 14 students, then take the students who are 14, 28, 42, . . . places down the list. (c) Choose SRSs of size 200 and 50 from each group.

2.89   Answers will vary.

2.91   (a) All adults who live in or near West Lafayette, Indiana. (b) Larger, since this voluntary response poll probably overrepresents those who are opposed to the one-way street.

2.93   Use a stratified random sample, with each distinct section in the library as a stratum.

CHAPTER 3

3.1   It is an experiment because the researcher attempts to affect the outcome by showing different price histories. The explanatory variable is whether the student sees a steady price or temporary price cuts. The response variable is the price the student would expect to pay.

3.3   A mother's prior attitude toward her baby, which probably influenced her choice of feeding method, is the lurking variable; the prior attitude is confounded with the effect of the nursing.

3.5   (a) Therapeutic touch practitioners; hand position (left or right); choice of hand (correct or incorrect). (b) The setting lacked realism; conditions were not similar to those typically faced by therapeutic touch practitioners.

3.7   Students registering for a course should be randomly assigned to a classroom or an online version of the course. Scores on a standardized test can then be compared.

3.9   Randomly allocate the 200 rooms into two groups of 100 rooms. First group gets the flat rate; second group gets the varied rates. Compare Internet usage. Rooms 119, 033, 199, 192, and 148 are the first five selected.

3.11   The difference in blood pressures was so great that it was unlikely to have occurred by chance (if calcium is not effective).

3.13   The effects of factors other than the nonphysical treatment have been eliminated or accounted for, so the differences in improvement observed between the subjects can be attributed to the differences in treatments.

3.15   If the economy worsened due to a recession during the five-year period, then unemployment could rise even if the training program was effective. Consumer spending is a possible lurking variable that would be confounded with the effectiveness of the training program.

3.17   Time of day would be a lurking variable. Students in the 8:30 A.M. class may be more motivated than those in the 2:30 P.M. class.

3.19   "Significant" means that a difference as large as the one we observed is unlikely to occur by chance.

3.21   Randomly assign 25 students to Group 1 (Brochure A) and 25 students to Group 2 (Brochure B). Have every student complete the questionnaire and compare responses.

3.23   "Placebo-controlled" means that the control group received a placebo. "Double-blind" means that neither the patients nor those who interacted with them during the experiment knew who received hydroxyurea and who received placebo.

3.25   (a) The placebo effect. (b) Use a three-treatment, completely randomized design. (c) No. (d) Since the patient assesses the effectiveness of the treatment, the experiment does not need to be double-blind.

3.27   For example, do rat tumors arising from high exposures for a relatively short period indicate that humans would get tumors from low doses for a longer period?

3.29   (a) Difference in score (after minus before) on the Beck Depression Inventory. (b) Give the Beck Depression Inventory to all subjects before and after. Randomly assign subjects into three groups of 110. Group 1 gets Saint-John's-wort, Group 2 gets a placebo, and Group 3 gets Zoloft. (c) Use a double-blind design.

3.31   (a) Subjects may learn how to squeeze the scale more efficiently after the first attempt. (b) Treat each subject as a block to reduce unwanted variability. (c) Answers will vary.

3.33   (b) For each block, go to a line in Table A and start reading across. First digit between 1 and 5 means put Variety A in corresponding plot. Next digit between 1 and 5 that is different from the first digit selected gets Variety B, etc.

3.35   (a) A block, because the diagnosis is an existing difference between subjects. (b) A treatment that is being randomly assigned to the subjects.

3.37   Use a matched pairs (block) design with each dummy as a block. Randomize the order of use of the two air bags. Repeat several times if possible with each dummy.

3.39   (a) The factors are storage and cooking. Storage is either fresh, a month at room temperature, or a month refrigerated. Cooking is either cooked immediately or cooked after an hour at room temperature. The response variables are the ratings of the color and flavor. (b) Use a completely randomized design, assigning an equal number of potatoes to each of the six treatment groups. (c) The different combinations of treatments will be presented in a random order.

3.41   (a) Nine treatments: 500°-front, 500°-middle,..., 1000°-back. (b) Randomly assign five converters to each of the nine groups. (c) Converters 19, 22, 39, 34, and 05 get 500°-front.

3.43   (a) This is a matched pairs design. (b) To avoid bias that may result from giving all subjects the same treatment first. (c) Yes.

3.45   Answers will vary.

3.47   This offers anonymity, since names are never revealed.

3.49   Answers will vary.

3.51   Answers will vary.

3.53   (a) The subjects should be told what kinds of questions the survey will ask and about how much time it will take. (b) For example, respondents may wish to contact the organization if they feel they have been treated unfairly by the interviewer. (c) Respondents should not know the poll's sponsor (responses might be affected), but sponsors should be revealed when results are published.

3.55   Answers will vary.

3.57   Answers will vary.

3.59   Answers will vary.

3.61   Answers will vary.

3.63   Answers will vary.

3.65   Answers will vary.

3.67   (a) Use a completely randomized design with equal numbers of patients assigned to each of the six possible treatments. (b) By using more subjects, any differences between treatments are more likely to be detected. (c) Block by gender, then randomly assign in about equal numbers to the six treatments within each block.

3.69   (a) Type of treatment; survival time. (b) No; we are examining available data. (c) The effect of the surgery is confounded with the initial health of the patients. (d) Randomly assign half the patients to surgery and the other half to nonsurgical treatment.

3.71   Matched pairs design. Randomize the order of left- and right-hand thread for each subject.

3.73   The 4561 healthy, working adults; flu treatment administered; number of days of lost work, number of days with health-care provider visits.

3.75   Institutional review board, informed consent, and confidentiality.

CHAPTER 4

4.1   A pie chart could be used since each number in the table represents a part of the total.

4.3   (a) The given percents add to 77.6%, so 22.4% were in other fields. (b) Either a bar graph or a pie chart would be appropriate.

4.5   (a) Adding the "Never Married," "Widowed," and "Divorced" groups gives 43,148,000. (c) A pie chart could be used since each number in the table represents a part of the total.

4.7   This is a pictogram and therefore misleadingly labeled. The CDs used to represent each company's share are scaled both horizontally and vertically.

4.9   (b) More electricity is used in January, when heating costs are high, and the least is used in the spring, typically May, when weather is mild. Usage increases somewhat over the summer, due to air conditioning, and is relatively high in August, the hottest part of the year. (c) February has fewer days (28) than January (31). (**d**) Perhaps many cold days in December 2002 increased heating (heat pump) costs. (**e**) Heating the house in the winter, by far.

4.11 A line graph shows a long-term increasing trend with sharp drops in 1981 and 1989.

4.13 (**a**) This is a pictogram and therefore misleadingly scaled. (**b**) A pie chart would not be appropriate since we don't know the total value of all exports.

4.15 (**a**) The transplant percents are hearts, 9%; heart-lung, 0%; lung, 4%; liver, 22%; kidney, 59%; other, 6%. (**b**) A pie chart is optimal, but one could also draw a histogram.

4.17 There will be a 12-month repeating pattern, rising in the summer and falling in winter.

4.19 Yes.

4.23 The distribution is strongly right-skewed. It has a prominent right tail. The lowest salary is \$211,400 and the highest is \$14.6 million.

4.25 The distribution is strongly right-skewed, trailing off rapidly from the peak at 0 through 4. The distribution spreads from 0 to 54, with few universities awarding more than 20 doctorates to minorities.

4.27 Answers will vary.

4.29 Right-skewed. Alabama (12.5), West Virginia (13.9), and Mississippi (16.0) might be considered unusual observations. They are all considered southern states. Common characteristics: lack of industry and jobs and low education rates.

4.31 The distribution is irregular in shape. There are two distinct groups, plus a low outlier (the veal group).

4.33 (**a**) 186 and 190 might be called unusual. (**b**) Strongly right-skewed. (**c**) The shape is similar to the original stemplot (right-skewed).

4.35 Both stemplots show a symmetric distribution with center in the upper 60s. (**c**) The stemplot shows the overall shape and the individual data values. The shape of the histogram depends on how the classes are defined.

4.37 Use a histogram or a stemplot. The distribution is roughly symmetric, perhaps slightly left-skewed, spread from 12.7% to 22.9%, and centered around 17% or 18%.

4.39 Use a histogram or a stemplot. The distribution is right-skewed, with peak around 12.

4.41 Half of all households make more than the median, and the other half make less.

4.43 Median = 153. About half the brands have more than 153 calories, and about half the brands have less than 153 calories. Q_1 = 138.5. Twenty-five percent of the brands have less than 138.5 calories. Q_3 = 180.5. Seventy-five percent of the brands have less than 180.5 calories.

4.45 Median = 7.8, $Q_1 = 6.5$, $Q_3 = 10.3$.

4.47 Bonds's 73 homers was an outlier.

4.49 (**a**) Minimum = –26, $Q_1 = 4$, median = 16.5, Q_3 = 27, maximum = 50. (**b**) The distribution is fairly symmetric.

4.51 Mean = 36.06, standard deviation = 13.45. The mean is Bonds's average number of home runs in a season. The standard deviation is a measure of the average deviation of Bonds's number of home runs from the center.

4.53 (**a**) Juan is correct. If the standard deviation is zero, then none of the observations deviate from the center. (**b**) Letishia is wrong. Consider $\{-1, 0, 1\}$ and $\{-1.37, -0.3509, 0, 0.3509, 1.37\}$. Both data sets have mean $\bar{x} = 0$ and standard deviation $s = 1$.

4.55 (**a**) Variable 2. (**b**) Variable 3. (**c**) Variable 1.

4.57 (**a**) Mean = 12, median = 11. (**c**) Most of the times are around the center (11 to 12), falling off to either side. But there is a relatively tall bar on the extreme right. Perhaps these times occurred when the subject was tired, and his reaction times were slow.

4.59 (**a**) Range is 14. (**b**) The range can be misleading. (**c**) Answers will vary.

4.61 Here's one solution: {1, 2, 3, 4, 5, 6, 7, 8, 9, 45}. Half the numbers are above the median.

4.63 Answers will vary.

4.65 The bar graph makes it easier to see percents (without writing them in or next to the wedges, as was done with the pie chart). It is also easier to compare sizes of bars than wedge angles.

4.67 The horizontal scale is not uniform, and so the graph has been distorted.

4.69 The average number of roller coaster fatalities per year is 1.9. (Notice that several years have no recorded fatalities, which suggests incomplete data.) There is no apparent trend.

4.71 (**a**) The distribution is strongly right-skewed. (**b**) Management prefers to quote the mean salary because the few extremely large salaries will pull the mean salary up. The players would prefer to cite the median salary since it would be smaller than the mean.

4.73 (**a**) Approximately 16%. (**b**) About 34%. (**c**) For students with student loans, about 16% of their indebtedness is from credit card charges; about 34% of their monthly payment will be applied to credit card debt; and 35% of the interest they pay will go to credit card companies.

CHAPTER 5

5.1 (**a**) The bars in the histogram from Activity 5.1A should theoretically be the same height. The curve forms a 1×1 square, which has area 1. (**b**) Mean = 0.5. (**c**) Median = 0.5. The mean and median are the same since the distribution is symmetric. (**d**) 40%. (**e**) A 0–9 spinner with 10 equal sectors. In all of these settings, the outcomes are equally likely to occur.

5.7 The middle 95% of IQs is between 89 and 133—that is, 111 ±2(11).

5.9 0.15%, because 144 is three standard deviations above the mean, so only half of the outer 0.3% have IQs above 144. It is not surprising that none of these 74 students had such a high IQ, since 0.15% of 74 is only 0.111 students.

5.11 (**a**) 327 to 345 days. (**b**) 16%.

5.13 Sixty-eight percent of the young women are between 62.5 and 67.5 inches tall, and so forth.

5.15 $x < 720$ means that $z < -1.42$, so 7.78% of all SAT scores are less than 720.

5.17 $N(495, 109)$. $x > 531$ is standardized to $z > (531 - 495)/109 = 0.33$. Using the TI-83, 37.07% of women scored higher than 531.

5.21 The area under the standard normal curve between $z = -1$ and $z = +1$ is 68%. Since the curve is symmetric about the center, 16% of the area lies below $z = -1$, and 16% lies above $z = 1$.

5.23 (**a**) An IQ score of 130 is two standard deviations above the mean, so about 2.5% of 1932 children had very superior scores (2.28% if using Table B or your calculator). (**b**) For a present-day child a score of 130 corresponds to a standard score of 0.6667. The standard score 0.7 is the 75.80 percentile, so about 24.2% of present-day children would have "very superior" scores.

5.25 (**a**) $x < 10.25$ has a standard score of $z = 0.5$ or about 69%. (**b**) $9.5 < x < 10.25$ is the same as $-1 < z < 0.5$, or about 53.3%.

5.27 (**a**) $115 < x < 295$ becomes $-1 < z < 1$, or about 68%. (**b**) $z = -0.2$. 42% spent less. (**c**) The assumption is that the distribution is approximately normal.

5.29 (**a**) 50%; 2.5%. (**b**) 0.37 to 0.43.

5.31 Price index = 126.

5.33 (**a**) The trend is generally down from 1983 to 1987 and generally up from 1987 to 2003. (**b**) Anticipating the two military conflicts in Iraq. Note that Iraq was a principal supplier of crude oil. (**c**) Price index = 120.

5.35 Answers will vary.

5.37 (**a**) Tuition rose by a factor of 3.26 from the base period to January 2000. Equivalently, tuition costs increased by 226%. (**b**) The tuition CPI is almost twice as large as the overall CPI.

5.39 About $61,500. A 1980 income of $30,000 is equivalent in buying power to 30,000(168/82.4) ≈ 61,456 in the year 2000.

5.41 Assuming the CPI continues to increase, the answer should be less than $14.90.

5.43 By asking the same questions, the GSS can be used to track how the population is changing over time.

5.45 Answers will vary.

5.47 (**a**) The New York CPI is greater than the Los Angeles CPI. (**b**) We do not know how prices compared in the base period.

5.49 Expressing the 1976 cost in 1999 dollars gives 13,500(166.6/56.9) ≈ 39,527. The actual 1999 cost is more than twice this amount, so the cost of a Steinway has gone up in real terms.

5.51 A salary of $143,760 in 1980 is equivalent to $322,060 in 2003. Baseball players are well ahead of inflation.

5.53 (**a**) Only to the extent that new funds are created, and some funds that were once popular go out of favor. (**b**) Less than.

5.55 (**a**) 500. (**b**) 68%.

5.57 About the 76th percentile.

5.59 Density curve (a).

5.61 About $38,300.

5.63 Tiger Woods's winnings are worth $6,620,970 in 1999 dollars. Sam Snead's 1938 winnings amount to $233,274(166.6/14.1) ≈ $2.64 million in 1999. Tom Watson's 1980 winnings amount to $1,041,002(166.6/82.4) ≈ $2.10 million in 1999. Snead's and Watson's earnings are very similar, while Woods earned 2.5 to 3 times as much (in real terms) as the other two.

CHAPTER 6

6.1 (**a**) Grade would be response, and time of study would be explanatory. (**b**) Relationship. (**c**) Rain would be explanatory, and yield would be response. (**d**) Relationship. (**e**) Income would be explanatory, and education would be response.

6.3 (**a**) There is a moderate, positive, linear relationship between IQ score and GPA. (**b**) About 103 and 0.5. (**c**) A, B: moderate IQ but low GPA. C: low IQ but high GPA.

6.5 (**a**) When individual income is higher, household income will also be higher. Household income will be higher when there are multiple wage earners in the family. Also, income per person is diluted by non–wage earners. (**b**) Some high outliers could cause the mean income per person to be greater than the median household income. (**c**) Except for the District of Columbia, the relationship is fairly linear with a positive slope. (**d**) Large households raise household income relative to personal income. Many very high earners (Wall Street) raise mean personal income.

6.7 (**a**) Stocks: highest, 50%; lowest, −28%. Treasury bills: highest, 15%; lowest, 1%. (**b**) There is a very weak relationship between interest rates and percent return on stocks.

6.9 (**b**) Linear, positive, fairly strong.

6.11 The strength of the linear relationship between GPA and first exam score.

6.13 (a) −1 to 1. (b) Any positive number.

6.15 We would expect a strong association between a woman's height and her height as a girl, a moderate association between a man's height and his adult son's height, and a weak association between a man's height and the height of his wife.

6.17 (a) Negative. (b) Negative. (c) Positive. (d) Small. (e) Moderately positive.

6.19 (a) Negative; slightly curved; yes—at about (4, 2.9). (b) No.

6.21 (a) The improvement seems to be rather linear except for the last data point, which shows a very minor improvement for the time period involved. (b) There is a very strong, negative, linear relationship between year and record time for women in the 10,000-meter run. (c) Women's times have decreased much more rapidly than men's times. (d) Yes.

6.23 (a) MA is explanatory. (b) Weak, positive, linear relationship; (12, 50) is an outlier. (c) There is a positive association, but it's not very strong. (d) Not really.

6.25 Close to −1.

6.27 −56.1 grams; prediction outside the range of the available data is risky.

6.29 Inactive girls are more likely to be obese. 3.2%.

6.31 False. No. 70% of the variation in y is explained by the linear regression model.

6.33 (a) x = number of beers consumed; y = blood alcohol content (BAC). (b) $r = 0.894$. Yes, since the relationship appears very linear. (c) Predicted BAC = −0.013 + 0.018(number of beers). Slope: for every additional beer consumed, a student's BAC increases by 0.018, according to our model. For the y intercept, our model predicts that a student who consumes no beers will have a BAC of −0.013. (d) 0.077. (e) No.

6.35 For example, motivation level, intelligence, socioeconomic status.

6.37 Stronger students are more likely to choose such math courses; weaker students may avoid them.

6.39 (a) The line shows a general decrease, but not a strong relationship. (b) The general pattern is concave upward (bowl shaped). This might strengthen the conclusion to avoid hospitals that treat few heart attacks.

6.41 (a) A strong, positive, linear relationship between age and height. (b) Predicted height = 64.928 + 0.635(age). Our model predicts a 0.635 cm increase in height per month for Kalama children; according to this model, a 0-month-old child would be 64.298 cm tall. (c) 85.247 cm; fairly confident due to the strong linear relationship. (d) 217.32 cm (over 7 feet!); not confident, since we are predicting beyond the range of the data.

6.43 Time spent standing is a confounding variable.

6.45 (a) For women, Record = 41,373 − 19.9(Year). For men, Record = 8166 − 3.29(Year). (b) The record times are decreasing more rapidly for women (19.9 per year) than for men (3.29 per year). (c) The two times should be equal in about 1999.

6.47 (a) Longer crickets should weigh more than shorter crickets. (b) It would not change.

6.49 Dolphins and hippos are outliers in the plot. Dolphins have unusually large brain weights for their body weights. Hippos are heavy but have unusually low brain weights.

6.51 74% of the variation in brain weight is explained by the linear relationship.

6.53 (b).

6.55 (a) HAV = 19.723 + 0.339(MA) (b) 28.194. (c) Not really. $r^2 = 0.091$.

6.57 (c) The outlier at (10, 1).

CHAPTER 7

7.1 Long trials of this experiment suggest about 40% heads.

7.3 The proportion from Table A is 0.105.

7.5 Answers will vary.

7.7 Answers will vary.

7.9 If two people talk at length, they will eventually discover something in common.

7.11 For example, people are often eager to assume that short-run success will continue indefinitely (gamblers are reluctant to quit after winning; people justify risky behavior because it has not killed them yet; etc.). Additionally, people tend to believe that tragedies happen to others.

7.13 (a) The wheel is not affected by its past outcomes, so on any one spin, black and red remain equally likely. (b) Wrong. If you hold 5 red cards, the deck now contains 5 fewer red cards, so your chance of another red decreases.

7.15 (a) Opinions will vary; maybe because airplane crashes are more sensational and "interesting." (b) The (lower) risk of flying is emphasized more than the risk of driving. Also, people feel more control behind the wheel of a car than sitting in a seat on an airplane.

7.17 As more flips are accumulated, the proportion of heads gets closer and closer to 0.5. This exercise illustrates probability as long-term relative frequency.

7.19 0.68; 0.32.

7.21 No, since the sum of Sheridan's probabilities is greater than 1.

7.23 Possible totals: 2 through 8. Probabilities: 1/16, 2/16, 3/16, 4/16, 3/16, 2/16, 1/16. The probability of a total of 5 is 4/16 = 1/4 = 0.25.

7.25 0.62%.

7.27 (a) About 50%. (b) 0.68. (c) 0.32.

7.29 Each face value has probability 1/13.

7.31   (a) This is fairly unusual, but it isn't that surprising to see a run of 16 consecutive male births at some hospital in the United States. (b) Not really. The law of averages suggests that over many births, the proportion of boys that are born will be very close to 0.5.

7.33   (a) All probabilities are between 0 and 1, and they add to 1. (And everyone must fall into exactly one category.) (b) 0.83. (c) 0.24.

7.35   Statement (b) is true.

7.37   (a) Answers will vary. (b) A personal probability might take into account specific information about your driving habits. (c) Most people believe that they are better-than-average drivers.

7.39   (a) 0.68. (b) 0.95.

7.41   (a) 4. (b) and (c) Answers will vary.

7.43   (a) 0.1. (b) 0.5. (c) 0.7.

7.45   About 0.21.

CHAPTER 8

8.1   (a) 0 to 4 for Democrats, 5 to 9 for Republicans. (b) 0 to 5 for Democrats, 6 to 9 for Republicans. (c) 0 to 3 for Democrats, 4 to 7 for Republicans, 8 to 9 for undecided. (d) 00 to 52 for Democrats, 53 to 99 for Republicans.

8.3   (a) 4 chose Democrats, 6 chose Republicans. (b) 3 Democrats; 7 Republicans. (c) 2 Democrats; 4 Republicans; 4 undecided. (d) 6 Democrats; 4 Republicans.

8.5   (a) Use `randInt(1,2,2)` where 1 represents a boy and 2 represents a girl. Discard any trials in which the second number is a 2. (b) Use `randInt(1,2,2)` as in (a). Discard any trials in which both numbers are 2s.

8.7   (a) `randInt(0,1)`. (b) `randInt(0,1,100)` → L1. (c) Answers will vary.

8.9   (a) `randInt(1,365,30)` → L_1. (b) No. The probability of duplicate birthdays among 30 nonrelated students is more than 0.5.

8.11   (a) 0.1. (b) One possibility: Use 0 through 2 to represent the top 10%, 3 through 5 to represent the rest of the top quarter, 6 through 8 to represent the rest of the top half, and 9 for the bottom half.

8.13   Results will vary. The theoretical probability that no more than 3 of the 8 are in the bottom half of their class is about 0.995—very high.

8.15   (a) 0, 1: no female offspring; 2, 3, 4: one female offspring; 5, 6, 7, 8, 9: two female offspring. (b) 0, 1, 2, 3, or 4 beetles in the third generation. (c) Answers will vary.

8.17   About $0.33.

8.19 $0.60.

8.21 $0.4996.

8.23   (a) Let 01 to 05 represent a box having a $1 bill and 06 to 00 represent a box with no $1 bill.

(b) `randInt(1,100)` with the same labeling as in (a). (c) Answers will vary.

8.25   3, 5.

8.27   (a) 0.2. (b) Let 0, 1 represent A; 2, 3, 4 represent B; 5, 6, 7 represent C; 8, 9 represent D or F. (c) Use `randInt(0,9)` with the same correspondence as (b).

8.29   Answers will vary. The theoretical probability is 0.32768.

8.31   (a) 0.699. (b) 0.301. (c) 0.609.

8.33   (a) Since the probability of someone winning on a single play of the game is 2/3, use `randInt(1,3)`, where 1 and 2 represent that there is a winner and 3 represents that there is no winner. Continue until there is a winner. Record the number of games played. Repeat several times. (b) The theoretical answer is 1.5 games.

8.35   (a) A single digit simulates one roll of the die, with 1, 2, 3, 4, 5, 6 standing for their respective faces (other digits ignored). Two successive digits (skipping 0, 7, 8, 9) simulate the roll of two dice. (b) The theoretical probability of winning is 0.4929.

8.37   0 to 5 means the van is available. The theoretical probability that one or more passengers is stranded is about 0.12.

8.39   (a) $49.70. (b) Although the probability that you will have to pay $100,000 is very small, if this were to happen, it would be financially disastrous. (c) The law of large numbers says that the average profit on many policies will be close to the expected value. So, on the average, the insurance company will earn about $50 per person insured.

8.41   5 heads. Simulation gives 502 heads in 100 trials, or an average of 5.02 heads per trial.

8.43   The psychic makes an average profit of $0.20 per customer (ignoring any postage costs).

CHAPTER 9

9.1   (a) The 175 seniors in Tonya's school. (b) The proportion who plan to attend the prom. (c) 0.72.

9.3   This interval, 10.2% to 13.8%, was calculated from sample data by a procedure that is guaranteed to capture the true population proportion 95% of the time.

9.5   (a) Proportion of all women who say they don't get enough time for themselves. (b) 0.439 to 0.501.

9.7   (a) `randInt(0,9,200)`. (b) $(L_1=0)$ → L_2. (c) `sum(L₂)/200`. (d) Answers will vary.

9.9   (a) (0.593, 0.847). (b) (0.6155, 0.8245). (c) This interval was produced by a method that gives correct results 95% of the time.

9.11   Answers will vary.

9.13   Answers will vary.

9.15   0.0125, or 1.25%.

9.17   (a) Normal with mean 0.14 and standard deviation 0.0142. (b) Above 18.2% is unlikely (0.15%); above 11.2% is likely (97.5%).

9.19   (a) The proportion of adults who feel they have achieved the good life. $\hat{p} = 0.09$. (b) 0.0064.

9.21   (a) No. No treatment was imposed. (b) This is not at all clear from the report. (c) No. There are several lurking variables.

9.23   (a) Response bias. Some teens may lie about this risky behavior. (b) With a sample size this large (6903), even small differences will be significant.

9.25 H_0: $p = 0.75$, H_a: $p > 0.75$. The standard value is 2.60 and the P-value is 0.005. This is strong evidence that more than 3 out of 4 middle school students have engaged in bullying behavior, according to the researchers' definition of bullying.

9.27   (a) Significant at the $\alpha = 0.05$ level means that the critical area is less than 0.05. It may or may not be less than 0.01. (b) No; this statement means that if H_0 is true, then we have observed a result that occurs less than 5% of the time.

9.29   (a) p is the proportion of all people with body temperature below 98.6°F. (b) H_0: $p = 0.5$, H_a: $p > 0.5$.

9.31   (a) About 0.67. (b) Standard score is 9.85, and the P-value is 0.0000. There is sufficient evidence to conclude that a majority of Hispanics have experienced no discrimination.

9.33 H_0: $p = 0.112$, H_a: $p \neq 0.112$. The standard score is -1.62, and the P-value is 0.105. There is insufficient evidence to conclude that the number of middle school smokers has declined.

9.35   Answers will vary.

9.37   No. There are too many lurking variables to be very confident in this conclusion. This was not a carefully controlled experiment. Those taking the statin drugs may have adopted other lifestyles that helped them reduce their risk of breast cancer.

9.39   False. But you can say that for the same data, a 95% confidence interval is about 1.97 times as long as a 68% interval.

9.41   (b) and (d) are false.

9.43   Joe is working with the population, not a sample.

9.45   It is essentially correct.

9.47   (a) The width decreases. (b) The P-value will decrease.

9.49   (a) In a sample of size 500, we expect to see about 5 with $P < 0.01$. (b) Test these four again.

9.51   Were these random samples? How big were the samples?

9.53   0.669 to 0.750.

9.55   About 66% to 74%.

9.57   They lied.

9.59   You have information about all states, not just a sample.

9.61   Mornings: 0.550 to 0.590. Evenings: 0.733 to 0.767. The evening proportion is quite a bit higher.

CHAPTER 10

10.1   In each group, the percent with C or better is 55%, 74.7%, and 37.5%. Some (but not too much) time spent in extracurricular activities seems to be beneficial.

10.3   (a) Hatched: 16, 38, 75. Did not hatch: 11, 18, 29. (b) Cold: 59.3%. Neutral: 67.9%. Hot: 72.1%. Cold water did not prevent hatching but made it less likely.

10.5   (a) 13% for Alaska Airlines and 10.9% for America West. (c) Both airlines do best at Phoenix and worst at Seattle. Alaska has most of its flights at Seattle, while America West has few flights there.

10.7   (a) H_0: Students' smoking behavior is not related to their parents' smoking behavior. H_a: Students' smoking behavior is related to their parents' smoking behavior. (b) 1447.51.

10.9   (a) H_0: There is no association between alcohol consumption and relapse. H_a: Alcohol consumption and relapse are associated. (b) 12.68, 20.32, 55.32, 88.68.

10.11   (a) 6.93; the largest contribution is from the lower right, with slightly less from the middle column. (b) df = 2. (c) $P = 0.03$. There is strong evidence of an association between extracurricular activities and success in a required course.

10.13 $X^2 = 1.703$, df = 2, $P = 0.427$. There is insufficient evidence that temperature and hatching are related.

10.15 $X^2 = 42.6$, df = 3, $P = 0.0000$. There is strong evidence that type of shooting is related to type of gun.

10.17   (a) All expected counts are greater than 5. (b) $X^2 = 37.6$, df = 2, $P = 0.0000$. There is strong evidence that students' smoking status is related to parents' smoking status.

10.19   (a) All expected counts are greater than 5. (b) $X^2 = 8.44$, df = 1, $P = 0.0037$. There is strong evidence that smoking relapse is related to alcohol consumption.

10.21   (a) 1,238,000. (b) 57.2%, 58.0%, 43.8%, and 44.4%. Women earn a majority of bachelor's and master's degrees but smaller percents of professional and doctoral degrees.

10.23   It appears that long guns are used more often for suicides than homicides. Handguns accounted for about 89% of the homicides but only about 71% of the suicides.

10.25   If an expected count were 0, then in the corresponding term in the formula for X^2, you would be dividing by 0.

10.27   (a) White defendant: 19 yes, 141 no. Black defendant: 17 yes, 149 no. (b) Overall death penalty: 11.9% of white defendants, 10.2% of black defendants. For white victims, 12.6% and 17.5%; for black victims, 0% and 5.8%. (c) The death penalty was more likely when the victim was white (14%) rather than black (5.4%). White defendants killed whites 94.3% of the time but were less likely to get the death penalty than blacks who killed whites.

10.31 (a) Normal with mean 115 and standard deviation 6. (b) 118.6 is fairly close to the middle of the curve and would not be too surprising if H_0 were true. However, 125.7 lies out toward the high tail of the curve and would rarely occur when $\mu = 115$.

10.33 (a) The mean is 0.496, which is close to 0.5. (b) We estimate the standard deviation to be 0.3251.

10.35 (a) Fairly normal, but with two low outliers. $\bar{x} = 105.84$ and $s = 14.27$. (b) 100.8 to 110.9. (c) Confidence interval methods assume that the numbers represent an SRS.

10.37 (a) H_0: $\mu = 100$; H_a: $\mu \neq 100$. (b) The sampling distribution is approximately N(100, 2.563). (c) Yes. (d) Standard score is 2.28. P-value = 0.0226. The mean IQ of this population is not 100.

10.39 (a) 271.5 to 278.5. (b) 90% confidence interval: 272.1 to 277.9; 99% confidence interval: 270.4 to 279.6. The margins of error are 2.90, 3.46, and 4.56. Margin of error goes up with increasing confidence.

10.41 H_0: $\mu = 2.6$ hours; H_a: $\mu \neq 2.6$ hours, where μ is the mean time to respond to a service call this year.

10.43 Answers will vary.

10.45 Standard score is -2.77; significant at $\alpha = 0.005$.

10.47 White students tend to perform better on the NAEP than minority students, so the overall average score is lower in a state (like New Jersey) with a higher percent of minorities.

10.49 (a) 2.96%, 13.07%, 6.36%. (b) The rows are 721, 173, 412; 22, 26, 28. (c) Expected counts: 702.14, 188.06, 415.80; 40.86, 10.94, 24.20. All expected counts are greater than 5. (d) H_0 : There is no relationship between a member's complaining and leaving the HMO. H_a: There is some relationship. df = 2; this is very significant.

10.51 (a) H_0: $\mu = 0$; H_a: $\mu < 0$, where μ = mean increase in systolic blood pressure. (b) Standard score is -2.2, $P = 0.035$. Reject H_0. (c) There is sufficient evidence to conclude that eating a three-ounce dark-chocolate bar daily will lower one's blood pressure. (d) The number of subjects was very small. Increase the sample size. (Note: We're available!)

10.53 (b) $\bar{x} = 15.59$ ft and $s = 2.550$ ft; 15.0 to 16.2 ft. (c) What population are we examining: full-grown sharks, male sharks?

10.55 (a) H_0: $\mu = 0$ seconds; H_a: $\mu > 0$ seconds. (b) $\bar{x} = 0.96$ seconds, $s = 12.55$ seconds. On average, they were slightly faster but did not show a big enough improvement to be important. (d) The standard score is 0.35. This gives no reason to doubt the null hypothesis.

Index